Android开发
范例实战宝典

武永亮　编著

清华大学出版社
北　京

内 容 简 介

本书全面、系统地介绍了 200 多个常用的 Android 应用开发实例,这些实例紧跟技术趋势,内容基本覆盖了 Android 开发的方方面面,几乎涉及 Android 开发的所有重要知识。书中给出了每个实例的实现过程,并精讲了每个实例的重点代码。作者专门为每一个实例都录制了配套的教学视频(共 14.2 小时),以帮助读者更好地学习,这些教学视频和书中的完整实例源代码一起收录于配书光盘中。另外,光盘中还赠送了大量的 Android 开发教学视频及其他资料。

本书共分为 9 章。主要内容包括:Android 简介及平台架构知识;Android 开发者必备利器——搭建开发环境;Android 中基本控件、常见布局及高级组件的使用;Android 中回调函数的事件处理、监听器的事件处理及多线程处理;Android 中系统 Intent 的使用及自定义 Intent 的使用;Android 的数据存储知识,重点介绍文件操作、ContentProvider 及资源文件;Android 中的服务和广播;Android 网络编程;Android 中的多媒体开发。

本书适合有一定 Java 基础,想快速提高 Android 开发水平的人员阅读。对于 Android 开发爱好者及经常使用 Android 做开发的程序员,本书更是一本不可多得的案头必备参考书。

本书封面贴有清华大学出版社防伪标签,无标签者不得销售。
版权所有,侵权必究。侵权举报电话:010-62782989　13701121933

图书在版编目(CIP)数据

Android 开发范例实战宝典 / 武永亮编著. —北京:清华大学出版社,2014(2015.4 重印)
ISBN 978-7-302-36701-7

Ⅰ. ①A… Ⅱ. ①武… Ⅲ. ①移动终端–应用程序–程序设计 Ⅳ. ①TN929.53

中国版本图书馆 CIP 数据核字(2014)第 116930 号

责任编辑:夏兆彦
封面设计:欧振旭
责任校对:徐俊伟
责任印制:刘海龙

出版发行:清华大学出版社
　　　　网　　　址:http://www.tup.com.cn,http://www.wqbook.com
　　　　地　　　址:北京清华大学学研大厦 A 座　　　邮　　编:100084
　　　　社 总 机:010-62770175　　　　　　　　　　邮　　购:010-62786544
　　　　投稿与读者服务:010-62776969,c-service@tup.tsinghua.edu.cn
　　　　质 量 反 馈:010-62772015,zhiliang@tup.tsinghua.edu.cn
印 刷 者:清华大学印刷厂
装 订 者:三河市新茂装订有限公司
经　　销:全国新华书店
开　　本:185mm×260mm　　印　张:39　　字　数:971 千字
　　　　(附光盘 1 张)
版　　次:2014 年 9 月第 1 版　　　　　　　　印　次:2015 年 4 月第 2 次印刷
印　　数:3501~5500
定　　价:89.00 元

产品编号:059781-01

前　言

2003年有"Android之父"之称的Andy Rubin在美国创建了Android科技公司。当时他的想法就是使移动设备更好地服务于人类。直到2005年，Google公司收购了Android科技公司，这才真正吹响了Google进军移动领域的号角。随后几年，Android一发而不可收拾，一跃成为了当前炙手可热的智能手机操作系统。

自2009年发布的第一个Android系统以来，仅仅几年时间，Android已经成为了使用最多的智能手机操作系统。这是与Android具有的以下几个特点分不开的。

（1）Android支持多种硬件设备，包括照相机、录像机和陀螺仪等，还有各种传感器。

（2）Android支持各种移动设备的网络，包括GSM/EDGE、IDEN、CDMA、EV-DO、UMTS、Bluetooth、Wi-Fi、LTE、NFC和WiMAX等。

（3）Android内置的网页浏览器基于WebKit内核，并且采用了Chrome引擎。Android 2.2版及之后的版本能原生支持Flash，在Android 4.0版内置的浏览器测试中，HTML 5和Acid 3故障处理，均获得了满分。

（4）Android支持多种媒体格式，包括WebM、H.263、H.264（in 3GP or MP4 container）、MPEG-4 SP、AMR、AMR-WB（in 3GP container）、AAC、HE-AAC（in MP4 or 3GP container）、MP3、MIDI、Ogg Vorbis、FLAC、WAV、JPEG、PNG、GIF和BMP等。如果用户需要播放更多格式的媒体，可以安装其他第三方应用程序。

这些特点使得Android系统在智能手机领域中具有不可动摇的地位。

目前，图书市场上的Android图书非常多，但也非常同质化，都以罗列Android开发技术为主，鲜见一本详细介绍Android常见开发实例的书。为了帮助读者更好地学习Android开发，笔者结合自己近几年的Android客户端开发经验和心得体会，花费了一年多的时间编写了本书。在本书中给出了笔者学习Android开发的各种问题总结及开发过程中遇到的各种问题的解决方案。希望读者能在本书的引领下跨入Android开发大门，并成为一名合格的Android开发人员。

本书重点讲解了200多个常见的Android开发实例，并对每个实例专门录制了配套多媒体教学视频，以辅助读者学习，这些教学视频和书中的完整实例源代码一起收录于配书光盘中。学习完本书后，相信读者应该可以具备较好的Android开发能力。

本书特色

1．实例丰富，代码精讲

本书详细讲解了200多个常用的Android开发实例，并对重点代码做了大量注释和讲解，以便于读者更加轻松地学习。通过对这些实例的演练，可以快速提高读者的开发水平。

2．内容全面，涵盖广泛

本书介绍了 Android 开发的环境搭建、界面开发、事件处理、信息传递、数据存储、网络编程、服务和广播及多媒体开发等内容，覆盖了 Android 开发的方方面面，几乎涉及 Android 开发的所有重要知识。

3．由浅入深，循序渐进

本书中的实例安排遵循从基础到高级的学习梯度，从 Android 开发的基础开始讲解，逐步深入到 Android 开发的高级技术及应用。讲解由浅入深，循序渐进，适合不同层次的读者阅读。

4．教学视频，高效直观

作者专门为每一个实例都录制了详细的配套多媒体教学视频（总长达 14.2 小时），以便让读者更加轻松、直观地学习本书内容，提高学习效率。这些视频与本书源代码一起收录于配书光盘中。

5．技术支持，答疑解惑

读者阅读本书时若有疑问可发 E-mail 到 bookservice2008@163.com 以获得帮助，也可以在本书的技术论坛（http://www.wanjuanchina.net）上留言，会有专人负责答疑。

本书内容及体系结构

第1章　打开Android世界的大门

本章主要介绍了 Android 的发展历史及框架结构。通过本章的学习，读者可以了解 Android 的发展历史及 Android 的平台架构知识。

第2章　Android开发者必备利器

本章主要介绍了 Android 开发环境的搭建，并给出了第一个 Android 程序，还介绍了 Android 开发过程中常用的调试工具，包括 Logcat、DDMS 和 ADB 等。通过学习本章内容，读者可以搭建好 Android 开发环境，并了解最简单的 Android 程序的开发过程。

第3章　让你的程序变成美女

本章涵盖 53 个开发实例，介绍了 Android 中常见的界面开发技术，其中重点介绍了基本控件的使用和常见的高级控件的使用。通过学习本章内容，读者可以构建出各种各样的 Android 程序界面。

第4章　让你的程序和用户说话

本章涵盖 38 个开发实例，介绍了 Android 中的事件处理机制及多线程处理机制。通过学习本章内容，读者可以实现 Android 程序与用户的各种交互。

第5章　Android程序内部的信息传递者

本章涵盖24个开发实例，介绍了Android中的Intent的使用，其中包括调用系统的Intent和自定义Intent的使用方法。通过学习本章内容，读者可以掌握在Android中通过Intent启动内部或外部应用程序。

第6章　Android的数据存储

本章涵盖16个开发实例，介绍了Android中的数据存储方式，主要有文件存储、ContentProvider和SQLite存储。通过学习本章内容，读者可以掌握Android中数据存储的具体方式。

第7章　Android的服务与广播

本章涵盖37个开发实例，介绍了Android中的Service和BroadCastReceiver的使用方法。通过学习本章内容，读者可以全面了解Android服务和广播所能实现的具体应用。

第8章　Android的网络编程

本章涵盖14个开发实例，介绍了Android中网络编程的相关知识，主要包括网络数据的请求获取和常见数据格式的解析。通过学习本章内容，读者可以让自己的应用具有网络访问的能力，并且可以对得到的数据进行数据解析。

第9章　Android中的多媒体开发

本章涵盖19个开发实例，介绍了Android中的多媒体开发技术，包括相机、录音机和播放器等。通过学习本章内容，读者可以轻松实现常见的与硬件相关功能的开发。

本书超值DVD光盘内容

- 本书各章涉及的实例源文件；
- 14.2小时本书配套教学视频；
- 8.7小时Android开发入门教学视频；
- 13.8小时Android开发实战教学视频；
- 11小时Android项目案例开发教学视频。

本书读者对象

- 有一定基础而想提高Android开发水平的人员；
- 想全面学习Android开发技术的人员；
- Android专业开发人员；
- 利用Android做开发的工程技术人员；
- Android开发爱好者；

- 大中专院校的学生；
- 社会培训班的学员；
- 需要一本案头必备手册的程序员。

本书作者

本书由武永亮主笔编写，其他参与编写的人员有陈晓建、陈振东、程凯、池建、崔久、崔莎、邓凤霞、邓伟杰、董建中、耿璐、韩红轲、胡超、黄格力、黄缙华、姜晓丽、李学军、刘娣、刘刚、刘宁、刘艳梅、刘志刚、司其军、滕川、王连心、沃怀凯、闫玉宝、欧振旭。

作者致谢

时光荏苒，转眼间从我想写一本 Android 开发方面的图书，到今天这本书的完成，历时大概一年有余，经过了很多个不眠之夜。当然，在此期间也有很多人在默默地帮助我。在此。我要特别感谢这些人。

感谢我的爱人！她把家里整理的井井有条，每当我遇到挫折的时候她都默默地陪在我身边。

感谢我的两位老师！王顶老师是我的启蒙恩师，是他把我领进了计算机的世界，让我看到了计算机世界的美妙之处。还有李文斌老师，他是我的领导，也是给我帮助最大的人，正是由于他对我严格的要求和帮助，我才能在人生和职业的道路上走的更远。

感谢我的学生！在我教学的过程中他们给了我很多启发，正是在解答他们的很多困惑时，我也有了进一步的理解，这让我多了很多进步和成长的机会。

感谢马翠翠！她是一个非常务实的好朋友，正是因为她的鼓励，本书才按时完成。

虽然笔者对本书中所述内容都尽量核实，并多次进行文字校对，但因时间所限，可能还存在疏漏和不足之处，恳请读者批评指正。

武永亮

目 录

第1章 打开 Android 世界的大门 ... 1
1.1 Android 的来龙去脉 ... 1
1.1.1 Android 的发展简介 ... 1
1.1.2 Android 的平台架构 ... 2
1.2 本书的目的及范例应用范围 ... 4
1.3 本书范例的使用方式 ... 5
1.4 参考网站 ... 6

第2章 Android 开发者必备利器——开发环境搭建（教学视频：14 分钟）... 7
2.1 搭建 Android 开发环境 ... 7
2.1.1 准备工作 ... 7
2.1.2 安装 JDK，配置基本 Java 环境 ... 7
2.1.3 安装 Eclipse ... 10
2.1.4 安装 Eclipse 的 ADT 插件 ... 11
2.1.5 获取 Android SDK ... 13
2.1.6 在 Eclipse 中配置 Android SDK ... 14
2.1.7 管理 AVD ... 15
2.2 建立第一个 Android 程序 ... 17
2.2.1 建立一个 Android 工程 ... 17
2.2.2 Android 程序的目录结构 ... 19
2.3 开发必备利器 ... 20
2.3.1 Logcat 的使用 ... 20
2.3.2 DDMS（Dalvik Debug Monitor Service）的使用 ... 21
2.3.3 ADB（Android Debug Bridge）的使用 ... 22
2.3.4 The Hierarchy Viewer 的使用 ... 23
2.3.5 Draw9-Patch 的使用 ... 23
2.3.6 真机测试 ... 23
2.4 Android 程序的基本组件 ... 24
2.4.1 Activity 组件介绍 ... 24
2.4.2 ContentProvider 组件介绍 ... 24
2.4.3 Service 组件介绍 ... 24
2.4.4 BroadcastReceiver 组件介绍 ... 25
2.4.5 Intent 组件介绍 ... 25
2.5 小结 ... 25

第3章 让你的程序变成美女（教学视频：247 分钟）... 26
3.1 Android 中基本控件的使用 ... 26

	范例 001	更改文字标签的内容	26
	范例 002	更改手机页面的背景色	29
	范例 003	文字超链接	30
	范例 004	让你的文字标签更加丰富多彩	32
	范例 005	用户名密码输入框	35
	范例 006	电话号码输入框	36
	范例 007	更改输入框的文字字体	38
	范例 008	我同意上述条款的页面	40
	范例 009	爱好调查页面	41
	范例 010	政治面貌调查表	44
	范例 011	IT 人员测试应用	45
	范例 012	应用中的关闭声音的按钮	49
	范例 013	应用中的音量调节效果	50
	范例 014	服务星级评价效果	51
	范例 015	页面加载中效果	52
	范例 016	日期获取框效果	54
	范例 017	时间获取框效果	56
	范例 018	日期时间弹出框效果	57
	范例 019	钟表显示效果	60
	范例 020	秒表应用	61
	范例 021	圆角按钮效果	63
3.2	Android 中常见布局的使用		65
	范例 022	用户注册页面的制作	65
	范例 023	学生成绩列表页面的制作	67
	范例 024	登录页面的制作	69
	范例 025	开发模型图的页面	71
	范例 026	图片相框效果	73
	范例 027	商城专区效果	74
	范例 028	三字经阅读程序	75
	范例 029	计算器程序的页面设计	77
3.3	Android 中高级组件的使用		80
	范例 030	单词搜索补全效果	80
	范例 031	多匹配补全效果	82
	范例 032	用户使用的操作系统调查表	84
	范例 033	电影票预售表格效果	86
	范例 034	文件表格列表效果	88
	范例 035	学生名单表	92
	范例 036	手机联系人列表效果	95
	范例 037	画廊图片浏览器	99
	范例 038	仿 iPhone 的 CoverFlow 效果	102
	范例 039	菜单弹出效果	108
	范例 040	打开文件的子菜单效果	111
	范例 041	文本框的复制粘贴全选菜单	114
	范例 042	仿 UC 浏览器的伪菜单效果	116

		范例 043	PopupMenu 效果	122
		范例 044	PopupWindow 效果	124
		范例 045	QQ 客户端的标签栏效果	127
		范例 046	仿新浪微博的主页效果	136
		范例 047	程序退出的对话框	140
		范例 048	程序的关于对话框	142
		范例 049	电话服务评价对话框	143
		范例 050	数据加载成功的提示	146
		范例 051	网络图片加载成功的提示	148
		范例 052	模拟收到短信的状态栏提示	151
		范例 053	模拟数据下载的状态栏提示	153
3.4	小结			157

第 4 章 让你的程序和用户说话（ 教学视频：149 分钟） 158

4.1	Android 中基于回调函数的事件处理			158
		范例 054	Activity 的声明周期回调	158
		范例 055	用户名长度检测效果	161
		范例 056	打字游戏实现	163
		范例 057	长按播放 TextView 动画	165
		范例 058	按钮的快捷键	168
		范例 059	屏幕单击测试器	170
		范例 060	Activity 内容加载完毕提示	172
		范例 061	横竖界面自动切换	173
		范例 062	动态添加联系人列表	175
4.2	Android 中基于监听器的事件处理			178
		范例 063	宝宝看图识字软件	178
		范例 064	控件的拖动效果	182
		范例 065	Email 格式的检测	186
		范例 066	隐藏导航栏	188
		范例 067	屏幕多点触摸测试器	190
		范例 068	图片的平移、缩放和旋转	194
		范例 069	图片浏览器滑动切换图片	199
		范例 070	简易画板	201
		范例 071	登录和注册页面的 ViewFlipper 效果	203
		范例 072	神庙逃亡的操作模拟效果	207
		范例 073	手势库的创建及手势识别	211
		范例 074	滑动切换 Activity 的背景效果	226
		范例 075	按钮控制小人儿移动	229
4.3	Android 中多线程处理			233
		范例 076	异步请求广告图片	233
		范例 077	本地三国演义文本的异步加载	236
		范例 078	应用程序的启动动画	239
		范例 079	NBA 球星信息介绍的网格视图	242
		范例 080	NBA 球星信息介绍的列表视图	248
		范例 081	文件下载	255

范例 082	中断文件下载	259
范例 083	线程间通讯	265
范例 084	本地图片加载速度测试器	267
范例 085	Surface 的读写刷新	270
范例 086	按两次物理返回键退出程序	274
范例 087	线程嵌套	276
范例 088	异步任务加载网络图片	279
范例 089	网站源代码查看器	282
范例 090	终止异步任务操作	286
范例 091	异步任务进度展示	289

4.4 小结 ... 292

第 5 章 Android 程序内部的信息传递者（教学视频：81 分钟）... 293

5.1 Android 中系统 Intent 的使用 ... 293

范例 092	Google 搜索内容	293
范例 093	打开浏览器浏览网页	296
范例 094	电话拨号软件	298
范例 095	分享短信	301
范例 096	短信发送客户端	303
范例 097	彩信分享客户端	306
范例 098	Email 发送客户端	309
范例 099	启动多媒体播放	312
范例 100	安装指定的应用程序	314
范例 101	卸载指定的应用程序	317
范例 102	打开照相机获取图片	319
范例 103	打开系统图库获取图片	322
范例 104	打开录音程序录音	325
范例 105	打开已安装的应用程序信息	328
范例 106	打开软件市场搜索应用	330
范例 107	选择联系人功能	333
范例 108	添加联系人功能	335
范例 109	程序内部启动外部程序	339
范例 110	启动 Google 地图显示某个位置	341
范例 111	启动 Google 地图进行路径规划	343

5.2 Android 中自定义 Intent 使用 ... 346

范例 112	登录页面功能	346
范例 113	注册页面功能	350
范例 114	获取随机验证码功能	355
范例 115	模拟站内搜索	359

5.3 小结 ... 362

第 6 章 Android 的数据存储（教学视频：70 分钟）... 363

6.1 Android 中的文件操作 ... 363

| 范例 116 | 可记住用户名密码的登录界面 | 363 |
| 范例 117 | 系统的设置界面 | 369 |

范例 118	系统图片剪裁	374
范例 119	SDCard 信息查询	379
范例 120	图片旋转保存	382
范例 121	学生成绩管理系统	385

6.2 Android 中的 ContentProvider ……393

范例 122	音乐播放器	393
范例 123	系统图片选择预览	399
范例 124	系统的联系人	402
范例 125	得到系统的音频文件	405

6.3 Android 中的资源文件 ……408

范例 126	全屏界面	408
范例 127	小图堆积背景	409
范例 128	自定义 EditText 样式	411
范例 129	透明背景的 Activity	413
范例 130	圆角控件的制作	415
范例 131	程序的国际化	417

6.4 小结 ……419

第 7 章 Android 中的服务和广播（教学视频：159 分钟）……420

7.1 Android 中的服务的使用 ……420

范例 132	查看手机运行的进程列表	420
范例 133	得到系统的唤醒服务	424
范例 134	定时任务启动	427
范例 135	发送状态栏信息	431
范例 136	得到屏幕状态	433
范例 137	程序中得到经纬度	435
范例 138	振动器应用	438
范例 139	获得当前网络状态	441
范例 140	获得手机 SIM 卡信息	444
范例 141	WiFi 管理器	446
范例 142	系统软键盘显示	450
范例 143	打开系统行车模式	451
范例 144	音量控制器	454
范例 145	短信群发软件	456
范例 146	电池状态查看器	459

7.2 Android 中的广播的使用 ……461

范例 147	飞行模式的切换	461
范例 148	创建桌面快捷方式	464
范例 149	程序开机自动启动	467
范例 150	拍照物理键的功能定制	469
范例 151	锁屏广播接收器	471
范例 152	系统设置信息改变的广播	474
范例 153	系统内存不足提醒	476
范例 154	接受耳机插入广播	478
范例 155	手机区域设置更改监听器	480

范例 156	SDCard 插入的广播	482
范例 157	SDCard 移除的广播	484
范例 158	APK 安装完成的广播	486
范例 159	APK 卸载完成的广播	488
范例 160	外部电源接入的广播	490
范例 161	重启系统的广播	492
范例 162	断开电源的广播	494
范例 163	墙纸改变的广播	496
范例 164	电话黑名单	498
范例 165	短信接收的广播	502
范例 166	短信发送的广播	504
范例 167	电池电量低的广播	507
范例 168	音乐播放器	509

7.3 小结 513

第 8 章 Android 的网络编程（教学视频：61 分钟） 514

8.1 网络请求 514

范例 169	在线天气查询	514
范例 170	在线百度搜索	517
范例 171	网络图片下载器	519
范例 172	文件上传	524
范例 173	异步图片加载	527
范例 174	UDP 网络通信	529
范例 175	在线音乐播放	533
范例 176	在线视频播放	537
范例 177	应用程序在线更新	542

8.2 数据格式解析 547

范例 178	DOM 方式解析 XML	547
范例 179	SAX 方式解析 XML	550
范例 180	PULL 方式解析 XML	554
范例 181	内置 JSON 解析	557
范例 182	Gson 解析 JSON	560

8.3 小结 563

第 9 章 Android 中的多媒体开发（教学视频：71 分钟） 564

9.1 Android 中多媒体应用开发 564

范例 183	屏幕方向改变	564
范例 184	调用系统相机拍照	566
范例 185	录音机	569
范例 186	录像机	571
范例 187	手电筒应用	574
范例 188	计时器	576
范例 189	语音识别功能	579
范例 190	语音转换文本	581
范例 191	TTS 文字朗读	583

范例192　本地音频播放···586
　　范例193　音效播放···588
　　范例194　播放本地视频···590
　　范例195　加速度传感器应用···592
　　范例196　光强度查看器···594
　　范例197　微信摇一摇功能···596
9.2　桌面插件开发···598
　　范例198　切换壁纸插件···598
　　范例199　倒计时插件···600
　　范例200　日期插件···603
　　范例201　电池状态显示插件···604
9.3　小结··607

第 1 章　打开 Android 世界的大门

如果有人问我现在最火的手机操作系统是什么？我一定会说是 Android。如果有人问我现在智能手机操作系统中占有率最高的是什么？我一定会说是 Android。如果有人问我今后最看好的行业是什么？我也一定会说是 Android。

我之所以这么肯定 Android 的市场，是因为 Android 现在确实很火。2011 年 5 月，当时的 Android 设备激活量达到了 1 亿，2012 年 Android 的设备激活量达到了 4 亿，截止到 2013 年，Android 设备的激活量已经达到 9 亿，也就是说现在每天都会有 1500 万台 Android 设备被激活，预计到今年底 Android 设备的总激活数将达到 10 亿，这是一个多么惊人的数字。当然 Android 设备的大卖，不但给各大手机生产厂商带来了不小的收益，也给我们程序开发人员提供了一片新的土地。Google Play 应用商店的 APP 下载数量已经突破 485 亿次，现在平均每月的下载量为 25 亿次，而且还在稳步增长。这样就为我们 Android 开发者提供了无限的商机。所以现在国内外对于 Android 手机开发者的需求量在逐年增大。

本书所介绍的所有实例都是在 Android 4.2.2 平台运行通过的，该版本是现阶段为止最新的 Android 操作系统，其中相对于之前的版本加入了许多特有的功能，以及特殊的修改。这个版本不但功能强大，而且稳定、高效，在 Android 各个操作系统的占有比率中也是名列前茅。本章是全书的基础，将会简要介绍 Android 的历史、现状，重点向读者介绍本书的应用场景及分布结构。

1.1　Android 的来龙去脉

1.1.1　Android 的发展简介

Android 是一个以 Linux 为基础的开源操作系统，其主要应用于嵌入式设备。这里需要给大家澄清一点，之前大家了解的 Android 可能仅仅局限于手机，其实 Android 应用的嵌入式设备种类很多，其涉及的领域也很多。例如，车载领域中的导航系统，医疗领域中的电子诊断设备，在智能监控领域的智能摄像头等，这些已经在各个领域中占有一定的市场，当然现在家用的很多设备也有 Android 的身影。例如，Android 系统的电视机，Android 系统的电脑等。所以 Android 的应用领域不仅仅是手机。Android 是由 Google 成立的 OHA（Open Handset Alliance，开发手持设备联盟）领导和开发的。

Android 系统最初是由 Andy Rubin 开发的，最早开发这个系统的目的是为了打造一个能与 PC 互动的智能相机网络，但后来智能手机市场开始爆棚，Android 被改造成手机的操作系统。其 2005 年被 Google 收购，2007 年 Google 与 80 余家硬件制造厂商、软件开发厂商和电信运营厂商成立 OHA，共同改良 Android 系统。随后 Google 开发了 Android 的源代码，让各大生产厂商推出搭在 Android 系统的智能手机，再后来 Android 系统扩展到平板电脑、电脑领域。与此同时 Google 通过官方商店 Google Play，向用户提供应用程序和游戏的下载,就这样在 2010 年末,根据有关数据显示,推出两年的 Android 已经超于称霸十年的诺基亚 Symbian 系统而成为全球第一大智能手机操作系统。

Android 系统之所以这么流行,个人总结有以下特点:

(1) Android 操作系统几乎支持所有的网络制式,包括 GSM/EDGE、IDEN、CDMA、BlueTooth 和 Wifi 等等,这是任何一个手机操作系统都无法做到的。

(2) Android 操作系统由于是在 Linux 基础上发展而来的,所以它更像一个电脑操作系统。它几乎可以做电脑可以做的所有事,如 Android 原生系统就支持短信、邮件、网络访问、多语言功能、内置浏览器、支持 Java、支持多种媒体格式的图片和视频等,凡是你能想到的电脑的功能都可以在 Android 的系统中找到缩影。

(3) Android 操作系统支持的硬件种类繁多,由于 Android 是由 Google 联合各大硬件厂商共同维护开发的,所以 Android 的操作系统支持的硬件种类繁多,如摄像头、电容电阻屏幕、GPS、加速器、陀螺仪、气压计、磁强计、体感控制器、游戏手柄、蓝牙设备、无线设备、感应和压力传感器等。

(4) Android 操作系统有强大的 Google 支撑,在原生的 Android 系统中会带有 Google 提供的各项服务,如 Google 地图服务、Google 读书服务和 Google 语音服务等。这就相当于 Android 已经有了很强大的服务团队。

有了以上的几点,你还不动心吗?想做一个合格的 Android 开发工程师吗?安心的看完本书吧。

1.1.2　Android 的平台架构

Android 系统是基于 Linux 系统发展而来的,使用的开发语言一共涉及到两种:底层采用 C/C++来进行开发,上层应用采用 Java 语言来开发。Android 系统的主要组成部分,如图 1.1 所示。

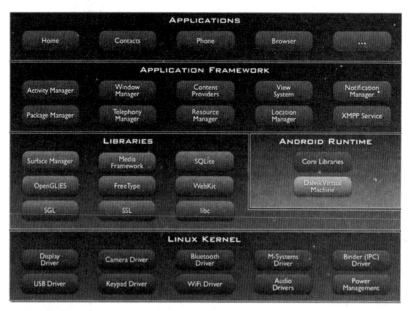

图 1.1　Android 层次结构图

从图 1.1 中明显看到 Android 的系统分为五个层次，由下至上分别为 Linux Kernel、Android Runtime、Libraries、Application Framework 和 Applications。它被叫做为一种"软件层叠结构"的方式进行构建，这种方式使得 Android 的各个层次之间相互分离，每个层次的分工明确，保证层次之间的关系，但是彼此都相互独立，降低了耦合性。对于这五个部分的功能介绍分别如下：

1. Linux Kernel层

Android 基于 Linux 2.6 提供核心系统服务，例如，安全、内存管理、进程管理、网络堆栈和驱动模型。Linux Kernel 也作为硬件和软件之间的抽象层，它隐藏具体硬件细节而为上层提供统一的服务。由于它的开发偏向于底层硬件，所以主要的开发语言为 C/C++。

如果你只是希望做 Android 的应用开发，那暂时不需要深入了解 Linux Kernel 层。

2. Android Runtime层

Android 包含一个核心库的集合，提供大部分在 Java 编程语言核心类库中可用的功能。每一个 Android 应用程序是 Dalvik 虚拟机中的实例，运行在他们自己的进程中。Dalvik 虚拟机设计成在一个设备可以高效地运行多个虚拟机。Dalvik 虚拟机可执行文件格式是.dex，dex 格式是专为 Dalvik 设计的一种压缩格式，适合内存和处理器速度有限的系统。

大多数虚拟机包括 JVM，都是基于栈的，而 Dalvik 虚拟机则是基于寄存器的。两种架构各有优劣，一般而言，基于栈的机器需要更多指令，而基于寄存器的机器指令更大。dx 是一套工具，可以将 Java .class 转换成 .dex 格式。一个 dex 文件通常会有多个.class。由于 dex 有时必须进行最佳化，会使文件大小增加 1～4 倍，以 ODEX 结尾。

Dalvik 虚拟机依赖于 Linux 内核提供的基本功能，如线程和底层内存管理。

3. Libraries层

Android 包含一个 C/C++库的集合，供 Android 系统的各个组件使用。这些功能通过 Android 的应用程序框架（application framework）暴露给开发者。下面列出一些核心库：

- 系统 C 库——标准 C 系统库（libc）的 BSD 衍生，调整为基于嵌入式 Linux 设备。
- 媒体库——基于 PacketVideo 的 OpenCORE。这些库支持播放和录制许多流行的音频和视频格式，以及静态图像文件，包括 MPEG4、H.264、MP3、AAC、AMR、JPG 和 PNG。
- 界面管理——管理访问显示子系统和无缝组合多个应用程序的二维和三维图形层。
- LibWebCore——新式的 Web 浏览器引擎，驱动 Android 浏览器和内嵌的 Web 视图。
- SGL——基本的 2D 图形引擎。
- 3D 库——基于 OpenGL ES 1.0 APIs 的实现。库使用硬件 3D 加速或包含高度优化的 3D 软件光栅。
- FreeType——位图和矢量字体渲染。
- SQLite——所有应用程序都可以使用的强大而轻量级的关系数据库引擎。

4．Application Framework层

通过提供开放的开发平台，Android 使开发者能够编制极其丰富和新颖的应用程序。开发者可以自由地利用设备硬件优势、访问位置信息、运行后台服务、设置闹钟、向状态栏添加通知等等。

开发者可以完全使用核心应用程序所使用的框架 APIs。应用程序的体系结构旨在简化组件的重用，任何应用程序都能发布它的功能且任何其他应用程序可以使用这些功能（需要服从框架执行的安全限制）。这一机制允许用户替换组件。

所有的应用程序其实是一组服务和系统，包括：

- 视图（View）——丰富的、可扩展的视图集合，可用于构建一个应用程序。包括包括列表、网格、文本框和按钮，甚至是内嵌的网页浏览器。
- 内容提供者（Content Providers）——使应用程序能访问其他应用程序（如通讯录）的数据，或共享自己的数据。
- 资源管理器（Resource Manager）——提供访问非代码资源，如本地化字符串、图形和布局文件。
- 通知管理器（Notification Manager）——使所有的应用程序能够在状态栏显示自定义警告。
- 活动管理器（Activity Manager）——管理应用程序生命周期，提供通用的导航回退功能。

5．Applications层

Android 包括一个核心应用程序集合，包括电子邮件客户端、SMS 程序、日历、地图、浏览器、联系人和其他设置。所有应用程序都是用 Java 编程语言写的，也是本书所介绍的主要内容所在。

以上所介绍的是 Android 的层次结构，在每个层次大家可能都会找到一些合适的职位。本书主要的内容在于讲解 Android 的上层应用开发，也就是使用已有的 API 构造常见的应用程序。希望各位读者能明确自己的开发定位。

1.2　本书的目的及范例应用范围

截止到目前为止，现在市面上已经有很多很多 Android 开发的相关书籍，其中不乏有一些经典之作，但是有很多同学会向我反映：老师，我们看完了 Android 的书，但是还是无法去完成一款专业的应用程序，怎么办呢？我的答案就是：编程不是历史，它不但需要你有坚实的理论基础，最主要的是要去练习，其中的练习包括这么几个阶段：

（1）能够了解 Android 程序开发的基础知识，如 Android 中是使用 Activity 来进行页面展示的。

（2）能够看懂别人的程序，如遇到某个功能能够看懂类似的实例，然后可以进行修改完成。

（3）能够看懂系统的代码，这个阶段相当于进阶阶段了，也就是你可以读懂 Android 系统的实现原理，不再停留在上层的 API 使用的阶段。

（4）设计自己的应用，独立开发想要的各种需求。这个阶段你就已经把 Android 融会贯通了。

本书以 200 个使用范例组成，其中很多都是常见的应用程序中的某些固定功能模块，例如，微信的摇一摇功能，安卓市场软件的下载安装功能等。本书把常见的功能模块拆解成最简单的单元，大家看完这些事例后，会发现他们很简单的就可以出现在你的应用程序中。

本书针对熟悉 Java 程序语言，并且了解 Android 程序开发基础知识的读者。当然，前者是必须的，如果对于 Android 一窍不通，但是你有一颗真心学会 Android 开发的心，我想你也能从本书获益良多。本书内容主要分为 9 章，每章内容都有自己的侧重点所在：

第 1 章打开 Android 世界的大门，主要介绍 Android 的发展史，以及本书的应用范围。

第 2 章 Android 开发者必备利器，主要介绍 Android 开发环境的搭建。

第 3 章让你的程序变成美女，主要介绍如何构造 Android 中漂亮的界面。

第 4 章让你的程序和用户交流，主要介绍如何让用户更好的和 Android 应用程序进行交互。

第 5 章 Android 程序内部的信息传递者，主要介绍如何在 Android 内部的组件之间传递消息。

第 6 章 Android 的数据存储，主要介绍 Android 中如何长期保存应用程序的数据。

第 7 章 Android 的服务与广播，主要介绍 Android 中的服务类软件如何开发及如何监听系统广播。

第 8 章 Android 的网络编程，主要介绍 Android 如何请求网络上的数据并进行解析。

第 9 章 Android 的多媒体开发，主要介绍 Android 的硬件相关的一些开发。

本书的所有范例采用统一编号，你从本书的目录就可看到，这些范例大多数都是根据我开发 Android 的应用程序的过程中总结而出的。现在在市面上很少有以范例书的整体思路贯穿的，本书在范例的讲解上统一了讲解风格，以减少大家理解程序的时间，对于范例的结构，基本分为以下几点。

- 实例简介：介绍本实例的应用场合；
- 运行效果：通过截图的形式看到实例的运行效果；
- 实例程序讲解：讲解实例中详细介绍了关键的代码和实现步骤；
- 实例扩展：本部分主要介绍此实例可扩展的实例，或者需要大家在开发本实例中注意的地方。

1.3 本书范例的使用方式

本书中所有的范例都经过作者在模拟机或者真机上反复测试，因为有些实例需要硬件的支持，在 AVD 上无法进行效果展示，如照相机效果、录音效果等。本书附带随书光盘，其中所有的 200 个范例都是在 Eclipse 中编译测试通过的。大家只需打开 Eclipse 导入工程即可运行，前提是你要参照第 2 章的内容配置好了 Android 的开发环境。个人建议大家在进行范例阅读的时候不要仅仅看范例的执行效果和代码，要尽量去想一下，在实际应用中什么地方可以使用此范例，是否可以进行改进，这样你的编程水平才可以迅速提高。

1.4 参考网站

大家在阅读本书的过程中可能还需要一些知识的查询和翻阅，最主要的参考网站就是 Google 的开发者网站：http://developer.android.com/index.html。此网站是 Android 的开发者网站，其中包括了最新的 Google 提供的 Android 的开发者文档，以及最新的 Android 咨询。当然还有一些网站大家可以去参考，如下所示：

Eclipse 的下载安装网站：http://www.eclipse.org/，在此网站下载安装 Eclipse，并且可以进行 Eclipse 的进阶学习。

Java SE 中 JDK 的参考文档：http://www.oracle.com/technetwork/java/javase/downloads/index.html，此网站提供了 JDK 的下载网址，并且提供了基本的 JDK 的开发者文档。

中国软件开发联盟网站：http://www.csdn.net/，这就是中国开发者联盟网站也就是通常所说的 CSDN。

全球 IT 界最受欢迎的技术问答网站：http://stackoverflow.com/，此网站中大家可以发表问题并且接受全世界的高手的指点。

以下两个是我们开发人员最常用的实例搜索网站。

谷歌的开源代码搜索站：https://code.google.com/intl/zh-CN/，在此网站大家可以搜索到你想要的开源代码实例。

GitHub 网站：https://github.com/，在此网站上大家可以托管共享你的代码工程，而且也可以看到别人托管的开源工程。

合理使用以上网站会给你的开发道路奠定坚实的基础。那让我们开始学习 Android 的旅途吧。

第 2 章 Android 开发者必备利器——开发环境搭建

要想进行 Android 应用的开发，需要提前进行一些准备工作，本章就来学习在 Android 开发前需要进行的准备工作。其中主要包括 Android 开发所需要的软件、硬件及开发环境的配置。

本章主要带领大家一起进行 Android 的开发环境的搭建，以及在开发过程中常用工具的使用。希望大家阅读完本章内容后，可以在自己的电脑上独立搭建自己的 Android 开发环境，熟练掌握 Android 常用工具的使用。

2.1 搭建 Android 开发环境

2.1.1 准备工作

Android 是 Google 推出的智能手机操作系统，其开发环境总体也分为三大平台。
- Windows 开发平台：Eclipse+ADT+Android SDK for Windows。
- Linux 开发平台：Eclipse+ADT+Android SDK for Linux。
- MAC OS 开发平台：Eclipse+ADT+Android SDK for MAC。

其中每个平台的开发环境的搭建思路大致都是一致的。大体分为，首先搭建 Java 开发环境，然后下载安装 Eclipse 开发工具，然后下载对应的 Eclipse 版本的 ADT 插件，然后下载对应的 Android SDK，最后进行配置即可。本书的开发环境是在 Windows 平台下进行搭建的，所以本章节仅介绍在 Windows 平台下如何进行 Android 开发环境的搭建。如果使用的是其他的操作系统平台的话，请参考谷歌官网为开发者提供的开发环境搭建步骤，地址是 http://developer.android.com/intl/zh-CN/sdk/index.html。

在 Windows 下搭建 Android 开发环境首先要具有一台安装了 Windows XP（32-bit）或者 Vista（32- or 64-bit）或者 Windows 7（32- or 64-bit）的电脑。然后才可以进行下面的操作。

2.1.2 安装 JDK，配置基本 Java 环境

首先，下载安装 JDK，打开 Java 官网，效果如图 2.1 所示。

地址为：http://www.oracle.com/technetwork/java/javase/downloads/index.html?ssSourceSiteId=ocomen，单击图中红色方框中的内容，进行 JDK 的下载。截止本书写作时间为止，Java 官网提供的最新的 JDK 为 jdk-7u17-windows-i586.exe。大家可以进行下载安装。安装过程如下：

图 2.1 JDK 官网

（1）双击 JDK7 安装程序，单击"下一步"按钮，如图 2.2 所示。

显示安装的组件和路径页面，如果需要修改路径的话可以单击"更改"按钮进行安装路径的修改，组件也可进行修改，这里就是用默认的组件，安装路径修改为 C:\Program Files\Java\jdk1.7\，单击"下一步"按钮，如图 2.3 所示。

图 2.2　JDK 安装界面

图 2.3　JDK 路径修改界面

（2）向导提示 jre 的安装路径，默认即可，如图 2.4 所示。单击"下一步"按钮提示完成，如图 2.5 所示。

图 2.4　jre 安装界面

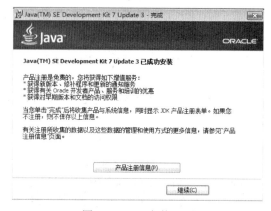

图 2.5　JDK 安装界面

（3）安装完 JDK 后，还要进行环境变量的配置。

（4）右击桌面上的"计算机"图标，选择"属性"命令，选择"高级系统设置"选项，在弹出的对话框中选择"高级"标签，并单击"环境变量"按钮，弹出"环境变量"对话框，如图 2.6 所示。

图 2.6　环境变量配置界面

（5）选择"系统变量"，单击"新建"按钮，依次输入如下"变量名"、"变量值"的内容。

- JAVA_HOME：C:\Program Files\Java\jdk1.7（JDK 安装路径）。
- PATH：%JAVA_HOME%\bin; %JAVA_HOME%\jre\bin。
- CLASSPATH：.;%JAVA_HOME%\lib;%JAVA_HOME%\lib\tools.jar。

（6）单击"确定"按钮，配置结束，如图 2.7 所示。

图 2.7　JAVA_HOME 环境变量设置界面

（7）然后通过 Win+R 调出系统的运行对话框，输入 cmd 打开系统命令行。

（8）在命令行窗口下键入以下命令，查看是否配置正确：

java –version：查看安装的 JDK 版本信息。

如果可以查看如图 2.8 所示的 java 信息，表示安装正常，可以使用。

图 2.8　Java 信息界面

2.1.3　安装 Eclipse

接下来就是安装 Eclipse 了。Eclipse 是一个开源的开发工具，在 Windows 平台上安装步骤很简单，只要登录 http://www.eclipse.org/downloads/，选择版本进行下载，然后进行解压缩即可，截止到本书的著作时间，最新的 Eclipse 版本为 4.2.2，如图 2.9 所示。

图 2.9　Eclipse 下载界面

解压缩完毕后，打开 Eclipse 提示如图 2.10 所示。

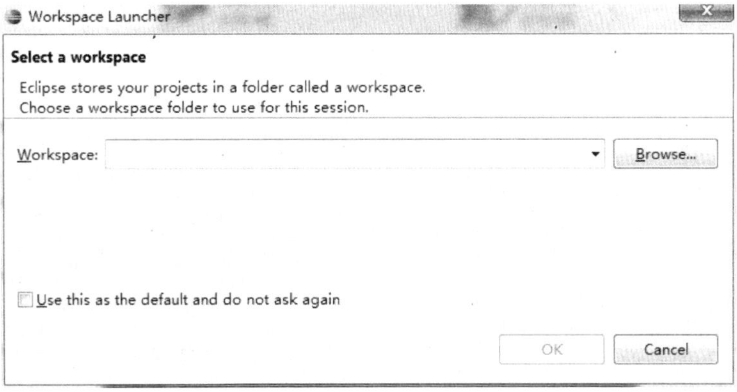

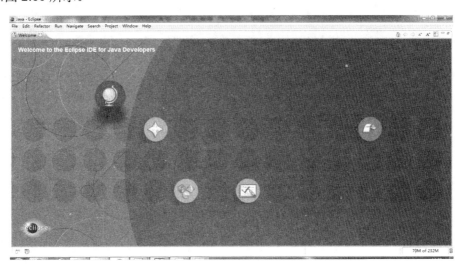

图 2.10　Eclipse 工作目录选择界面

大家可以在此向导界面中设定你的工作空间的目录。然后即可看到 Eclipse 的欢迎页了，如图 2.11 所示。

图 2.11　Eclipse 欢迎界面

恭喜大家，Eclipse 安装成功了。

2.1.4　安装 Eclipse 的 ADT 插件

Eclipse 安装完成后，还不能开发 Android 的应用，需要安装 ADT（Android Development Tools）插件才可以进行 Android 应用的开发，安装方式有两种，如下所示。

1．在线安装

Google 提供了 ADT 插件的在线安装和更新的地址，大家可以通过网络进行插件的安装和更新，具体步骤在：http://developer.android.com/intl/zh-CN/sdk/installing/installing-adt.html。

单击 Help 菜单，选择下拉菜单的 Install New Software 选项，如图 2.12 所示。

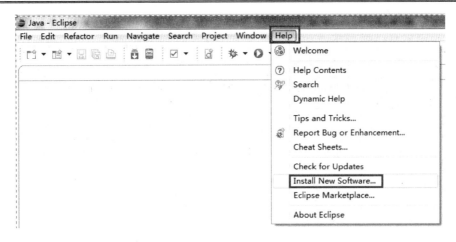

图 2.12 Eclipse 添加插件界面

然后弹出 Eclipse 安装插件的向导框，输入 https://dl-ssl.google.com/android/eclipse/，获取 ADT 插件然后选择下一步安装即可，如图 2.13 所示。

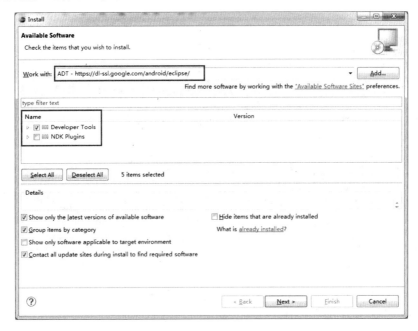

图 2.13 在线安装 ADT 界面

推荐大家使用在线安装的方式进行 ADT 下载安装，因为今后有了更新的话可以直接升级。

2．离线安装

Google 还提供了 ADT 插件离线安装包，大家可以通过下载离线安装包，然后进行离线安装，地址在 http://developer.android.com/intl/zh-CN/sdk/installing/installing-adt.html 中，截止到当前时间为止，ADT 的最新版本为 ADT-21.1.0。下载完成后，在安装插件的页面选择本地安装即可，如图 2.14 所示。

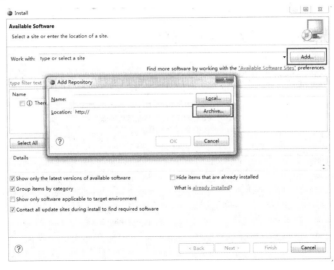

图 2.14　本地安装 ADT 界面

离线安装的方式，在安装的时候不需要电脑联网，但是后期无法自动进行 ADT 的更新，只能再次下载新的 ADT 离线安装文件进行更新。

两种 ADT 的安装方式都可以进行 ADT 的安装，安装完毕后，在 Eclipse 的状态栏会多出 ADT 的按钮，但是这两个按钮还不能单击，单击会出错，如图 2.15 所示。

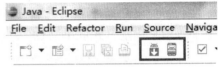

图 2.15　ADT 状态栏按钮

这两个按钮的介绍分别如下所示。

❑ Android SDK Manager：用来管理 Android 的 SDK，包括升级和下载等。

❑ Android Virtual Device Manager：用来管理 Android 的 AVD。

2.1.5　获取 Android SDK

要开发 Android 的应用就要下载 Android 的 SDK，谷歌官方提供了 Android SDK 的下载方式，地址：http://developer.android.com/intl/zh-CN/sdk/index.html。下载对应版本的 SDK Tools 解压缩即可，如图 2.16 所示。

图 2.16　Android SDK 下载网址

2.1.6 在 Eclipse 中配置 Android SDK

打开 Eclipse，选择 Windows 菜单下的 Preferences 选项，如图 2.17 所示。

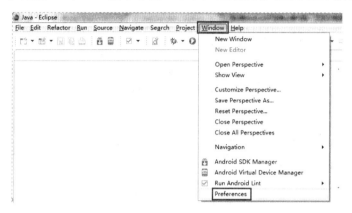

图 2.17　选择 Eclipse 的配置界面

（1）弹出配置对话框，然后选择 Android 选项下，配置 Android SDK 的解压缩目录，在下面就会识别出 Android SDK 中已经安装的 Android SDK 的版本了，如图 2.18 所示。

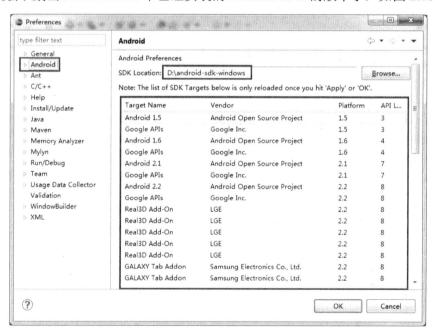

图 2.18　配置 Android SDK 界面

（2）当然如果是新下载的 Android SDK 的话，里面有的 SDK 的版本应该没有图 2.18 那么多。想要进行更新的话，单击 Android SDK Manager 按钮，会弹出 SDK 的更新对话框，如图 2.19 所示。

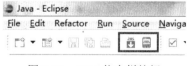

图 2.19　ADT 状态栏按钮

（3）在更新对话框中选择需要更新的 SDK 的版本，进行更新即可，如图 2.20 所示。

由于这个更新的时间很漫长，所以大家可以选择需要的 SDK 版本进行更新。本书中所有的实例都是在 Android 4.2.2 版本测试通过的，所以希望大家先更新 Android 4.2.2 版本即可，也就是 API Level 17，剩下的大家可以在有时间的时候再进行更新即可。

2.1.7 管理 AVD

当然我们搭建好 Android 的开发环境后，就可以创建一个 AVD（Android Virtual Device）了，也就是 Android 的虚拟设备，我们叫做 Android 的虚拟机。因为 Android 的应用都要在 Android 的环境下才可以安装调试，所以我们需要创建一个 Android 的虚拟设备方便我们看到程序的效果及调试。创建 AVD 的方法如下。

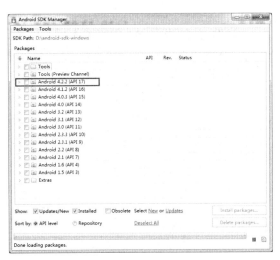

图 2.20　Android SDK Manager 界面

（1）单击 Android Virtual Device Manager 按钮，弹出 AVD 的管理窗口，如图 2.21 所示。

（2）打开后，单击 New 来创建一个新的 AVD。因为本书中所有的例子都是在 Android 4.2.2 上测试通过的，这里就创建 Android 4.2.2 的 AVD，如图 2.22 所示。

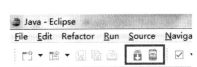

图 2.21　ADT 状态栏按钮

（3）创建 AVD 的界面，如图 2.23 所示。

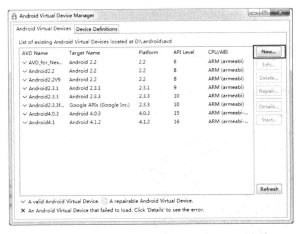

图 2.22　Android Virtual Device Manager 界面

图 2.23　创建 AVD 的界面

其中一些选项需要大家进行填写，具体如下。

❑ AVD Name：就是 Android 模拟器的名字，这个可以随意的取，本人习惯采用 Android

的版本号作为名字以方便区别各个 AVD。本例填写 Android4.2.2。
- Device：代表 AVD 的屏幕大小，例如：这里选择 3.2 寸的 HVGA 的分辨率。
- Target：代表 AVD 的系统版本，例如：这里选择 Android 4.2.2-API Level 17。
- Internal Storage：代表 AVD 的手机内部存储空间，这里默认 200M。
- SD Card：代表 AVD 的 SDCard 存储空间，这里为 256M。

（4）单击 OK 按钮，就创建好 AVD 了。

（5）在之前的 AVD Manager 的页面就会有新建的 AVD 了，如图 2.24 所示。

（6）选择新建的 AVD，然后单击 Start 按钮。启动选择默认即可，如图 2.25 所示。

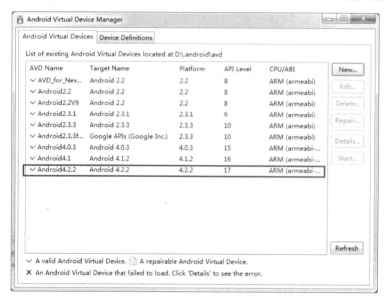

图 2.24　AVD 列表页面　　　　　图 2.25　启动 AVD 的界面

（7）然后，就可以启动建立的 AVD 了，如图 2.26 所示。

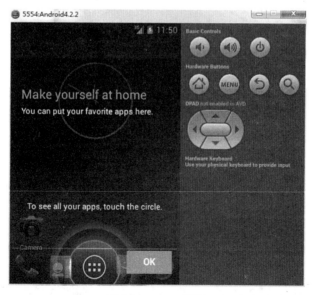

图 2.26　AVD 界面

当看到如图 2.26 所示后，恭喜你，说明 AVD 已经建立成功了。

2.2 建立第一个 Android 程序

2.2.1 建立一个 Android 工程

（1）单击 File 菜单中的 New 选项，选择 Android Application Project，创建一个 Android 工程，如图 2.27 所示。

图 2.27　创建 Android 工程

（2）然后弹出建立 Android 工程的向导对话框，如图 2.28 所示。

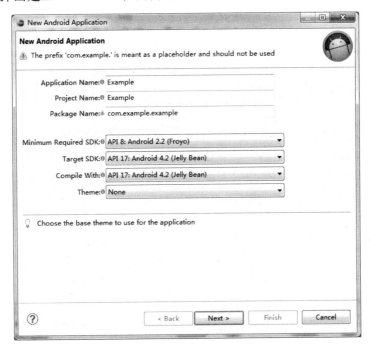

图 2.28　选择工程的 SDK 版本

其中有一些选项希望用户填写，具体含义如下所示。
- Application Name：是应用的名字。
- Project Name：是工程的名字。
- Package Name：是工程的基础包名，本书中所有的实例的包名都为 com.wyl,example，工程名为案例的编号。
- Minimun Require SDK：代码程序可运行的最低 SDK 的版本，这里选择 Android 2.2 API8。

- ❑ Target SDK：代码工程可运行的最高 SDK 版本，一般选择最新的，这里选择 Android 4.2.2 API17。
- ❑ Compile With：是工程用来编译的 SDK 版本，一般也是最新的，这里选择 Android 4.2.2 API17。
- ❑ Theme：是创建工程的 UI 风格，根据用户的需要进行选择即可，这里默认选择 None。

（3）填写完毕后单击 Next 按钮。弹出"创建工程"对话框，如图 2.29 所示。

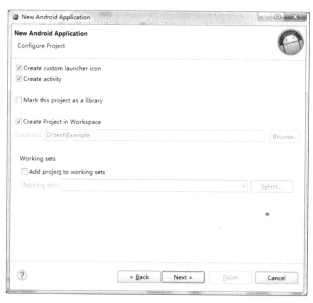

图 2.29　创建工程选项页面

其中 Create custom launcher icon 选项是用来设置程序的图标的，其他选项基本默认即可。然后弹出"创建基本 Activity"的对话框，如图 2.30 所示。

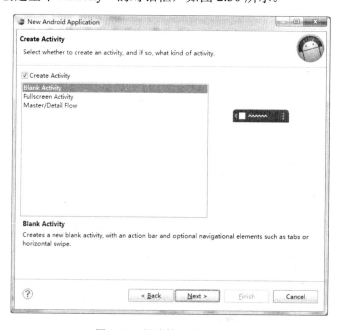

图 2.30　创建第一个 Activity

（4）这里设置创建基本的 Activity，然后单击 Next 按钮，如图 2.31 所示。

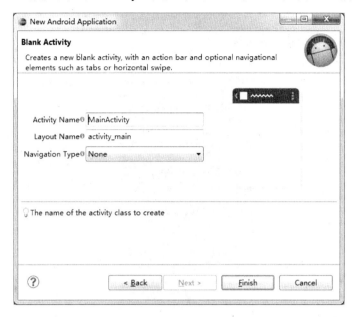

图 2.31　设置第一个 Activity 的名字及布局名

（5）默认单击 Finish 按钮即可。

2.2.2　Android 程序的目录结构

建立好工程后，大家会发现 Eclipse 里面就有了我们创建的工程了，这个工程直接可以运行。想要运行此工程的话，可以右击工程选择 Run AS，然后选择 Android Application 命令，如图 2.32 所示。

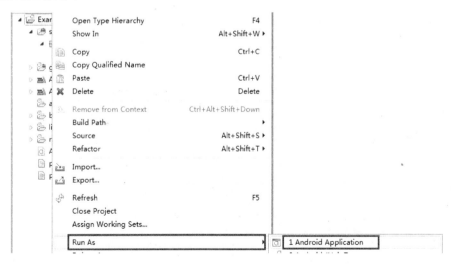

图 2.32　运行 Android 工程的界面

可以在 AVD 中看到程序的运行效果，如图 2.33 所示是一个"Hello world！"字样的效果。大家可以看到工程的目录如图 2.34 所示。

图 2.33　Hello world 实例效果　　　图 2.34　Android 工程目录

在工程中有很多文件夹，其中的功能希望大家有所了解，具体如下所示。
- src 目录：源代码目录，在此目录下是包名，对应包名中包含对应的 java 源代码。
- gen 目录：是 ADT 自动生成的资源 ID 目录。此目录是不需要大家修改的，是 ADT 自动产生的。
- assets 目录：是工程中用到的外部资源的文件夹所在。这里暂时为空。
- bin 目录：是编译工程成功后的 apk 所在的目录。
- libs 目录：是外部的 jar 所在的文件夹。
- res 目录：为工程内部的资源目录，其中包括图片资源、字符串资源和布局资源等。
- AndroidManifest.xml 文件：是工程的配置文件，在其中设置工程 apk 的权限及组件注册等。

对于我们开发 Android 的应用来说 src 目录、res 目录和 AndroidManifest 文件，这些是我们在今后做实例的时候使用最多的，当然其他目录也可能会使用到，等我们讲到对应实例的时候再做具体讲解。

2.3　开发必备利器

2.3.1　Logcat 的使用

Logcat 是我们在 Android 应用开发过程中调试的一个重要的工具，它可以捕捉 AVD 中的一些 Log 信息，其中 Log 的级别分为 error、waring、info 和 debug 等，这些我们都可以通过 Logcat 进行捕捉。打开 Eclipse 的 Logcat 的方法，如图 2.35 所示。

然后选择：Android 文件夹中的 LogCat 窗口，如图 2.36 所示。

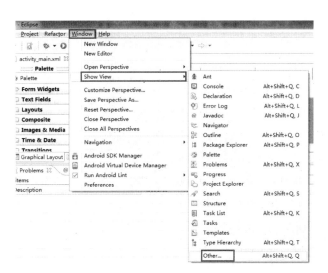

图 2.35　显示窗口

图 2.36　显示 LogCat 窗口

这样在 Eclipse 中就可以看到 LogCat 窗口了，如图 2.37 所示。

图 2.37　Logcat 窗口显示效果

今后我们的 Log 就会在这个窗口中显示出来了。

2.3.2　DDMS（Dalvik Debug Monitor Service）的使用

DDMS 是 Eclipse 中的调试控制器，在今后程序调试的过程中有很多方面要用到 DDMS。在 Eclipse 中打开 DDMS 的方法如图 2.38 所示。

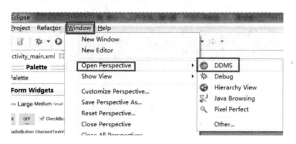

图 2.38　打开 DDMS 窗口

在 DDMS 页面中我们可以看到很多调试的窗口如图 2.39 所示。

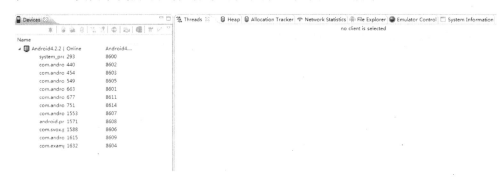

图 2.39　DDMS 窗口显示效果

例如：
- Devices 窗口显示连接当前电脑的所有设备。
- File Explorer 窗口显示某个设备的文件列表。
- Emulator Control 窗口模拟对于 AVD 的电话、短信或者经纬度的模拟。

当然在 DDMS 中我们今后还会用到很多窗口，今后在使用过程中再给大家讲解。

2.3.3　ADB（Android Debug Bridge）的使用

ADB 是 Android 手机连接电脑调试的最基本的进程，在我们的命令行中也可以使用 ADB 的命令进行调试和对手机的各种操作，在 cmd 框中输入 adb，可以看到所有 adb 的命令，如图 2.40 所示。

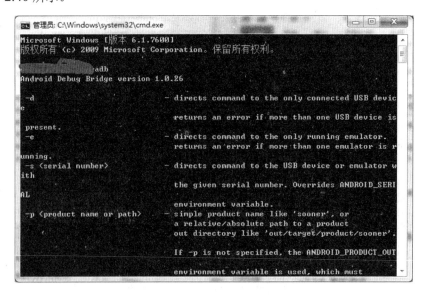

图 2.40　ADB 命令窗口

其主要的命令参数，请看 http://developer.android.com/intl/zh-CN/tools/help/adb.html，在此页面中列举出了 ADB 命令的所有的参数及用法。大家可以根据自己的情况来使用。

2.3.4　The Hierarchy Viewer 的使用

Hierarchy Viewer 是在 Android SDK 中提供的视图检查工具，位置在 Android SDK 的解压后的 tools 文件夹下，名为 hierarchyviewer.bat。它是 Android 自带的非常有用而且使用简单的工具，可以更好地检视和设计用户界面（UI）。使用方法就是双击 hierarchyviewer.bat，选择需要查看的应用包名即可，如图 2.41 所示。

图 2.41　Hierarchy Viewer 窗口显示效果

在图 2.41 所示中左边的树形结构就代表了我们当前的 Hello world!应用的页面布局效果。这样就可以很直观的看到页面中的布局层次了。

2.3.5　Draw9-Patch 的使用

Draw 9-Patch 是一个九宫格的画图器。它可以对每张图分布成九个块儿，这样的话，就可以实现同一张图片在不同屏幕的手机上都可以用了。具体使用方法是打开 tools 目录下的 draw9patch.bat 文件，然后打开对应图片进行编辑即可，如图 2.42 所示。

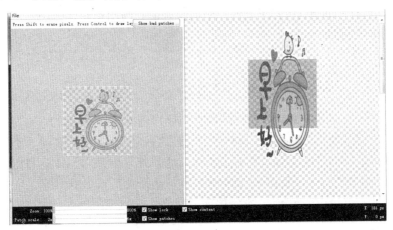

图 2.42　Draw9-Patch 窗口显示效果

2.3.6　真机测试

对于我们做 Android 开发的话，大部分时候是使用 AVD 进行测试和调试。但是对于某些特殊的功能 AVD 无法进行模拟，例如：手机的摇一摇功能，还有手机的摄像头功能，这些都需要有一个 Android 的真实设备。

电脑连接 Android 手机的步骤有如下几个：

（1）保证手机驱动在电脑上已经安装完成。一般第三方的手机软件都可以安装。

（2）进入 Eclipse 的 DDMS，查看 Devices，如果列出了手机，说明手机和 Eclipse 已经正常连接了，然后大家就可以在真机上进行开发和调试了，如图 2.43 所示。

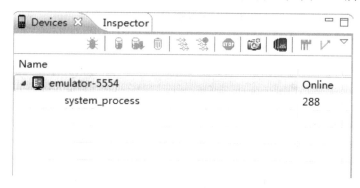

图 2.43　Devices 窗口显示效果

2.4　Android 程序的基本组件

2.4.1　Activity 组件介绍

Activity 是 Android 中最基本也是最重要的一个组件，它主要负责 Android 中的页面展示，所有大家看到的 Android 中的界面，都是 Activity。所以这部分内容是和用户直接接触的内容，负责用户的 UI 和 UE，所以你的应用是否能够吸引用户，界面起了绝大部分的作用。

2.4.2　ContentProvider 组件介绍

Android 中每个应用程序都有自己的内存空间，而且应用程序之间的内存空间是无法相互访问的，这就带来了一个问题，如果几个应用程序之间希望共享同一份数据的话，将很难实现。例如，我们手机上有可能有多个短信客户端，但是它们访问的短信数据确是同一份库。Android 中通过 ContentProvider 来实现应用程序间的数据共享。所以应用程序间的数据只有通过 ContentProvider 来进行共享。

2.4.3　Service 组件介绍

Service 运行在 Android 的后台，它不和用户直接交互，但是它却能够为用户提供一些服务。例如：后台的音乐播放、后台的任务下载等。当然 Android 系统中大部分与硬件相关的一些功能也都是通过服务来实现的，如电话服务、短信服务和 GPS 服务等。所以如果当你希望开发的功能是在后台运行的，那么你就应该考虑使用 Service 实现了。

2.4.4 BroadcastReceiver 组件介绍

Broadcast 是 Android 中各个应用程序之间传输消息的最基本机制，也是唯一的机制。而我们在应用中就可以通过 BroadcastReceiver 来截获系统所发出的广播消息，从而获取系统所要传达的消息。例如，接收短信广播，可以实现短信的拦截功能，接收电话广播可以实现电话的黑名单功能等。所以如果你想要实现的功能是通过系统的广播，来实现一些功能，那你就要考虑使用 BroadcastReceiver 了。

2.4.5 Intent 组件介绍

之前我们介绍的几个组件，是 Android 中的基本组件，但是这些组件之间想要进行交互，就一定要使用 Intent 了。例如，通过 Activity 去启动另一个 Activity，通过 Activity 去发送广播等。这些都要用到 Intent 组件。而且很多与系统的功能交互也要使用 Intent，所以如果你希望通过一个组件去启动另一个组件的话，就要使用 Intent 了。

2.5 小 结

在本章节中主要介绍了 Android 的开发环境的搭建还有 Android 开发过程中常见的开发调试工具，通过这些工具，我们就可以开发 Android 应用了。当然本章还介绍了 Android 中最基本组件的功能，希望通过本章的学习大家都能够配置好 Android 的开发环境。下一章我们会讲述如何开发 Android 中的界面应用。

第 3 章　让你的程序变成美女

对于 Android 应用开发最基本的就是用户界面（GUI，Graphics User Interface）的开发。如果一个应用没有好的界面，那么将很难吸引最终用户。所以用户界面的开发对于 Android 应用开发是很重要的，也是我们首先要掌握的。

Android 系统中提供了大量的 UI 组件，这些组件小到简单的文本框 TextView，大到浏览器核心控件 WebView，都可以给用户提供不同的功能感受。我们开发者只要根据用户的需求将这些 UI 组件组合在一起，就像拼装一辆汽车。尽量在有限的手机屏幕中给用户带来无限的美感体验，那么何愁你的应用没有人用呢？

本章主要通过各种应用界面的实例介绍，来带领大家一起学习 Android 的界面开发。希望大家阅读完本章内容后，可以根据自己的需求独立完成各种界面的开发。

3.1　Android 中基本控件的使用

范例 001　更改文字标签的内容

1．实例简介

在上一章中我们搭建完成 Android 的开发环境，新建立了一个 Android 工程，在 AVD 中运行可以看到在一个界面中显示 Hello world 的文字标签。这是我们的第一个 Android 程序，但是这个程序过于死板，如何让文字标签显示我们想让它显示的文字内容呢？这个实例会带领我们通过两种方式修改文字标签的文字内容，方式 1：通过控件的 xml 布局中的 text 属性修改 TextView 的文字，方式 2：通过在 Java 代码中得到 TextView 对象，然后通过对象的 setText 方法来设置 TextView 的文字。

2．运行效果

该实例运行效果如图 3.1 所示。

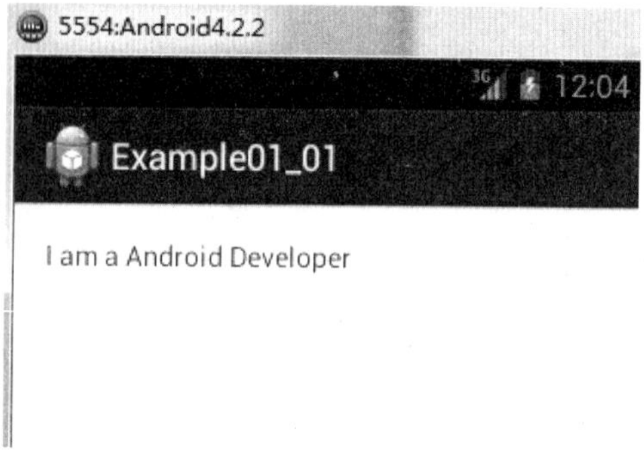

3．实例程序讲解

方式 1：通过修改 xml 布局文件中 TextView 控件的 text 属性来完成如上效果，主要修改的地方在我们建立的工程下的 res/layout/activity_main.xml。代码如下：

图 3.1　在界面中显示 I am a Android Developer

```
01 <RelativeLayout xmlns:android="http://schemas.android.com/apk/res/android"
02   xmlns:tools="http://schemas.android.com/tools"
03   android:layout_width="match_parent"
04   android:layout_height="match_parent"
05   android:paddingBottom="@dimen/activity_vertical_margin"
06   android:paddingLeft="@dimen/activity_horizontal_margin"
07   android:paddingRight="@dimen/activity_horizontal_margin"
08   android:paddingTop="@dimen/activity_vertical_margin"
09   tools:context=".MainActivity" >
10   <!-- 定义 TextView 控件，在 text 节点中设置文本标签的文字 -->
11   <TextView
12       android:layout_width="wrap_content"
13       android:layout_height="wrap_content"
14       android:text="I am a Android Developer" />
15
16 </RelativeLayout>
```

这是我们的 Activity 的布局文件，其中第 11~14 行构造了一个 TextView 控件，在 TextView 控件中 text 属性就代表这个文本标签上显示的文字，所以只要修改 text 节点的值为你想输入的字符串即可，如第 14 行的修改。

方式 2：在 Java 代码中得到 TextView 对象，然后通过对象的 setText 方法来设置 TextView 的文字。要通过这种方式修改 TextView 的内容，步骤如下。

（1）在 xml 布局文件的 TextView 控件中加上 id 字段。

```
01 <RelativeLayout xmlns:android="http://schemas.android.com/apk/res/android"
02   xmlns:tools="http://schemas.android.com/tools"
03   android:layout_width="match_parent"
04   android:layout_height="match_parent"
05   android:paddingBottom="@dimen/activity_vertical_margin"
06   android:paddingLeft="@dimen/activity_horizontal_margin"
07   android:paddingRight="@dimen/activity_horizontal_margin"
08   android:paddingTop="@dimen/activity_vertical_margin"
09   tools:context=".MainActivity" >
10   <!-- 定义 TextView 控件，在 id 节点中设置文本标签的 id -->
11   <TextView
12       android:id="@+id/Tv"
13       android:layout_width="wrap_content"
14       android:layout_height="wrap_content"
15       android:text="@string/hello_world" />
16
17 </RelativeLayout>
```

如上面中代码的第 12 行，通过 id 节点给 TextView 对象加上唯一标示的 id。这里需要注意的是 id 的值是自定义 id，所以加入的方式为@+id/Tv。其中的 Tv 是我们的 TextView 的 id。

（2）在代码中获得此 TextView 对象，通过 setText 方法修改此 TextView 的值。

主要修改的地方在我们建立的工程下的 src/com.wyl.example/MainActivity.java，代码如下。

```
01 package com.wyl.example;                              //当前包名
02 //导入必备的包
03 import android.os.Bundle;
04 import android.app.Activity;
05 import android.view.Menu;
06 import android.widget.TextView;
07
08 public class MainActivity extends Activity {          //定义 MainActivity 继承
                                                         自 Activity
```

```
09
10    private TextView Tv;                                //定义TextView的对象
11
12    @Override
13    protected void onCreate(Bundle savedInstanceState) {
14       super.onCreate(savedInstanceState);              //调用父类的onCreate方法
15       setContentView(R.layout.activity_main);          //通过setContentView方
                                                          法设置当前页面的布局文件为activity_main
16       Tv = (TextView)findViewById(R.id.Tv);            //通过findViewById得到
                                                          对应的TextView对象
17       Tv.setText("I am a Android Developer"); //通过TextView对象的setText
                                                          设置文本标签的内容
18    }
19    @Override
20    public boolean onCreateOptionsMenu(Menu menu) { //当前Activity的菜单
                                                          创建,本例没有用途
21       //Inflate the menu; this adds items to the action bar if it is present.
22       getMenuInflater().inflate(R.menu.main, menu);
23       return true;
24    }
25 }
```

如上代码第 10 行定义了一个 TextView 对象, 在第 16 行我们通过 findViewById 拿到了刚才定义了那个 TextView 的对象, 在第 17 行通过 TextView 中的 setText 方法来修改 TextView 的值。

通过上面两种方法我们都可以达到修改文本标签内容的目的, 相对来说第一种方法, 是在程序加载的时候就确定了 TextView 的内容。第二种方法是在程序运行的时候确定了 TextView 的内容, 所以如果你的文本标签的内容要根据程序运行过程中某些状态来变化的话, 要选择第二种方法。例如, 用户名标签, 一般使用第一种方式, 因为它一旦确定基本不再修改; 如果是显示网络数据的文本标签, 那么就要采用第二种方法了, 这样才能根据程序的运行状态修改标签的内容。

4. 实例扩展

扩展 1: 在 xml 布局文件中 android:text 的内容可以是字符串, 也可以是系统的资源 Id。

```
01 <TextView
02    android:id="@+id/Tv"
03    android:layout_width="wrap_content"
04    android:layout_height="wrap_content"
05    android:text="@string/str" />
```

如上面代码的第 5 行, 其中@string/str 就代表工程的 str 字符串资源, 工程的字符串资源一般保存在 res/values/strings.xml 中。

```
01 <?xml version="1.0" encoding="utf-8"?>
02 <resources>
03
04    <string name="app_name">Example01_01</string>
05    <string name="action_settings">Settings</string>
06    <string name="hello_world">Hello world!</string>
07    <string name="str">hi Android</string>
08
09 </resources>
```

在 strings.xml 文件中的第 7 行，你可以看到 str 资源的值是 hi Android，这就是你设置给 TextView 的真实内容了。

扩展 2：在 Java 代码中修改 TextView 的值的话，setText 方法有多种重载形式：

```
public final void setText (CharSequence text)              //设置文本标签内容值为
text 变量的值
public final void setText (int resid)                      //设置文本标签内容值为资
源 resid 的值
public void setText (CharSequence text, TextView.BufferType type)
                                                           //设置内容值为 Text 的值 type 代表缓冲类型
public final void setText (int resid, TextView.BufferType type)
                                                           //设置内容值资源 resid 的值 type 代表缓冲类型
public final void setText (char[] text, int start, int len)
                                 //设置内容为 text 数组的从第 start 位开始的后 len 个字符
```

范例 002　更改手机页面的背景色

1．实例简介

到目前为止，我们现在看到的页面的颜色都是系统默认的颜色值，如果我们想要在程序的执行过程中显示与众不同的页面，更改页面的背景颜色是最基本的思路。本实例就带领大家一起来学习如何更改页面的背景颜色。

2．运行效果

该实例运行效果如图 3.2 所示。

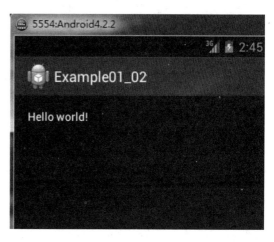

图 3.2　更改界面的背景为黑色，字体颜色为白色

3．实例程序讲解

想要实现更改页面的背景，只需修改 res/layout/activity_main.xml 即可。代码如下：

```
01  <RelativeLayout xmlns:android="http://schemas.android.com/apk/res/android"
02    xmlns:tools="http://schemas.android.com/tools"
03    android:layout_width="match_parent"
04    android:layout_height="match_parent"
05    android:paddingBottom="@dimen/activity_vertical_margin"
06    android:paddingLeft="@dimen/activity_horizontal_margin"
07    android:paddingRight="@dimen/activity_horizontal_margin"
08    android:paddingTop="@dimen/activity_vertical_margin"
09    android:background="@android:color/background_dark"
10    tools:context=".MainActivity" >
11    <!-- 在上面 RelativeLayout 节点中，添加 background 属性 -->
12    <TextView
13        android:layout_width="wrap_content"
14        android:layout_height="wrap_content"
15        android:textColor="@android:color/white"
16        android:text="@string/hello_world" />
17
18  </RelativeLayout>
```

这是我们的 Activity 的布局文件，其中第一个节点 RelativeLayout 代表当前页面布局效果为相对布局。第 9 行添加了 android:background 节点，并且其值设置成了 @android:color/background_dark，其代表 Android 系统中的颜色资源 background_dark。当然这里的值也可以是一个颜色值，如下：

```
android:background=" #FF0000"
```

其中，#FF0000 代表红色的颜色值，这样页面背景就被更改为红色。

最优的一种方式是在工程中自定义颜色资源文件，将各种颜色值加入。

在 res/ralues/目录中建立 colors.xml 资源文件，内容为：

```
<?xml version="1.0" encoding="utf-8"?>
<resources>
    <color name="red">#FF0000</color>
    <color name="green">#00FF00</color>
    <color name="blue">#0000FF</color>
</resources>
```

这样在布局文件中就可以通过如下代码来设置自定义的颜色资源了。

```
android:background="@color/green"
```

4．实例扩展

扩展 1：在 Android 中设置任何一种控件的背景方式都一样，可以在对应的 xml 布局文件中设置也可以在 Java 代码中设置。在 Java 代码中设置控件背景的步骤如下：

（1）得到需要修改背景颜色的控件，通过 findViewById 方法。

（2）通过 setBackgroundColor 方法设置控件的背景颜色。

扩展 2：对于 Android 中控件的背景的修改，不仅仅可以设置成单一颜色值，而且还可以将一张图片设置成控件的背景。实现方法与设置背景颜色相同。就是在对应的 xml 布局文件中给相应的控件加上如下代码：

```
android:background="@drawable/ic_launcher"
```

其中@drawable/ic_launcher 代表工程目录中 res/drawable/的 ic_launcher.png 图片。这样你的页面就以此图片为背景了。

范例 003　文字超链接

1．实例简介

在 Android 系统中默认情况下 TextView 仅仅用来显示文字内容，可我们经常会看到一些应用中的 TextView 不但可以显示文字，而且当它显示特殊意义的文字时具有特殊的功能。例如，当 TextView 中显示网址的时候可以单击跳转到网址；当 TextView 中显示电话的时候可以单击打电话。其实这些功能实现起来也很简单，本实例就带领大家一起来学习如何制作具有超链接效果的文字标签。

2．运行效果

该实例运行效果如图 3.3 所示。

第 3 章 让你的程序变成美女

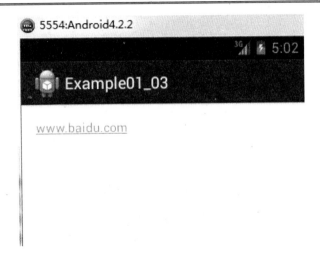

图 3.3 显示超链接的文字内容

3．实例程序讲解

想要实现具有网址超链接功能的文字标签,只需修改 res/layout/activity_main.xml 即可。代码如下：

```
01  <TextView
02      android:layout_width="wrap_content"
03      android:layout_height="wrap_content"
04      android:autoLink="web"
05      android:text="www.baidu.com" />     <!-- 定义 TextView 标签，加入
                      autoLink 节点值为 web，text 节点值为 www.baidu.com -->
```

这是我们的 Activity 的布局文件，其中在 TextView 标签中添加 autoLink 字段，并且设置值为 web，则可以实现文字超链接的功能。

4．实例扩展

对于 TextView 的 autoLink 属性，其可选择的属性值如下所示。
- none：不匹配任何类型的文字，默认为此选项。
- web：匹配 URL 地址，单击后打开浏览器显示地址。
- email：匹配邮箱地址，单击后打开邮箱发送邮件。
- phone：匹配电话号码，单击后打开拨号界面。
- map：匹配地图地址，单击后打开地图选项。
- all：匹配所有的格式，自动检测 web、phone、email 和 map 四种格式。

例如：

```
01  <TextView
02      android:layout_width="wrap_content"
03      android:layout_height="wrap_content"
04      android:autoLink="email"
05      android:text="squallwu_2006@qq.com" />     <!-- 定义 TextView 标签，
          加入 autoLink 节点值为 email，text 节点值为 squallwu_2006@qq.com -->
06  <TextView
07      android:layout_width="wrap_content"
```

```
08          android:layout_height="wrap_content"
09          android:autoLink="phone"
10          android:text="10086" />           <!-- 定义 TextView 标签,加入 autoLink
                                                   节点值为 phone, text 值为 10086 -->
11    <TextView
12          android:layout_width="wrap_content"
13          android:layout_height="wrap_content"
14          android:autoLink="map"
15          android:text="1812 Avenue K Plano,Texas 75074" />
                                              <!-- 定义 TextView 标签,加入 autoLink 节点值为 map,
                                                   text 节点值为 1812 Avenue K Plano,Texas 75074-->
```

其中,单击 autoLink 为 email 的文字标签,程序会打开手机上的邮件客户端(确保你的手机安装了邮件客户端)。单击 autoLink 为 phone 的文字标签,程序会打开手机上的电话拨号界面。单击 autoLink 为 map 的文字标签,程序会打开手机上的谷歌地图客户端(确保你的手机安装了谷歌地图客户端),而且对于现在来说支持美国的地图。

范例 004 让你的文字标签更加丰富多彩

1. 实例简介

在 Android 系统中的 TextView 不但可以实现文字展示和几种特殊形式的超链接,其实在 TextView 中还可以实现更加绚丽的效果。例如,可以在 TextView 中显示 Html 样式。这些效果对于大家在做实际项目的时候是非常实用的,本实例就带领大家一起来学习如何制作具有 Html 样式的文字标签。

2. 运行效果

该实例运行效果如图 3.4 所示。

3. 实例程序讲解

想要实现在文字标签中显示 Html 样式,一定要在 Java 代码中对 TextView 进行设置,

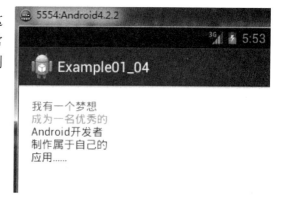

图 3.4 显示网页格式的文字内容

所以步骤就是先设置 TextView 的 id,在 Java 中通过 findViewById 方法得到对应的 TextView 对象,然后再通过 settext 方法进行设置。首先修改 res/layout/activity_main.xml 即可。代码如下:

```
01  <LinearLayout xmlns:android="http://schemas.android.com/apk/res/android"
02      xmlns:tools="http://schemas.android.com/tools"
03      android:layout_width="match_parent"
04      android:layout_height="match_parent"
05      android:paddingBottom="@dimen/activity_vertical_margin"
06      android:paddingLeft="@dimen/activity_horizontal_margin"
07      android:paddingRight="@dimen/activity_horizontal_margin"
08      android:paddingTop="@dimen/activity_vertical_margin"
09      android:orientation="vertical"
10      tools:context=".MainActivity" >
```

```
11      <!-- 给 TextView 设置 id 字段 -->
12      <TextView
13          android:id="@+id/Tv"
14          android:layout_width="wrap_content"
15          android:layout_height="wrap_content"
16          android:text="@string/hello_world"/>
17
18  </LinearLayout>
```

这是我们的 Activity 的布局文件,其中在第 13 行给 TextView 标签添加 id 字段,并且设置控件 id 为 Tv,这样就可以在 Activity 中通过 id 得到此 TextView 了。

在 src/com.wyl.example/MainActivity.java 中得到 TextView 的对象,然后通过 setText 方法设置文字内容。代码如下:

```
01  package com.wyl.example;                        //当前包名
02
03  import android.os.Bundle;
04  import android.app.Activity;
05  import android.text.Html;
06  import android.widget.TextView;
07
08  public class MainActivity extends Activity {    //定义 MainActivity
                                                    //继承自 Activity
09
10
11      private TextView Tv;                        //定义 TextView 对象 Tv
12
13      @Override
14      protected void onCreate(Bundle savedInstanceState) {
15          super.onCreate(savedInstanceState);     //调用父类的 onCreate 方法
16          setContentView(R.layout.activity_main); //setContentView 方法设
                                                    //置页面布局为 activity_main
17          Tv = (TextView)findViewById(R.id.Tv);   //通过 findViewById 得到
                                                    //TextView 对象
18
19          String str = "<font color=\"#FF0000\">我有一个梦想</font><br>"
20          +"<font color=\"#00FF00\">成为一名优秀的</font><br>"
21          +"<font color=\"#0000FF\">Android 开发者</font><br>"
22          +"<font color=\"#0F0FFF\">制作属于自己的</font><br>"
23          +"<font color=\"#0F0F0F\">应用……</font>";  //定义 html 的字符串 str
24          Tv.setText(Html.fromHtml(str));         //通过 setText 设置文字标
                                                    //签内容
25      }
26  }
```

在代码的第 17 行通过 findViewById 得到 TextView 的对象,第 19~23 行定义了一段 Html 语法的字符串文字。在第 24 行将 str 字符串通过 Html 类的 fromHtml 方法转成对应字符串,然后设置给 Tv。

4.实例扩展

当然 TextView 不但可以简单的识别 Html 中的颜色标签,还可以识别<img>标签和<a href>标签等。下面我们看一段代码。

```
01  package com.wyl.example;                        //当前包名
02
```

```
03  import android.app.Activity;
04  import android.graphics.drawable.Drawable;
05  import android.os.Bundle;
06  import android.text.Html;
07  import android.text.Html.ImageGetter;
08  import android.widget.TextView;
09
10  public class MainActivity extends Activity {        //定义MainActivity
                                                         继承自Activity
11
12    private TextView Tv;                              //定义TextView对象Tv
13
14    @Override
15    protected void onCreate(Bundle savedInstanceState) {
16      super.onCreate(savedInstanceState);      //调用父类的onCreate方法
17      setContentView(R.layout.activity_main); //通过setContentView方法
                                   设置当前页面的布局文件为activity_main
18      Tv = (TextView) findViewById(R.id.Tv);
19      //定义带有图片的html字符串str,图片为本地资源id R.drawable.ic_launcher
20      String str = "<h1>测试图片</h1><p><img src="+R.drawable.ic_launcher+"></p>";
21      Tv.setText(Html.fromHtml(str, imgGetter, null));//通过settext 设
                                                  置文字标签的内容
22    }
23
24    ImageGetter imgGetter = new Html.ImageGetter() {        //定义ImageGetter对象
25      public Drawable getDrawable(String source) { //当遇到img标签时调用
                                                     此回调方法
26        int id = Integer.parseInt(source);
27        Drawable drawable = getResources().getDrawable(id);
                                            //根据资源id得到图片对象
28                                          drawable.setBounds(0, 0,
drawable.getIntrinsicWidth(),drawable.getIntrinsicHeight());
29        return drawable;
30      }
31    };
32  }
```

在代码的第 20 行我们定义了一个带有<img>标签的字符串 str，通过 Tv 的 setText 方法将 str 的内容按照 Html 的形式展示出来，但是此时使用的方法变成：

Spanned android.text.Html.fromHtml(String source, ImageGetter imageGetter, TagHandler tagHandler)

这个方法的第一个参数是我们如上的 str，第二个参数是 ImageGetter 的类的对象，也就是我们 24～31 行定义的 imgGetter 对象，在 ImageGetter 类中有接口方法 getDrawable，此方法在 str 中检测到<img>标签后调用，并且把识别到的图片值传给 getDrawable 的参数 source，然后你就根据得到的图片值来进行图片的显示，此实例中图片值显示了一个本地的资源 id 为 R.drawable.ic_launcher，所以在 getDrawable 中通过 getResources().getDrawable 方法来获得对应的 Drawable 对象，然后返回出去。

TextView 控件对于 Html 语言的支持，官方文档原话为 Not all HTML tags are supported.，并不是所有的 Html 标签都支持，经过测试还支持的标签有：<a href=""> 超链接；<b>文字粗体；<h1>标题文字，同理 h 的其他级别；<strong> 文字强调等。当然随着 API 的升级更新可能有些支持的标签不再支持，或者添加了新的标签，那么当大家用到某种标签的时候自己进行测试即可。

范例 005 用户名密码输入框

1. 实例简介

在 Android 系统中的 TextView 主要负责内容的显示,对于内容的输入我们最常用的控件是 EditText。EditText 可以用来接收用户录入的信息,当然为了能给用户提示输入的内容,可以在输入前加入相应的提示。本实例就带领大家一起来做一个常见的用户密码输入界面。

2. 运行效果

该实例运行效果如图 3.5 所示。

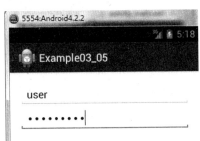

图 3.5 用户名密码输入框

3. 实例程序讲解

想要实现在页面中显示用户名和密码的输入框,只需对我们程序的 res/layout/activity_main.xml 进行修改即可。代码如下:

```xml
01  <LinearLayout xmlns:android="http://schemas.android.com/apk/res/android"
02      xmlns:tools="http://schemas.android.com/tools"
03      android:layout_width="match_parent"
04      android:layout_height="match_parent"
05      android:paddingBottom="@dimen/activity_vertical_margin"
06      android:paddingLeft="@dimen/activity_horizontal_margin"
07      android:paddingRight="@dimen/activity_horizontal_margin"
08      android:paddingTop="@dimen/activity_vertical_margin"
09      android:orientation="vertical">
10      <!-- 设置用户名的输入框,hint 代表输入前的提示语 -->
11      <EditText
12          android:layout_width="match_parent"
13          android:layout_height="wrap_content"
14          android:hint="请输入用户名"/>
15
16      <!-- 设置密码的输入框,hint 代表输入前的提示语 -->
17      <EditText
18          android:layout_width="match_parent"
19          android:layout_height="wrap_content"
20          android:password="true"
21          android:hint="请输入密码"/>
22
23  </LinearLayout>
```

这是我们的 Activity 的布局文件,其中第 11 行在当前布局中添加了一个 EditText 控件代表用户名的输入框,为了能够在用户输入之前对用户提示输入信息,加上了 android:hint 属性代表提示信息。在第 17~21 行添加了代表密码输入的 EditText,并且相对于上面的用户输入框加入了 android:password 属性代表当前为一个密码框,即用户输入过程中输入的字符是不可见的。

4. 实例扩展

当然对于用户的密码可能还会有一些限制,例如:密码长度最长 10 个字符,而且只能是数字。这样的话就要对密码的 EditView 再进行一些设置了，代码如下：

```
01  <!-- 设置密码的输入框, hint 代表输入前的提示语  -->
02  <EditText
03      android:layout_width="match_parent"
04      android:layout_height="wrap_content"
05      android:maxLength="10"
06      android:numeric="integer"
07      android:password="true"
08      android:hint="请输入密码"/>
```

在上面的第 5 行加入了 maxLength 属性，代表此输入框最多能够接收的字符的个数。在第 6 行加入了 numeric 属性代表此输入框只能输入数字。

对于 EditText 的其他限制条件也可以同样的方法进行设定，例如：maxLines 代表最多输入的行数，digits 属性代表能够接收的字符的范围等。这些属性大家在使用的时候去查看官方文档即可。

范例 006 电话号码输入框

1. 实例简介

在 Android 系统中 EditText 可以接收用户的输入，也可以进行简单的校验，但是我们通常使用的 EditText 接收的输入有一定的约束的。例如，邮件的输入框和电话号码的输入框等。本实例就带领大家一起来做自己要求的约束格式的输入框。

2. 运行效果

该实例运行效果如图 3.6 所示。

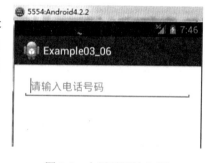

图 3.6 电话号码输入框

3. 实例程序讲解

图 3.6 想要实现的效果是当在电话的输入框中输入正确的电话号码时输入框的文字是绿色，如果输入不是电话号码的格式，输入框的颜色为红色。要想实现这样的效果有以下三个步骤。

（1）修改 res/layout/activity_main.xml 文件添加 EditText 控件并设置 id。代码如下：

```
01  <LinearLayout xmlns:android="http://schemas.android.com/apk/res/
        android"
02      xmlns:tools="http://schemas.android.com/tools"
03      android:layout_width="match_parent"
04      android:layout_height="match_parent"
05      android:paddingBottom="@dimen/activity_vertical_margin"
06      android:paddingLeft="@dimen/activity_horizontal_margin"
07      android:paddingRight="@dimen/activity_horizontal_margin"
08      android:paddingTop="@dimen/activity_vertical_margin"
09      android:orientation="vertical">
10      <!-- 设置电话号码输入框, hint 代表输入前的提示语, id 代表控件的 id -->
11      <EditText
12          android:id="@+id/EtPhone"
```

```
13        android:layout_width="match_parent"
14        android:layout_height="wrap_content"
15        android:hint="请输入电话号码"/>
16
17  </LinearLayout>
```

这是我们的 Activity 的布局文件,其中,第 11~15 行在当前布局中添加了一个 EditText 控件代表电话号码的输入框,为了能够在用户输入之前对用户提示输入信息,加上了 android:hint 属性代表提示信息,为了在 java 代码中能够得到此 EditText,设置的 id 属性。

(2) 修改 src/com.wyl.example/MainActivity.java 文件,得到 EditText 控件,然后设置内容的监视者查看 EditText 内容的变化,最后设置相应的字体颜色。主要代码如下:

```
01  public class MainActivity extends Activity {        //定义 MainActivity 继承
                                                        自 Activity
02
03      private EditText EtPhone;                       //定义 EditText 对象
04
05      @Override
06      protected void onCreate(Bundle savedInstanceState) {
07          super.onCreate(savedInstanceState);         //调用父类的 onCreate 方法
08
09          //通过 setContentView 方法设置当前页面的布局文件为 activity_main
10          setContentView(R.layout.activity_main);
11          //通过 findviewById 方法得到布局中的 Edittext
12          EtPhone = (EditText)findViewById(R.id.EtPhone);
13          //设置 EtPhone 的监听器
14          EtPhone.addTextChangedListener(new TextWatcher() {
15
16              //文字改变时的回调方法
17              @Override
18              public void onTextChanged(CharSequence s, int start, int before,
                 int count) {
19                  //TODO Auto-generated method stub
20
21              }
22              //文字改变前的回调方法
23              @Override
24              public void beforeTextChanged(CharSequence s, int start, int
                 count,
25                  int after) {
26                  //TODO Auto-generated method stub
27
28              }
29              //文字改变后的回调方法
30              @Override
31              public void afterTextChanged(Editable s) {
32                  //TODO Auto-generated method stub
33                  //得到 Editable 对象的 string
34                  String phoneStr = s.toString();
35                  //判断输入的内容是否为 phone
36                  boolean b = isPhoneNumber(phoneStr);
37                  if (b) {
38                      //如果 b 为 true,设置 EtPhone 的字体颜色为绿色
39                      EtPhone.setTextColor(Color.rgb(0, 255,0));
40                  } else {
41                      //如果 b 为 true,设置 EtPhone 的字体颜色为红色
42                      EtPhone.setTextColor(Color.rgb(255, 0,0));
```

```
43              }
44          }
45      });
46  }
47
48  /*
49   * 方法名：isPhoneNumber
50   * 参数：String
51   * 返回值：boolean
52   * 方法作用：判断参数字符串是否为电话格式
53   *
54   */
55  public boolean isPhoneNumber(String str) {
56      //定义电话格式的正则表达式
57      String regex = "^((13[0-9])|(15[^4,\\D])|(18[0,5-9]))\\d{8}$";
58      //设定查看模式
59      Pattern p = Pattern.compile(regex);
60      //判断 str 是否匹配，返回匹配结果
61      Matcher m = p.matcher(str);
62      return m.find();
63  }
64  }
65
```

在此代码中第 12～14 行通过 findViewById 得到 EditText 对象，并设置了文字改变监听器 TextWatcher 的对象，这样的话，等 EditText 的文字发生改变的时候，程序会自动调用第 29～44 行的文字监听器 TextWatcher 的 afterTextChanged 方法，然后修改 EditText 的文字颜色。在其中修改颜色之前会进行电话号码条件的检查，我们这里使用正则表达式来实现，在代码中的第 48～63 行。

4．实例扩展

基于如上思路我们可以制作邮箱的验证对话框和邮编的验证对话框等，只要大家实现不同的正则表达式即可。大家可以自己进行练习。

范例 007　更改输入框的文字字体

1．实例简介

在 Android 系统中 EditText 具有默认的字体格式，但是这些字体格式相对比较单调。如果你想要使 EditText 的字体与众不同的话，就要改变字体格式了。本实例就带领大家一起来做一个不同字体的输入框。

2．运行效果

该实例运行效果如图 3.7 所示。

图 3.7　不同字体的输入框

3．实例程序讲解

图 3.7 想要实现不同字体的 EditText 输入框。要想实现这样的效果只要修改当前页面

的布局文件 res/layout/activity_main.xml 中 EditText 控件属性即可。代码如下：

```xml
01 <LinearLayout xmlns:android="http://schemas.android.com/apk/res/android"
02     xmlns:tools="http://schemas.android.com/tools"
03     android:layout_width="match_parent"
04     android:layout_height="match_parent"
05     android:paddingBottom="@dimen/activity_vertical_margin"
06     android:paddingLeft="@dimen/activity_horizontal_margin"
07     android:paddingRight="@dimen/activity_horizontal_margin"
08     android:paddingTop="@dimen/activity_vertical_margin"
09     android:orientation="vertical">
10     <!-- 设置输入框，typeface 属性是 normal -->
11     <EditText
12         android:layout_width="match_parent"
13         android:layout_height="wrap_content"
14         android:typeface="normal"
15         android:hint="我的字体是 normal"/>
16     <!-- 设置输入框，typeface 属性是 sans -->
17     <EditText
18         android:layout_width="match_parent"
19         android:layout_height="wrap_content"
20         android:typeface="sans"
21         android:hint="我的字体是 sans"/>
22     <!-- 设置输入框，typeface 属性是 serif -->
23     <EditText
24         android:layout_width="match_parent"
25         android:layout_height="wrap_content"
26         android:typeface="serif"
27         android:hint="我的字体是 serif"/>
28     <!-- 设置输入框，typeface 属性是 monospace -->
29     <EditText
30         android:layout_width="match_parent"
31         android:layout_height="wrap_content"
32         android:typeface="monospace"
33         android:hint="我的字体是 monospace"/>
34     <!-- 设置输入框，在 java 文件中设置字体 -->
35     <EditText
36         android:id="@+id/Et"
37         android:layout_width="match_parent"
38         android:layout_height="wrap_content"
39         android:hint="我的字体是华文行楷"/>
40
41 </LinearLayout>
```

在这个布局中的第 14、20、26、32 行都对 EditText 控件设置了不同的字体属性。这样在看到输入框的时候，它就具有不同的字体效果了。

4．实例扩展

对于 Android 系统中只提供了三种字体，而且这三种字体效果也不是特别好看，如果你想要实现一些比较个性的字体的话，就可以引入外部的字体格式了。常见的字体格式的后缀为.ttf。但是引入外部字体的话就一定要在 java 文件中设置字体格式了。步骤如下：

（1）通过 findViewById 得到对应的 EditText 控件。

```
EditText Et = (EditText)findViewById(R.id.Et);
```

（2）通过 Typeface 的 createFromAsset 方法获得 typeface 对象，前提是在你的工程中的

asset/fonts 目录下有对应的字体文件。我这里为了方便直接是用 Windows 中的华文行楷的字体。

```
Typeface typeFace =Typeface.createFromAsset(getAssets(),"fonts/ STXINGKA.TTF ");
Et.setTypeface(typeFace);
```

通过如上步骤大家就可以定义属于自己的格式的 EditText 对话框了。

范例 008　我同意上述条款的页面

1. 实例简介

在我们使用 Android 的一些应用的时候经常会看到一个页面，尤其是在注册用户信息的时候，会有一个我同意上述条款的页面。本实例就带领大家一起来做一个条款的显示页面。

2. 运行效果

该实例运行效果如图 3.8 所示。

3. 实例程序讲解

想要实现图 3.8 效果需要使用 CheckBox 控件。此控件主要有两种状态，选中和未选中。一般在我们程序中想要对某个功能的选项进行是否选择的时候使用。要想实现这样的效果只要修改当前页面的布局文件 res/layout/activity_main.xml，在其中添加 CheckBox 控件即可。代码如下：

图 3.8　我同意条款页面

```
01  <LinearLayout xmlns:android="http://schemas.android.com/apk/res/android"
02      xmlns:tools="http://schemas.android.com/tools"
03      android:layout_width="match_parent"
04      android:layout_height="match_parent"
05      android:paddingBottom="@dimen/activity_vertical_margin"
06      android:paddingLeft="@dimen/activity_horizontal_margin"
07      android:paddingRight="@dimen/activity_horizontal_margin"
08      android:paddingTop="@dimen/activity_vertical_margin"
09      android:orientation="vertical">
10
11      <!-- 定义 TextView 控件 -->
12      <TextView
13          android:layout_width="wrap_content"
14          android:layout_height="wrap_content"
15          android:text="@string/str" />
16
17      <!-- 定义 CheckBox 控件 -->
18      <CheckBox
19          android:layout_width="wrap_content"
20          android:layout_height="wrap_content"
21          android:text="我同意如上条款" />
22
23  </LinearLayout>
```

在上面代码的第 17~21 行定义了 CheckBox 控件，这样你在页面中就可以看到复选框的效果了。

4．实例扩展

如果 Android 系统中的 CheckBox 的样式不能满足你的要求。你可以实现一个个性的 CheckBox。这样的话就需要定义样式。首先准备好两张 png 的图片，用于 CheckBox 的选中和未选中的图片，然后在 res/drawable 中定义 checkboxstyle.xml 文件。其内容为：

```xml
01  <?xml version="1.0" encoding="UTF-8"?>
02  <selector xmlns:android="http://schemas.android.com/apk/res/android">
03      <!-- CheckBox 选中时的图片 -->
04      <item android:state_checked="true" android:drawable="@drawable/select" />
05      <!-- CheckBox 未选中时的图片 -->
06  <item android:state_checked="false" android:drawable="@drawable/unselect" />
07  </selector>
```

然后在 CheckBox 控件中加上 button 属性即可。

```xml
01      <!-- 定义 CheckBox 控件 -->
02      <CheckBox
03          android:layout_width="wrap_content"
04          android:layout_height="wrap_content"
05          android:button="@drawable/checkboxstyle"
06          android:text="我同意如上条款" />
```

其中 button 属性代表 CheckBox 的选项背景，使用@drawable/checkboxstyle，也就是之前建立的 checkboxstyle.xml 文件，这样你的 CheckBox 在选择的过程中就会显示关联的图片了。

范例009 爱好调查页面

1．实例简介

对于 CheckBox 控件，在一个页面中不但可以代表一个是否的逻辑，而且也经常用在多选的情况。例如，对于学生的爱好调查，有些学生可能喜欢篮球，而且喜欢足球。本实例就带领大家一起来做一个爱好调查页面。

2．运行效果

该实例运行效果如图 3.9 所示。

3．实例程序讲解

在图 3.9 效果中选中三个复选框中的任意一个，在下方的 TextView 中都会进行选择结果的显示。要想实现这样的效果，步骤如下所示。

图 3.9　爱好调查页面

（1）修改当前页面的布局文件 res/layout/activity_main.xml，在其中添加三个 CheckBox 控件，分别代表篮球、乒乓球和足球的复选框。代码如下：

```xml
01 <LinearLayout xmlns:android="http://schemas.android.com/apk/res/android"
02     xmlns:tools="http://schemas.android.com/tools"
03     android:layout_width="match_parent"
04     android:layout_height="match_parent"
05     android:paddingBottom="@dimen/activity_vertical_margin"
06     android:paddingLeft="@dimen/activity_horizontal_margin"
07     android:paddingRight="@dimen/activity_horizontal_margin"
08     android:paddingTop="@dimen/activity_vertical_margin"
09     android:orientation="vertical">
10
11     <!-- 定义 CheckBox 控件，代表篮球选项-->
12     <CheckBox
13         android:id="@+id/CbBasketball"
14         android:layout_width="wrap_content"
15         android:layout_height="wrap_content"
16         android:text="篮球" />
17
18     <!-- 定义 CheckBox 控件，代表乒乓球选项-->
19     <CheckBox
20         android:id="@+id/CbPingpangball"
21         android:layout_width="wrap_content"
22         android:layout_height="wrap_content"
23         android:text="乒乓球" />
24
25     <!-- 定义 CheckBox 控件，代表足球选项-->
26     <CheckBox
27         android:id="@+id/CbFootball"
28         android:layout_width="wrap_content"
29         android:layout_height="wrap_content"
30         android:text="足球" />
31
32     <!-- 定义 TextView 控件，来显示选中结果 -->
33     <TextView
34         android:id="@+id/TvResult"
35         android:layout_width="wrap_content"
36         android:layout_height="wrap_content"
37         android:text="@string/str" />
38
39 </LinearLayout>
```

在上面代码的第 11～16 行、第 18～23 行和第 25～30 行分别定义了三个 CheckBox 控件。在第 32～37 行定义了一个结果文本标签。而且给这四个控件分别设置的 id 属性，方便我们在 java 文件中获取相应的控件对象。

（2）在 src/com.wyl.example/MainActivity.java 代码中进行 CheckBox 的取值操作。代码如下：

```java
01 package com.wyl.example;                              //当前包名
02
03 //导入必备的包
04 import android.app.Activity;
05 import android.os.Bundle;
06 import android.widget.CheckBox;
07 import android.widget.CompoundButton;
08 import android.widget.TextView;
09 import android.widget.CompoundButton.OnCheckedChangeListener;
10
11 public class MainActivity extends Activity {          //定义 MainActivity 继承
                                                        自 Activity
```

```java
12
13      private CheckBox CbBasketball;           //定义篮球的复选框对象
14      private CheckBox CbPingpangball;         //定义乒乓球的复选框对象
15      private CheckBox CbFootball;             //定义足球的复选框对象
16      private TextView TvResult;               //定义结果文本便签对象
17
18      @Override
19      protected void onCreate(Bundle savedInstanceState) {
20          super.onCreate(savedInstanceState);      //调用父类的onCreate方法
21
22          //通过setContentView方法设置当前页面的布局文件为activity_main
23          setContentView(R.layout.activity_main);
24          findView();                              //获取页面中的控件
25          setListener();                           //设置控件的监听器
26      }
27
28      private void setListener() {
29          //TODO Auto-generated method stub
30          //设置所有CheckBox的状态改变监听器
31          CbBasketball.setOnCheckedChangeListener(myCheckChangelistener);
32          CbPingpangball.setOnCheckedChangeListener(myCheckChangelistener);
33          CbFootball.setOnCheckedChangeListener(myCheckChangelistener);
34      }
35
36      OnCheckedChangeListener          myCheckChangelistener         =         new
        OnCheckedChangeListener() {
37
38          @Override
39          public void onCheckedChanged(CompoundButton buttonView, boolean isChecked) {
40              //TODO Auto-generated method stub
41              //设置TextView的内容显示CheckBox的选择结果
42              setText();
43          }
44      };
45
46
47      private void findView() {
48          //TODO Auto-generated method stub
49          //通过findViewById得到对应的控件对象
50          CbBasketball = (CheckBox)findViewById(R.id.CbBasketball);
51          CbPingpangball = (CheckBox)findViewById(R.id.CbPingpangball);
52          CbFootball = (CheckBox)findViewById(R.id.CbFootball);
53          TvResult = (TextView)findViewById(R.id.TvResult);
54      }
55
56      private void setText(){
57          String str;
58          TvResult.setText("");                    //清空TextView的内容
59          //如果CbBasketball被选中,则加入TvResult内容显示
60          if (CbBasketball.isChecked()) {
61              str = TvResult.getText().toString()+CbBasketball.getText().toString()+",";
62              TvResult.setText(str);
63          }
64          //如果CbPingpangball被选中,则加入TvResult内容显示
65          if (CbPingpangball.isChecked()) {
66              str = TvResult.getText().toString()+CbPingpangball.getText().toString()+",";
```

```
67          TvResult.setText(str);
68        }
69        //如果CbFootball被选中,则加入TvResult内容显示
70        if (CbFootball.isChecked()) {
71          str = TvResult.getText().toString()+CbFootball.getText().
              toString();
72          TvResult.setText(str);
73        }
74    }
75  }
76
```

在如上的代码中第 49～53 行分别得到了三个 CheckBox 控件及一个 TextView 对象。在第 31～34 行给三个 CheckBox 对象设置了一个监听器。在 36～44 行定义了 OnCheckedChangeListener，所以当 CheckBox 的选中状态发生改变的时候会调用 onCheckedChanged，去执行 57～73 行的代码设置 TvResult 对象的显示内容。这样就可以实现当任意一个 CheckBox 控件的状态改变的时候 TextView 的内容随之改变了。

4．实例扩展

对于 CheckBox 控件来说，如果一个页面中有多个 CheckBox，那么他们之间是没有逻辑联系的，如果希望一个页面中的一些 CheckBox 有联系的话，必须使用代码人为进行关联了。例如，在一个页面中有六个 CheckBox，其中三个是学生的爱好，足球、篮球和乒乓球，还有三个是大家精通的语言，C 语言、C++和 Java。那么在进行选择的时候就需要人工的对于这六个 CheckBox 进行区分了。

范例 010　政治面貌调查表

1．实例简介

对于 CheckBox 控件适用于多选的情况，而在某些情况下需要在某些选项中选择其中之一，这个时候就需要使用 RadioButton 了。例如，对于政治面貌的选择：党员、团员和群众。本实例就带领大家一起来做一个政治面貌选择页面。

2．运行效果

该实例运行效果如图 3.10 所示。

3．实例程序讲解

在图 3.10 效果中选中三个选项中当选中一项时，其他两项自动取消选择。要想实现这样的效果，只需修改当前页面的布局文件 res/layout/activity_main.xml，在其中添加一个 RadioGroup 控件，来代表一个单选框组，然后在 RadioGroup 中添加三个 RadioButton 选项。代码如下：

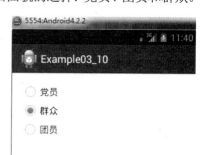

图 3.10　政治面貌调查页面

```
01  <LinearLayout xmlns:android="http://schemas.android.com/apk/res/android"
02      xmlns:tools="http://schemas.android.com/tools"
03      android:layout_width="match_parent"
04      android:layout_height="match_parent"
05      android:orientation="vertical"
```

```
06        android:paddingBottom="@dimen/activity_vertical_margin"
07        android:paddingLeft="@dimen/activity_horizontal_margin"
08        android:paddingRight="@dimen/activity_horizontal_margin"
09        android:paddingTop="@dimen/activity_vertical_margin" >
10
11        <!-- 定义RadioGroup控件，代表政治面貌选择组 -->
12        <RadioGroup
13            android:id="@+id/RgPolitical"
14            android:layout_width="wrap_content"
15            android:layout_height="wrap_content" >
16            <!-- 定义RadioButton控件，代表党员选项 -->
17            <RadioButton
18                android:layout_width="wrap_content"
19                android:layout_height="wrap_content"
20                android:text="党员" />
21            <!-- 定义RadioButton控件，代表群众选项 -->
22            <RadioButton
23                android:layout_width="wrap_content"
24                android:layout_height="wrap_content"
25                android:text="群众" />
26            <!-- 定义RadioButton控件，代表团员选项 -->
27            <RadioButton
28                android:layout_width="wrap_content"
29                android:layout_height="wrap_content"
30                android:text="团员" />
31        </RadioGroup>
32
33    </LinearLayout>
```

在上面代码的第11~31行定义了一个RadioGroup控件，代表一个单选框组，在一个组中的单选项是具有关联的，也就是说在一个选项组中的选项只可以选择一个，其他的默认不选择。在RadioGroup中添加三个RadioButton，分别代表党员、群众和团员，但是这三项只可以选择其中一项。

4．实例扩展

对于RadioGroup来说，大家可以把他当做是一个单选框组，它就给我们的单选框进行逻辑的划分，而且不但RadioGroup可以进行逻辑的划分，它可以简单的布局其中的RadioButton。我们上例中的效果是纵向列表选择，可以通过RadioGroup控件的android:orientation属性修改为横向显示。代码如下所示：

```
01    <RadioGroup
02        android:id="@+id/RgPolitical"
03        android:layout_width="wrap_content"
04        android:layout_height="wrap_content"
05        android:orientation="horizontal" >
06
07        …  …
08    </RadioGroup>
```

范例011　IT人员测试应用

1．实例简介

对于RadioGroup和RadioButton，其实在我们生活中应用很广泛。例如，现在比较流行的测试软件,通过对给出问题进行回答,应用中会计算用户得到的积分,并且根据积分得到用户的性格。本实例就带领大家一起来做一个简易的 IT 人员测试应用。

2. 运行效果

该实例运行效果如图 3.11 所示。

3. 实例程序讲解

在图 3.11 效果中有三道题目,并且每道题目有三个选项,每个题目只可以进行一个选项的选择,而且三个题目之间的选项互不影响。要想实现这样的效果,步骤如下所示。

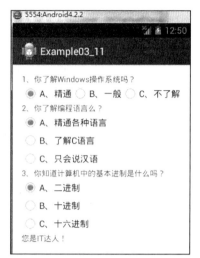

图 3.11 IT 人员测试应用

(1)由于我们本例中用到的固定字符串相对比较多,所以在 res/values/strings.xml 中建立相应字符串节点,方便我们在 xml 布局文件中使用。代码如下:

```
01  <?xml version="1.0" encoding="utf-8"?>
02  <resources>
03      <string name="app_name">Example03_11</string>
04      <string name="subject1">1、你了解 Windows 操作系统吗? </string>
05      <string name="subject2">2、你了解编程语言么? </string>
06      <string name="subject3">3、你知道计算机中的基本进制是什么吗? </string>
07
08      <string name="subject1_opt1">A、精通</string>
09      <string name="subject1_opt2">B、一般</string>
10      <string name="subject1_opt3">C、不了解</string>
11
12      <string name="subject2_opt1">A、精通各种语言</string>
13      <string name="subject2_opt2">B、了解 C 语言</string>
14      <string name="subject2_opt3">C、只会说汉语</string>
15
16      <string name="subject3_opt1">A、二进制</string>
17      <string name="subject3_opt2">B、十进制</string>
18      <string name="subject3_opt3">C、十六进制</string>
19  </resources>
```

(2)定义界面的布局,修改 res/layout/activity_main.xml 布局文件。代码如下:

```
01  <LinearLayout xmlns:android="http://schemas.android.com/apk/res/android"
02      xmlns:tools="http://schemas.android.com/tools"
03      android:layout_width="match_parent"
04      android:layout_height="match_parent"
05      android:orientation="vertical"
06      android:paddingBottom="@dimen/activity_vertical_margin"
07      android:paddingLeft="@dimen/activity_horizontal_margin"
08      android:paddingRight="@dimen/activity_horizontal_margin"
09      android:paddingTop="@dimen/activity_vertical_margin" >
10
11      <TextView
12          android:layout_width="match_parent"
13          android:layout_height="wrap_content"
14          android:text="@string/subject1"
```

```
15          />
16      <!-- 定义 RadioGroup 控件，代表选择题 1 -->
17      <RadioGroup
18          android:id="@+id/RgSubject1"
19          android:layout_width="wrap_content"
20          android:layout_height="wrap_content"
21          android:orientation="horizontal" >
22          <!-- 定义 RadioButton 控件，选项 1 -->
23          <RadioButton
24              android:id="@+id/RbSubject1_opt1"
25              android:layout_width="wrap_content"
26              android:layout_height="wrap_content"
27              android:text="@string/subject1_opt1" />
28          <!-- 定义 RadioButton 控件，选项 2 -->
29          <RadioButton
30              android:id="@+id/RbSubject1_opt2"
31              android:layout_width="wrap_content"
32              android:layout_height="wrap_content"
33              android:text="@string/subject1_opt2" />
34          <!-- 定义 RadioButton 控件，选项 3 -->
35          <RadioButton
36              android:id="@+id/RbSubject1_opt3"
37              android:layout_width="wrap_content"
38              android:layout_height="wrap_content"
39              android:text="@string/subject1_opt3" />
40      </RadioGroup>
41      <!-- 定义选项二，选项三的代码同选项一，这里省略 -->
42      ......
43      <!-- 定义 TextView 控件，代表测试结果 -->
44      <TextView
45          android:id="@+id/TvResult"
46          android:layout_width="match_parent"
47          android:layout_height="wrap_content"
48          />
49
50  </LinearLayout>
```

其中第 17～40 行定义了题目一的题目和选项，并且设置了相应的 id，方便在 java 文件中进行对象的获取。

（3）在 java 文件中获取相应的控件对象，然后设置监听器，当用户选择完毕后计算出用户的得分。代码如下：

```
001  //import 代码省略
002  public class MainActivity extends Activity {
003      private RadioGroup RgSubject1;          //定义题目 1 的选项
004      private RadioGroup RgSubject2;          //定义题目 2 的选项
005      private RadioGroup RgSubject3;          //定义题目 3 的选项
006      private TextView TvResult;              //定义显示结果的文本标签
007
008      //定义 MainActivity 继承自 Activity
009      @Override
010      protected void onCreate(Bundle savedInstanceState) {
011          super.onCreate(savedInstanceState);   //调用父类的 onCreate 方法
012
013          //通过 setContentView 方法设置当前页面的布局文件为 activity_main
014          setContentView(R.layout.activity_main);
015          findView();                              //得到控件对象
```

```
016            setListener();                          //设置 RadioGroup 的监听器事件
017        }
018
019    private void setListener() {
020        //TODO Auto-generated method stub
021        //给三组 RadioGroup 设置 CheckChangeListener
022        RgSubject1.setOnCheckedChangeListener(mylistener);
023        RgSubject2.setOnCheckedChangeListener(mylistener);
024        RgSubject3.setOnCheckedChangeListener(mylistener);
025    }
026
027    private void findView() {
028        //TODO Auto-generated method stub
029        //通过 findViewById 得到对应控件的对象
030        RgSubject1 = (RadioGroup) findViewById(R.id.RgSubject1);
031        RgSubject2 = (RadioGroup) findViewById(R.id.RgSubject2);
032        RgSubject3 = (RadioGroup) findViewById(R.id.RgSubject3);
033        TvResult = (TextView) findViewById(R.id.TvResult);
034    }
035
036    //自定义 RadioGroup 的 OnCheckedChangeListener 对象
037    RadioGroup.OnCheckedChangeListener   mylistener
                    =new RadioGroup.OnCheckedChangeListener() {
038        //当 RadioGroup 对象状态发生改变时的回调函数
039        @Override
040        public void onCheckedChanged(RadioGroup group, int checkedId) {
041            //TODO Auto-generated method stub
042            //判断是否三组题目都进行了选择
043            if (RgSubject1.getCheckedRadioButtonId() != -1
044                    && RgSubject2.getCheckedRadioButtonId() != -1
045                    && RgSubject3.getCheckedRadioButtonId() != -1 ) {
046                int score = 0;            //记录用户的分数
047                //加上第一题的选择得分
048                switch (RgSubject1.getCheckedRadioButtonId()) {
049                case R.id.RbSubject1_opt1:
050                    score +=3;
051                    break;
052                case R.id.RbSubject1_opt2:
053                    score +=2;
054                    break;
055                case R.id.RbSubject1_opt3:
056                    score +=1;
057                    break;
058                default:
059                    break;
060                }
061                //加上第二题的选择得分
062                switch (RgSubject2.getCheckedRadioButtonId()) {
063                case R.id.RbSubject2_opt1:
064                    score +=3;
065                    break;
066                case R.id.RbSubject2_opt2:
067                    score +=2;
068                    break;
069                case R.id.RbSubject2_opt3:
070                    score +=1;
071                    break;
072                default:
073                    break;
074                }
```

```
075                    //加上第三题的选择得分
076                    switch (RgSubject3.getCheckedRadioButtonId()) {
077                    case R.id.RbSubject3_opt1:
078                        score +=3;
079                        break;
080                    case R.id.RbSubject3_opt2:
081                        score +=2;
082                        break;
083                    case R.id.RbSubject3_opt3:
084                        score +=1;
085                        break;
086                    default:
087                        break;
088                    }
089                    //对于用户的得分给出评价结果
090                    if (score >= 8) {
091                        TvResult.setText("您是IT达人！");
092                    }else if (score >= 4) {
093                        TvResult.setText("您是一般电脑用户！");
094                    }else{
095                        TvResult.setText("您需要提高电脑知识哦！");
096                    }
097                }
098            }
099        };
100    }
```

在上面代码的第 29～33 行获得 RadioGroup 控件。在第 22～24 行为 RadioGroup 控件设置监听器。在第 36～99 行自定义了 RadioGroup 的 OnCheckedChangeListener，其中的 onCheckedChanged 方法是当 RadioGroup 状态改变后的回调方法，本例中在此方法中得到了三个 RadioGroup 的选项，并且根据一定的公式计算出了用户对于电脑了解程度的积分，并在 TextView 控件中给出提示。

4．实例扩展

RadioGroup 中 RadioButton 控件只能选一个，而且通过 RadioGroup 控件的 getCheckedRadioButtonId()可以得到选择的 RadioButton 的 id，如果没有选择的话，此函数返回-1，大家可以通过此返回值判断此 RadioGroup 中的选项是否被用户选择。

范例 012　应用中的关闭声音的按钮

1．实例简介

在我们使用 Android 应用或者玩儿游戏的时候经常会用到一个功能，就是关闭程序的声音，负责开关声音的按钮，不同于我们之前讲到的控件，一般情况是一个可以按下的按钮，有按下的状态，有弹起的状态。本实例就带领大家一起来做一个声音开关按钮的效果。

2．运行效果

该实例运行效果如图 3.12 所示。

图 3.12　声音开启关闭效果

3. 实例程序讲解

在图 3.12 效果中有一个开关按钮，此开关按钮有两个状态，当按下的时候显示声音关闭，当弹起的时候显示声音开启。要想实现这样的效果，只需要在 res/values/strings.xml 中建立一个 ToggleButton 节点即可。代码如下：

```
01  <!-- 定义一个 ToggleButton 的控件，
02      设置 texton，按钮开启时显示的文字，
03      设置 textoff，按钮关闭时显示的问题 -->
04  <ToggleButton
05      android:layout_width="wrap_content"
06      android:layout_height="wrap_content"
07      android:textOn="声音开启"
08      android:textOff="声音关闭"
09  />
```

在如上代码的第 4~9 行建立了一个 ToggleButton 节点，加上了 textOn 属性，就代表当开关开启时显示的文字，加上了 textOff 属性，就代表当开关关闭时显示的文字，这样即可以实现一个开关按钮了。

4. 实例扩展

在 Android 4.0 版本之后，Android SDK 中加入了一个新的表示开关的控件 Switch，此控件不但能够表示开关的功能，而且可以实现滑动开关。要实现 Switch 控件，只需要在布局文件中加入如下代码：

```
01  <!-- 定义一个 Switch 的控件，
02      设置 texton，按钮开启时显示的文字，
03      设置 textoff，按钮关闭时显示的问题 -->
04  <Switch
05      android:layout_width="wrap_content"
06      android:layout_height="wrap_content"
07      android:textOn="声音开启"
08      android:textOff="声音关闭"
09  />
```

需要注意的是如果要使用 Switch 控件的话，我们建立的 Android 工程一定要保证在 Android 4.0 版本以上，所以在工程的 AndroidManifest.xml 文件中一定要保证 minSdkVersion 和 targetSdkVersion 大于等于 14，这样才可以保证程序会在 Android 4.0 版本以上的手机运行。

范例 013 应用中的音量调节效果

1. 实例简介

在我们使用 Android 应用或者玩儿游戏的时候，可能不仅仅是开启关闭声音，有时候会希望应用能有声音但是降低一些声音的音量，这时候我们就会用到 Android 中的另一个控件 SeekBar。本实例就带领大家一起来做滑动调节音量的效果。

2. 运行效果

该实例运行效果如图 3.13 所示。

图 3.13 音量调节效果

3．实例程序讲解

在图 3.13 效果中有一个 TextView 控件来显示文字标签，还有一个 SeekBar 控件来实现滑动的进度效果。要想实现这样的效果，只需要在 res/values/strings.xml 中建立一个 SeekBar 节点即可。代码如下：

```
01    <!-- 定义一个 SeekBar 的控件,
02        设置 max，代表此 SeekBar 最大时的数值,
03        设置 progress 属性，代表在滑动过程中最小的滑动距离 -->
04    <SeekBar
05        android:layout_width="match_parent"
06        android:layout_height="wrap_content"
07        android:max="100"
08        android:progress="2"
09        />
```

在如上代码的第 4～9 行建立了一个 SeekBar 节点，加上了 max 属性，代表当滑动到最大值的时候代表的数字，加上了 progress 属性，在滑动过程中每次最小的滑动距离，这样即可以实现一个滑动的效果了。等我们今后学习了如何设置系统的音量，那就可以真正的实现通过滑动来控制声音了。

4．实例扩展

对于 SeekBar 控件，Android 中提供了丰富的属性控制，在我们做应用的时候大家可以直接来使用，我这里列出 Seekbar 控件的常用属性如下。

❑ progressDrawable：设置拖动条的样式。
❑ thumb：设置 SeekBar 中滑块的图片。
❑ secondaryProgress：代表第二进度的大小。

除了上述的属性以外，SeekBar 还有很多可以设置的属性，大家使用的时候查一下官方文档即可。

范例 014 服务星级评价效果

1．实例简介

在我们去 Android 的市场上搜索应用的时候，最关注的一项指标就是这个应用其他用

户的评分如何,当然如果你使用了一款应用你感觉这款应用很好,你也会给出一个比较好的评价,在 Android 中给出了方便用户评价的控件就是 RatingBar。本实例就带领大家一起使用 RatingBar 来实现一个评分的效果。

2. 运行效果

该实例运行效果如图 3.14 所示。

3. 实例程序讲解

在图 3.14 中用一个 RatingBar 控件来实现滑动进行星级评价的效果。要想实现这样的效果,只需要在 res/values/strings.xml 中建立一个 RatingBar 节点即可。代码如下:

图 3.14 服务星级评价效果

```
01    <!-- 定义一个 RatingBar 的控件,
02         设置 numStars 属性,代表显示最大的星级数量,
03         设置 stepSize 属性,代表在评价过程中最小的移动单位 -->
04    <RatingBar
05        android:layout_width="wrap_content"
06        android:layout_height="wrap_content"
07        android:numStars="5"
08        android:stepSize="0.5"
09        />
```

在如上代码的第 4~9 行建立了一个 RatingBar 节点,加上了 numStars 属性,就代表当前评价控件最多的星级数量;加上了 setpSize 属性,就表示在滑动过程中每次最小的滑动星级该变量,这样就可以实现滑动进行等级评价的效果了。

4. 实例扩展

对于 RatingBar 控件其实和 SeekBar 控件的效果比较相似,但是 RatingBar 更侧重于表示星级评价的时候使用,同样 Android 对于 RatingBar 也提供了丰富的属性控制,我这里列出 RatingBar 控件的常用属性如下。

❑ style:设置 RatingBar 的样式。
❑ rating:设置 RatingBar 的默认评分。
❑ isIndicator:设置 RatingBar 是否是一个指示器,如果是的话,用户就无法进行滑动修改了。

除了上述的属性以外,RatingBar 还有很多可以设置的属性,大家使用的时候查一下官方文档即可。

范例 015 页面加载中效果

1. 实例简介

在开发 Android 应用的时候,经常会遇到一些比较耗时的操作,例如:网络文件下载、在线视频加载和程序中图片音频资源的加载等。在这样的过程中如果你不做处理的话,用户会以为你的程序反应比较慢，或者是不是程序死掉了，所以我们在做比较耗时的操作的时候一般通过一个 ProgressBar 来给出用户提示，或者显示加载进度。本实例就带领大家一起使用 ProgressBar 来实现页面加载中的效果。

2．运行效果

该实例运行效果如图 3.15 所示。

3．实例程序讲解

在图 3.15 中用一个 ProgressBar 控件来实现正在加载页面内容的效果。要想实现这样的效果，只需要在 res/values/strings.xml 中建立一个 ProgressBar 节点即可。代码如下：

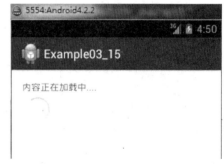

图 3.15　页面加载中效果

```xml
<!-- 定义一个 ProgressBar 的控件   -->
<ProgressBar
    android:layout_width="wrap_content"
    android:layout_height="wrap_content"
    />
```

在如上代码中建立了一个 ProgressBar 节点，这样就可以实现正在加载的效果了。但是我们通常使用的 ProgressBar 不仅仅只有这一种样式，我们可以通过 style 属性给 ProgressBar 设置相应的样式效果，代码如下：

```xml
<!-- 定义一个 ProgressBar 的控件   -->
<ProgressBar
    android:layout_width="wrap_content"
    android:layout_height="wrap_content"
    style="@android:style/Widget.ProgressBar"
    />
```

其中 style 属性就设置了 ProgressBar 控件的显示样式，Android 系统中提供了集中系统默认的样式，大家可以进行选择。

- Widget.ProgressBar.Horizontal：水平 ProgressBar 的进度条效果。
- Widget.ProgressBar.Small：小圆形的进度条效果。
- Widget.ProgressBar.Large：大圆形的进度条效果。
- Widget.ProgressBar.Inverse：普通进度条翻转。
- Widget.ProgressBar.Small.Inverse：小进度条翻转。
- Widget.ProgressBar.Large.Inverse：大进度条翻转。

当然还可以通过 max 属性设置横向进度条的最大进度，通过 progress 属性设置当前进度，通过 visibility 属性来设置这个 ProgressBar 的可见性。

4．实例扩展

对于应用加载进度的提示，还有一个控件是我们经常使用的，就是 ProgressDialog。ProgressDialog 不同于 ProgressBard 的一点是，在加载过程中，ProgressBar 是通过按键无法取消的，而 ProgressDialog 运行过程中按手机物理返回键，ProgressDialog 就消失了。而且 ProgressDialog 不是一个控件类，而是作为 android.app 包中的一个应用类来使用，如果想要使用 ProgressDialog 的话，在程序的 java 代码中定义一个 ProgressDialog 对象，然后调用其 show 方法即可，代码如下：

```
//定义 ProgressDialog 对象 pd
ProgressDialog pd = new ProgressDialog(this);
//设置 pd 的标题信息
pd.setTitle("程序正在加载中.....");
//调用 pd 的 show 方法显示此 ProgressDialog
pd.show();
```

当然，对于 ProgressBar 和 ProgressDialog 更加深入的用法，我们会在讲到异步请求和加载的时候再次讲到。

范例 016 日期获取框效果

1. 实例简介

在开发 Android 应用的时候，经常会遇到输入日期的情况，例如：请填入您的生日，请填入您工作的日期等。如果我们使用之前例子中的 EditText 来接受日期的话，我们需要写复杂的正则表达式来控制用户输入的格式，而且也无法快速的进行日期的调整，所以 Android 中提供了获取日期的控件 DatePicker。本实例就带领大家一起使用 DatePicker 来实现获取日期的效果。

2. 运行效果

该实例运行效果如图 3.16 所示。

3. 实例程序讲解

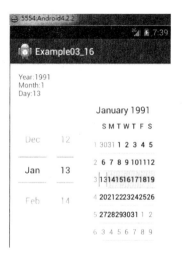

图 3.16 获取日期效果

在图 3.16 效果中用一个 DatePicker 控件来实现，当单击 DatePicker 时就可以修改日期，并且 TextView 中显示对应的日期值。要想实现这样的效果，只需要在 res/values/strings.xml 中建立一个 DatePicker 节点。代码如下：

```
01  <TextView
02      android:id="@+id/Tv"
03      android:layout_width="fill_parent"
04      android:layout_height="wrap_content"
05      android:text="请输入您的生日...." />
06  <!-- 定义一个 DatePicker 的控件，设置 id 属性方便在 java 文件中获得对象-->
07  <DatePicker
08      android:id="@+id/Dp"
09      android:layout_width="fill_parent"
10      android:layout_height="wrap_content" />
```

在如上代码中 7～10 行建立了一个 DatePicker 节点，这样就可以显示日期获取的效果了。在 DatePicker 和 TextView 中分别设置了 id 属性，方便在 java 文件中获取相应的对象进行操作。

然后在 src/com.wyl.example/MainActivity.java 中获取 DatePicker 对象并且进行监听器

的设置，这样当 DatePicker 的值发生改变时 TextView 的值也就随时改变了。代码如下：

```
01  /*import 代码省略*/
02  public class MainActivity extends Activity {
03      private TextView Tv;                    //定义结果文本标签
04      private DatePicker Dp;                  //定义日期获取控件
05
06  //定义 MainActivity 继承自 Activity
07      @Override
08      protected void onCreate(Bundle savedInstanceState) {
09          super.onCreate(savedInstanceState);     //调用父类的 onCreate 方法
10
11          //通过 setContentView 方法设置当前页面的布局文件为 activity_main
12          setContentView(R.layout.activity_main);
13          findView();                             //获取控件对象
14          setListener();                          //设置 datePicker 的监听器
15      }
16
17  private void setListener() {
18      //TODO Auto-generated method stub
19      //初始化 DatePicker 对象，并设置日期改变的监听器
20      Dp.init(1990, 10, 12, new OnDateChangedListener() {
21          //当 Dp 的日期改变时回调 onDateChanged 方法
22          @Override
23          public void onDateChanged(DatePicker view, int year, int monthOfYear,
24                  int dayOfMonth) {
25              //TODO Auto-generated method stub
26              //获取 Dp 的年月日的值，在 TextView 中显示
27              Tv.setText("Year:"+Dp.getYear()+
28                      "\nMonth:"+(Dp.getMonth()+1)+
29                      "\nDay:"+Dp.getDayOfMonth());
30          }
31      });
32  }
33
34  private void findView() {
35      //TODO Auto-generated method stub
36      //通过 findViewById 得到对应的控件对象
37      Tv = (TextView)findViewById(R.id.Tv);
38      Dp = (DatePicker)findViewById(R.id.Dp);
39      }
40  }
```

在代码的第 37～38 行获取了 DatePicker 和 TextView 的控件对象，在第 19～31 行初始化 DatePicker 对象并且绑定了 OnDateChangedListener 类的一个无名对象，其中实现了 onDateChanged 方法，当 DatePicker 对象发生改变时自动回调此方法。在 onDateChanged 方法中通过 Dp 的 getYear、getMonth 和 getDayOfMonth 分别获取到 DatePicker 的年月日，然后显示在 TextView 控件中。

4. 实例扩展

对于 DatePicker 来说，我们还有很多可以设置的属性，在这里我列出常见的 DatePicker 的属性。

- calendarViewShown：是否显示日历栏。
- android:endYear：用于设置最终的年份。

- android:maxDate：用于设置最大的日期。
- android:minDate：用于设置最小的日期。
- android:spinnersShown：用于设置 spinner 是否显示。
- android:startYear：用于设置开始年份。

范例 017　时间获取框效果

1．实例简介

在开发 Android 应用的时候，有时候也会遇到输入时间的情况，例如：设定时间闹钟和填写工作日志等。如果我们使用之前例子中的 EditText 来接受时间的话，我们也需要写复杂的正则表达式来控制用户输入的格式，而且也无法快速的进行时间的调整，所以 Android 中提供了获取日期的控件 TimePicker。本实例就带领大家一起使用 TimePicker 来实现获取时间的效果。

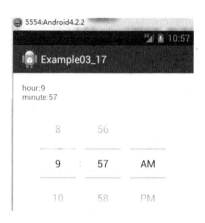

2．运行效果

该实例运行效果如图 3.17 所示。

3．实例程序讲解

图 3.17　获取时间效果

在图 3.17 效果中用一个 TimePicker 控件来实现，当单击 TimePicker 时就可以修改时间，并且 TextView 中显示对应的时间值。要想实现这样的效果，只需要在 res/values/strings.xml 中建立一个 TimePicker 节点。代码如下：

```
01  <TextView
02      android:id="@+id/Tv"
03      android:layout_width="fill_parent"
04      android:layout_height="wrap_content"
05      android:text="请输入闹钟时间...." />
06  <!-- 定义一个TimePicker的控件,设置id属性方便在java文件中获得对象-->
07  <TimePicker
08      android:id="@+id/Tp"
09      android:layout_width="fill_parent"
10      android:layout_height="wrap_content" />
```

在如上代码中 7～10 行建立了一个 TimePicker 控件，这样就可以显示获取时间的效果了。在 TimePicker 和 TextView 中分别设置了 id 属性，方便在 java 文件中获取相应的对象进行操作。

然后在 src/com.wyl.example/MainActivity.java 中获取 TimePicker 对象并且进行监听器的设置，这样当 TimePicker 的值发生改变时 TextView 的值也就随时改变了。代码如下：

```
01  /*import 代码省略*/
02  public class MainActivity extends Activity {
03      private TextView Tv;                    //定义结果文本标签
04      private TimePicker Tp;                  //定义内容获取控件
05
```

```
06      //定义MainActivity继承自Activity
07      @Override
08      protected void onCreate(Bundle savedInstanceState) {
09          super.onCreate(savedInstanceState);      //调用父类的onCreate方法
10
11          //通过setContentView方法设置当前页面的布局文件为activity_main
12          setContentView(R.layout.activity_main);
13          findView();                              //获取控件对象
14          setListener();                           //设置TimePicker的监听器
15      }
16
17      private void setListener() {
18          //TODO Auto-generated method stub
19          //初始化TimePicker对象,并设置日期改变的监听器
20          Tp.setOnTimeChangedListener(new OnTimeChangedListener() {
21
22              @Override
23              public void onTimeChanged(TimePicker view, int hourOfDay, int minute) {
24                  //TODO Auto-generated method stub
25                  Tv.setText("hour:"+hourOfDay+
26                          "\nminute:"+minute);
27              }
28          });
29      }
30
31      private void findView() {
32          //TODO Auto-generated method stub
33          //通过findViewById得到对应的控件对象
34          Tv = (TextView)findViewById(R.id.Tv);
35          Tp = (TimePicker)findViewById(R.id.Tp);
36      }
37  }
```

在代码的第 32~35 行获取了 TimePicker 和 TextView 的控件对象,在第 19~31 行初始化 TimePicker 对象并且绑定了 OnTimeChangedListener 类的一个无名对象,其中实现了 onTimeChanged 方法,当 TimePicker 对象发生改变时自动回调此方法。在 onTimeChanged 方法中的 hourOfDay 和 minute 分别代表修改后的小时和分钟,然后显示在 TextView 控件中。

4. 实例扩展

对于 TimePicker 来说就是 Android 中提供的专门用来处理时间的控件,所以基本关于时间的设置在 TimePicker 中都可以得到设置,例如:时间是否是 24 小时制,是否显示 AM/PM 等。

范例018 日期时间弹出框效果

1. 实例简介

在开发 Android 应用的时候,一般会在输入日期的时候也输入时间,例如:请输入您的具体出生时间,请输入您的购车时间,请打印我这个月的话费清单等。这时候我们不但要得到日期,而且要得到时间。使用之前讲到的 DatePicker 和 TimePicker 的话我们页面的

空间就会被挤压的很小了，所以 Android 中提供了更方便的获取时间和日期的方法，就是使用 DatePickerDialog 和 TimePickerDialog。它使用弹出对话框的方式让用户进行日期和时间的选择，这样就避免了日期控件和时间控件比较占用页面空间的问题。本实例就带领大家一起使用 DatePickerDialog 和 TimePickerDialog 来设置日期时间的效果。

2．运行效果

该实例运行效果如图 3.18 所示。

3．实例程序讲解

在图 3.18 效果中用一个 TextView 控件来显示设置的日期和时间的具体内容，定义两个按钮，当单击这两个按钮的时候，分别显示获取日期和时间的对话框来得到日期和时间，并且当日期获取对话框和时间获取对话框得到数据后，修改 TextView 的内容。要想实现这样的效果，步骤如下所示。

图 3.18　获取日期时间效果

（1）需要在 res/values/strings.xml 中建立一个 TextView 控件和两个 Button 控件，并且设置 id，方便在 java 文件中获取这些控件。代码如下：

```
01  <TextView
02      android:id="@+id/Tv"
03      android:layout_width="fill_parent"
04      android:layout_height="wrap_content"
05      android:text="请输入准确的日期时间...." />
06  <!-- 定义两个 Button 的控件，设置 id 属性方便在 java 文件中获得对象-->
07  <Button
08      android:id="@+id/BtnDate"
09      android:layout_width="fill_parent"
10      android:layout_height="wrap_content"
11      android:text="设置日期" />
12  <Button
13      android:id="@+id/BtnTime"
14      android:layout_width="fill_parent"
15      android:layout_height="wrap_content"
16      android:text="设置时间" />
```

在如上代码建立了一个 TextView 控件和两个 Button 控件，在第 2 行、第 8 行和第 13 行分别设置了对应的 id，方便在 java 文件中获取相应的对象进行操作。

（2）修改 src/com.wyl.example/MainActivity.java 文件中的代码。代码如下：

```
001  /*import 代码省略*/
002
003
004  public class MainActivity extends Activity {
005      private TextView Tv;              //定义结果文本标签
006      private Button BtnDate;           //定义获取日期按钮
007      private Button BtnTime;           //定义获取时间按钮
008      private final int DATEDIALOG = 0; //创建 DatePickerDialog 的标示
009      private final int TIMEDIALOG = 1; //创建 TimePickerDialog 的标示
010
011      //定义 MainActivity 继承自 Activity
```

```
012    @Override
013    protected void onCreate(Bundle savedInstanceState) {
014        super.onCreate(savedInstanceState);    //调用父类的 onCreate 方法
015
016        //通过 setContentView 方法设置当前页面的布局文件为 activity_main
017        setContentView(R.layout.activity_main);
018        findView();                            //获取控件对象
019        setListener();                         //设置 button 的监听器
020    }
021
022    private void setListener() {
023        //TODO Auto-generated method stub
024        //设置 BtnDate 和 BtnTime 的单击监听器
025        BtnDate.setOnClickListener(mylistener);
026        BtnTime.setOnClickListener(mylistener);
027    }
028
029    //自定义 Button 的 OnClickListener 对象
030    OnClickListener mylistener = new OnClickListener() {
031        //按钮单击时的 onClick 回调方法
032        @Override
033        public void onClick(View v) {
034            //TODO Auto-generated method stub
035            switch (v.getId()) {
036            case R.id.BtnDate:
037                showDialog(DATEDIALOG);
038                break;
039            case R.id.BtnTime:
040                showDialog(TIMEDIALOG);
041                break;
042            default:
043                break;
044            }
045        }
046    };
047
048    /* (non-Javadoc)
049     * @see android.app.Activity#onCreateDialog(int)
050     */
051    //当调用 showDialog 方法是系统的回调方法
052    @Override
053    protected Dialog onCreateDialog(int id) {
054        //TODO Auto-generated method stub
055        //根据传入的 id，初始化不同的 Dialog
056        switch (id) {
057        case DATEDIALOG:
058            //定义 DatePickerDialog，并进行初始化
059            DatePickerDialog dpd = new DatePickerDialog(
060                this, new OnDateSetListener() {
061                    //设置日期改变的监听器
062                    @Override
063                    public void onDateSet(DatePicker view, int year,
064                            int monthOfYear, int dayOfMonth) {
065                        //TODO Auto-generated method stub
066                        //设置 TextView 的内容
067                        Tv.setText("year:" + year + "\nmonth:"
068                                + monthOfYear + "\ndayOfMonth:"
069                                + dayOfMonth);
070                    }
```

```
071                     }, 1990, 10, 1);
072             return dpd;
073         case TIMEDIALOG:
074             //定义TimePickerDialog,并进行初始化
075             TimePickerDialog tpd = new TimePickerDialog(
076                     this, new OnTimeSetListener() {
077                         //设置时间改变的监听器
078                         @Override
079                         public void onTimeSet(TimePicker view,
080                                 int hourOfDay, int minute) {
081                             //TODO Auto-generated method stub
082                             Tv.setText(Tv.getText()+
083                                     "\nhourOfDay:" + hourOfDay
084                                     + "\nminute:"+ minute);
085                         }
086                     }, 12, 0, true);
087             return tpd;
088         default:
089             break;
090         }
091         return super.onCreateDialog(id);
092     }
093
094     private void findView() {
095         //TODO Auto-generated method stub
096         //通过findViewById得到对应的控件对象
097         Tv = (TextView) findViewById(R.id.Tv);
098         BtnDate = (Button) findViewById(R.id.BtnDate);
099         BtnTime = (Button) findViewById(R.id.BtnTime);
100     }
101 }
```

在代码第 97~99 行获取对应控件的对象,在第 25~26 行设置了第 29~46 行定义的按钮的监听器,当单击按钮时调用单击监听器的 onclick 方法,然后调用 showDialog 方法,然后回调 onCreateDialog 方法,执行 Dialog 的初始化,在第 59~71 行定义了日期选择对话框,在第 75~86 行定义了时间选择对话框。并分别设定了日期改变和时间改变的监听器。这样就实现了当获得时间或者日期的时候 TextView 的内容跟随变化了。

4. 实例扩展

对于 DatePickerDialog 和 TimePickerDialog 都是 android.app 包下的类,所以在 Android 中会把它们当做一个 app 来进行处理,所以使用这两个控件的时候和之前的控件不太相同,其实它们更像是弹出的另一个页面,而其当这些 Dialog 弹出后按下手机上的物理返回键,这些对话框也会消失。

范例 019 钟表显示效果

1. 实例简介

在开发 Android 应用的时候,可能会遇到在应用中显示时间的情况。例如,在用户全屏播放视频时显示时间和在用户全屏游戏时显示时间等。这时候我们需要在页面中展示当前的时间,Android 中提供 AnalogClock 和 DigitalClock 两个控件可以显示系统时间了。本实例就带领大家一起使用 AnalogClock 和 DigitalClock 来显示系统时间。

2．运行效果

该实例运行效果如图 3.19 所示。

3．实例程序讲解

在图 3.19 中用两种形式显示了当前的系统时间，用到两个系统控件 AnalogClock 和 DigitalClock，想要实现图 3.19 效果，需要在 res/values/strings.xml 中添加 AnalogClock 和 DigitalClock 控件即可。代码如下：

图 3.19　显示系统时间效果

```
01    <!-- 定义AnalogClock的控件   -->
02    <AnalogClock
03        android:layout_width="wrap_content"
04        android:layout_height="wrap_content" />
05    <!-- 定义DigitalClock的控件   -->
06    <DigitalClock
07        android:layout_width="wrap_content"
08        android:layout_height="wrap_content"/>
```

在如上代码中 AnalogClock 代表模拟时钟控件，DigitalClock 代表数字时钟控件，默认都显示系统时间。

4．实例扩展

对于系统时间的显示可以使用 AnalogClock 和 DigitalClock 控件来显示，当然你也可以定义自己的显示形式，前提是你可以得到系统的时间，在 Android 中得到系统时间的方法是：

```
//定义获取的时间格式
SimpleDateFormat formatter = new SimpleDateFormat(
        "yyyy:MM:dd HH:mm:ss");
//通过系统的时间戳得到date对象
Date curDate = new Date(System.currentTimeMillis());
//格式化Date对象，改成我们要求的格式
String str = formatter.format(curDate);
```

然后系统时间就存储在字符串 str 中了，大家可以通过自定义显示时间的 View 来显示时间了。

范例 020　秒表应用

1．实例简介

在 Android 中，一些应用需要计时功能，例如，游戏的过程中是否在规定时间内完成了谜题等。Android 中提供了 Chronometer 计时器控件，使我们可以很方便的实现计时功能。本实例就带领大家一起使用 Chronometer 来完成秒表应用。

2．运行效果

该实例运行效果如图 3.20 所示。

图 3.20　秒表应用

3. 实例程序讲解

在图 3.20 中包含了一个 Chronometer 控件用来显示秒表，三个按钮用来控制 Chronometer 的开始、停止和重置。想要实现上述效果，步骤如下所示。

（1）修改 res/layout/activity_main.xml 文件，代码如下：

```xml
01  <!-- 定义Chronometer的控件，定义计时器控件  -->
02  <Chronometer
03      android:id="@+id/Chron"
04      android:layout_width="wrap_content"
05      android:layout_height="wrap_content"/>
06  <Button
07      android:id="@+id/BtnStart"
08      android:layout_width="match_parent"
09      android:layout_height="wrap_content"
10      android:text="开始计时"
11      />
12  <Button
13      android:id="@+id/BtnStop"
14      android:layout_width="match_parent"
15      android:layout_height="wrap_content"
16      android:text="停止计时"
17      />
18  <Button
19      android:id="@+id/BtnReset"
20      android:layout_width="match_parent"
21      android:layout_height="wrap_content"
22      android:text="重置计时器"
23      />
```

在如上代码中第 2～5 行定义了一个 Chronometer 控件，第 6～23 行定义了三个按钮，并且分别设置了相应的 id，方便在 java 文件中获取控件对象。

（2）修改 src/com.wyl.example/MainActivity.java 文件，代码如下：

```java
01  /*import 代码省略*/
02  //定义MainActivity继承自Activity
03  public class MainActivity extends Activity {
04      private Button BtnStart;    //定义开始计时按钮
05      private Button BtnStop;     //定义停止计时按钮
06      private Button BtnReset;    //定义重置计时器按钮
07      private Chronometer Chron;  //定义计时器变量
08
09      @Override
10      protected void onCreate(Bundle savedInstanceState) {
11          super.onCreate(savedInstanceState);     //调用父类的onCreate方法
12
13          //通过setContentView方法设置当前页面的布局文件为activity_main
14          setContentView(R.layout.activity_main);
15          findView();                             //获取控件对象
16          setListener();                          //设置button的监听器
17      }
18
19      private void setListener() {
20          //TODO Auto-generated method stub
21          //设置按钮的单击监听器
22          BtnStart.setOnClickListener(mylistener);
23          BtnStop.setOnClickListener(mylistener);
```

```
24        BtnReset.setOnClickListener(mylistener);
25    }
26
27    //自定义 Button 的 OnClickListener 对象
28    OnClickListener mylistener = new OnClickListener() {
29        //按钮单击时的 onClick 回调方法
30        @Override
31        public void onClick(View v) {
32            //TODO Auto-generated method stub
33            switch (v.getId()) {
34            case R.id.BtnStart:
35                //开始计时
36                Chron.start();
37                break;
38            case R.id.BtnStop:
39                //停止计时
40                Chron.stop();
41                break;
42            case R.id.BtnReset:
43                //重置计时器
44                Chron.setBase(SystemClock.elapsedRealtime());
45                break;
46            default:
47                break;
48            }
49        }
50    };
51
52    private void findView() {
53        //TODO Auto-generated method stub
54        //通过 findViewById 得到对应的控件对象
55        BtnStart = (Button) findViewById(R.id.BtnStart);
56        BtnStop = (Button) findViewById(R.id.BtnStop);
57        BtnReset = (Button) findViewById(R.id.BtnReset);
58        Chron = (Chronometer)findViewById(R.id.Chron);
59    }
60 }
```

在上面代码的第 54~58 行得到相应的控件对象,在第 21~24 行给三个按钮设置了自定义的监听器,在第 27~50 行自定了一个单击监听器,当单击三个按钮的时候分别调用 Chronometer 对象的 start、stop 和 setBase 方法,实现计时器的开始、停止和重置。这里需要指出的是 setBase 方法设置计时器的及时开始基准点,设置为 SystemClock.elapsedRealtime(),也就是从当前系统时间开始计时,也就相当于清零了计时器。

4. 实例扩展

Chronometer 控件是 TextView 控件的一个子类,其实原理很简单就是每隔一秒钟调用一次更新内容的方法来更新 Chronometer 控件的显示内容。当然如果大家想制作毫秒的秒表的话,最常用的方法是通过记录当前时间的时间戳,然后用时间相减的方式来实现。

范例 021 圆角按钮效果

1. 实例简介

在 Android 中,肯定会用到按钮这个控件,这个控件我们在之前的实例中也看到了,相对也比较简单，但是现在我们应用中使用原始的按钮的效果很少。例如，QQ 软件的登录框已经是一个圆角的按钮，购物网站的登录按钮是自定义颜色的按钮等。本实例就带领大家一起来完成一个圆角按钮的美化效果。

2．运行效果

该实例运行效果如图 3.21 所示。

3．实例程序讲解

在图 3.21 中包含了一个 Button 控件，但是又和我们之前看到的按钮效果不同，主要是这个按钮的四个角是圆角效果。想要实现这样的效果就要定义按钮的背景样式了，步骤如下所示。

图 3.21 圆角按钮效果

（1）在 res/drawable 目录下新建名为 btn_bg.xml 的文件，代码如下：

```xml
01  <?xml version="1.0" encoding="utf-8"?>
02  <shape xmlns:android="http://schemas.android.com/apk/res/android"
03      android:shape="rectangle" >
04      <!-- 定义按钮的背景的颜色 -->
05      <solid android:color="#00CCFF" />
06      <!-- 设置按钮的每个角的弧形角度 -->
07      <corners android:radius="10dip" />
08      <!-- 按钮文字与边界的距离 -->
09      <padding
10          android:bottom="5dp"
11          android:left="5dp"
12          android:right="5dp"
13          android:top="5dp" />
14  </shape>
```

在此 xml 中定义了按钮的背景样式，其中第 5 行定义了按钮的背景颜色，第 7 行指定了按钮的四个角的弧形角度，第 9~13 行定义了按钮中的文字相对于按钮边缘的距离。

（2）修改 res/layout/activity_main.xml 文件，代码如下：

```xml
01  <LinearLayout xmlns:android="http://schemas.android.com/apk/res/android"
02      xmlns:tools="http://schemas.android.com/tools"
03      android:layout_width="match_parent"
04      android:layout_height="match_parent"
05      android:orientation="vertical"
06      android:paddingBottom="@dimen/activity_vertical_margin"
07      android:paddingLeft="@dimen/activity_horizontal_margin"
08      android:paddingRight="@dimen/activity_horizontal_margin"
09      android:paddingTop="@dimen/activity_vertical_margin" >
10
11      <!-- 定义Chronometer的控件，定义计时器控件 -->
12
13      <Button
14          android:layout_width="match_parent"
15          android:layout_height="wrap_content"
16          android:background="@drawable/btn_bg"
17          android:text="圆角按钮"
18          />
19
20  </LinearLayout>
```

在如上代码中第 11～18 行定义了一个 Button 控件，其中第 16 行将之前定义好的按钮的背景样式 btn_bg 设置给按钮的 background 属性。这样的一个圆角的按钮效果就出来了。

4．实例扩展

其实 Android 中对于美化按钮的控件很多，如 ImageButton，这个控件具有按钮的所有功能，而且它可以设置一张图片显示在按钮上面，或者通过 style 的属性能够给 Button 快速的进行美化，常见的 Button 的美化 style 有如下几种。

- @android:attr/button：普通按钮的效果。
- @android:attr/buttonBarButtonStyle：按钮栏的按钮效果。
- @android:attr/buttonBarStyle：按钮栏效果。
- @android:attr/buttonStyle：按钮的效果。
- @android:attr/buttonStyleInset：插入按钮的效果。
- @android:attr/buttonStyleSmall：小型按钮的效果。
- @android:attr/buttonStyleToggle：选项按钮的效果。

对于如上效果，大家可以自己进行练习，当然有些效果是需要高级的 Android SDK 版本支持的。剩下的就是大家在做自己的应用的时候选择自己喜欢的效果就可以了。

3.2　Android 中常见布局的使用

范例 022　用户注册页面的制作

1．实例简介

我们在使用 Android 应用的时候，免不了要进行应用账号的注册，尤其是网站的应用客户端，如 QQ 软件、淘宝客户端和京东客户端等，某些功能只能注册后才可使用。本实例就带领大家通过 Android 中最常用的布局 LinearLayout 来完成一个常见的注册页面。

2．运行效果

该实例运行效果如图 3.22 所示。

图 3.22　用户注册页面的制作

3．实例程序讲解

在上面的例子效果中有很多控件，但是这些控件的摆放规则总体是以线性的形式放置的，这就需要使用 LinearLayout 布局了。想要实现如上效果，只要修改 res/layout/activity_main.xml 文件，代码如下：

```
01  <LinearLayout xmlns:android="http://schemas.android.com/apk/res/android"
02      xmlns:tools="http://schemas.android.com/tools"
03      android:layout_width="match_parent"
04      android:layout_height="match_parent"
05      android:orientation="vertical"
```

```
06      android:paddingBottom="@dimen/activity_vertical_margin"
07      android:paddingLeft="@dimen/activity_horizontal_margin"
08      android:paddingRight="@dimen/activity_horizontal_margin"
09      android:paddingTop="@dimen/activity_vertical_margin" >
10
11      <!-- 定义用户名的输入控件   -->
12      <EditText
13          android:layout_width="match_parent"
14          android:layout_height="wrap_content"
15          android:hint="请输入注册用户名"
16          android:ems="10" >
17      </EditText>
18      <!-- 定义密码的输入控件   -->
19      <EditText
20          android:layout_width="match_parent"
21          android:layout_height="wrap_content"
22          android:hint="请输入密码"
23          android:ems="10"
24          android:inputType="textPassword" />
25      <!-- 定义再次输入密码的控件   -->
26      <EditText
27          android:layout_width="match_parent"
28          android:layout_height="wrap_content"
29          android:hint="请再次输入密码"
30          android:ems="10"
31          android:inputType="textPassword" />
32      <!-- 定义输入邮箱的控件   -->
33      <EditText
34          android:layout_width="match_parent"
35          android:layout_height="wrap_content"
36          android:hint="请输入您的邮箱"
37          android:ems="10"
38          android:inputType="textEmailAddress" />
39      <!-- 定义输入手机号码的控件   -->
40      <EditText
41          android:layout_width="match_parent"
42          android:layout_height="wrap_content"
43          android:hint="请输入您的手机"
44          android:ems="10" />
45      <!-- 定义注册按钮控件   -->
46      <Button
47          android:layout_width="match_parent"
48          android:layout_height="wrap_content"
49          android:background="@drawable/btn_bg"
50          android:text="注册"
51          />
52  </LinearLayout>
```

在如上代码中第 1~9 行定义了一个 LinearLayout——线性布局，在第 52 行结束了这个布局（线性布局），顾名思义就是在此布局中的所有控件都以一条线的形式来排列。在第 5 行，设置了 orientation 属性的值为 vertical，代表此线性布局为垂直线性布局，也就是从上到下逐一排列。这样一个通用的注册页面就完成了。

4．实例扩展

LinearLayout 指的是线性布局，我们此实例中展示的是垂直布局，当然也可以水平布局，这样的话，在 LinearLayout 布局中的控件就会从左到右水平排列了，方法就是设置

orientation 属性的值为 horizontal。

范例 023　学生成绩列表页面的制作

1．实例简介

我们在使用 Android 应用的时候，经常会看到类似的表格布局，它主要以展现系统中的数据列表为主，如学生成绩列表和购买商品列表等。这样我们就用到了 Android 中另一个布局 TableLayout。本实例就带领大家通过 Android 中的 TableLayout 来完成一个常见的学生列表页面。

2．运行效果

该实例运行效果如图 3.23 所示。

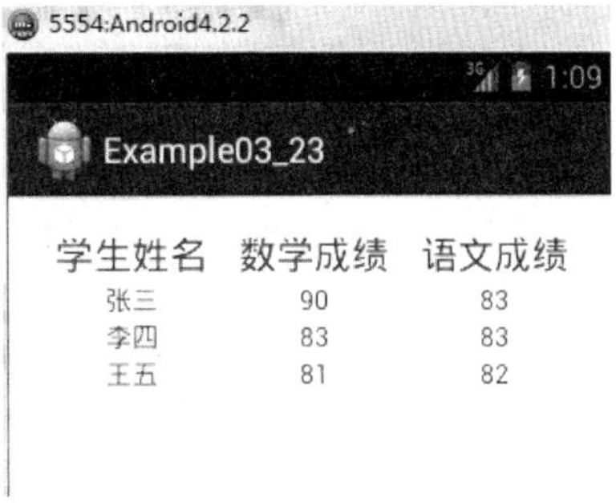

图 3.23　学生成绩列表页面的制作

3．实例程序讲解

在上面的例子效果中所有控件都好像在一个表格中，在 Android 中实现表格的布局效果就要使用到另外的一种布局 TableLayout，其中的每一行是一个 TableRow，TableRow 也可以设置 orientation 属性的。想要实现 TableLayout 效果，只要修改 res/layout/activity_main.xml 文件，代码如下：

```xml
001 <!-- 定义基础布局 TableLayout -->
002 <TableLayout xmlns:android="http://schemas.android.com/apk/res/android"
003     xmlns:tools="http://schemas.android.com/tools"
004     android:layout_width="match_parent"
005     android:layout_height="match_parent"
006     android:orientation="vertical"
007     android:paddingBottom="@dimen/activity_vertical_margin"
008     android:paddingLeft="@dimen/activity_horizontal_margin"
009     android:paddingRight="@dimen/activity_horizontal_margin"
010     android:paddingTop="@dimen/activity_vertical_margin" >
011     <!-- 定义 TableRow 控件，代表第一行 -->
012     <TableRow android:orientation="horizontal" >
013
014         <TextView
015             android:layout_width="wrap_content"
016             android:layout_height="match_parent"
017             android:gravity="center"
018             android:textSize="20dp"
019             android:layout_weight="1"
020             android:text="学生姓名" />
021
022         <TextView
023             android:layout_width="match_parent"
024             android:layout_height="match_parent"
025             android:layout_weight="1"
026             android:textSize="20dp"
027             android:gravity="center"
028             android:text="数学成绩" />
029
030         <TextView
```

```xml
031            android:layout_width="match_parent"
032            android:layout_height="match_parent"
033            android:layout_weight="1"
034            android:textSize="20dp"
035            android:gravity="center"
036            android:text="语文成绩" />
037    </TableRow>
038    <!-- 定义 TableRow 控件，代表第二行 -->
039    <TableRow android:orientation="horizontal" >
040
041        <TextView
042            android:layout_width="wrap_content"
043            android:layout_height="match_parent"
044            android:layout_weight="1"
045            android:gravity="center"
046            android:text="张三" />
047
048        <TextView
049            android:layout_width="wrap_content"
050            android:layout_height="match_parent"
051            android:layout_weight="1"
052            android:gravity="center"
053            android:text="90" />
054
055        <TextView
056            android:layout_width="wrap_content"
057            android:layout_height="match_parent"
058            android:layout_weight="1"
059            android:gravity="center"
060            android:text="83" />
061    </TableRow>
062    <!-- 定义 TableRow 控件，代表第三行 -->
063    <TableRow android:orientation="horizontal" >
064
065        <TextView
066            android:layout_width="match_parent"
067            android:layout_height="match_parent"
068            android:layout_weight="1"
069            android:gravity="center"
070            android:text="李四" />
071
072        <TextView
073            android:layout_width="wrap_content"
074            android:layout_height="match_parent"
075            android:layout_weight="1"
076            android:gravity="center"
077            android:text="83" />
078
079        <TextView
080            android:layout_width="wrap_content"
081            android:layout_height="match_parent"
082            android:layout_weight="1"
083            android:gravity="center"
084            android:text="83" />
085    </TableRow>
086    <!-- 定义 TableRow 控件，代表第四行 -->
087    <TableRow android:orientation="horizontal" >
088
089        <TextView
090            android:layout_width="match_parent"
```

```
091            android:layout_height="match_parent"
092            android:layout_weight="1"
093            android:gravity="center"
094            android:text="王五" />
095
096        <TextView
097            android:layout_width="wrap_content"
098            android:layout_height="match_parent"
099            android:layout_weight="1"
100            android:gravity="center"
101            android:text="81" />
102
103        <TextView
104            android:layout_width="wrap_content"
105            android:layout_height="match_parent"
106            android:layout_weight="1"
107            android:gravity="center"
108            android:text="82" />
109    </TableRow>
110
111 </TableLayout>
```

在如上代码中第 1～9 行定义了一个 TableLayout，表格布局，在第 12～37 行、第 39～61 行，第 63～85 行和第 87～109 行分别定义了四个 TableRow，代表此表格中的四行，并且这四个 TableRow 都设置了 orientation 属性为 horizontal，代表每行中都以水平形式来排列。这样一个学生成绩列表页面就完成了。

4．实例扩展

TableLayout 一般的使用场景就是信息的列表展示，或者当你的数据来源于网络或者数据库时动态进行数据显示的时候。所以大家可以根据自己的应用需要来进行选择是否使用 TableLayout 布局了。

范例 024　登录页面的制作

1．实例简介

我们在使用 Android 应用的时候，尤其是一些网站的客户端，我们必须要进行登录才可以使用一些功能，如 qq 登录和百度音乐的登录等。这样我们就用到了 Android 中另一个布局 RelativeLayout。本实例就带领大家通过 Android 中的 RelativeLayout 来完成一个登录页面。

2．运行效果

该实例运行效果如图 3.24 所示。

图 3.24　登录页面

3．实例程序讲解

在上面的例子效果中，我们的所有控件都是以相对位置来进行布局的。例如，退出按钮在输入密码按钮的下面，在屏幕的最右端；而登录按钮在退出按钮的左边，并且也在输入密码按钮的下面。想要实现所有控件都以相对位置布局，那就要使用 RelativeLayout 来进行布局了。对于我们这个实例，我们只要修改 res/layout/activity_main.xml 文件，代码如下：

```xml
01  <!-- 定义基础布局RelativeLayout -->
02  <RelativeLayout xmlns:android="http://schemas.android.com/apk/res/android"
03      xmlns:tools="http://schemas.android.com/tools"
04      android:layout_width="match_parent"
05      android:layout_height="match_parent"
06      android:paddingBottom="@dimen/activity_vertical_margin"
07      android:paddingLeft="@dimen/activity_horizontal_margin"
08      android:paddingRight="@dimen/activity_horizontal_margin"
09      android:paddingTop="@dimen/activity_vertical_margin" >
10
11      <!-- 定义用户名输入框-->
12      <EditText
13          android:id="@+id/EtUsername"
14          android:layout_width="match_parent"
15          android:layout_height="wrap_content"
16          android:hint="请输入用户名" />
17      <!-- 定义用户密码输入框-->
18      <EditText
19          android:id="@+id/EtPasswrod"
20          android:layout_width="match_parent"
21          android:layout_height="wrap_content"
22          android:layout_below="@id/EtUsername"
23          android:hint="请输入密码" />
24      <!-- 定义退出按钮-->
25      <Button
26          android:id="@+id/BtnExit"
27          android:layout_width="wrap_content"
28          android:layout_height="wrap_content"
29          android:layout_below="@id/EtPasswrod"
30          android:layout_alignRight="@id/EtPasswrod"
31          android:hint="退出" />
32      <!-- 定义登录按钮-->
33      <Button
34          android:id="@+id/BtnLogin"
35          android:layout_below="@id/EtPasswrod"
36          android:layout_toLeftOf="@id/BtnExit"
37          android:layout_width="wrap_content"
38          android:layout_height="wrap_content"
39          android:hint="登录" />
40  </RelativeLayout>
```

在如上代码中第 1～9 行定义了一个 RelativeLayout（相对布局），在第 22、29、30、35 和 36 行分别使用了只有相对布局中的控件才具有的属性。这样你就可以以相对位置来进行页面的布局了。

4．实例扩展

RelativeLayout 的使用场景很多，我们经常会遇到某个控件是相对父控件右对齐这样的需求，这时候你只有使用 RelativeLayout 了。在 RelativeLayout 中的控件会多出一些属性，这是我们程序的软件，常见的属性如下所示。

- ❑ android:layout_above：该控件的位置在给定 ID 的控件的上面。
- ❑ android:layout_below：该控件的位置在给定 ID 的控件的下面。
- ❑ android:layout_toLeftOf：该控件的位置在给定 ID 的控件的左面。

- android:layout_toRightOf：该控件的位置在给定 ID 的控件的右面。
- android:layout_alignBaseline：该控件与给定 ID 的控件 baseline 对齐。
- android:layout_alignTop：该控件与给定 ID 的控件顶部对齐。
- android:layout_alignBottom：该控件与给定 ID 的控件底部对齐。
- android:layout_alignLeft：该控件与给定 ID 的控件左对齐。
- android:layout_alignRight：该控件与给定 ID 的控件右对齐。
- android:layout_alignParentTop：该控件与其父控件的顶部对齐。
- android:layout_alignParentBottom：该控件与其父控件的底部对齐。
- android:layout_alignParentLeft：该控件与其父控件的左对齐。
- android:layout_alignParentRight：该控件与其父控件的右对齐。
- android:layout_centerHorizontal：将该控件水平居中。
- android:layout_centerVertical：将该控件的垂直居中。
- android:layout_centerInParent：将该控件相对于父控件居中。

当然还有一些属性我这里没有提到，当大家使用到的时候去 Android 的开发文档中查询即可。

范例 025　开发模型图的页面

1．实例简介

我们在使用 Android 应用的时候，也有可能遇到这样一种效果，就是控件之间的位置相对独立，而且是固定不变的。例如，固定图表模板的制作、固定文档模板的制作等。这就要求我们在做这样的页面的时候只能够以屏幕坐标为基准设置控件的位置，这就用到了 AbsoluteLayout（绝对布局）。本实例就带领大家使用 Android 中的 AbsoluteLayout 来完成一个软件开发模型图的页面。

2．运行效果

该实例运行效果如图 3.25 所示。

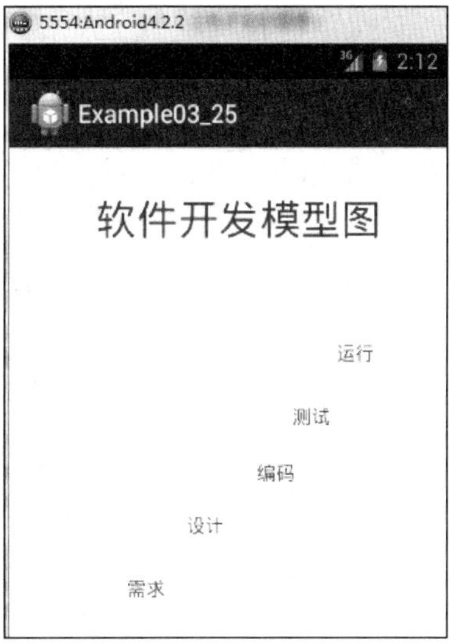

图 3.25　软件开发模型图页面

3．实例程序讲解

在上面的例子效果中，我们可以看到，其中有 6 个 TextView，标题 TextView，还有 5 个位置飘忽的 TextView，这就要使用 Android 中的 AbsoluteLayout 来进行布局了。对于我们这个实例，只要修改 res/layout/activity_main.xml 文件，代码如下：

```
01  <!-- 定义基础布局 AbsoluteLayout -->
02  <AbsoluteLayout xmlns:android="http://schemas.android.com/apk/res/android"
03      xmlns:tools="http://schemas.android.com/tools"
04      android:layout_width="match_parent"
05      android:layout_height="match_parent"
06      android:paddingBottom="@dimen/activity_vertical_margin"
07      android:paddingLeft="@dimen/activity_horizontal_margin"
```

```
08        android:paddingRight="@dimen/activity_horizontal_margin"
09        android:paddingTop="@dimen/activity_vertical_margin" >
10
11        <!-- 定义标题 TextView -->
12        <TextView
13            android:layout_width="wrap_content"
14            android:layout_height="wrap_content"
15            android:layout_x="46dp"
16            android:layout_y="14dp"
17            android:text="软件开发模型图"
18            android:textSize="30sp" />
19
20        <TextView
21            android:layout_width="wrap_content"
22            android:layout_height="wrap_content"
23            android:layout_x="223dp"
24            android:layout_y="120dp"
25            android:text="运行" />
26
27        <TextView
28            android:layout_width="wrap_content"
29            android:layout_height="wrap_content"
30            android:layout_x="190dp"
31            android:layout_y="165dp"
32            android:text="测试" />
33
34        <TextView
35            android:layout_width="wrap_content"
36            android:layout_height="wrap_content"
37            android:layout_x="164dp"
38            android:layout_y="205dp"
39            android:text="编码" />
40
41        <TextView
42            android:layout_width="wrap_content"
43            android:layout_height="wrap_content"
44            android:layout_x="112dp"
45            android:layout_y="242dp"
46            android:text="设计" />
47
48        <TextView
49            android:layout_width="wrap_content"
50            android:layout_height="wrap_content"
51            android:layout_x="69dp"
52            android:layout_y="287dp"
53            android:text="需求" />
54
55    </AbsoluteLayout>
```

在如上代码中第 1～9 行定义了一个绝对布局 AbsoluteLayout，在第 15～16 行、第 23～24 行、第 30～31 行、第 37～38 行、第 44～45 行和第 51～52 行分别使用了只有绝对布局中的控件才具有的属性 layout_x 和 layout_y，通过这两个属性可以定位此控件在屏幕上的绝对位置。这样你就可以实现以控件的绝对位置来进行页面的布局了。

4．实例扩展

对于现在市面上 Android 的设备数不胜数，而且其中手机中的分辨率也是各不相同，所以如果使用 AbsoluteLayout 来进行页面布局的话，会造成在不同的分辨率下的效果差异，所以现在 AbsoluteLayout 使用的频率相对比较少。但是任何一种布局都有它擅长的地方存在，如果在做应用的时候发现需要以屏幕的绝对坐标来固定控件的位置的话，那么 AbsoluteLayout 是你的必备之选，所以大家在做应用的时候要根据自己的需要进行选择。

范例 026　图片相框效果

1．实例简介

我们在使用 Android 应用的时候，也有可能遇到这样一种效果，就是把屏幕当做画布，把所有的控件当做画布上的颜料，这样一层一层的画上去，下层的被后放上去的控件所覆盖，如图片合成效果和相框效果等。这就要求我们在做这样的页面的时候是一层一层的放上控件，而且之前放上去的控件会自动被覆盖，这就用到了 Android 中的 FrameLayout（帧布局）。本实例就带领大家使用 Android 中的 FrameLayout 来完成一个图片相框的效果。

2．运行效果

该实例运行效果如图 3.26 所示。

图 3.26　图片相框效果

3．实例程序讲解

在上面的例子效果中，我们可以看到，其中有两个 ImageView，分别显示我的相框和我需要显示的图片，而且这两个 ImageView 是层叠的，这就要使用 Android 中的 FrameLayout 来进行布局了。对于我们这个实例，只要修改 res/layout/activity_main.xml 文件，代码如下：

```xml
01  <!-- 定义基础布局 FrameLayout -->
02  <FrameLayout xmlns:android="http://schemas.android.com/apk/res/android"
03      xmlns:tools="http://schemas.android.com/tools"
04      android:layout_width="match_parent"
05      android:layout_height="match_parent"
06      android:paddingBottom="@dimen/activity_vertical_margin"
07      android:paddingLeft="@dimen/activity_horizontal_margin"
08      android:paddingRight="@dimen/activity_horizontal_margin"
09      android:paddingTop="@dimen/activity_vertical_margin" >
10
11      <ImageView
12          android:layout_width="wrap_content"
13          android:layout_height="wrap_content"
14          android:src="@drawable/imgbg" />
15
16      <ImageView
17          android:layout_width="wrap_content"
18          android:layout_height="wrap_content"
19          android:padding="30dp"
20          android:src="@drawable/img" />
21
22  </FrameLayout>
```

在如上代码中第 1～9 行定义了一个帧布局 FrameLayout，在第 14 和 20 行，分别对

ImageView 设置了不同的图片资源（图片资源大家请复制到工程的 res/drawable 目录下，本例中我的图片资源名为 imgbg 和 img）。这样你就可以实现以控件在放入页面中的时候像画布一样一层层的放入了。

4．实例扩展

帧布局 FrameLayout 的布局方式是用户在绘制每一个控件的时候，都当做页面的每一帧来画到屏幕上，对于这种方式来说，多数用在特殊的控件布局，或者一些游戏的开发中使用。

范例 027 商城专区效果

1．实例简介

我们在使用 Android 应用的时候，也有可能遇到这样一种效果，就是把屏幕分成了很多的矩形框，然后每部分矩形框中加入相应的内容，显得比较个性。例如，商品的促销列表和商城的分区列表等。这就要求我们在做这样的页面的时候将屏幕分成一个一个的区域，然后填入相应的控件，这就用到了在 Android 4.0 SDK 中的新加入的一种布局 GridLayout（网格布局）。本实例就带领大家使用 Android 中的GridLayout 来完成一个商城的促销分区效果。

2．运行效果

该实例运行效果如图 3.27 所示。

3．实例程序讲解

在上面的例子效果中，我们可以看到，其中有 5 个 TextView，分别显示我们的个类服装的分区，但是这 5 个 TextView 又不是固定的几种关系，而是相对服装的网格关系，这就要使用 Android 中的 GridLayout 来进行布局了。对于我们这个实例，只要修改 res/layout/activity_main.xml 文件，代码如下：

图 3.27 商城的促销分区效果

```xml
01  <!-- 定义基础布局 GridLayout -->
02  <GridLayout xmlns:android="http://schemas.android.com/apk/res/android"
03      xmlns:tools="http://schemas.android.com/tools"
04      android:layout_width="match_parent"
05      android:layout_height="match_parent"
06      android:paddingBottom="@dimen/activity_vertical_margin"
07      android:paddingLeft="@dimen/activity_horizontal_margin"
08      android:paddingRight="@dimen/activity_horizontal_margin"
09      android:paddingTop="@dimen/activity_vertical_margin"
10      android:columnCount="4">
11
12      <TextView
13          android:layout_width="150dp"
14          android:layout_height="60dp"
15          android:layout_columnSpan="2"
16          android:layout_gravity="left"
17          android:background="@color/blue"
18          android:gravity="center"
19          android:text="男装"
20          android:width="50dp" />
```

```
21
22      <TextView
23          android:layout_width="150dp"
24          android:layout_height="60dp"
25          android:layout_columnSpan="2"
26          android:background="@color/red"
27          android:gravity="center"
28          android:text="女装"
29          android:width="50dp" />
30
31      <TextView
32          android:layout_width="75dp"
33          android:layout_height="60dp"
34          android:layout_gravity="top"
35          android:background="@color/red"
36          android:gravity="center"
37          android:text="童装"
38          android:width="25dp" />
39
40      <TextView
41          android:layout_width="150dp"
42          android:layout_height="60dp"
43          android:layout_columnSpan="2"
44          android:background="@color/green"
45          android:gravity="center"
46          android:text="流行装"
47          android:width="50dp" />
48
49      <TextView
50          android:layout_width="75dp"
51          android:layout_height="60dp"
52          android:background="@color/blue"
53          android:gravity="center"
54          android:text="嘻哈"
55          android:width="25dp" />
56
57  </GridLayout>
```

在如上代码中第 1~9 行定义了一个网格布局 GridLayout，在第 15、25 和 43 行，分别对 TextView 设置了他们所跨的列数，默认为 1 列，在本实例中为了能方便区分各个分区的位置，我给每个 TextView 都设置了不同颜色的背景。这样就实现网格的布局效果了。

4．实例扩展

网格布局 GridLayout 的布局方式是可以将屏幕分成大小不同的一个一个的格子，然后把控件一个一个的放进去，这样方便进行布局显示，对于一些有网格布局要求的页面使用这种布局，格外轻松。需要注意的是想要使用 GridLayout，你的工程的 android:minSdkVersion 和 android:targetSdkVersion 都必须大于等于 API Level 14。

范例 028 三字经阅读程序

1．实例简介

我们在使用 Android 应用的时候，经常会遇到这样的情况，就是一个页面显示的内容较多，在一屏幕无法显示完。例如，应用的帮助文档和应用中的软件注册须知页面等。这就要求我们在做这样的页面的时候能够让一个屏幕滚动显示更多的内容，而我们之前讲到的所有的布局类型都只能显示一屏，这就用到了 Android 中的另一种布局 ScrollView（滚动视图）。本实例就带领大家使用 Android 中的 ScrollView 来完成一个三字经的阅读程序。

2．运行效果

该实例运行效果如图 3.28 所示。

3．实例程序讲解

在上面的例子效果中，明显三字经的内容对于我们的屏幕来说太多了，一屏幕是无法完整显示的，而我们的实例中屏幕可以随着 TextView 中内容的多少而进行滚动，

图 3.28　三字经阅读程序

也就是一屏幕显示不下可以滚动两屏幕，甚至更多。这就要使用 Android 中的 ScrollView 来进行布局了。对于我们这个实例，只要修改 res/layout/activity_main.xml 文件，代码如下：

```xml
01  <!-- 定义基础布局 ScrollView -->
02  <ScrollView xmlns:android="http://schemas.android.com/apk/res/android"
03      xmlns:tools="http://schemas.android.com/tools"
04      android:layout_width="match_parent"
05      android:layout_height="match_parent"
06      android:paddingBottom="@dimen/activity_vertical_margin"
07      android:paddingLeft="@dimen/activity_horizontal_margin"
08      android:paddingRight="@dimen/activity_horizontal_margin"
09      android:paddingTop="@dimen/activity_vertical_margin"
10      >
11      <!-- 定义 TextView，显示三字经的内容 -->
12      <TextView
13          android:layout_width="match_parent"
14          android:layout_height="match_parent"
15          android:text="@string/tv_content"/>
16  
17  </ScrollView>
```

在如上代码中第 1~10 行定义了一个滚动布局 ScrollView，在 ScrollView 中包含了一个 TextView，TextView 的内容为一个常量字符串，需要大家修改 res/values/strings.xml 文件，添加 string 节点，代码如下：

```xml
<string name="tv_content"> 人之初，性本善。………..</string>
```

这样就可以设置 TextView 的 text 属性为@string/tv_content，来使用此常量字符串了。这样就实现一个可以滚动显示的布局效果了。

4．实例扩展

对于 ScrollView 来说 Android 对它有一个使用的限制，就是 ScrollView 中只能有一个 View。这令我们在使用的时候有些不便，本例只包含了一个 TextView 控件没问题，但是如果要求页面中包含多个控件的话怎么办呢？所以我们一般在使用 ScrollView 的时候，在其中直接放一个 LinearLayout 布局，然后把我们需要放置的多个控件放到 LinearLayout 中，这样就可以了。所以大家可以去看 Android 实现代码，我们之前讲过的各种布局都是 View 类的子类，Android 中的 View 和 ViewGroup 的关系是明显的设计模式中的组合模式。

范例 029 计算器程序的页面设计

1．实例简介

我们在使用 Android 应用的时候，经常会发现一般 Android 应用中的一个页面不会单纯只用一种布局来完成，有可能在一个页面中包含了多种布局的效果，这些布局之间存在嵌套的关系。例如，一个页面的整体布局为线性布局，而其中又包含了相对布局；一个页面的整体是相对布局，而局部又使用的是绝对布局等。这就要求我们需要灵活掌握这几种布局的使用方法，结合具体需求来使用。本实例就带领大家使用 Android 中的嵌套 LinearLayout 来完成一个计算器程序页面的设计。

2．运行效果

该实例运行效果如图 3.29 所示。

图 3.29 计算器页面设计

3．实例程序讲解

在上面的例子效果中，我们看到了很多个 Button 控件，而且这些控件之间的位置之前是基本遵循线性布局的，但是其中有横向的线性布局，也有纵向的线性布局。这就要求我们要在一个页面中进行布局的嵌套了。对于我们这个实例，只要修改 res/layout/activity_main.xml 文件，代码如下：

```
001  <!-- 定义基础布局 LinearLayout -->
002  <LinearLayout xmlns:android="http://schemas.android.com/apk/res/android"
003      xmlns:tools="http://schemas.android.com/tools"
004      android:layout_width="match_parent"
005      android:layout_height="match_parent"
006      android:paddingBottom="@dimen/activity_vertical_margin"
007      android:paddingLeft="@dimen/activity_horizontal_margin"
008      android:paddingRight="@dimen/activity_horizontal_margin"
009      android:paddingTop="@dimen/activity_vertical_margin"
010      android:orientation="vertical">
011      <EditText
012          android:layout_width="match_parent"
013          android:layout_height="wrap_content"
014          android:text="0"
015          />
016      <!-- 定义第一行的 LinearLayout -->
017      <LinearLayout
018          android:layout_width="match_parent"
019          android:layout_height="wrap_content"
020          android:orientation="horizontal" >
021
022          <Button
023              android:layout_width="wrap_content"
024              android:layout_height="wrap_content"
025              android:layout_weight="1"
026              android:text="7" />
027
```

```
028        <Button
029            android:layout_width="wrap_content"
030            android:layout_height="wrap_content"
031            android:layout_weight="1"
032            android:text="8" />
033
034        <Button
035            android:layout_width="wrap_content"
036            android:layout_height="wrap_content"
037            android:layout_weight="1"
038            android:text="9" />
039
040        <Button
041            android:layout_width="wrap_content"
042            android:layout_height="wrap_content"
043            android:layout_weight="1"
044            android:text="/" />
045
046        <Button
047            android:layout_width="wrap_content"
048            android:layout_height="wrap_content"
049            android:layout_weight="1"
050            android:text="%" />
051    </LinearLayout>
052    <!-- 定义第二行的 LinearLayout -->
053    <LinearLayout
054        android:layout_width="match_parent"
055        android:layout_height="wrap_content"
056        android:orientation="horizontal" >
057
058        <Button
059            android:layout_width="wrap_content"
060            android:layout_height="wrap_content"
061            android:layout_weight="1"
062            android:text="4" />
063
064        <Button
065            android:layout_width="wrap_content"
066            android:layout_height="wrap_content"
067            android:layout_weight="1"
068            android:text="5" />
069
070        <Button
071            android:layout_width="wrap_content"
072            android:layout_height="wrap_content"
073            android:layout_weight="1"
074            android:text="6" />
075
076        <Button
077            android:layout_width="wrap_content"
078            android:layout_height="wrap_content"
079            android:layout_weight="1"
080            android:text="*" />
081
082        <Button
083            android:layout_width="wrap_content"
084            android:layout_height="wrap_content"
085            android:layout_weight="1"
086            android:text="1/x" />
087    </LinearLayout>
088    <!-- 定义第三行的 LinearLayout -->
```

```xml
089  <LinearLayout
090      android:layout_width="match_parent"
091      android:layout_height="wrap_content"
092      android:orientation="horizontal" >
093
094      <LinearLayout
095          android:layout_width="wrap_content"
096          android:layout_height="wrap_content"
097          android:layout_weight="4"
098          android:orientation="vertical" >
099
100          <LinearLayout
101              android:layout_width="match_parent"
102              android:layout_height="wrap_content" >
103
104              <Button
105                  android:layout_width="wrap_content"
106                  android:layout_height="wrap_content"
107                  android:layout_weight="1"
108                  android:text="1" />
109
110              <Button
111                  android:layout_width="wrap_content"
112                  android:layout_height="wrap_content"
113                  android:layout_weight="1"
114                  android:text="2" />
115
116              <Button
117                  android:layout_width="wrap_content"
118                  android:layout_height="wrap_content"
119                  android:layout_weight="1"
120                  android:text="3" />
121
122              <Button
123                  android:layout_width="wrap_content"
124                  android:layout_height="wrap_content"
125                  android:layout_weight="1"
126                  android:text="-" />
127          </LinearLayout>
128          <LinearLayout
129              android:layout_width="match_parent"
130              android:layout_height="wrap_content" >
131
132              <Button
133                  android:layout_width="wrap_content"
134                  android:layout_height="wrap_content"
135                  android:layout_weight="2"
136                  android:text="0" />
137                  />
138              <Button
139                  android:layout_width="wrap_content"
140                  android:layout_height="wrap_content"
141                  android:layout_weight="1"
142                  android:text="." />
143
144              <Button
145                  android:layout_width="wrap_content"
146                  android:layout_height="wrap_content"
147                  android:layout_weight="1"
148                  android:text="+" />
149          </LinearLayout>
```

```
150            </LinearLayout>
151
152        <Button
153            android:layout_width="wrap_content"
154            android:layout_height="match_parent"
155            android:layout_weight="1"
156            android:text="=" />
157     </LinearLayout>
158
159 </LinearLayout>
```

在如上代码中第 1~10 行定义了基础的 LinearLayout 布局，在第 18~20 行定义了第一行的 LinearLayout，在第 53~56 行定义了第二行的 LinearLayout 布局，在第 88~92 行又定义了第三行的 LinearLayout，而且立刻在第 94~98 行又嵌套了一层 LinearLayout。这样就实现一个在布局之中嵌套布局了。

4．实例扩展

在 Android 中所有的布局都可以当做一个 View 来使用，所以各种布局之间都可以进行嵌套，在我们的实际应用中，可能会更频繁的嵌套布局，但是在嵌套布局的过程中，无形的加大了系统对于布局文件的开销，所以我们在做页面的时候要使用尽量简单的布局来完成。例如，本实例可以使用我们之前讲过的 GridLayout 来实现，这样的话操作步骤及布局的嵌套情况会比我们现在要少很多，大家可以自行练习。

3.3 Android 中高级组件的使用

范例 030　单词搜索补全效果

1．实例简介

我们在使用 Android 应用的时候，经常会遇到搜索的情况，这是由于我们的手机大小的限制，所以用户的输入相对于电脑的输入来说受到很大的限制，这时候用户就希望当输入想要搜索的内容的前一部分时，应用能够自动提示可能的后部分内容，然后用户通过的提示进行选择，这样的话给用户的使用体验是非常好的。例如，在百度搜索中，输入"放假"两个字，在下拉列表中就提示出了与放假有关的热门搜索词等。这就要求在用户的输入过程中我们的输入框能够随时和我们的库进行比较，如果有类似的词语就立刻返回给用户。本实例就带领大家使用 Android 中的 AutoCompleteTextView 来完成单词查询页面的设计。

2．运行效果

该实例运行效果如图 3.30 所示。

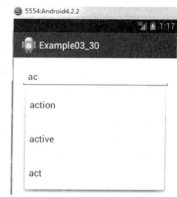

图 3.30　单词补全输入框

3. 实例程序讲解

在上面例子的效果中,在我们输入过程中,输入框根据输入的前几个字符去查询我们的库,如果有匹配上,那么就以下拉列表的形式显示出来,然后通过单击可以快速输入。这就是 Android 提供的自动补全控件 AutoCompleteTextView 的作用了,它可以自动的补全内容。想要实现我们上例的效果,步骤如下所示。

(1) 修改 res/layout/activity_main.xml 文件,代码如下:

```xml
01  <!-- 定义基础布局 LinearLayout -->
02  <LinearLayout xmlns:android="http://schemas.android.com/apk/res/android"
03      xmlns:tools="http://schemas.android.com/tools"
04      android:layout_width="match_parent"
05      android:layout_height="match_parent"
06      android:paddingBottom="@dimen/activity_vertical_margin"
07      android:paddingLeft="@dimen/activity_horizontal_margin"
08      android:paddingRight="@dimen/activity_horizontal_margin"
09      android:paddingTop="@dimen/activity_vertical_margin"
10      android:orientation="vertical">
11      <!-- 定义自动补全控件 -->
12      <AutoCompleteTextView
13          android:id="@+id/Actv"
14          android:layout_width="fill_parent"
15          android:layout_height="wrap_content" />
16  </LinearLayout>
```

在如上代码中第 12~15 行定义了 AutoCompleteTextView 控件,并且设置其 id 属性为 Actv,方便我们在 java 文件中获取此控件。

(2) 修改 src/com.wyl.example/MainActivity.java 文件,代码如下:

```java
01  /*import 代码省略*/
02  //定义 MainActivity 继承自 Activity
03  public class MainActivity extends Activity {
04
05      private AutoCompleteTextView Actv;          //定义 AutoCompleteTextView 对象
06
07      //定义单词查询库
08      private static final String[] words = { "abbreviation", "action", "active","act",
09              "ally ", "ball", "bask" };
10
11      @Override
12      protected void onCreate(Bundle savedInstanceState) {
13          super.onCreate(savedInstanceState);         //调用父类的 onCreate 方法
14          //通过 setContentView 方法设置当前页面的布局文件为 activity_main
15          setContentView(R.layout.activity_main);
16
17          //获取 AutoCompleteTextView 对象
18          Actv = (AutoCompleteTextView)findViewById(R.id.Actv);
19          //自定义 ArrayAdapter,设置了 simpleitem 样式
20          ArrayAdapter<String> adapter = new ArrayAdapter<String>(this,
21                  android.R.layout.simple_dropdown_item_1line, words);
22          //给 AutoCompleteTextView 对象设置 adapter
23          Actv.setAdapter(adapter);
24      }
25  }
```

在上面代码中第 18 行得到了页面中的 AutoCompleteTextView 对象,在第 20 行通过 words 字符串数组生成了一个 ArrayAdapter 对象,在第 23 行给 AutoCompleteTextView 对象设置了 adapter。这样当用户在输入的过程中 Android 系统就会自动去 words 字符串数组中查找相关的单词了。

4．实例扩展

一般在 Android 中我们要进行比对的库不是单纯的字符串数组,一般会是从网络获取,或者从数据库读取,这样的话才能使我们比对的库相对来说比较大。还有,如果比对的库的数量相对来说比较多的话,当你输入一个字母对于数据的查询量会比较大,所以一般我们会设置 android:completionThreshold 属性,也就是当用户输入一定数量的字符后,再开启我们的比对提示功能。

范例 031　多匹配补全效果

1．实例简介

我们在使用 Android 应用的时候,不但会遇到搜索的情况,而且有时候搜索框的内容会有多个搜索的词在其中。例如,在发送邮件的时候我希望给张三、张十、李四和李六同时发送邮件,这时候我不但希望能够进行自动的提示和补全,而且还希望能够多次进行。这就要求在用户的输入过程中我们的输入框能够随时和我们的库进行比较,如果有类似的词语就立刻返回给用户,而且可以多次进行此操作,结果在同一个显示框中显示。本实例就带领大家使用 Android 中的 MultiAutoCompleteTextView 控件来完成一个发送邮件时收件人的输入框效果。

2．运行效果

该实例运行效果如图 3.31 所示。

3．实例程序讲解

在上面例子的效果中,在我们输入过程中,输入框根据输入的前几个字符去查询我们的库,如果有匹配上,那么就以下拉列表的形式显示出来,然后通过单击可以快速输入,然后在输入框的最后面会自动添加我们定义好的分隔符,本例为逗号,然后用户可以继续输入。这就是

图 3.31　收件人输入框

Android 提供的多重自动补全控件 MultiAutoCompleteTextView 的作用了,它可以多次进行自动补全内容。想要实现我们上例的效果,步骤如下所示。

(1) 修改 res/layout/activity_main.xml 文件,代码如下:

```
01    <!-- 定义基础布局 LinearLayout -->
02    <LinearLayout xmlns:android="http://schemas.android.com/apk/res/android"
03        xmlns:tools="http://schemas.android.com/tools"
04        android:layout_width="match_parent"
05        android:layout_height="match_parent"
06        android:paddingBottom="@dimen/activity_vertical_margin"
```

```
07      android:paddingLeft="@dimen/activity_horizontal_margin"
08      android:paddingRight="@dimen/activity_horizontal_margin"
09      android:paddingTop="@dimen/activity_vertical_margin"
10      android:orientation="vertical">
11      <!-- 定义多次自动补全控件 -->
12      <MultiAutoCompleteTextView
13          android:id="@+id/Mactv"
14          android:layout_width="fill_parent"
15          android:layout_height="wrap_content" />
16  </LinearLayout>
```

在如上代码中第 12~15 行定义了 MultiAutoCompleteTextView 控件，并且设置其 id 属性为 Mactv，方便我们在 java 文件中获取此控件。

（2）修改 src/com.wyl.example/MainActivity.java 文件，代码如下：

```
01  /*import 代码省略*/
02  //定义 MainActivity 继承自 Activity
03  public class MainActivity extends Activity {
04      //定义 MultiAutoCompleteTextView 对象
05      private MultiAutoCompleteTextView Mactv;
06  
07      //定义收件人查询库
08      private static final String[] names = { "zhangsan", "zhangshi", "lisi","liliu",
09              "liushasha", "wangli", "wangzhengsan" };
10  
11      @Override
12      protected void onCreate(Bundle savedInstanceState) {
13          super.onCreate(savedInstanceState);      //调用父类的 onCreate 方法
14          //通过 setContentView 方法设置当前页面的布局文件为 activity_main
15          setContentView(R.layout.activity_main);
16  
17          //获取 MultiAutoCompleteTextView 对象
18          Mactv = (MultiAutoCompleteTextView)findViewById(R.id.Mactv);
19          //自定义 ArrayAdapter，设置了 simpleitem 样式
20          ArrayAdapter<String> adapter = new ArrayAdapter<String>(this,
21                  android.R.layout.simple_dropdown_item_1line, names);
22          //给 MultiAutoCompleteTextView 对象设置 adapter
23          Mactv.setAdapter(adapter);
24          //给 MultiAutoCompleteTextView 对象设置分隔符号
25          Mactv.setTokenizer(new MultiAutoCompleteTextView.CommaTokenizer());
26      }
27  }
```

在上面代码中第 18 行得到了页面中的 MultiAutoCompleteTextView 对象，在第 20 行通过 names 字符串数组生成了一个 ArrayAdapter 对象，在第 23 行给 MultiAutoCompleteTextView 对象设置了 adapter。这样当用户在输入的过程中 Android 系统就会自动去 names 字符串数组中查找相关的单词了。第 25 行给 MultiAutoCompleteTextView 设置分隔符，这里我们使用的是默认的分割符号。

4．实例扩展

一般在 Android 中的 AutoCompleteTextView 和 MultiAutoCompleteTextView 其实功能上没有太大差异，只是一个为一次匹配，一个为多次匹配。大家在使用的过程中根据自己应用的需求进行选择即可。

范例 032 用户使用的操作系统调查表

1．实例简介

我们在使用 Android 应用的时候，也经常看到类似 Windows 中的下拉菜单控件，如城市的选择、地区的选择和信息的调查等应用。这就要求在某些场景下一些信息是由用户选择得到的，而不是用户随意输入的。本实例就带领大家使用 Android 中的 Spinner 控件来完成一个计算机用户操作系统调查表的效果。

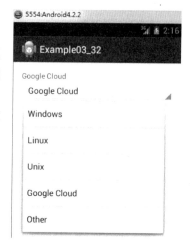

2．运行效果

该实例运行效果如图 3.32 所示。

3．实例程序讲解

图 3.32 用户使用系统调查表

在上面例子的效果中，我们可以通过单击 Spinner 控件显示用户可选择的系统的列表，当用户进行选择的同时，上面的 TextView 控件的内容随时改变。想要实现我们上例的效果，步骤如下所示。

（1）添加 res/values/arrays.xml 文件，代码如下：

```xml
<?xml version="1.0" encoding="utf-8"?>
<resources>
    <string-array name="System">
        <item>Windows</item>
        <item>Linux</item>
        <item>Unix</item>
        <item>Google Cloud</item>
        <item>Other</item>
    </string-array>
</resources>
```

在 arrays.xml 文件中定义了一个常量字符串数组，资源名字为 System。我们可以在 java 文件中通过此名字来引用此数组。

（2）修改 res/layout/activity_main.xml 文件，代码如下：

```
01  <!-- 定义基础布局 LinearLayout -->
02  <LinearLayout xmlns:android="http://schemas.android.com/apk/res/android"
03      xmlns:tools="http://schemas.android.com/tools"
04      android:layout_width="match_parent"
05      android:layout_height="match_parent"
06      android:paddingBottom="@dimen/activity_vertical_margin"
07      android:paddingLeft="@dimen/activity_horizontal_margin"
08      android:paddingRight="@dimen/activity_horizontal_margin"
09      android:paddingTop="@dimen/activity_vertical_margin"
10      android:orientation="vertical">
11      <!-- 定义 TextView 控件 -->
12      <TextView
13          android:id="@+id/Tv"
14          android:layout_width="match_parent"
15          android:layout_height="wrap_content" />
```

```
16  <!-- 定义选择框 Spinner 控件 -->
17  <Spinner
18      android:id="@+id/Sp"
19          android:layout_width="match_parent"
20          android:layout_height="wrap_content"
21      />
22  </LinearLayout>
```

在如上代码中第 16～21 行定义了 Spinner 控件，并且设置其 id 属性为 Sp，方便我们在 java 文件中获取此控件。

（3）修改 src/com.wyl.example/MainActivity.java 文件，代码如下：

```
01  /*import 代码省略*/
02  //定义 MainActivity 继承自 Activity
03  public class MainActivity extends Activity {
04      //定义 TextView 对象
05      private TextView Tv;
06      //定义 Spinner 对象
07      private Spinner Sp;
08      //定义字符序列数组用来存储 Spinner 的数据源
09      ArrayAdapter<CharSequence> adapter;
10
11      @Override
12      protected void onCreate(Bundle savedInstanceState) {
13          super.onCreate(savedInstanceState);      //调用父类的 onCreate 方法
14          //通过 setContentView 方法设置当前页面的布局文件为 activity_main
15          setContentView(R.layout.activity_main);
16          findView();
17          setSp();
18      }
19
20      private void setSp() {
21          //TODO Auto-generated method stub
22          //将可选内容与 ArrayAdapter 连接起来
23          adapter = ArrayAdapter.createFromResource(
24                  this, R.array.System, android.R.layout.simple_spinner_item);
25
26          //设置下拉列表的风格
27          adapter.setDropDownViewResource(android.R.layout.simple_spinner_dropdown_item);
28          //设置 Sp 的 adapter
29          Sp.setAdapter(adapter);
30
31          //添加事件 Sp 的选择事件监听
32          Sp.setOnItemSelectedListener(new OnItemSelectedListener() {
33
34              @Override
35              public void onItemSelected(AdapterView<?> arg0, View arg1,
36                      int arg2, long arg3) {
37                  //TODO Auto-generated method stub
38                  //当用户选择了某项的时候，Tv 显示用户的选项
39                  Tv.setText(adapter.getItem(arg2).toString());
40              }
41
42              @Override
43              public void onNothingSelected(AdapterView<?> arg0) {
44                  //TODO Auto-generated method stub
45                  //当用户没有选择任何项的时候，Tv 显示 selected Nothing
```

```
46                  Tv.setText("Selected Noting");
47              }
48          });
49      }
50
51      private void findView() {
52          //TODO Auto-generated method stub
53          //通过findViewById得到对应的控件对象
54          Tv = (TextView) findViewById(R.id.Tv);
55          Sp = (Spinner) findViewById(R.id.Sp);
56      }
57  }
```

在上面代码中第 52～55 行得到了页面中的控件对象，在第 23～29 行定义的 ArrayAdapter 并且把它设置给了 Sp 对象，作为 Sp 对象的下拉菜单的内容。在第 31～49 行定义了 Sp 对象的选择监听器，其中两个回调方法分别在用户进行选择和没有选择时进行回调。这样当用户单击 Spinner 对象的时候就会显示出 adapter 中的数据内容了。

4．实例扩展

Android 中的 Spinner 有一个问题，大家可能都发现了，就是在第一次默认加载的时候，Spinner 中的第一项默认选中，但是有时候你又不希望它进行默认选中，这个时候只要在 OnItemSelectedListener 选中的回调方法中加入一个判断即可。在我们的类中设置一个标志变量，开始为 false，第一次过后，在 OnItemSelectedListener 的回调方法中修改标志变量为 true 即可。当然大家在使用的过程中可能还会遇到各种要求，大家只要了解其原理，然后进行定义修改即可。

范例 033　电影票预售表格效果

1．实例简介

我们在使用 Android 应用的时候，会遇到列表的情况，这些列表是有规律排列的。例如，当我们通过应用来预订电影票的时候，为了让用户能够看到票的位置则采用列表的形式显示剩余票的状态，预订列车座位时也是同样情况。这就要求在我们进行列表时，如果应用需要根据表格显示列表。本实例就带领大家使用 Android 中的 GridView 控件来完成一个显示电影票座位的效果。

2．运行效果

该实例运行效果如图 3.33 所示。

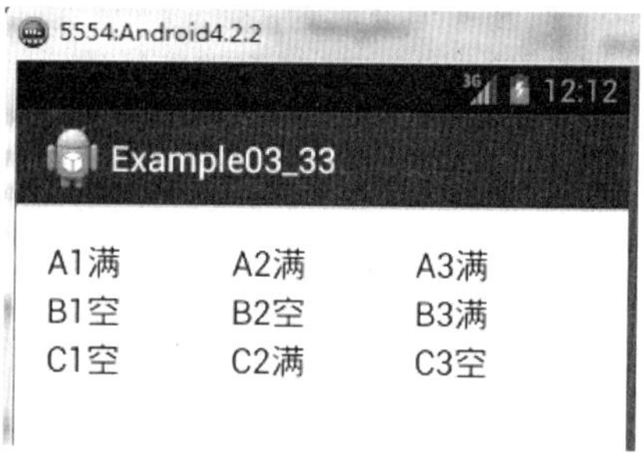

图 3.33　电影票预售表格效果

3．实例程序讲解

在上面例子的效果中，我们可以通过单击每个选项进行选择，当然现在仅仅是一个列表效果。想要实现我们上例的效果，步骤如下所示。

（1）添加 res/values/arrays.xml 文件，代码如下：

```
01  <?xml version="1.0" encoding="utf-8"?>
02  <resources>
```

```
03  <string-array name="seat">
04      <item>A1 满</item>
05      <item>A2 满</item>
06      <item>A3 满</item>
07      <item>B1 空</item>
08      <item>B2 空</item>
09      <item>B3 满</item>
10      <item>C1 空</item>
11      <item>C2 满</item>
12      <item>C3 空</item>
13  </string-array>
14  </resources>
```

在 arrays.xml 文件中定义了一个常量字符串数组，资源名字为 seat。我们可以在 java 文件中通过此名字来引用此数组。

（2）修改 res/layout/activity_main.xml 文件，代码如下：

```
01  <!-- 定义基础布局 LinearLayout -->
02  <LinearLayout xmlns:android="http://schemas.android.com/apk/res/android"
03      xmlns:tools="http://schemas.android.com/tools"
04      android:layout_width="match_parent"
05      android:layout_height="match_parent"
06      android:paddingBottom="@dimen/activity_vertical_margin"
07      android:paddingLeft="@dimen/activity_horizontal_margin"
08      android:paddingRight="@dimen/activity_horizontal_margin"
09      android:paddingTop="@dimen/activity_vertical_margin"
10      android:orientation="vertical">
11      <!-- 定义 GridView 控件 -->
12      <GridView
13          android:id="@+id/Gv"
14          android:layout_width="match_parent"
15          android:layout_height="match_parent"
16          android:numColumns="3" />
17
18  </LinearLayout>
```

在如上代码中第 12～16 行定义了 GridView 控件，并且设置其 id 属性为 Gv，方便我们在 java 文件中获取此控件。还设置了 numColumns 属性，代表 GridView 的列数。

（3）修改 src/com.wyl.example/MainActivity.java 文件，代码如下：

```
01  /*import 代码省略*/
02  //定义 MainActivity 继承自 Activity
03  public class MainActivity extends Activity {
04      //定义 GridView 对象
05      private GridView Gv;
06      //定义字符序列数组用来存储 GridView 的数据源
07      ArrayAdapter<CharSequence> adapter;
08
09
10      @Override
11      protected void onCreate(Bundle savedInstanceState) {
12          super.onCreate(savedInstanceState);      //调用父类的 onCreate 方法
13          //通过 setContentView 方法设置当前页面的布局文件为 activity_main
14          setContentView(R.layout.activity_main);
15          findView();
16          setGv();
17      }
```

```
18
19    private void setGv() {
20        //TODO Auto-generated method stub
21        //将可选内容与ArrayAdapter连接起来
22        adapter = ArrayAdapter.createFromResource(
23            this, R.array.seat, android.R.layout.simple_gallery_item);
24
25        //设置Gv的adapter
26        Gv.setAdapter(adapter);
27    }
28
29    private void findView() {
30        //TODO Auto-generated method stub
31        //通过findViewById得到对应的控件对象
32        Gv = (GridView) findViewById(R.id.Gv);
33    }
34 }
```

在上面代码中第 4 行定义了 GridView 控件对象，在第 32 行得到了布局中的 GridView 控件。在第 22～26 行创建 adapter，并且给 Gv 对象设置。这样一个表格列表的样式就显示出来了，在这里的 adapter 是通过得到 seat 数组中的数据显示的，而且每个 item 设置了 simplegalleryitem 的样式。

4．实例扩展

Android 中的 GridView 还有很多常用的属性如下所示。
- verticalSpacing：设置两行之间的距离。
- horizontalSpacing：设置两列之间的距离。
- stretchMode：设置缩放的形式。

当然希望了解 GridView 的所有属性的话，请查看 Android 开发的官方文档。

范例 034 文件表格列表效果

1．实例简介

我们在使用 Android 应用的时候，遇到的列表有可能不是单纯的文字列表，也有可能在列表中有图片也有文字。例如，查看手机中的文件的时候，会发现在每个文件的 itm 用一种特殊的图片来表示文件的类型，或者大家可以看到 Android 系统的功能列表，其中每个应用程序都是有程序图标和程序名称的。这就要求在显示表格列表的时候需要自定义每一个 item 的样式。本实例就带领大家根据自己的需要来自定义 GridView 控件的每一个 item，实现一个带图片的文件列表效果。

图 3.34 文件表格列表效果

2．运行效果

该实例运行效果如图 3.34 所示。

3. 实例程序讲解

在上面例子的效果中，我们可以通过单击每个选项进行文件的选择，当然现在仅仅是一个列表效果。其中用到了四张资源图片分别是 dir.png、doc.png、img.png 和 video.png，分别代表目录、文档、图片和视频这四种类型，并且将这四张图片复制到工程目录的 /res/dreawable 目录中。然后实现我们上例的效果步骤如下所示。

（1）修改 res/layout/activity_main.xml 文件，代码如下：

```xml
01 <!-- 定义基础布局 LinearLayout -->
02 <LinearLayout xmlns:android="http://schemas.android.com/apk/res/android"
03     xmlns:tools="http://schemas.android.com/tools"
04     android:layout_width="match_parent"
05     android:layout_height="match_parent"
06     android:paddingBottom="@dimen/activity_vertical_margin"
07     android:paddingLeft="@dimen/activity_horizontal_margin"
08     android:paddingRight="@dimen/activity_horizontal_margin"
09     android:paddingTop="@dimen/activity_vertical_margin"
10     android:orientation="vertical">
11     <!-- 定义 GridView 控件 -->
12     <GridView
13         android:id="@+id/Gv"
14         android:layout_width="match_parent"
15         android:layout_height="match_parent"
16         android:numColumns="3" />
17
18 </LinearLayout>
```

在如上代码中第 12～16 行定义了 GridView 控件，并且设置其 id 属性为 Gv，方便我们在 java 文件中获取此控件。还设置了 numColumns 属性，代表 GridView 的列数。

（2）在 res/layout/目录下添加 activity_filelist_item.xml 文件，代码如下：

```xml
01 <?xml version="1.0" encoding="utf-8"?>
02 <!-- 定义 GridView 的 item 样式 -->
03 <LinearLayout
04     xmlns:android="http://schemas.android.com/apk/res/android"
05     android:layout_width="match_parent"
06     android:layout_height="match_parent"
07     android:orientation="vertical">
08     <!-- 定义 item 中的图片控件 -->
09     <ImageView
10         android:id="@+id/Iv"
11         android:layout_width="wrap_content"
12         android:layout_height="wrap_content"
13         />
14     <!-- 定义 item 中的文字控件 -->
15     <TextView
16         android:id="@+id/Tv"
17         android:layout_width="match_parent"
18         android:layout_height="wrap_content"
19         android:gravity="center"
20         />
21 </LinearLayout>
```

这个 xml 布局文件是用来定义 GridView 中每个 item 的样式的。

（3）在 src/com.wyl.example/目录下添加 MyFile.java 文件，代码如下：

```
01  /*
02   * 文件名：MyFile.java
03   * 类型：实体类
04   * 功能：定义了 MyFile 实体类
05   */
06
07  package com.wyl.example;
08
09  public class MyFile {
10      public String FileName;        //文件名称属性
11      public int ImgId;              //文件的类型缩略图
12
13      //MyFile 的构造方法
14      public MyFile() {
15          super();
16          //TODO Auto-generated constructor stub
17      }
18      //MyFile 带参数的构造方法
19      public MyFile(String fileName, int imgId) {
20          super();
21          FileName = fileName;
22          ImgId = imgId;
23      }
24  }
```

此文件定义了一个实体类 MyFile，为了方便我们等给 GridView 设置数据。这里需要注意一点，在我们后面定义 adapter 时可以通过一个对象的 list，也可以通过 map 来传递数据，但是站在面向对象的角度还是通过传对象的列表更好，这样大家会发现如果把其中的每一个 item 对应一个对象，这样对程序编写和后期维护带来了很多方便。

（4）在 src/com.wyl.example/目录下添加 FileListAdapter.java 文件，代码如下：

```
01  //自定义 adapter
02  public class FileListAdapter extends BaseAdapter {
03
04      //定义 Context
05      private Context mContext;
06      //定义要显示的 MyFile 列表
07      private List<MyFile> fileList;
08
09      //FileListAdapter 的构造方法
10      public FileListAdapter(Context c,List<MyFile> fl) {
11          mContext = c;
12          fileList = fl;
13      }
14
15      //获取显示的条目数量
16      @Override
17      public int getCount() {
18          //TODO Auto-generated method stub
19          return fileList.size();
20      }
21
22      //获取列表中的单个对象
23      @Override
24      public Object getItem(int position) {
25          //TODO Auto-generated method stub
26          return fileList.get(position);
```

```
27  }
28
29  //获取列表中对象的id
30  @Override
31  public long getItemId(int position) {
32      //TODO Auto-generated method stub
33      return position;
34  }
35
36  //构造每一个item的View视图
37  @Override
38  public View getView(int position, View convertView, ViewGroup parent) {
39      //定义位置占位符类的对象
40      ViewHolder viewholder =new ViewHolder();
41      if (convertView == null) {
42          //初始化当前view的布局视图
43          convertView = LayoutInflater.from(mContext).inflate(
44                  R.layout.activity_filelist_item, null);
45      }
46      //获取到对应的控件对象
47      viewholder.fileImage = (ImageView) convertView
48              .findViewById(R.id.Iv);
49      viewholder.fileName = (TextView) convertView
50              .findViewById(R.id.Tv);
51      //给控件对象设置相应的内容
52      viewholder.fileImage.setBackgroundResource(fileList.get(position)
.ImgId);
53      viewholder.fileName.setText(fileList.get(position).FileName);
54
55      return convertView;
56  }
57
58  //定义内部类作为占位符组合
59  class ViewHolder {
60      ImageView fileImage;
61      TextView fileName;
62  }
63  }
```

此文件自定义了一个adapter，因为我们要自定义GridView的每一个item，所以这里为了方便定义了一个adapter，在此adapter中接受了一个MyFile的链表，通过第38~56行getView回调方法，加载了之前我们定义的activity_filelist_item布局文件，逐个生成每一个item的视图。

（5）修改src/com.wyl.example/MainActivity.java文件，代码如下：

```
01  /*import 代码省略*/
02  public class MainActivity extends Activity {
03      //定义GridView对象
04      private GridView Gv;
05      //定义用来存储GridView的数据源
06      FileListAdapter adapter;
07      //定义用来存储需要显示的对象的列表
08      private List<MyFile> fileList = new ArrayList<MyFile>();
09
10
11      @Override
12      protected void onCreate(Bundle savedInstanceState) {
13          super.onCreate(savedInstanceState);         //调用父类的onCreate方法
```

```
14        //通过setContentView方法设置当前页面的布局文件为activity_main
15        setContentView(R.layout.activity_main);
16        findView();
17        setData();
18        setGv();
19    }
20
21    private void setData() {
22        //TODO Auto-generated method stub
23        //构造模拟数据
24        fileList.add(new MyFile("test.txt",R.drawable.doc));
25        fileList.add(new MyFile("test.jpg",R.drawable.img));
26        fileList.add(new MyFile("test.avi",R.drawable.video));
27        fileList.add(new MyFile("dir1",R.drawable.dir));
28        fileList.add(new MyFile("test.doc",R.drawable.doc));
29        fileList.add(new MyFile("test.rmvb",R.drawable.video));
30        fileList.add(new MyFile("test.mp4",R.drawable.video));
31        fileList.add(new MyFile("test.rm",R.drawable.video));
32        fileList.add(new MyFile("test.png",R.drawable.img));
33        fileList.add(new MyFile("dir2",R.drawable.dir));
34    }
35
36    private void setGv() {
37        //TODO Auto-generated method stub
38        //将可选内容与ArrayAdapter连接起来
39        adapter = new FileListAdapter(this,fileList);
40
41        //设置Gv的adapter
42        Gv.setAdapter(adapter);
43    }
44
45    private void findView() {
46        //TODO Auto-generated method stub
47        //通过findViewById得到对应的控件对象
48        Gv = (GridView) findViewById(R.id.Gv);
49    }
50 }
```

在上面代码中第4行定义了GridView控件对象，第47~48行得到了GridView控件，在第21~34行加入了模拟数据，第40~42行定义了FileListAdapter对象，并且设置给了GridView对象。这样一个表格列表的样式就显示出来了，在这里的adapter是使用自定义的item的样式。

4. 实例扩展

在我们这个实例中通过setData方法进行了模拟定义了一些数据，当然在我们通常的应用中这些数据一般是从网络上获取的，或者从数据库中获取的，如果是这样的话，获取数据的过程有可能就会很漫长了，这时候你就要通过另外的线程来进行数据获取了，当数据获取完成之后再来设置adapter，然后再设置GridView了。

范例035 学生名单表

1. 实例简介

我们在使用Android应用的时候，表格式的列表是非常常见的，当然还有一种列表形式是比较常见的就是条目列表,这种列表是以横行为单位显示数据的。例如,学生名单表、新闻列表、微博好友列表、qq 好友列表和安装的软件列表等。本实例就带领大家使用 Android 中的 ListView 控件来完成一个学生名单表的效果。

2.运行效果

该实例运行效果如图 3.35 所示。

3.实例程序讲解

在上面例子的效果中,就需要用到 Android 中的 ListView 控件了。想要实现我们上例的效果,步骤如下所示。

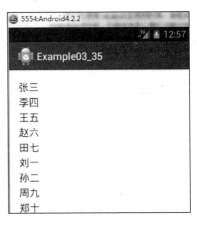

图 3.35 学生名单表

(1)添加 res/values/arrays.xml 文件,代码如下:

```xml
<?xml version="1.0" encoding="utf-8"?>
<resources>
    <string-array name="names">
        <item>张三</item>
        <item>李四</item>
        <item>王五</item>
        <item>赵六</item>
        <item>田七</item>
        <item>刘一</item>
        <item>孙二</item>
        <item>周九</item>
        <item>郑十</item>
    </string-array>
</resources>
```

在 arrays.xml 文件中定义了一个常量字符串数组,资源名字为 names。我们可以在 java 文件中通过此名字来引用此数组。

(2)修改 res/layout/activity_main.xml 文件,代码如下:

```
01  <!-- 定义基础布局 LinearLayout -->
02                                                                       <LinearLayout
xmlns:android="http://schemas.android.com/apk/res/android"
03  xmlns:tools="http://schemas.android.com/tools"
04  android:layout_width="match_parent"
05  android:layout_height="match_parent"
06  android:paddingBottom="@dimen/activity_vertical_margin"
07  android:paddingLeft="@dimen/activity_horizontal_margin"
08  android:paddingRight="@dimen/activity_horizontal_margin"
09  android:paddingTop="@dimen/activity_vertical_margin"
10  android:orientation="vertical">
11  <!-- 定义 ListView 控件 -->
12  <ListView
13      android:id="@+id/Lv"
14      android:layout_width="match_parent"
15      android:layout_height="match_parent"/>
16
17  </LinearLayout>
```

在如上代码中第 12～15 行定义了 ListView 控件，并且设置其 id 属性为 Lv，方便我们在 java 文件中获取此控件。

（3）修改 src/com.wyl.example/MainActivity.java 文件，代码如下：

```java
01  /*import 代码省略*/
02  //定义 MainActivity 继承自 Activity
03  public class MainActivity extends Activity {
04      //定义 ListView 对象
05      private ListView Lv;
06      //定义字符序列数组用来存储 ListView 的数据源
07      ArrayAdapter<CharSequence> adapter;
08
09
10      @Override
11      protected void onCreate(Bundle savedInstanceState) {
12          super.onCreate(savedInstanceState);        //调用父类的 onCreate 方法
13          //通过 setContentView 方法设置当前页面的布局文件为 activity_main
14          setContentView(R.layout.activity_main);
15          findView();
16          setGv();
17      }
18
19      private void setGv() {
20          //TODO Auto-generated method stub
21          //将可选内容与 ArrayAdapter 连接起来
22          adapter = ArrayAdapter.createFromResource(
23              this, R.array.names, android.R.layout.simple_gallery_item);
24
25          //设置 Lv 的 adapter
26          Lv.setAdapter(adapter);
27      }
28
29      private void findView() {
30          //TODO Auto-generated method stub
31          //通过 findViewById 得到对应的控件对象
32          Lv = (ListView) findViewById(R.id.Lv);
33      }
34  }
```

在上面代码中第 31～32 行得到了页面中的 ListView 控件对象，在第 22～26 行定义的 ArrayAdapter 并且把它设置给了 Lv 对象，作为 Lv 对象的列表内容。这样一个 ListView 的效果就显示出来了。

4．实例扩展

Android 中的 ListView 还有很多常用的属性本例没有设置，常见的属性列表如下所示。
- choiceMode：规定此 ListView 可以选中其中的一项或多项。
- divider：设置此 ListView 每个 item 之间的分割线的颜色。
- dividerHeight：设置此 ListView 每个 item 之间的距离。
- footerDividersEnabled：设置在最后一项之后是否画出分割线。
- headerDividersEnabledd：设置在第一项之前是否画出分割线。

如果大家在使用过程中用到其他属性，可以去 Android 的开发者文档中查询。

范例 036 手机联系人列表效果

1. 实例简介

我们在使用 Android 应用的时候，看到的 ListView 效果经常也会看到图片和文字共存的情况。例如，手机的联系人应用中不但可以看到联系人名字，还可以看到联系人电话及对应的联系人图片。这就要求在显示列表的时候自定义每一个 item 的样式。本实例就带领大家根据自己的需要来自定义 ListView 控件的每一个 item 效果，实现一个带图片的联系人列表效果。

2. 运行效果

该实例运行效果如图 3.36 所示。

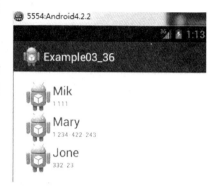

图 3.36 手机联系人列表效果

3. 实例程序讲解

在上面例子的效果中，ListView 的 item 是自定义的，而且 ListView 中的数据是取得自手机的联系人信息。想要实现我们上例的效果，步骤如下所示。

（1）修改 res/layout/activity_main.xml 文件，代码如下：

```
01  <!-- 定义基础布局 LinearLayout -->
02  <LinearLayout xmlns:android="http://schemas.android.com/apk/res/android"
03      xmlns:tools="http://schemas.android.com/tools"
04      android:layout_width="match_parent"
05      android:layout_height="match_parent"
06      android:paddingBottom="@dimen/activity_vertical_margin"
07      android:paddingLeft="@dimen/activity_horizontal_margin"
08      android:paddingRight="@dimen/activity_horizontal_margin"
09      android:paddingTop="@dimen/activity_vertical_margin"
10      android:orientation="vertical">
11      <!-- 定义 ListView 控件 -->
12      <ListView
13          android:id="@+id/Lv"
14          android:layout_width="match_parent"
15          android:layout_height="match_parent"/>
16  
17  </LinearLayout>
```

在如上代码中第 12~15 行定义了 ListView 控件，并且设置其 id 属性为 Lv，方便我们在 java 文件中获取此控件。

（2）在 res/layout/目录下添加 activity_list_item.xml 文件，代码如下：

```
01  <?xml version="1.0" encoding="utf-8"?>
02  <!-- 定义 ListView 的 item 样式 -->
03  <LinearLayout xmlns:android="http://schemas.android.com/apk/res/android"
04      android:layout_width="match_parent"
05      android:layout_height="match_parent"
06      android:orientation="horizontal" >
07
```

```xml
08    <!-- 定义联系人图片控件 -->
09    <ImageView
10        android:id="@+id/IvNews"
11        android:layout_width="wrap_content"
12        android:layout_height="match_parent" />
13
14    <LinearLayout
15        android:layout_width="match_parent"
16        android:layout_height="match_parent"
17        android:orientation="vertical" >
18
19        <!-- 定义联系人姓名的文字控件 -->
20        <TextView
21            android:id="@+id/TvNewsTitle"
22            android:layout_width="match_parent"
23            android:layout_height="wrap_content"
24            android:gravity="left"
25            android:textSize="20sp" />
26
27        <!-- 定义联系人电话的文字控件 -->
28        <TextView
29            android:id="@+id/TvNewsInfo"
30            android:layout_width="match_parent"
31            android:layout_height="wrap_content"
32            android:gravity="left"
33            android:textSize="10sp" />
34    </LinearLayout>
35
36 </LinearLayout>
```

这个 xml 布局文件是用来定义 ListView 中每个 item 的样式的。其中设计到了两个 LinearLayout，第一个横向的 LinearLayout 定义了基本的 item 布局，其中包括了一个 ImageView 和一个 LinearLayout，这个 LinearLayout 是纵向的，其中包含了显示联系人姓名和联系人电话的两个 TextView。

（3）在 src/com.wyl.example/目录下添加 MyFile.java 文件，代码如下：

```java
01 /*
02  * 文件名：People.java
03  * 类型：实体类
04  * 功能：定义了 People 联系人实体类
05  */
06
07 package com.wyl.example;
08
09 public class People {
10     public String PeopleName;      //联系人姓名
11     public int ImgId;              //联系人照片
12     public String PeopleNumber;    //联系人电话
13
14     //People 的构造方法
15     public People() {
16         super();
17         //TODO Auto-generated constructor stub
18     }
19     //People 带参数的构造方法
20     public People(String fileName, int imgId, String info) {
21         super();
22         PeopleName = fileName;
```

```
23        ImgId = imgId;
24        PeopleNumber = info;
25    }
26 }
```

此文件定义了一个实体类 People，为了方便我们给 ListView 设置数据使用的。ListView 中的每一个 item 对应了一个 People 对象。

（4）在 src/com.wyl.example/目录下添加 PeopleListAdapter.java 文件，代码如下：

```
01 //自定义 adapter
02 public class PeopleListAdapter extends CursorAdapter {
03
04     //定义 LayoutInflater 对象
05     private LayoutInflater mInflater;
06
07     //NewsListAdapter 的构造方法
08     public PeopleListAdapter(Context c, Cursor cursor) {
09         super(c, cursor);
10         mInflater = LayoutInflater.from(c);
11     }
12
13     //设置 item 页面的布局效果
14     @Override
15     public void bindView(View view, Context context, Cursor cursor) {
16         //TODO Auto-generated method stub
17         ViewHolder viewholder = new ViewHolder();
18         //获取到对应的控件对象
19         viewholder.PeopleImage = (ImageView) view.findViewById(R.id.IvNews);
20         viewholder.PeopleName = (TextView) view.findViewById(R.id.TvNewsTitle);
21         viewholder.PeopleNumber = (TextView) view.findViewById(R.id.TvNewsInfo);
22         //给控件对象设置相应的内容
23         viewholder.PeopleImage.setBackgroundResource(R.drawable.ic_launcher);
24         //通过 curso 的 getsting 方法得到对应的字段的值
25         viewholder.PeopleName
26             .setText(cursor.getString(cursor
27 .getColumnIndex(ContactsContract.CommonDataKinds.Phone.DISPLAY_NAME)));
28         viewholder.PeopleNumber
29           .setText(cursor.getString(cursor
30             .getColumnIndex(ContactsContract.CommonDataKinds.Phone.NUMBER)));
31     }
32
33     //初始化每个 item 的 view
34     @Override
35     public View newView(Context context, Cursor cursor, ViewGroup parent) {
36         //TODO Auto-generated method stub
37         return mInflater
38             .inflate(R.layout.activity_list_item, parent, false);
39     }
40
41     //定义内部类作为占位符组合
42     class ViewHolder {
43         ImageView PeopleImage;
44         TextView PeopleName;
45         TextView PeopleNumber;
46     }
47 }
```

此文件自定义了一个 adapter，继承自 CursorAdapter，因为我们 adatper 的数据源是一个 cursor。在第 15～31 行实现了 bindView 回调方法，在此方法中定义了每个 item 的显示样式。其中通过联系人的 cursor 得到联系人的姓名和电话号码。

（5）修改 src/com.wyl.example/MainActivity.java 文件，代码如下：

```java
01  /*import 代码省略*/    //定义 MainActivity 继承自 Activity
02  public class MainActivity extends Activity {
03      //定义 ListView 对象
04      private ListView Lv;
05      //定义用来存储 ListView 的数据源
06      private PeopleListAdapter adapter;
07      //定义用来存储需要显示的对象的 cursor
08      private Cursor cursor;
09
10
11      @Override
12      protected void onCreate(Bundle savedInstanceState) {
13          super.onCreate(savedInstanceState);       //调用父类的 onCreate 方法
14          //通过 setContentView 方法设置当前页面的布局文件为 activity_main
15          setContentView(R.layout.activity_main);
16          findView();
17          setData();
18          setLv();
19      }
20
21      private void setData() {
22          //TODO Auto-generated method stub
23          //通过 getContentResolver 获得手机中的联系人信息
24          cursor = getContentResolver().query(ContactsContract.CommonDataKinds.Phone.CONTENT_URI, null, null, null, null);
25      }
26
27      private void setLv() {
28          //TODO Auto-generated method stub
29          //将可选内容与 ArrayAdapter 连接起来
30          adapter = new PeopleListAdapter(this,cursor);
31
32          //设置 Lv 的 adapter
33          Lv.setAdapter(adapter);
34      }
35
36      private void findView() {
37          //TODO Auto-generated method stub
38          //通过 findViewById 得到对应的控件对象
39          Lv = (ListView) findViewById(R.id.Lv);
40      }
41  }
42
```

在上面代码中第 3～8 行定义了 ListView 对象、自定义的 adapter 对象和 cursor 对象，第 21～25 行通过 getContentResolver 得到了手机中的联系人的信息 cursor，在第 30～33 行通过 cursor 得到了 adapter 对象，并且设置给了 Lv 对象。这样一个自定义 item 的列表样式就完成了。

4．实例扩展

在我们这个实例中只获得了联系人的姓名和电话，大家可以获得更多的联系人信息。

例如，联系人的住址和联系人的座机等其他信息，这就需要在我们工程的 Manifest 文件中添加能够读取联系人信息的权限<uses-permission android:name="android.permission.READ_CONTACTS" />，这在 Android 中是通过 ContentProvider 来实现的，在后面的章节中会涉及到。

范例 037　画廊图片浏览器

1．实例简介

我们在使用 Android 应用的时候，图片的展示和浏览是经常用到的功能，现在的图片展示工具很多，如列表式的图片展示和表格式的图片展示等。在 Android 中，提供了一种专门用来展示图片的控件，叫做 Gallery（画廊）。本实例就带领大家使用 Android 中的 Gallery 和 ImageView 实现一个滚动图片浏览的效果。

2．运行效果

该实例运行效果如图 3.37 所示。

3．实例程序讲解

图 3.37　滚动图片浏览效果

在上面例子的效果中，上面是一个 Gallery 画廊效果，可以通过单击进行选择，下面是 ImageView 效果来显示画廊中选中效果的大图。此工程中需要六张图片资源 b1.png、b2.png、b3.png、b4.png、b5.png 和 b6.png，我已经复制到工程的/res/drawable 目录下了。想要实现我们上例的效果，步骤如下所示。

（1）修改 res/layout/activity_main.xml 文件，代码如下：

```xml
01 <!-- 定义基础布局 LinearLayout -->
02 <LinearLayout xmlns:android="http://schemas.android.com/apk/res/android"
03     xmlns:tools="http://schemas.android.com/tools"
04     android:layout_width="match_parent"
05     android:layout_height="match_parent"
06     android:paddingBottom="@dimen/activity_vertical_margin"
07     android:paddingLeft="@dimen/activity_horizontal_margin"
08     android:paddingRight="@dimen/activity_horizontal_margin"
09     android:paddingTop="@dimen/activity_vertical_margin"
10     android:orientation="vertical">
11     <!-- 定义 Gallery 控件 -->
12     <Gallery
13         android:id="@+id/Gal"
14         android:layout_width="fill_parent"
15         android:layout_height="wrap_content" />
16     <!-- 定义 ImageView 控件 -->
17     <ImageView
18         android:id="@+id/Iv"
19         android:layout_width="fill_parent"
20         android:layout_height="wrap_content"
21     />
```

```
22
23 </LinearLayout>
```

在如上代码中第 11~16 行定义了 Gallery 控件，并且设置其 id 属性为 Gal，方便我们在 java 文件中获取此控件。

（2）在 src/com.wyl.example/目录下添加 ImgsAdapter.java 文件，代码如下：

```
01  //自定义 adapter
02  public class ImgsAdapter extends BaseAdapter{
03
04      //定义 Context 对象
05      private Context context;
06      //存放图片 ID 的数组
07      private Integer[] imgsArray;
08
09
10      public ImgsAdapter(Context context,Integer[] mImageIds){
11          this.context = context;
12          imgsArray = mImageIds;
13      }
14
15
16      @Override
17      public int getCount() {
18          //TODO Auto-generated method stub
19          //返回数组的总数
20          return imgsArray.length;
21      }
22
23      @Override
24      public Object getItem(int arg0) {
25          //TODO Auto-generated method stub
26          return arg0;
27      }
28
29      @Override
30      public long getItemId(int arg0) {
31          //TODO Auto-generated method stub
32          return arg0;
33      }
34
35      @Override
36      public View getView(int arg0, View arg1, ViewGroup arg2) {
37          //TODO Auto-generated method stub
38       if (arg1 == null) {
39          //您需要返回的是 ImageView，因为您要实现的是相册
40              ImageView view = new ImageView(this.context);
41          //得到图片资源 id
42              int id = imgsArray[arg0];
43          //设置 imageview 的资源 id
44              view.setImageResource(id);
45          //对 ImageView 进行布局
46              view.setLayoutParams(new Gallery.LayoutParams(80,80));
47          //设置 ImageView 的图片显示类型为 fitxy
48              view.setScaleType(ImageView.ScaleType.FIT_XY);
49              arg1 = view;
50       }
51
52          return arg1;
```

```
53     }
54 }
```

此文件自定义了一个 adapter，继承自 BaseAdapter，实现了 getView 方法，在此方法中定义了每个 item 的显示样式，这里我们的 Gallery 只是显示图片，所以这里每个 item 都是一个 ImageView。其中设置了 ImageView 的资源及显示的放大缩小类型。

（3）修改 src/com.wyl.example/MainActivity.java 文件，代码如下：

```
00 /*import 代码省略*/
01 //定义 MainActivity 继承自 Activity
02 public class MainActivity extends Activity {
03     //定义 Gallery 对象
04     private Gallery Gal;
05     //定义 ImageView 对象
06     private ImageView Iv;
07     //定义用来存储 ListView 的数据源
08     private ImgsAdapter adapter;
09     //图片的资源 ID
10     private Integer[] imgsIds = {
11             R.drawable.b1,
12             R.drawable.b2,
13             R.drawable.b3,
14             R.drawable.b4,
15             R.drawable.b5,
16             R.drawable.b6 };
17
18
19     @Override
20     protected void onCreate(Bundle savedInstanceState) {
21         super.onCreate(savedInstanceState);        //调用父类的 onCreate 方法
22         //通过 setContentView 方法设置当前页面的布局文件为 activity_main
23         setContentView(R.layout.activity_main);
24         findView();
25         setGal();
26     }
27
28     private void setGal() {
29         //TODO Auto-generated method stub
30         //将可选内容与 ArrayAdapter 连接起来
31         adapter = new ImgsAdapter(this,imgsIds);
32
33         //设置 Gal 的 adapter
34         Gal.setAdapter(adapter);
35         //设置 Gal 的监听器为 myOnitemListener
36         Gal.setOnItemClickListener(myOnitemListener);
37     }
38     //自定义 OnItemClickListener 监听器
39     OnItemClickListener myOnitemListener = new OnItemClickListener() {
40
41         @Override
42         public void onItemClick(AdapterView<?> arg0, View arg1, int arg2,
43                 long arg3) {
44             //TODO Auto-generated method stub
45             //设置 imageview 的图片资源
46             Iv.setImageResource(imgsIds[arg2]);
47         }
48     };
49
```

```
50    private void findView() {
51        //TODO Auto-generated method stub
52        //通过findViewById得到对应的控件对象
53        Gal = (Gallery) findViewById(R.id.Gal);
54        Iv = (ImageView) findViewById(R.id.Iv);
55    }
56  }
```

在上面代码中第 4~6 行定义了 Gallery 和 ImageView 对象，第 51~54 行得到了布局中的控件对象，在第 29~36 行通过定义了 ImgsAdapter 对象，并设置给了 Gallery 对象。第 38~48 行，自定义了一个监听单击事件的监听器 OnItemClickListener。这样一个图片显示的画廊就完成了。

4．实例扩展

在我们这个实例中画廊只是为了展示图片，常见的属性如下所示。

❑ animationDuration：定义画廊切换的时间。
❑ spacing：定义画廊中图片之间的间距。
❑ unselectedAlpha：定义未选中的图片的透明度。

大家可以根据自己的需要进行设置。

范例 038 仿 iPhone 的 CoverFlow 效果

1．实例简介

我们在使用 Android 应用的时候，经常会和 iPhone 里面的一些效果作对比，例如 iPhone 中有一种图片展示的比较华丽的效果叫做 CoverFlow。在 Android 中没有直接提供 CoverFlow 控件，但是可以使用自定义 Gallery 来实现类似 CoverFlow 的效果。本实例就带领大家使用 Android 中的 Gallery 来实现仿 iPhone 中的 CoverFlow 的效果。

2．运行效果

该实例运行效果如图 3.38 所示。

3．实例程序讲解

在上面例子的效果中，上面是一个 CoverFlow 效果，可以通过单击进行选择某张图片突出显示，也可以进行滑动突出显示某张图片。此工程中需要六张图片资源 b1.png、b2.png、b3.png、b4.png、b5.png 和 b6.png，我已经拷贝到工程的/res/drawable 目录下了。想要实现我们上例的效果，步骤如下所示。

图 3.38 仿 iPhone 的 CoverFlow 效果

（1）在工程目录先建立 MyCoverFlow.java 文件，代码如下：

```
001  //import 代码省略
002  //自定义控件类, 继承自 Gallery
003  public class MyCoverFlow extends Gallery {
004
```

```
005    //定义Camera对象来实现Gallery的伪3d效果
006    private Camera camera = new Camera();
007    //标记Gallery旋转的最大角度
008    private int maxRotation = 60;
009    //当旋转角度改变是图片大小进行改变
010    private int zoom = -120;
011    //定义Coveflow的中心点
012    private int center;
013
014    //构造方法
015    public MyCoverFlow(Context context) {
016        super(context);
017        //设置图片可有倒影效果
018        this.setStaticTransformationsEnabled(true);
019    }
020    //构造方法
021    public MyCoverFlow(Context context, AttributeSet attrs) {
022        super(context, attrs);
023        this.setStaticTransformationsEnabled(true);
024    }
025    //构造方法
026    public MyCoverFlow(Context context, AttributeSet attrs, int defStyle) {
027        super(context, attrs, defStyle);
028        this.setStaticTransformationsEnabled(true);
029    }
030
031    //得到图片的中心位置
032    private int getCenter() {
033        return (getWidth() - getPaddingLeft() - getPaddingRight()) / 2
034                + getPaddingLeft();
035    }
036    //得到整个View的中心位置
037    private int getCenterOfView(View view) {
038        return view.getLeft() + view.getWidth() / 2;
039    }
040
041    protected boolean getChildStaticTransformation(View child, Transformation trans) {
042
043        //得到中心点
044        final int childCenter = getCenterOfView(child);
045        //得到item的宽度
046        final int childWidth = child.getWidth();
047        //默认角度为0
048        int rotationAngle = 0;
049
050        trans.clear();
051        //根据Matrix矩阵旋转
052        trans.setTransformationType(Transformation.TYPE_MATRIX);
053
054        if (childCenter == center) {
```

```
055                //设置旋转后的图片
056                setTransImage((ImageView) child, trans, 0);
057            } else {
058                //根据中心位置计算旋转角度
059                rotationAngle = (int) (((float) (center - childCenter) /
                      childWidth) * maxRotation);
060                if (Math.abs(rotationAngle) > maxRotation) {
061                    rotationAngle = (rotationAngle < 0) ? -maxRotation
062                            : maxRotation;
063                }
064                //设置旋转后的图片
065                setTransImage((ImageView) child, trans, rotationAngle);
066            }
067
068        return true;
069    }
070
071    //当Gallery的大小改变时回调
072    protected void onSizeChanged(int w, int h, int oldw, int oldh) {
073        center = getCenter();
074        super.onSizeChanged(w, h, oldw, oldh);
075    }
076
077    private void setTransImage(ImageView child, Transformation t,
078            int rotationAngle) {
079        //保存当前内容
080        camera.save();
081        //得到旋转矩阵
082        final Matrix imageMatrix = t.getMatrix();
083        //得到imageview的高度
084        final int imageHeight = child.getLayoutParams().height;
085        //得到imageview的宽度
086        final int imageWidth = child.getLayoutParams().width;
087        //得到imageview的角度
088        final int rotation = Math.abs(rotationAngle);
089
090        //根据camer的移动方向决定角度
091        camera.translate(0.0f, 0.0f, 100.0f);
092
093        //根据旋转的角度得到图片的放大倍数
094        if (rotation < maxRotation) {
095            float zoomAmount = (float) (zoom + (rotation * 1.5));
096            camera.translate(0.0f, 0.0f, zoomAmount);
097        }
098
099        //设置Y轴的旋转角度
100        camera.rotateY(rotationAngle);
101        //得到旋转矩阵
102        camera.getMatrix(imageMatrix);
103        //设置旋转矩阵的倒影宽高
104        imageMatrix.preTranslate(-(imageWidth / 2), -(imageHeight /
2));
```

```
105            imageMatrix.postTranslate((imageWidth / 2), (imageHeight / 2));
106            //保存视角效果
107            camera.restore();
108        }
109  }
```

在上面的代码中第 3 行定义 MyCoverFlow 类继承自 Gallery 类，其中实现了 getCenter 方法来得到整个 View 的中心位置，方便后面设定图片的位置。定义了 getChildStaticTransformation 方法来得到图片转换后的方法，其中使用到了图片的 Matrix 矩阵。类中还定义了 setTransImage 方法来得到图片的倒影处理。所以我们自定义的一个 Gallery 类，实现了基本 CoverFlow 的所有功能。

（2）修改 res/layout/activity_main.xml 文件，代码如下：

```
01  <?xml version="1.0" encoding="utf-8"?>
02  <!-- 定义基本的线性布局 -->
03  <LinearLayout xmlns:android="http://schemas.android.com/apk/res/android"
04      android:layout_width="fill_parent"
05      android:layout_height="fill_parent" >
06
07      <!-- 定义自定义的 CoverFlow 控件 -->
08      <com.wyl.example.MyCoverFlow
09          android:id="@+id/Mcf"
10          android:layout_width="fill_parent"
11          android:layout_height="wrap_content"/>
12
13  </LinearLayout>
```

在如上代码中第 7～11 行定义了 MyCoverFlow 控件，并且设置其 id 属性为 Mcf，方便我们在 java 文件中获取此控件。

（3）在 src/com.wyl.example/目录下添加 ImgsAdapter.java 文件，代码如下：

```
01  //自定义 Gallery 的 adapter
02  public class ImgsAdapter extends BaseAdapter {
03      //得到上下文环境
04      private Context context;
05      //设置显示的图片 id
06      private Integer[] ImagIds;
07      //保存最终需要绘制的 imageview 数组
08      private ImageView[] Images;
09
10      //构造方法
11      public ImgsAdapter(Context c, Integer[] ImageIds) {
12          context = c;
13          ImagIds = ImageIds;
14          Images = new ImageView[ImagIds.length];
15      }
16
17      //旋转图片
18      public boolean invertedImages() {
19          //倒影图片的高度
20          final int reflectionGap = 4;
```

```java
21          int index = 0;
22
23          //遍历每个imageview,生成倒影
24          for (int imageId : ImagIds) {
25              Bitmap originalImage = BitmapFactory.decodeResource(context
26                      .getResources(), imageId);
27              //得到图片的宽高
28              int width = originalImage.getWidth();
29              int height = originalImage.getHeight();
30
31              //通过matrix矩阵生成图片的旋转图片
32              Matrix matrix = new Matrix();
33              matrix.preScale(1, -1);
34
35              Bitmap reflectionImage = Bitmap.createBitmap(originalImage, 0,
36                      height / 2, width, height / 2, matrix, false);
37              //生成合并后的图片
38              Bitmap bitmapWithReflection = Bitmap.createBitmap(width,
39                      (height + height / 2), Config.ARGB_8888);
40              //通过canvas合并图片和图片的倒影
41              Canvas canvas = new Canvas(bitmapWithReflection);
42              canvas.drawBitmap(originalImage, 0, 0, null);
43              //用默认的画笔进行绘制原图片
44              Paint deafaultPaint = new Paint();
45              canvas.drawRect(0, height, width, height + reflectionGap,
46                      deafaultPaint);
47              canvas.drawBitmap(reflectionImage, 0, height + reflectionGap,
                    null);
48              //用LinearGradient画笔绘制倒影图片
49              Paint paint = new Paint();
50              LinearGradient shader = new LinearGradient(0, originalImage
51                      .getHeight(), 0, bitmapWithReflection.getHeight()
52                      + reflectionGap, 0x70ffffff, 0x00ffffff, TileMode.
                    CLAMP);
53
54              paint.setShader(shader);
55              paint.setXfermode(new PorterDuffXfermode(Mode.DST_IN));
56              canvas.drawRect(0, height, width, bitmapWithReflection.
                    getHeight()
57                      + reflectionGap, paint);
58
59              //将绘制完的图片生成imageview
60              ImageView imageView = new ImageView(context);
61              imageView.setImageBitmap(bitmapWithReflection);
62              imageView.setLayoutParams(new Gallery.LayoutParams(100, 200));
63              //加入imageview数组
64              Images[index++] = imageView;
65          }
66          return true;
67      }
68
69      public int getCount() {
```

```
70          return ImagIds.length;
71      }
72
73      public Object getItem(int position) {
74          return position;
75      }
76
77      public long getItemId(int position) {
78          return position;
79      }
80
81      public View getView(int position, View convertView, ViewGroup parent) {
82          return Images[position];
83      }
84  }
```

此文件自定义了一个 adapter，继承自 BaseAdapter，其中的 invertedImages 方法实现了对于 adapter 中的每一个图片数据源的翻转操作。在 getView 方法中将处理过的 image 效果返回给 gallery 来显示。

（4）修改 src/com.wyl.example/MainActivity.java 文件，代码如下：

```
01  //定义MainActivity继承自Activity
02  public class MainActivity extends Activity {
03      //定义显示的图片Id数组
04      private Integer[] imgs = { R.drawable.b1,
05              R.drawable.b2, R.drawable.b3,
06              R.drawable.b4, R.drawable.b5,
07              R.drawable.b6 };
08      //定义自定义的CoverFlow类的对象
09      private MyCoverFlow CoverFlow;
10
11      public void onCreate(Bundle savedInstanceState) {
12          super.onCreate(savedInstanceState);
13          //通过setContentView方法设置当前页面的布局文件为activity_main
14          setContentView(R.layout.activity_main);
15
16          setCoverFlow();
17      }
18
19      private void setCoverFlow() {
20          //定义自定义的ImgsAdapter对象
21          ImgsAdapter adapter = new ImgsAdapter(this, imgs);
22          //设置adapter的倒影图片
23          adapter.invertedImages();
24
25          //得到MyCoverFlow对象设置其adapter
26          CoverFlow = (MyCoverFlow) findViewById(R.id.Mcf);
27          CoverFlow.setAdapter(adapter);
28      }
29  }
```

在上面代码中第 16 行去初始化 CoverFlow 控件。第 19～28 行，定义了 ImgsAdapter，然后进行图片的翻转,然后设置给 CoverFlow 对象。这样在页面中就可以显示一个仿 iPhone 的 CoverFlow 效果了。

4．实例扩展

在这个实例中我们仅仅是为了仿造 iPhone 中的 CoverFlow 效果，所以设置图片的旋转及选中图片的放大倍数和未选中图片的旋转角度，当然大家如果希望能够制作自己个性的 Gallery 的话，只需要定义自己的 Gallery 类就可以了，其中的图片角度的旋转或者图片的放大缩小及其他特效，大家都可以在 gallery 中去实现。

范例 039　菜单弹出效果

1．实例简介

我们在使用 Android 应用的时候，由于手机的屏幕大小有限，并不能像网页那样展示很多的内容，但是我们希望 Android 应用放在一个页面的功能又很多，这该怎么办呢？在 Android 的手机中一般都会有 Menu 菜单键，不论是物理键，还是软键盘。所以在 Android 中如果一个页面的功能很多又希望能够尽量安排在一起的话，那我们就得用菜单键来实现。本实例就带领大家来开发 Android 中的菜单弹出效果。

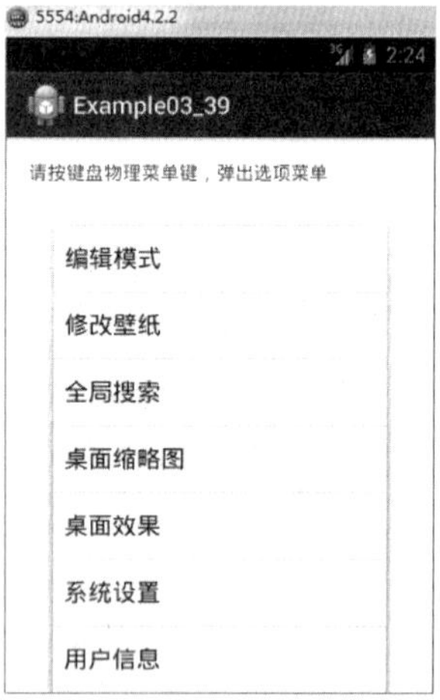

2．运行效果

该实例运行效果如图 3.39 所示。

图 3.39　菜单弹出效果

3．实例程序讲解

在上面例子的效果中，是一个菜单弹出的效果，可以通过单击物理的菜单键显示系统的菜单。想要实现我们上例的效果，步骤如下所示。

（1）修改 res/layout/activity_main.xml 文件，代码如下：

```
01  <!-- 定义基础布局 LinearLayout -->
02  <LinearLayout xmlns:android="http://schemas.android.com/apk/res/android"
03      xmlns:tools="http://schemas.android.com/tools"
04      android:layout_width="match_parent"
05      android:layout_height="match_parent"
06      android:paddingBottom="@dimen/activity_vertical_margin"
07      android:paddingLeft="@dimen/activity_horizontal_margin"
08      android:paddingRight="@dimen/activity_horizontal_margin"
09      android:paddingTop="@dimen/activity_vertical_margin"
10      android:orientation="vertical">
11      <!-- 定义 TextView 控件 -->
12      <TextView
13          android:layout_width="match_parent"
14          android:layout_height="match_parent"
15          android:text="请按键盘物理菜单键，弹出选项菜单"
```

```
16          />
17  </LinearLayout>
```

在上面的代码中第 12~16 行定义基本的提示 TextView。

(2) 修改 src/com.wyl.example/MainActivity.java 文件，代码如下：

```
01  //定义 MainActivity 继承自 Activity
02  public class MainActivity extends Activity {
03      //定义 Menu 中每个菜单选项的 Id
04      private final static int Menu_1 = Menu.FIRST;
05      private final static int Menu_2 = Menu.FIRST + 1;
06      private final static int Menu_3 = Menu.FIRST + 2;
07      private final static int Menu_4 = Menu.FIRST + 3;
08      private final static int Menu_5 = Menu.FIRST + 4;
09      private final static int Menu_6 = Menu.FIRST + 5;
10      private final static int Menu_7 = Menu.FIRST + 6;
11
12      @Override
13      protected void onCreate(Bundle savedInstanceState) {
14          super.onCreate(savedInstanceState); //调用父类的 onCreate 方法
15          //通过 setContentView 方法设置当前页面的布局文件为 activity_main
16          setContentView(R.layout.activity_main);
17      }
18
19      //创建 Menu 菜单的回调方法
20      public boolean onCreateOptionsMenu(Menu m) {
21          //参数 m 就是拿到的当前 Activity 菜单对象
22          //想要给当前页面添加方法的话就 add 进去即可
23          //add 方法的参数：add(分组 id,itemid, 排序, 菜单文字)
24          m.add(0, Menu_1, 0, "编辑模式");
25          m.add(0, Menu_2, 0, "修改壁纸");
26          m.add(0, Menu_3, 0, "全局搜索");
27          m.add(0, Menu_4, 0, "桌面缩略图");
28          m.add(0, Menu_5, 0, "桌面效果");
29          m.add(0, Menu_6, 0, "系统设置");
30          m.add(0, Menu_7, 0, "用户信息");
31          return super.onCreateOptionsMenu(m);
32      }
33
34      //Menu 菜单选项的选项选择的回调事件
35      public boolean onOptionsItemSelected(MenuItem item) {
36          //参数为用户选择的菜单选项对象
37          //根据菜单选项的 id 来执行相应的功能
38          switch (item.getItemId()) {
39          case 1:
40              Toast.makeText(this, "你单击了编辑模式选项", Toast.LENGTH_
                    SHORT).show();
41              break;
42          case 2:
43              Toast.makeText(this, "你单击了修改壁纸", Toast.LENGTH_
                    SHORT).show();
44              break;
```

```
45      case 3:
46          Toast.makeText(this, "你单击了全局搜索", Toast.LENGTH_
            SHORT).show();
47          break;
48      case 4:
49          Toast.makeText(this, "你单击了桌面缩略图", Toast.LENGTH_
            SHORT).show();
50          break;
51      case 5:
52          Toast.makeText(this, "你单击了桌面效果", Toast.LENGTH_
            SHORT).show();
53          break;
54      case 6:
55          Toast.makeText(this, "你单击了系统设置", Toast.LENGTH_
            SHORT).show();
56          break;
57      case 7:
58          Toast.makeText(this, "你单击了用户信息", Toast.LENGTH_
            SHORT).show();
59          break;
60      }
61      return super.onOptionsItemSelected(item);
62  }
63
64  //选项菜单关闭时的回调方法
65  public void onOptionsMenuClosed(Menu menu) {
66      Log.e("onOptionsMenuClosed","用户菜单关闭了");
67  }
68
69  //菜单显示之前的回调方法
70  public boolean onPrepareOptionsMenu(Menu menu) {
71      Log.e("onPrepareOptionsMenu","用户菜单准备好被显示了");
72      //方法返回 true，就会显示 Menu，否则 Menu 不会被显示
73      return true;
74  }
75  }
```

在上面代码中第 3~10 行定义了菜单的 id 常量。第 19~32 行，实现了 Activity 的 onCreateOptionsMenu 方法，此方法是 Activity 的回调方法，在物理的菜单键被按下的时候自动调用，在其中的 Menu 参数就代表菜单对象，然后在此 Menu 对象中通过 add 方法添加具体的菜单选项即可。在第 34~62 行，这里实现了 Activity 的 onOptionsItemSelected 方法，此方法也是回调方法，当菜单的某个选项被选中的时候自动调用，其中 MenuItem 参数为你选中的菜单对象，根据此对象的 id 的值可以得知用户选择的是哪个选项，然后就可以进行相应操作了。本例在 64~74 行，分别实现的 Activity 的 onOptionsMenuClosed 和 onPreOptionsMenu 方法，这两个方法也是回调方法，会在菜单关闭和菜单准备显示之前系统自动调用。

4. 实例扩展

在 Android 4.0 之前对于系统的菜单可以显示选项图片，但是在 Android 4.0 之后通常

无法直接显示菜单图片了,所以大家在 Android 4.0 的系统上很少能够见到带有图片选项的菜单。当然如果想要实现带图片的菜单选项的话,也是可以的,只是需要自定义菜单 item 的显示效果了。大家可以自己下载来进行练习。

范例 040　打开文件的子菜单效果

1. 实例简介

我们在使用 Android 应用的时候,还会遇到对于菜单有层级的现象。例如,显示操作菜单中的打开、保存、关闭以及单击保存弹出保存文件的菜单。所以在 Android 中如果希望菜单具有层级效果,就需要用到子菜单的功能。本实例就带领大家来开发 Android 中的打开文件的子菜单效果。

2. 运行效果

该实例运行效果如图 3.40 所示。

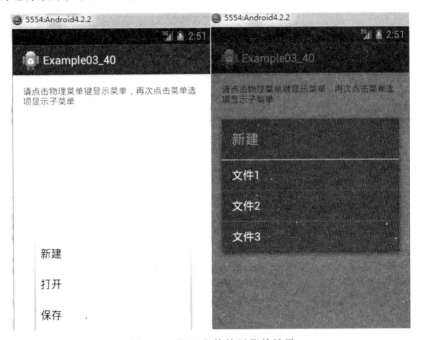

图 3.40　打开文件的子菜单效果

3. 实例程序讲解

在上面例子的效果中,是一个子菜单弹出的效果,可以通过单击物理的菜单键显示系统的菜单,然后选中菜单中的选项后,弹出子菜单效果。想要实现我们上例的效果,步骤如下所示。

(1) 修改 res/layout/activity_main.xml 文件,代码如下:

```
01  <!-- 定义基础布局 LinearLayout -->
02  <LinearLayout xmlns:android="http://schemas.android.com/apk/res/android"
```

```xml
03     xmlns:tools="http://schemas.android.com/tools"
04     android:layout_width="match_parent"
05     android:layout_height="match_parent"
06     android:paddingBottom="@dimen/activity_vertical_margin"
07     android:paddingLeft="@dimen/activity_horizontal_margin"
08     android:paddingRight="@dimen/activity_horizontal_margin"
09     android:paddingTop="@dimen/activity_vertical_margin"
10     android:orientation="vertical">
11     <!-- 定义 TextView 控件 -->
12     <TextView
13         android:layout_width="match_parent"
14         android:layout_height="match_parent"
15         android:text="请单击物理菜单键显示菜单，再次单击菜单选项显示子菜单"
16         />
17 </LinearLayout>
```

在上面的代码中第 11~16 行定义基本的提示 TextView。

（2）修改 src/com.wyl.example/MainActivity.java 文件，代码如下：

```java
01 //定义 MainActivity 继承自 Activity
02 public class MainActivity extends Activity {
03     //定义 Menu 中每个菜单选项的 Id
04     final int Menu_1 = Menu.FIRST;
05     final int Menu_2 = Menu.FIRST + 1;
06     final int Menu_3 = Menu.FIRST + 2;
07     final int Menu_4 = Menu.FIRST + 3;
08     final int Menu_5 = Menu.FIRST + 4;
09     final int Menu_6 = Menu.FIRST + 5;
10     final int Menu_7 = Menu.FIRST + 6;
11     final int Menu_8 = Menu.FIRST + 7;
12     final int Menu_9 = Menu.FIRST + 8;
13
14     @Override
15     protected void onCreate(Bundle savedInstanceState) {
16         super.onCreate(savedInstanceState);       //调用父类的 onCreate 方法
17         //通过 setContentView 方法设置当前页面的布局文件为 activity_main
18         setContentView(R.layout.activity_main);
19     }
20
21     //创建 Menu 菜单的回调方法
22     public boolean onCreateOptionsMenu(Menu m) {
23         //定义子菜单，然后添加到 Menu 中
24         SubMenu fileMenu = m.addSubMenu("新建");
25         //在菜单选项的子菜单中添加选项内容
26         //add 方法的参数：add(分组 id,itemid, 排序, 菜单文字)
27         fileMenu.add(0, Menu_1, 0, "文件1");
28         fileMenu.add(0, Menu_2, 0, "文件2");
29         fileMenu.add(0, Menu_3, 0, "文件3");
30
31         SubMenu openMenu = m.addSubMenu("打开");
32         //在菜单选项的子菜单中添加选项内容
33         //add 方法的参数：add(分组 id,itemid, 排序, 菜单文字)
```

```java
34        openMenu.add(0, Menu_4, 0, "文件4");
35        openMenu.add(0, Menu_5, 0, "文件5");
36        openMenu.add(0, Menu_6, 0, "文件6");
37
38        SubMenu saveMenu = m.addSubMenu("保存");
39
40        //在菜单选项的子菜单中添加选项内容
41        //add方法的参数：add(分组id,itemid, 排序, 菜单文字)
42        saveMenu.add(0, Menu_7, 0, "文件7");
43        saveMenu.add(0, Menu_8, 0, "文件8");
44        saveMenu.add(0, Menu_9, 0, "文件9");
45        return super.onCreateOptionsMenu(m);
46    }
47
48    //Menu 菜单选项的选项选择的回调事件
49    public boolean onOptionsItemSelected(MenuItem item) {
50        //参数为用户选择的菜单选项对象
51        //根据菜单选项的id来执行相应的功能
52        switch (item.getItemId()) {
53        case 1:
54            Toast.makeText(this, "新建文件1", Toast.LENGTH_SHORT).show();
55            break;
56        case 2:
57            Toast.makeText(this, "新建文件2", Toast.LENGTH_SHORT).show();
58            break;
59        case 3:
60            Toast.makeText(this, "新建文件3", Toast.LENGTH_SHORT).show();
61            break;
62        case 4:
63            Toast.makeText(this, "打开文件4", Toast.LENGTH_SHORT).show();
64            break;
65        case 5:
66            Toast.makeText(this, "打开文件5", Toast.LENGTH_SHORT).show();
67            break;
68        case 6:
69            Toast.makeText(this, "打开文件6", Toast.LENGTH_SHORT).show();
70            break;
71        case 7:
72            Toast.makeText(this, "保存文件7", Toast.LENGTH_SHORT).show();
73            break;
74        case 8:
75            Toast.makeText(this, "保存文件8", Toast.LENGTH_SHORT).show();
76            break;
77        case 9:
78            Toast.makeText(this, "保存文件9", Toast.LENGTH_SHORT).show();
79            break;
80        }
81        return super.onOptionsItemSelected(item);
82    }
83 }
```

在上面代码中第3～12行定义了菜单的id常量。第21～46行，实现了Activity的

onCreateOptionsMenu 方法，此方法是 Activity 的回调方法，在物理的菜单键被按下的时候自动调用，在其中的 Menu 参数就代表菜单对象，然后在此 Menu 对象中通过 addSubMenu 方法添加子菜单，然后再通过子菜单的 add 方法加入子菜单的内容。在第 49～82 行实现了 Activity 的 onOptionsItemSelected 方法，此方法也是回调方法，当菜单的某个选项包括子菜单的选项被选中的时候自动调用，其中 MenuItem 参数为你选中的菜单对象，根据此对象的 id 的值可以得知用户选择的是哪个选项，然后就可以进行相应操作了。

4．实例扩展

Android 中菜单底层的实现使用到了一个基本的设计模式，就是组合模式，也就是说 Android 中的菜单可以被表示成一棵树，然后通过忽略叶子节点和子分支节点进行添加。这样的话也就是 Android 中的菜单可以一直延续添加下去，不过用户对于手机的使用习惯来说一般菜单最好不要超过三级，这样用户会感觉软件的使用比较繁琐，所以需要权衡功能的展示和用户的体验，来合理的安排你的页面布局吧。

范例 041　文本框的复制粘贴全选菜单效果

1．实例简介

我们在使用 Android 应用的时候，经常会遇到这样的情况，对于某个控件相关的数据进行操作。例如，单击某一个选项，长按弹出删除此选项的菜单；在文本输入框的输入过程中，长按弹出复制文字、粘贴文字和全选文字的菜单。在 Android 中希望对于某个控件有菜单弹出的效果，就需要用到 Android 中的上下文菜单的功能。本实例就带领大家来开发一个 Android 中文本编辑框的长按弹出效果。

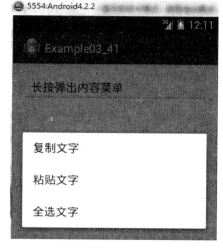

2．运行效果

该实例运行效果如图 3.41 所示。

图 3.41　文本框的复制粘贴全选菜单效果

3．实例程序讲解

在上面例子的效果中，页面有一个输入框，当对输入框长按的时候，弹出复制文字、粘贴文字和全选文字的上下文菜单。想要实现我们上例的效果，步骤如下所示。

（1）修改 res/layout/activity_main.xml 文件，代码如下：

```
01  <!-- 定义基础布局 LinearLayout -->
02  <LinearLayout xmlns:android="http://schemas.android.com/apk/res/android"
03      xmlns:tools="http://schemas.android.com/tools"
04      android:layout_width="match_parent"
05      android:layout_height="match_parent"
06      android:paddingBottom="@dimen/activity_vertical_margin"
07      android:paddingLeft="@dimen/activity_horizontal_margin"
08      android:paddingRight="@dimen/activity_horizontal_margin"
09      android:paddingTop="@dimen/activity_vertical_margin"
```

```xml
10 android:orientation="vertical">
11 <!-- 定义 TextView 控件 -->
12 <EditText
13     android:id="@+id/Et"
14     android:layout_width="match_parent"
15     android:layout_height="wrap_content"
16     android:text="长按弹出内容菜单"
17     />
18 </LinearLayout>
```

在上面的代码中第 11～16 行定义基本的输入框 EditText。

（2）修改 src/com.wyl.example/MainActivity.java 文件，代码如下：

```java
01 //定义 MainActivity 继承自 Activity
02 public class MainActivity extends Activity {
03 //定义 ContextMenu 中每个菜单选项的 Id
04 final int Menu_1 = Menu.FIRST;
05 final int Menu_2 = Menu.FIRST + 1;
06 final int Menu_3 = Menu.FIRST + 2;
07
08 @Override
09 protected void onCreate(Bundle savedInstanceState) {
10     super.onCreate(savedInstanceState);       //调用父类的 onCreate 方法
11     //通过 setContentView 方法设置当前页面的布局文件为 activity_main
12     setContentView(R.layout.activity_main);
13     //获得布局中的控件
14     EditText Et = (EditText)findViewById(R.id.Et);
15     //给 EditText 注册上下文菜单
16     registerForContextMenu(Et);
17 }
18
19 //创建 ContextMenu 菜单的回调方法
20 public void onCreateContextMenu(ContextMenu m, View v,
21         ContextMenuInfo menuInfo) {
22     super.onCreateContextMenu(m,v,menuInfo);
23
24     //在上下文菜单选项中添加选项内容
25     //add 方法的参数：add(分组 id,itemid, 排序, 菜单文字)
26     m.add(0, Menu_1, 0, "复制文字");
27     m.add(0, Menu_2, 0, "粘贴文字");
28     m.add(0, Menu_3, 0, "全选文字");
29 }
30
31 //ContextMenu 菜单选项的选项选择的回调事件
32 public boolean onContextItemSelected(MenuItem item) {
33     //参数为用户选择的菜单选项对象
34     //根据菜单选项的 id 来执行相应的功能
35     switch (item.getItemId()) {
36     case 1:
37         Toast.makeText(this, "复制文字", Toast.LENGTH_SHORT).show();
38         break;
39     case 2:
40         Toast.makeText(this, "粘贴文字", Toast.LENGTH_SHORT).show();
41         break;
42     case 3:
43         Toast.makeText(this, "全选文字", Toast.LENGTH_SHORT).show();
44         break;
45     }
```

```
46          return super.onOptionsItemSelected(item);
47     }
48 }
```

在上面代码中第 14~16 行定义了得到了 EditText 对话框,并且注册了上下文菜单。第 19~29 行,实现了 Activity 的 onCreateContextMenu 方法,此方法是 Activity 的回调方法,当长按某一个控件的时候弹出上下文菜单,其中有三个参数,第一个参数是上下文菜单对象,第二个参数是长按单击的 View 对象,第三个参数是关于菜单的菜单的信息,然后在此方法中添加了复制文字、粘贴文字和全选文字三个菜单选项。在第 31~47 行实现了 Activity 的 onContextItemSelected 方法,此方法也是回调方法,当上下文菜单的某个选项被选中的时候自动调用,其中 MenuItem 参数为你选中的上下文菜单对象,根据此对象的 id 的值可以得知用户选择的是哪个选项,然后就可以进行相应操作了。

4.实例扩展

Android 中的上下文菜单类似于 Windows 中的鼠标右键,针对不同的控件显示不同的菜单选项。在 Android 应用中上下文菜单最主要的一个应用就是在 ListView,通过长按 ListView 中的某一个 Item,就可以对某一个 Item 做修改和删除工作了。例如,对学生列表,通过单击其中的某一个 Item 进行删除,或者位置的移动,这在一些应用里面是比较炫的效果了,所以大家可以根据自己的需求添加这些效果。

范例 042 仿 UC 浏览器的伪菜单效果

1.实例简介

在 Android 中的菜单有很多种,除了我们之前了解的几种以外,自定义的菜单现在应用中的使用也是很多的。例如,UC 浏览器的菜单效果,其中单击 Menu 物理键会弹出菜单,但是这种菜单并不是系统的菜单,而是自定义菜单,其中包括了图片和文字,当然也可以设置固定的效果。本实例就带领大家一起来完成一个仿 UC 浏览器的伪菜单效果。

2.运行效果

该实例运行效果如图 3.42 所示。

3.实例程序讲解

在上面例子的效果中,页面有一个文本提示框,当用户单击手机上的物理菜单键时就会显示我们的菜单效果。在此工程中用到了 9 张外部图片,我已经提前拷贝到 res/drawable 目录下,名称为 m1、m2、m3、m4、m5、m6、m7、m8、m9,我们在下面的代码中会进行使用。想要实现我们上例的效果,步骤如下所示。

图 3.42 仿 UC 浏览器的伪菜单效果

(1)新建 res/layout/menu_item.xml 文件,代码如下:

```
01 <?xml version="1.0" encoding="utf-8"?>
02 <!-- 定义伪 Menu 的一个 item 样式 -->
03 <LinearLayout xmlns:android="http://schemas.android.com/apk/res/android"
04     android:layout_width="fill_parent"
```

```
05      android:layout_height="wrap_content"
06      android:orientation="vertical">
07      <!-- 定义伪 Menu 的 item 的图片控件 -->
08      <ImageView
09          android:id="@+id/Iv"
10          android:layout_width="40dip"
11          android:layout_height="40dip"
12          android:layout_gravity="center"/>
13      <!-- 定义伪 Menu 的 item 的文字控件 -->
14      <TextView
15          android:id="@+id/Tv"
16          android:layout_width="match_parent"
17          android:layout_height="wrap_content"
18          android:gravity="center"
19          android:textColor="@android:color/white"
20          android:text="选项" />
21
22  </LinearLayout>
```

此文件是我们定义的菜单的一个 item 的布局文件，其中基本布局是垂直线性布局，包含了一个 ImageView 和一个 TextView。在第 7～20 行定义了这两个控件，并且设置了相应的 id，方便我们在 java 文件中获取对应的控件对象。

（2）新建 res/layout/menu_layout.xml 文件，代码如下：

```
01  <?xml version="1.0" encoding="utf-8"?>
02  <!-- 定义基础的 Linearlayout 布局 -->
03  <LinearLayout xmlns:android="http://schemas.android.com/apk/res/android"
04      android:layout_width="fill_parent"
05      android:layout_height="fill_parent"
06      android:orientation="vertical" >
07
08      <!-- 定义 GridView 控件 -->
09      <GridView
10          android:id="@+id/Gv"
11          android:layout_width="fill_parent"
12          android:layout_height="fill_parent"
13          android:gravity="center"
14          android:horizontalSpacing="15dip"
15          android:numColumns="3"
16          android:stretchMode="columnWidth"
17          android:background="@android:color/black"
18          android:verticalSpacing="15dip" />
19
20  </LinearLayout>
```

此文件是我们定义的菜单的布局文件，其中基本布局是线性布局。第 8～18 行在线性布局中定义了一个 GridView 表格视图，并且设置了相应的 id 属性、gravity 属性和 stretchMode 属性等。

（3）修改 res/layout/activity_main.xml 文件，代码如下：

```
01  <!-- 定义基础布局 LinearLayout -->
02  <LinearLayout xmlns:android="http://schemas.android.com/apk/res/android"
03      xmlns:tools="http://schemas.android.com/tools"
04      android:layout_width="match_parent"
05      android:layout_height="match_parent"
06      android:paddingBottom="@dimen/activity_vertical_margin"
07      android:paddingLeft="@dimen/activity_horizontal_margin"
```

```
08      android:paddingRight="@dimen/activity_horizontal_margin"
09      android:paddingTop="@dimen/activity_vertical_margin"
10      android:orientation="vertical">
11      <!-- 定义 TextView 控件 -->
12      <TextView
13          android:layout_width="match_parent"
14          android:layout_height="wrap_content"
15          android:text="单击 Menu 物理键,弹出自定义菜单"
16          />
17  </LinearLayout>
```

这个文件定义了整个页面的基本布局,其中只有一个 TextView,提示用户单击 Menu 物理键。

(4) 新建 src/com.wyl.example/Item.java 文件,代码如下:

```
01  //定义实体类,代表 GridView 的选项实体类
02  public class Item {
03      public int itemImg;              //选项图片 id
04      public String itemTitle;         //选项文字
05
06      //构造方法
07      public Item(int itemImg, String itemTitle) {
08          super();
09          this.itemImg = itemImg;
10          this.itemTitle = itemTitle;
11      }
12  }
```

此文件为一个实体类,也就是大家以后需要做业务层时对应的对象类,在这里进行模拟数据,所以仅仅作为占位使用,大家今后可以根据自己的需要进行修改。

(5) 新建 src/com.wyl.example/GvAdapter.java 文件,代码如下:

```
01  //自定义 GridView 的 adapter
02  public class GvAdapter extends BaseAdapter {
03
04      // 定义 Context
05      private Context mContext;
06      // 定义要显示的 MyFile 列表
07      private List<Item> fileList;
08
09      //FileListAdapter 的构造方法
10      public GvAdapter(Context c,List<Item> fl) {
11          mContext = c;
12          fileList = fl;
13      }
14
15      // 获取显示的条目数量
16      @Override
17      public int getCount() {
18          // TODO Auto-generated method stub
19          return fileList.size();
20      }
21
22      // 获取列表中的单个对象
23      @Override
24      public Object getItem(int position) {
25          // TODO Auto-generated method stub
26          return fileList.get(position);
```

```
27  }
28
29  // 获取列表中对象的id
30  @Override
31  public long getItemId(int position) {
32      // TODO Auto-generated method stub
33      return position;
34  }
35
36  //构造每一个item的View视图
37  @Override
38  public View getView(int position, View convertView, ViewGroup parent) {
39      //定义位置占位符类的对象
40      ViewHolder viewholder =new ViewHolder();
41      if (convertView == null) {
42          //初始化当前view的布局视图
43          convertView = LayoutInflater.from(mContext).inflate(
44              R.layout.menu_item, null);
45      }
46      //获取到对应的控件对象
47      viewholder.fileImage = (ImageView) convertView
48              .findViewById(R.id.Iv);
49      viewholder.fileName = (TextView) convertView
50              .findViewById(R.id.Tv);
51      //给控件对象设置相应的内容
52      viewholder.fileImage.setBackgroundResource(fileList.get
        (position).itemImg);
53      viewholder.fileName.setText(fileList.get(position).itemTitle);
54
55      return convertView;
56  }
57
58  //定义内部类作为占位符组合
59  class ViewHolder {
60      ImageView fileImage;
61      TextView fileName;
62  }
63  }
12  }
```

此文件是自定义的一个 Adapter，为了等下给我们的自定义 Menu 中的 GridView 来使用。在其中的第 36~56 行实现了 getView 方法在此方法中通过 **LayoutInflater** 类，加载之前的 **menu_item** 的布局作为当前的视图布局，然后得到对应的控件设置相应的数据。

（6）修改 src/com.wyl.example/MainActivity.java 文件，代码如下：

```
001  // 定义MainActivity继承自Activity
002  public class MainActivity extends Activity {
003      private AlertDialog Dia;         //定义Dialog,当单击菜单是弹出菜单
004      private GridView Gv;             //得到菜单视图中的GridView对象
005      private View view;               //定义了菜单视图
006
007      private final int ITEM_1 = 0;    //搜索
008      private final int ITEM_2 = 1;    //文件管理
009      private final int ITEM_3 = 2;    //下载管理
010      private final int ITEM_4 = 3;    //全屏
011      private final int ITEM_5 = 4;    //网址
012      private final int ITEM_6 = 5;    //书签
```

```java
013    private final int ITEM_7 = 6;    //加入书签
014    private final int ITEM_8 = 7;    //分享
015    private final int ITEM_9 = 8;    //退出
016    //定义要显示在GridView中的item链表
017    private List<Item> li = new ArrayList<Item>();
018
019    @Override
020    protected void onCreate(Bundle savedInstanceState) {
021        // TODO Auto-generated method stub
022        super.onCreate(savedInstanceState);    //调用父类的onCreate方法
023        // 通过setContentView方法设置当前页面的布局文件为activity_main
024        setContentView(R.layout.activity_main);
025        //定义菜单的选项
026        setData();
027        setMenuView();
028    }
029    private void setData() {
030        // TODO Auto-generated method stub
031        //定义菜单的选项
032        li.add(new Item(R.drawable.m1, "搜索"));
033        li.add(new Item(R.drawable.m2, "文件管理"));
034        li.add(new Item(R.drawable.m3, "下载管理"));
035        li.add(new Item(R.drawable.m4, "全屏"));
036        li.add(new Item(R.drawable.m5, "网址"));
037        li.add(new Item(R.drawable.m6, "书签"));
038        li.add(new Item(R.drawable.m7, "加入书签"));
039        li.add(new Item(R.drawable.m8, "分享"));
040        li.add(new Item(R.drawable.m9, "退出"));
041    }
042    //初始化Menu的视图效果
043    private void setMenuView() {
044        // TODO Auto-generated method stub
045        //通过View的inflate方法,加载xml文件生成View
046        view = View.inflate(this, R.layout.menu_layout, null);
047        //定义AlertDialog,并且设置对话框的视图为之前的view
048        Dia = new AlertDialog.Builder(this).create();
049        Dia.setView(view);
050        //设置AlertDialog的键盘监听器事件
051        Dia.setOnKeyListener(new OnKeyListener() {
052            public boolean onKey(DialogInterface dialog, int keyCode,
053                    KeyEvent event) {
054                if (keyCode == KeyEvent.KEYCODE_MENU)    //监听按键
055                    dialog.dismiss();
056                return false;
057            }
058        });
059
060        //定义adapter
061        GvAdapter gva = new GvAdapter(this, li);
062
063        //得到菜单的视图中的GridView控件
064        Gv = (GridView) view.findViewById(R.id.Gv);
065        //设置GridView的adapter数据源
066        Gv.setAdapter(gva);
067        /** 监听menu选项 **/
068        Gv.setOnItemClickListener(new OnItemClickListener() {
069            public void onItemClick(AdapterView<?> arg0, View arg1, int arg2,
```

```
070                    long arg3) {
071                switch (arg2) {
072                case ITEM_1:
073                    Toast.makeText(MainActivity.this,"选择了搜索",
                           Toast.LENGTH_SHORT).show();
074                    break;
075                case ITEM_2:
076                    Toast.makeText(MainActivity.this,"选择了文件",
                           Toast.LENGTH_SHORT).show();
077                    break;
078                case ITEM_3:
079                    Toast.makeText(MainActivity.this,"选择了下载",
                           Toast.LENGTH_SHORT).show();
080                    break;
081                case ITEM_4:
082                    Toast.makeText(MainActivity.this,"选择了全屏",
                           Toast.LENGTH_SHORT).show();
083                    break;
084                case ITEM_5:
085                    Toast.makeText(MainActivity.this,"选择了网址",
                           Toast.LENGTH_SHORT).show();
086                    break;
087                case ITEM_6:
088                    Toast.makeText(MainActivity.this,"选择了书签",
                           Toast.LENGTH_SHORT).show();
089                    break;
090                case ITEM_7:
091                    Toast.makeText(MainActivity.this,"选择了书签",
                           Toast.LENGTH_SHORT).show();
092                    break;
093                case ITEM_8:
094                    Toast.makeText(MainActivity.this,"选择了分享",
                           Toast.LENGTH_SHORT).show();
095                    break;
096                case ITEM_9:
097                    Toast.makeText(MainActivity.this,"选择了退出",
                           Toast.LENGTH_SHORT).show();
098                    break;
099                }
100                //dialog取消
101                Dia.dismiss();
102            }
103        });
104    }
105    //选项菜单建立的回调函数
106    @Override
107    public boolean onCreateOptionsMenu(Menu menu) {
108        menu.add("menu");                    //必须创建一项
109        return super.onCreateOptionsMenu(menu);
110    }
111    //选项菜单打开的回调函数
112    @Override
113    public boolean onMenuOpened(int featureId, Menu menu) {
114        //如果dialog为空,就初始化菜单,否则显示菜单
115        if (Dia == null) {
116            Dia = new AlertDialog.Builder(this).setView(view).show();
117        } else {
118            Dia.show();
119        }
```

```
120         return false;             // 返回为 true 则显示系统 menu
121     }
122 }
```

在上面代码中第 7～15 行定义了菜单的选项 id。第 29～41 行，添加模拟数据，这里使用到了之前导入的 9 张图片。在第 46～50 行定义了一个 AlertDialog 对象，并且设置了它的视图布局，设置了键盘监听事件。在第 68 行设置了菜单中的 GridView 的菜单选项单击事件，通过单击菜单的 id 做出对应的操作。第 107 行实现了 onCreateOptionsMenu 方法，当单击 menu 的时候调用的方法。在第 113 行实现了 onMenuOpened 方法，当菜单打开时调用的方法。

这样你的伪菜单就应该能正常显示了。

4．实例扩展

本例实现的其实是通过对 Menu 创建的回调方法进行实现来完成的，其实本例子的实质是通过一个自定义的 View 来显示菜单。所以大家可以在做自己的应用的过程时根据需要来定义自己想要的菜单，对于本例子来说，大家只要修改第 46 行去加载自定义的 View 视图布局，然后对于选项的选择做自己的修改就可以了。

范例 043　PopupMenu 效果

1．实例简介

在 Android 中的 SDK 3.0 版本以后加入了一个特殊的菜单效果。它可以在任何的 View 上显示，而且会根据 View 位置显示菜单的效果。本实例就带领大家一起来完成一个 Android 中特有的 PopupMenu 菜单效果。

2．运行效果

该实例运行效果如图 3.43 所示。

图 3.43　PopupMenu 菜单效果

3．实例程序讲解

在上面例子的效果中，页面有一个文本提示框，当用户单击此文本框时就会弹出 PopupMenu 的菜单。想要实现我们上例的效果，步骤如下所示。

（1）新建 res/menu/main.xml 文件，代码如下：

```
01 <?xml version="1.0" encoding="utf-8"?>
02 <menu xmlns:android="http://schemas.android.com/apk/res/android" >
03
04     <item
05         android:id="@+id/send"
06         android:title="@string/send"/>
07     <item
08         android:id="@+id/look"
09         android:title="@string/look"/>
10     <item
```

```
11        android:id="@+id/delete"
12        android:title="@string/delete"/>
13
14  </menu>
```

此文件是我们定义菜单的布局文件,在此文件中定义了菜单的三个选项分别设置了对应的 id 和对应的文字常量。

(2) 修改 res/values/strings.xml 文件,代码如下:

```
01  <?xml version="1.0" encoding="utf-8"?>
02  <resources>
03      <string name="app_name">Example03_43</string>
04
05      <string name="send">发送邮件</string>
06      <string name="look">阅读邮件</string>
07      <string name="delete">删除邮件</string>
08  </resources>
```

此文件中定义工程时用到的文字常量。

(3) 修改 src/com.wyl.example/MainActivity.java 文件,代码如下:

```
01  //定义 MainActivity 继承自 Activity
02  public class MainActivity extends Activity {
03      //定义 TextView 控件
04      private TextView Tv;
05
06      @Override
07      protected void onCreate(Bundle savedInstanceState) {
08          // TODO Auto-generated method stub
09          super.onCreate(savedInstanceState);        //调用父类的 onCreate 方法
10          // 通过 setContentView 方法设置当前页面的布局文件为 activity_main
11          setContentView(R.layout.activity_main);
12          findView();
13          setListener();
14      }
15      private void findView() {
16          // TODO Auto-generated method stub
17          //得到视图中的控件
18          Tv = (TextView)findViewById(R.id.Tv);
19      }
20
21      private void setListener() {
22          // TODO Auto-generated method stub
23          //设置 TextView 的单击监听器
24          Tv.setOnClickListener(mylistener);
25      }
26      //自定义自己的单击监听器
27      OnClickListener mylistener = new OnClickListener() {
28          //单击后的 onClick 回调方法
29          @Override
30          public void onClick(View v) {
31              // TODO Auto-generated method stub
32              //定义 popupmenu 对象
33              PopupMenu popup = new PopupMenu(MainActivity.this, v);
34
35              //设置 popupmenu 对象的布局
36              popup.getMenuInflater().inflate(R.menu.main, popup.getMenu());
37
```

```
38            //设置 popupmenu 对象的菜单单击事件
39            popup.setOnMenuItemClickListener(new OnMenuItemClickListener() {
40               //菜单单击后的回调函数
41               @Override
42               public boolean onMenuItemClick(MenuItem item) {
43                     Toast.makeText(getBaseContext(), "你选择了: "
                        + item.getTitle(), Toast.LENGTH_SHORT).show();
44                     return true;
45               }
46            });
47
48            //显示 popupmenu 菜单
49            popup.show();
50         }
51      };
52   }
```

在上面代码中第 18 行得到 TextView 对象，在第 24 行设置自定义的监听器。在第 27 行定义了自定义的 OnClickListener 监听器。在第 35～49 行定义一个 PopupMenu 对象，并设置其选项菜单为 main.xml。然后设置 PopupMenu 的菜单选项单击的监听器。然后通过 popup.show()方法显示菜单。

4．实例扩展

Android 中的 PopupMenu 基本上都是结合 View 来使用的，而且对于 Android 来说最常用的一个用法就是在 ActionBar 上添加 PopupMenu，现在基本国外的流行应用都采用这种方式来弹出菜单，一方面因为它不会占用屏幕的控件，另一方面它会根据 View 的位置自动调整菜单弹出的方向。

范例 044　PopupWindow 效果

1．实例简介

在 Android 中的 SDK 3.0 版本不但加入了 PopupMenu 效果可以弹出菜单，而且还加入 PopupWinodw 效果，可以弹出一个可交互的页面，就像我们在做网站开发的时候弹出登录框的样子。本实例就带领大家使用 Android 中 PopupWindow 实现弹出页面的效果。

2．运行效果

该实例运行效果如图 3.44 所示。

3．实例程序讲解

在上面例子的效果中，页面有一个文本提示框，当用户单击此文本框时就会弹出 PopupWindow 窗口。想要实现我们上例的效果，

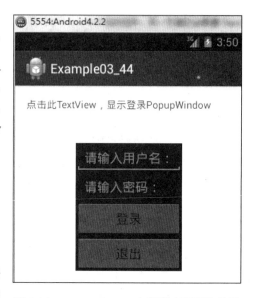

图 3.44　PopupWindow 实现弹出页面的效果

步骤如下所示。

（1）新建 res/layout/popup.xml 文件，代码如下：

```xml
01 <?xml version="1.0" encoding="utf-8"?>
02 <!-- 定义基础布局 LinearLayout -->
03 <LinearLayout xmlns:android="http://schemas.android.com/apk/res/android"
04     android:layout_width="wrap_content"
05     android:layout_height="wrap_content"
06     android:gravity="center_horizontal"
07     android:background="@android:color/background_dark"
08     android:orientation="vertical" >
09     <!-- 定义输入用户名的 EditText -->
10     <EditText
11         android:layout_width="match_parent"
12         android:layout_height="wrap_content"
13         android:textColor="@android:color/white"
14         android:hint="请输入用户名："
15         />
16     <!-- 定义输入密码的 EditText -->
17     <EditText
18         android:layout_width="match_parent"
19         android:layout_height="wrap_content"
20         android:textColor="@android:color/white"
21         android:hint="请输入密码："
22         />
23 <!-- 定义登录按钮 -->
24     <Button
25         android:id="@+id/BtnLogin"
26         android:layout_width="match_parent"
27         android:layout_height="wrap_content"
28         android:text="登录" />
29 <!-- 定义退出按钮 -->
30     <Button
31         android:id="@+id/BtnExit"
32         android:layout_width="match_parent"
33         android:layout_height="wrap_content"
34         android:text="退出"/>
35
36 </LinearLayout>
```

此文件是我们定义的弹出窗口的布局文件，此文件有用户名输入框、密码输入框、登录按钮和退出按钮，并且设置了对应的 id。

（2）修改 res/layout/activity_main.xml 文件，代码如下：

```xml
01 <!-- 定义基础布局 LinearLayout -->
02 <LinearLayout xmlns:android="http://schemas.android.com/apk/res/android"
03     xmlns:tools="http://schemas.android.com/tools"
04     android:id="@+id/LlMain"
05     android:layout_width="match_parent"
06     android:layout_height="match_parent"
07     android:paddingBottom="@dimen/activity_vertical_margin"
08     android:paddingLeft="@dimen/activity_horizontal_margin"
09     android:paddingRight="@dimen/activity_horizontal_margin"
10     android:paddingTop="@dimen/activity_vertical_margin"
11     android:orientation="vertical">
12     <!-- 定义 TextView 控件 -->
13     <TextView
14         android:id="@+id/Tv"
```

```
15        android:layout_width="match_parent"
16        android:layout_height="wrap_content"
17        android:text="单击此TextView,显示登录PopupWindow"
18      />
19  </LinearLayout>
```

此文件中定义了 activity 的布局,只显示了一个 TextView 的提示信息文本框。

(3) 修改 src/com.wyl.example/MainActivity.java 文件,代码如下:

```
01  // 定义 MainActivity 继承自 Activity
02  public class MainActivity extends Activity {
03      //定义 TextView 对象
04      private TextView Tv;
05
06      @Override
07      protected void onCreate(Bundle savedInstanceState) {
08          // TODO Auto-generated method stub
09          super.onCreate(savedInstanceState);       //调用父类的 onCreate 方法
10          // 通过 setContentView 方法设置当前页面的布局文件为 activity_main
11          setContentView(R.layout.activity_main);
12          findView();
13          setListener();
14      }
15
16
17      private void findView() {
18          // TODO Auto-generated method stub
19          //得到控件对象
20          Tv = (TextView)findViewById(R.id.Tv);
21      }
22
23      private void setListener() {
24          // TODO Auto-generated method stub
25          //设置单击监听器
26          Tv.setOnClickListener(mylistener);
27      }
28
29      // 自定义自己的单击监听器
30      OnClickListener mylistener = new OnClickListener() {
31          // 单击后的 onClick 回调方法
32          @Override
33          public void onClick(View v) {
34              // TODO Auto-generated method stub
35              //得到 layoutInflater 对象
36              LayoutInflater lif = (LayoutInflater) (MainActivity.this)
37                      .getSystemService(LAYOUT_INFLATER_SERVICE);
38              // 获取自定义布局文件 activity_popup.xml 的视图
39              View poplayout = lif.inflate(R.layout.activity_popup, null);
40              //根据 poplayout 得到 popwindow 对象
41              PopupWindow popwindow = new PopupWindow(poplayout,
42                      LayoutParams.WRAP_CONTENT, LayoutParams.WRAP_CONTENT);
43              //设置 popwindow 的背景
44              popwindow.setBackgroundDrawable(new BitmapDrawable());
45              //设置可获取焦点
46              popwindow.setFocusable(true);
47              //设置 popwindow 之外,popwindow 就消失
48              popwindow.setOutsideTouchable(true);
49              //设定 popwindow 的显示位置
```

```
50          popwindow.showAtLocation(findViewById(R.id.LlMain), Gravity.CENTER,
51                  0, 0);
52      }
53  };
54 }
```

在上面代码中第 20 行得到了 TextView 对象，在第 26 行设置自定义的监听器。在第 30 行定义了自定义的 OnClickListener 监听器。在第 36~51 行定义一个 PopupWindow 对象，并设置其布局为 activity_popup.xml。然后设置 PopupWindow 的相关属性。

4．实例扩展

Android 中的 Window 指的是最上层显示给用户的界面。Android 中通过窗口管理器来管理程序中的所有的界面，也就是上述代码中通过 getSystemService 方法得到系统的窗口管理器对象，然后再来使用。

在 Android 中我们到现在为止介绍了三个比较相近的概念：View、Activity 和 window，他们的概念分别如下所示。

- View：指的是控件和视图，其包括了 ViewGroup。
- Activity：它的主要功能是得到页面布局，产生对应的窗口，及用户的动作接收。
- Window：是真正用户看到的最顶层的界面效果。

所以我们之前认为看到的界面都是 Activity，其实是不太准确的，确切的说我们所能看到的都是 Window。这三个概念希望大家能够理解。

范例 045 QQ 客户端的标签栏效果

1．实例简介

在 Android 中，现在的应用经常会遇到同时显示几个页面的情况，这几个页面之间是并列的关系，而且用户希望能够快速的来回切换。对于这样的要求在 Android 中经常会使用一个控件叫做 TabHost。它可以实现同时加载几个页面，而且方便进行几个页面之间的切换。本实例就带领大家使用 Android 中的 TabHost 来实现仿 QQ 客户端的标签栏效果。

2．运行效果

该实例运行效果如图 3.45 所示。

图 3.45 QQ 客户端的标签栏效果

3. 实例程序讲解

在上面例子的效果中，显示了一个 TabHost 控件，其中包含了三个页面，一个是好友列表页面，一个是群组列表页面，一个是讨论组列表页面。想要实现我们上例的效果，步骤如下所示。

（1）先定义三个单独的页面的页面布局，新建 res/layout/ activity_mydiscussion.xml 文件，代码如下：

```xml
01  <?xml version="1.0" encoding="utf-8"?>
02  <LinearLayout xmlns:android="http://schemas.android.com/apk/res/android"
03      android:layout_width="match_parent"
04      android:layout_height="match_parent" >
05  <!-- 定义 ListView 的布局 -->
06      <ListView
07          android:id="@+id/LvDiscussion"
08          android:layout_width="match_parent"
09          android:layout_height="match_parent"
10          />
11
12  </LinearLayout>
```

此文件是我们定义了讨论组页面的布局，其中包含 ListView 控件，设置 id 为 LvDiscussion。

新建 res/layout/ activity_myfriend.xml 文件，代码如下：

```xml
01  <?xml version="1.0" encoding="utf-8"?>
02  <LinearLayout xmlns:android="http://schemas.android.com/apk/res/android"
03      android:layout_width="match_parent"
04      android:layout_height="match_parent" >
05  <!-- 定义 ListView 的布局 -->
06      <ListView
07          android:id="@+id/ LvFirend"
08          android:layout_width="match_parent"
09          android:layout_height="match_parent"
10          />
11
12  </LinearLayout>
```

此文件是我们定义了好友页面的布局文件，其中包含 ListView 控件，设置 id 为 LvFriend。

新建 res/layout/ activity_mygroup.xml 文件，代码如下：

```xml
01  <?xml version="1.0" encoding="utf-8"?>
02  <LinearLayout xmlns:android="http://schemas.android.com/apk/res/android"
03      android:layout_width="match_parent"
04      android:layout_height="match_parent" >
05  <!-- 定义 ListView 的布局 -->
06      <ListView
07          android:id="@+id/LvGroup"
08          android:layout_width="match_parent"
09          android:layout_height="match_parent"
10          />
11
12  </LinearLayout>
```

此文件是我们定义了群组页面的布局文件，其中包含 ListView 控件，设置 id 为 LvGroup。

（2）创建三个页面中 ListView 的 Item 的布局。新建 res/layout/ activity_mydiscussion_item.xml 文件，代码如下：

```xml
<?xml version="1.0" encoding="utf-8"?>
<!-- 定义ListView的item样式 -->
<LinearLayout xmlns:android="http://schemas.android.com/apk/res/android"
    android:layout_width="match_parent"
    android:layout_height="match_parent"
    android:orientation="horizontal" >

    <!-- 定义讨论组图片 -->
    <ImageView
        android:id="@+id/Iv"
        android:layout_width="wrap_content"
        android:layout_height="match_parent"
        android:scaleType="fitXY" />

    <LinearLayout
        android:layout_width="match_parent"
        android:layout_height="match_parent"
        android:layout_marginLeft="10dip"
        android:orientation="vertical" >

        <!-- 定义讨论组名称文字控件 -->
        <TextView
            android:id="@+id/TvTitle"
            android:layout_width="wrap_content"
            android:layout_height="wrap_content"
            android:gravity="left"
            android:textColor="@color/blue"
            android:textSize="20sp" />

        <!-- 定义讨论组简介文字的文字控件 -->
        <TextView
            android:id="@+id/TvInfo"
            android:layout_width="match_parent"
            android:layout_height="wrap_content"
            android:gravity="left"
            android:textColor="@color/blue"
            android:textSize="10sp" />
    </LinearLayout>

</LinearLayout>
```

此文件中定义了讨论组 Item 的布局，其中定义了讨论组的图片控件、讨论组名称和讨论组简介的文本标签。

新建 res/layout/ activity_myfriend_item.xml 文件，代码如下：

```xml
<?xml version="1.0" encoding="utf-8"?>
<!-- 定义ListView的item样式 -->
<LinearLayout xmlns:android="http://schemas.android.com/apk/res/android"
    android:layout_width="match_parent"
    android:layout_height="match_parent"
    android:orientation="horizontal" >

```

```xml
08        <!-- 定义箭头图片 -->
09        <ImageView
10            android:id="@+id/Iv"
11            android:layout_width="wrap_content"
12            android:layout_height="match_parent" />
13
14        <!-- 定义好友文字控件 -->
15        <TextView
16            android:id="@+id/TvTitle"
17            android:layout_width="wrap_content"
18            android:layout_height="wrap_content"
19            android:layout_marginLeft="10dip"
20            android:gravity="left"
21            android:textColor="@color/blue"
22            android:textSize="20sp" />
23
24        <!-- 定义好友人数的文字控件 -->
25        <TextView
26            android:id="@+id/TvInfo"
27            android:layout_width="match_parent"
28            android:layout_height="wrap_content"
29            android:gravity="right"
30            android:textColor="@color/blue"
31            android:textSize="10sp" />
32
33  </LinearLayout>
```

此文件中定义了好友组 Item 的布局，其中定义了好友组的箭头控件，还有分组名称和分组中的人数的文本标签。

新建 res/layout/ activity_mygroup_item.xml 文件，代码如下：

```xml
01  <?xml version="1.0" encoding="utf-8"?>
02  <!-- 定义 ListView 的 item 样式 -->
03  <LinearLayout xmlns:android="http://schemas.android.com/apk/res/android"
04      android:layout_width="match_parent"
05      android:layout_height="match_parent"
06      android:orientation="horizontal" >
07
08      <!-- 定义分组图片 -->
09      <ImageView
10          android:id="@+id/Iv"
11          android:layout_width="wrap_content"
12          android:layout_height="match_parent"
13          android:scaleType="fitXY" />
14
15      <LinearLayout
16          android:layout_width="match_parent"
17          android:layout_height="match_parent"
18          android:layout_marginLeft="10dip"
19          android:orientation="vertical" >
20
21          <!-- 定义群组名文字控件 -->
22          <TextView
23              android:id="@+id/TvTitle"
24              android:layout_width="wrap_content"
25              android:layout_height="wrap_content"
26              android:gravity="left"
27              android:textColor="@color/blue"
28              android:textSize="20sp" />
```

```
29
30              <!-- 定义群组简介文字控件 -->
31              <TextView
32                  android:id="@+id/TvInfo"
33                  android:layout_width="match_parent"
34                  android:layout_height="wrap_content"
35                  android:gravity="left"
36                  android:textColor="@color/blue"
37                  android:textSize="10sp" />
38      </LinearLayout>
39
40  </LinearLayout>
```

此文件中定义了好友群页面的 Item 的布局，其中定义了群组的图片控件，还有群组名称和群组简介的文本标签。

（3）修改 res/layout/ activity_main.xml 文件，代码如下：

```
01  <?xml version="1.0" encoding="utf-8"?>
02  <!-- 定义基础的 TabHost 控件，显示分页标签 -->
03  <TabHost xmlns:android="http://schemas.android.com/apk/res/android"
04      android:id="@android:id/tabhost"
05      android:layout_width="fill_parent"
06      android:layout_height="fill_parent" >
07
08      <!-- 定义 TabHost 的标签 -->
09      <LinearLayout
10          android:layout_width="fill_parent"
11          android:layout_height="fill_parent"
12          android:orientation="vertical" >
13          <!-- 定义 TabWidget 控件 -->
14          <TabWidget
15              android:id="@android:id/tabs"
16              android:layout_width="fill_parent"
17              android:layout_height="wrap_content" />
18          <!-- 定义 FrameLayout 控件 -->
19          <FrameLayout
20              android:id="@android:id/tabcontent"
21              android:layout_width="fill_parent"
22              android:layout_height="fill_parent"
23              android:padding="5dp" />
24      </LinearLayout>
25
26  </TabHost>
```

此文件是本例子中主界面的布局文件，其中包含了 TabHost 控件，确定在主页面中的显示内容是以多页标签为显示框架。

（4）此程序相对我们之前建立的程序都比较复杂，其中涉及到了 Activity 页面、实体类和页面中三个 ListView 的自定义 Adapter 类。所以把工程中的 src 目录分成了三个包，分别如下所示。

- com.wyl.example：程序的 UI 包。
- com.wyl.example.adapter：程序的 adapter 包。
- com.wyl.example.moudle：程序中用到的实体类的包。

在程序的 com.wyl.example.moudle 包中建立三个实体类，如下所示。

新建 src/com.wyl.example.moudle/MyDiscussion.java 文件，代码如下：

```
01  //定义讨论组的实体类
02  public class MyDiscussion {
03  public String Name;        //item名字
04  public int ImgId;          //item图片
05  public String Info;        //item信息
06
07  //MyDiscussion 的构造方法
08  public MyDiscussion() {
09      super();
10      // TODO Auto-generated constructor stub
11  }
12  //MyDiscussion 带参数的构造方法
13  public MyDiscussion(String n, int imgId, String info) {
14      super();
15      Name = n;
16      ImgId = imgId;
17      Info = info;
18  }
19  }
```

此文件定义了讨论组的实体类。

新建 src/com.wyl.example.moudle/MyFriend.java 文件，代码如下：

```
01  //定义了好友的实体类
02  public class MyFriend {
03  public String Name;        //item名字
04  public int ImgId;          //item图片
05  public String Info;        //item信息
06
07  //MyFriend 的构造方法
08  public MyFriend() {
09      super();
10      // TODO Auto-generated constructor stub
11  }
12  //MyFriend 带参数的构造方法
13  public MyFriend(String n, int imgId, String info) {
14      super();
15      Name = n;
16      ImgId = imgId;
17      Info = info;
18  }
19  }
```

此文件定义了好友组的实体类。

新建 src/com.wyl.example.moudle/MyGroup.java 文件，代码如下：

```
01  //定义讨论组的实体类
01  //定义分组的实体类
02  public class MyGroup {
03  public String Name;        //item名字
04  public int ImgId;          //item图片
05  public String Info;        //item信息
06
07  //MyGroup 的构造方法
08  public MyGroup() {
09      super();
10      // TODO Auto-generated constructor stub
```

```
11  }
12  //MyGroup 带参数的构造方法
13  public MyGroup(String n, int imgId, String info) {
14      super();
15      Name = n;
16      ImgId = imgId;
17      Info = info;
18  }
19  }
```

此文件定义了群组的实体类。

（5）在程序的 com.wyl.example.adapter 包中建立三个类，如下所示。

新建 src/com.wyl.example.adapter/MyDiscussionListAdapter.java 文件，代码如下：

```
01  //自定义 adapter
02  public class MyDiscussionListAdapter extends BaseAdapter {
03
04      //定义 LayoutInflater 对象
05      private LayoutInflater mInflater;
06      private List<MyDiscussion> list = new ArrayList<MyDiscussion>();
07      private Context context;
08
09      // NewsListAdapter 的构造方法
10      public MyDiscussionListAdapter(Context c, List<MyDiscussion> l) {
11          context = c;
12          list = l;
13      }
14
15      // 定义内部类作为占位符组合
16      class ViewHolder {
17          ImageView Image;
18          TextView Name;
19          TextView Number;
20      }
21
22      @Override
23      public int getCount() {
24          // TODO Auto-generated method stub
25          return list.size();
26      }
27
28      @Override
29      public Object getItem(int position) {
30          // TODO Auto-generated method stub
31          return list.get(position);
32      }
33
34      @Override
35      public long getItemId(int position) {
36          // TODO Auto-generated method stub
37          return position;
38      }
39
40      @Override
41      public View getView(int position, View convertView, ViewGroup parent) {
42          // TODO Auto-generated method stub
43          ViewHolder viewholder = null;
44          if (convertView == null) {
45              //初始化 friend 的 item 视图
```

```
46            convertView = LayoutInflater.from(context).inflate(
47                R.layout.activity_mydiscussion_item, null);
48            viewholder = new ViewHolder();
49            //获得视图中的对象控件
50            viewholder.Image = (ImageView) convertView.findViewById
              (R.id.Iv);
51          viewholder.Name = (TextView) convertView.findViewById
            (R.id.TvTitle);
52          viewholder.Number = (TextView) convertView.findViewById
            (R.id.TvInfo);
53
54            convertView.setTag(viewholder);
55          } else {
56            viewholder = (ViewHolder) convertView.getTag();
57          }
58          //设置控件的属性
59          viewholder.Image.setBackgroundResource(list.get(position).ImgId);
60          viewholder.Name.setText(list.get(position).Name);
61          viewholder.Number.setText(list.get(position).Info);
62
63        return convertView;
64      }
65   }
```

在上面代码中第 41 行实现了 BaseAdapter 的 getView 方法,来得到对应的讨论组的 Item 的布局,然后设置对应的数据来源。

然后新建 src/com.wyl.example.adapter/MyFriendListAdapter.java 文件,代码的基本逻辑和上述 Adapter 类似这里就不再展示代码了。

然后新建 src/com.wyl.example.adapter/ MyGroupListAdapter.java 文件,代码的基本逻辑和上述 Adapter 类似这里就不再展示代码了。

(6) 在程序的 com.wyl.example 包中建立三个类,分别是三个标签页的 Activity,如下所示。

新建 src/com.wyl.example /MyDiscussionActivity.java 文件,代码如下:

```
01  //定义MyDiscussion继承自Activity
02  public class MyDiscussionActivity extends Activity {
03   // 定义ListView对象
04   private ListView LvFriend;
05   // 定义adapter对象
06   private MyDiscussionListAdapter adapter;
07   // 定义adapter的数据
08   private List<MyDiscussion> l = new ArrayList<MyDiscussion>();
09
10   public void onCreate(Bundle savedInstanceState) {
11     super.onCreate(savedInstanceState);
12     setContentView(R.layout.activity_mygroup);
13     // 定义模拟数据
14     getData();
15
16     // 得到控件中的ListView对象
17     LvFriend = (ListView) findViewById(R.id.LvGroup);
18     // 定义adapter
19     adapter = new MyDiscussionListAdapter(MyDiscussionActivity.this,
     l);
20     // 设置ListView的数据adapter
```

```
21       LvFriend.setAdapter(adapter);
22   }
23   //加入模拟数据
24   private void getData() {
25       // TODO Auto-generated method stub
26       l.add(new MyDiscussion("临时讨论组", R.drawable.dicussion, "张三,
         李四,王五"));
27       l.add(new MyDiscussion("项目研讨组", R.drawable.dicussion, "赵六,
         刘八,池塘的花"));
28       l.add(new MyDiscussion("自定义讨论组", R.drawable.dicussion,"shune,
         黑暗骑士,白雪"));
29       l.add(new MyDiscussion("课程讨论组", R.drawable.dicussion, "A1,
         blue,李经理"));
30   }
31   }
```

其中就是得到对应 ListView 控件然后进行 adapter 的设置。

新建 src/com.wyl.example /MyFriendActivity.java 文件,代码逻辑和 MyDiscussionActivity.java 代码逻辑类似,这里就不再展示代码。

新建 src/com.wyl.example /MyGroupActivity.java 文件,代码逻辑和 MyDiscussionActivity.java 代码逻辑类似,这里就不再展示代码。

(7) 在程序的 com.wyl.example 包中,修改 src/com.wyl.example /MainActivity.java 文件,代码如下:

```
01   //定义 MainActivity 继承自 TabActivity
02   public class MainActivity extends TabActivity {
03   /** Called when the activity is first created. */
04   @Override
05   public void onCreate(Bundle savedInstanceState) {
06       super.onCreate(savedInstanceState);
07       setContentView(R.layout.activity_main);
08       //得到当前 activity 中的 tabhost 对象
09       TabHost tabHost = getTabHost();
10       //定义 tabhost 中的 tabspec 对象
11       TabHost.TabSpec spec;
12       //定义 intent 对象
13       Intent i;
14       //设置第一个标签页的布局
15       i = new Intent(this, MyFriendActivity.class);
16       spec = tabHost.newTabSpec("0")
17               .setIndicator("好友")
18               .setContent(i);
19       //添加到 tabHost 中
20       tabHost.addTab(spec);
21       //设置第二个标签页的布局
22       i = new Intent(this, MyGroupActivity.class);
23       spec = tabHost.newTabSpec("1")
24               .setIndicator("群")
25               .setContent(i);
26       //添加到 tabHost 中
27       tabHost.addTab(spec);
28       //设置第三个标签页的布局
29       i = new Intent(this, MyDiscussionActivity.class);
30       spec = tabHost.newTabSpec("2")
31               .setIndicator("讨论组")
```

```
32                .setContent(i);
33        //添加到 tabHost 中
34        tabHost.addTab(spec);
35        //设置当前的 tabHost 的选中标签
36        tabHost.setCurrentTab(1);
37    }
38 }
```

在上面代码中第 9 行得到 TabHost 对象,在第 20～25 行添加好友标签页,然后依次添加群组标签页和讨论组标签页。在第 35～36 行设置当前 tabHost 的当前标签为标签 id 为 1 的标签页。

4．实例扩展

Android 中的 TabHost 默认都是在屏幕的上端的,而且在本例中设计到了三个列表标签页,其中三个标签页的数据我这里都是写死的模拟数据,在真是应用的开发过程中,这些数据一般都是从数据库获得,或者从网络申请下来的 Json 或 xml 信息,解析获得,今后我们会再次讲解更加高级的使用方法。

范例 046 仿新浪微博的主页效果

1．实例简介

Android 中 TabHost 默认是在页面最上面,但是我们看到的应用中有些页面的 TabHost 是在页面最下面的,而且也美化了很多,如新浪微博和 QQ 客户端等。本实例就带领大家一起来完成一个仿 Android 的新浪微博的主页效果。

2．运行效果

该实例运行效果如图 3.46 所示。

3．实例程序讲解

在上面例子的效果中,有一个 TabHost 在页面的最下端,而且每个标签都有图片,当用户单击某个标签页的时候,显示对应的页面。想要实现我们上例的效果,步骤如下所示。

图 3.46 仿新浪微博的主页效果

(1)修改/res/layout/activity_main.xml 文件,代码如下:

```
01 <?xml version="1.0" encoding="UTF-8"?>
02 <!--定义基础的 TabHost 控件   -->
03 <TabHost xmlns:android="http://schemas.android.com/apk/res/android"
04     android:id="@android:id/tabhost"
05     android:layout_width="fill_parent"
06     android:layout_height="fill_parent" >
07     <!-- 模拟 TabHost 的底层 RadioGroup 样式 -->
08     <LinearLayout
```

```xml
09      android:orientation="vertical"
10      android:layout_width="fill_parent"
11      android:layout_height="fill_parent">
12        <FrameLayout
13          android:id="@android:id/tabcontent"
14          android:layout_width="fill_parent"
15          android:layout_height="0.0dip"
16          android:layout_weight="1.0" />
17        <TabWidget
18          android:id="@android:id/tabs"
19          android:visibility="gone"
20          android:layout_width="fill_parent"
21          android:layout_height="wrap_content"
22          android:layout_weight="0.0" />
23        <!-- 定义RadioGroup模拟选项卡效果 -->
24        <RadioGroup
25          android:gravity="center_vertical"
26          android:layout_gravity="bottom"
27          android:orientation="horizontal"
28          android:id="@+id/main_radio"
29          android:background="@drawable/maintab_toolbar_bg"
30          android:layout_width="fill_parent"
31          android:layout_height="wrap_content">
32            <RadioButton
33              android:id="@+id/radio_button0"
34              android:tag="radio_button0"
35              android:layout_marginTop="2.0dip"
36              android:text="@string/friend"
37              android:drawableTop="@drawable/icon_1"
38              style="@style/main_tab_bottom" />
39            <RadioButton
40              android:id="@+id/radio_button1"
41              android:tag="radio_button1"
42              android:layout_marginTop="2.0dip"
43              android:text="@string/cast"
44              android:drawableTop="@drawable/icon_2"
45              style="@style/main_tab_bottom" />
46            <RadioButton
47              android:id="@+id/radio_button2"
48              android:tag="radio_button2"
49              android:layout_marginTop="2.0dip"
50              android:text="@string/main"
51              android:drawableTop="@drawable/icon_3"
52              style="@style/main_tab_bottom" />
53            <RadioButton
54              android:id="@+id/radio_button3"
55              android:tag="radio_button3"
56              android:layout_marginTop="2.0dip"
57              android:text="@string/msg"
58              android:drawableTop="@drawable/icon_4"
59              style="@style/main_tab_bottom" />
60            <RadioButton
61              android:id="@+id/radio_button4"
62              android:tag="radio_button4"
63              android:layout_marginTop="2.0dip"
64              android:text="@string/more"
65              android:drawableTop="@drawable/icon_5"
66              style="@style/main_tab_bottom" />
67        </RadioGroup>
68    </LinearLayout>
69 </TabHost>
```

在此页面中主要以 TabHost 为基础布局,其中使用了 RadioGroup 和 RadioButton 模拟 TabHost 的切换标签的效果。

(2) 建立五个标签页面的布局文件如下。

新建 activity_one.xml 文件,代码如下:

```xml
01 <?xml version="1.0" encoding="utf-8"?>
02 <!-- 定义基础的 LinearLayout 布局 -->
03 <LinearLayout xmlns:android="http://schemas.android.com/apk/res/android"
04     android:layout_width="fill_parent"
05     android:layout_height="fill_parent"
06     android:gravity="top|center"
07     android:orientation="vertical" >
08
09 <!-- 定义 textview,标示当前页面 -->
10 <TextView
11     android:layout_width="fill_parent"
12     android:layout_height="wrap_content"
13     android:gravity="center"
14     android:text="这是时间 Activity"/>
15
16 </LinearLayout>
```

这里只显示了一个 TextView 的提示信息。

新建 activity_two.xml 文件、activity_three.xml 文件、activity_four.xml 文件和 activity_five.xml 文件,代码和 activity_one.xml 的代码相同,只是提示语不同。这里不再显示代码。

(3) 新建五个 Activity 页面:

新建 src/com.wyl.example/OneActivity.java,其继承自 Activity,在 oncreate 方法中通过 setcontentView 设置当前页面的布局为 activity_one.xml 布局文件。

然后建立 src/com.wyl.example/TwoActivity.java、src/com.wyl.example/ThreeActivity.java、src/com.wyl.example/FourActivity.java 和 src/com.wyl.example/FiveActivity.java,四个文件分别加载对应的布局文件。

(4) 修改 src/com.wyl.example/MainActivity.java 文件,代码如下:

```java
01 //定义 activity 继承自 TabActivity,并且实现了 OnCheckedChangeListener 接口
02 public class MainActivity extends TabActivity implements OnCheckedChangeListener{
03     /** Called when the activity is first created. */
04 //定义的 tabhost 对象
05 private TabHost mHost;
06 //定义 RadioGroup 对象
07 private RadioGroup radioderGroup;
08
09     @Override
10     public void onCreate(Bundle savedInstanceState) {
11         super.onCreate(savedInstanceState);
12         setContentView(R.layout.activity_main);
13         //实例化 TabHost
14         mHost=this.getTabHost();
15
16         //添加选项卡,并且设置跳转 intent
17         mHost.addTab(mHost.newTabSpec("ONE").setIndicator("ONE")
18             .setContent(new Intent(this,OneActivity.class)));
19         mHost.addTab(mHost.newTabSpec("TWO").setIndicator("TWO")
```

```
20              .setContent(new Intent(this,TwoActivity.class)));
21          mHost.addTab(mHost.newTabSpec("THREE").setIndicator("THREE")
22              .setContent(new Intent(this,ThreeActivity.class)));
23          mHost.addTab(mHost.newTabSpec("FOUR").setIndicator("FOUR")
24              .setContent(new Intent(this,FourActivity.class)));
25          mHost.addTab(mHost.newTabSpec("FIVE").setIndicator("FIVE")
26              .setContent(new Intent(this,FiveActivity.class)));
27          //得到 radioGroup 对象
28          radioderGroup = (RadioGroup) findViewById(R.id.main_radio);
29          //设置 radioGroup 对象的切换监听器
30          radioderGroup.setOnCheckedChangeListener(this);
31      }
32
33      //实现 OnCheckedChangeListener 中的 RadioGroup 的选项切换回调函数
34      @Override
35      public void onCheckedChanged(RadioGroup group, int checkedId) {
36          //根据所选中的 RadioGroup 的选项 id, 设置 tabhost 的选项卡
37          switch(checkedId){
38          case R.id.radio_button0:
39              mHost.setCurrentTabByTag("ONE");
40              break;
41          case R.id.radio_button1:
42              mHost.setCurrentTabByTag("TWO");
43              break;
44          case R.id.radio_button2:
45              mHost.setCurrentTabByTag("THREE");
46              break;
47          case R.id.radio_button3:
48              mHost.setCurrentTabByTag("FOUR");
49              break;
50          case R.id.radio_button4:
51              mHost.setCurrentTabByTag("FIVE");
52              break;
53          }
54      }
55  }
```

在上面代码中第 14 行得到 TabHost 对象，在第 16～26 行添加 TabHost 的标签页。在第 30 行给我们布局中的 RadioGroup 设置选项切换监听器。在第 35 行实现 OnCheckedChangeListener 中的回调方法，当单击 RadioButton 时，切换 TabHost 标签。这样就可以实现美化的 TabHost 了。

4．实例扩展

Android 中的 TabHost 是现在最常用的一个控件，包括自定义的 TabHost 样式，但是对于 TabHost 来说有很多的属性我们在实例中没有使用到，在这里给大家总结一下。

- clearAllTabs 方法：可以清楚所有的标签页。
- getCurrentTab 方法：可以得到当前所选中的标签页的 id。
- setCurrentTab 方法：可以设置当前 TabHost 所选中的标签页。

这些方法在我们平时使用 TabHost 的过程中都是很常用到的。大家可以记忆一下，如果需要更多的方法，请查询 Android 的官方开发文档。

范例 047 程序退出的对话框

1．实例简介

我们在使用 Android 应用的时候，如果想要关闭对话框，一般都是单击手机物理键盘的返回键，但是如果当单击就退出程序的话，用户在操作过程中有可能误单击返回，这样程序就关闭了，数据也就无法保存了，所以现在 Android 应用的退出功能一般都是用户单击返回键，然后弹出确认对话框，让用户确认退出，然后再退出程序。在 Android 中想要实现这种效果就要去创建 AlertDialog 对象了。本实例就带领大家来使用 Android 中 AlertDialog 来完成程序退出的对话框。

2．运行效果

该实例运行效果如图 3.47 所示。

图 3.47 程序退出的对话框

3．实例程序讲解

在上面例子的效果中，当用户单击手机物理返回键时弹出退出确认对话框，当用户单击"确认"按钮，退出程序，当用户单击"取消"按钮，关闭对话框。想要实现我们上例的效果，步骤如下所示。

（1）修改 res/layout/activity_main.xml 文件，代码如下：

```
01  <?xml version="1.0" encoding="UTF-8"?>
02  <!--定义基础的 LinearLayout 布局   -->
03  <LinearLayout xmlns:android="http://schemas.android.com/apk/res/android"
04      android:layout_width="fill_parent"
05      android:layout_height="fill_parent" >
06
07  <!-- 定义 TextView 控件 -->
08  <TextView
09          android:layout_width="fill_parent"
10          android:layout_height="fill_parent"
11          android:text="单击物理返回键弹出退出对话框" />
12
13  </LinearLayout>
```

在上面的代码中只定义了一个标记页面的 TextView 控件。

（2）修改 src/com.wyl.example/MainActivity.java 文件，代码如下：

```
01  //定义 MainActivity 继承自 Activity
02  public class MainActivity extends Activity {
03    /** Called when the activity is first created. */
04
05    @Override
06    public void onCreate(Bundle savedInstanceState) {
07        super.onCreate(savedInstanceState);
08        // 设置页面的布局文件
```

```
09      setContentView(R.layout.activity_main);
10  }
11
12  //在此方法中创建dialog
13  protected void creatdialog() {
14      //初始化AlertDialog构建器对象
15      AlertDialog.Builder b = new Builder(MainActivity.this);
16      //设置dialog的信息
17      b.setMessage("确认退出吗？");
18      //设置dialog的标题
19      b.setTitle("提示");
20      //添加确认和取消按钮
21      b.setPositiveButton("确认", new OnClickListener() {
22          @Override
23          public void onClick(DialogInterface dialog, int which) {
24              //如果用户单击确认退出，则对话框消失，程序关闭
25              dialog.dismiss();
26              MainActivity.this.finish();
27          }
28      });
29      b.setNegativeButton("取消", new OnClickListener() {
30          @Override
31          public void onClick(DialogInterface dialog, int which) {
32              //如果用户单击取消退出，则对话框消失
33              dialog.dismiss();
34          }
35      });
36      //创建对话框并且显示
37      b.create().show();
38  }
39
40  //在Activity中的键盘监听回调事件
41  public boolean onKeyDown(int keyCode, KeyEvent event) {
42      if (keyCode == KeyEvent.KEYCODE_BACK && event.getRepeatCount() == 0) {
43          creatdialog();
44      }
45      return false;
46  }
47  }
```

在上面代码中第 13 行定义 creatdialog 方法创建 AlertDialog，在第 15～37 行根据 AlertDialog.Builder 创建 AlertDialog，然后设置标题，设置提示信息，设置确认按钮的单击效果，取消按钮的单击效果，然后调用 show 方法，显示 AlertDialog。在第 41 行监听键盘的按键回调是方法，当用户按下返回键的时候弹出对话框。

4. 实例扩展

Android 中 Activity 具有很多方法，其中 finish 方法就是把当前的 Activity 从程序的栈中取出来，结束掉，显示程序栈的下面的页面。所以这样就可以实现程序退出了。当然如果在这个程序中有多个 Activity 已经打开，想要结束掉程序的话，需要使用结束程序的进程 id 或者终止当前程序的虚拟机的方式来实现了，代码如下所示。

结束程序的进程 id：

```
android.os.Process.killProcess(android.os.Process.myPid());
```

关闭当前程序的虚拟机：

```
System.exit(0);
```

范例 048　程序的关于对话框

1．实例简介

我们在使用 Android 应用的时候，一般都会看到程序的版本信息或者作者简介。在 Android 中完成这种提示性质的信息，一般就需要自定义 AlertDialog 的某些视图。本实例就带领大家来使用 Android 中 AlertDialog 来完成程序的关于对话框。

2．运行效果

该实例运行效果如图 3.48 所示。

3．实例程序讲解

在上面例子的效果中，当用户单击手机物理菜单键时弹出关于对话框，显示程序版本信息。想要实现我们上例的效果，步骤如下所示。

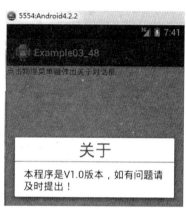

图 3.48　程序的关于对话框

（1）修改 res/layout/activity_main.xml 文件，代码如下：

```
01  <?xml version="1.0" encoding="UTF-8"?>
02  <!--定义基础的 LinearLayout 布局  -->
03  <LinearLayout xmlns:android="http://schemas.android.com/apk/res/android"
04      android:layout_width="fill_parent"
05      android:layout_height="fill_parent" >
06
07  <!-- 定义 TextView 控件 -->
08  <TextView
09      android:layout_width="fill_parent"
10      android:layout_height="fill_parent"
11      android:text="单击物理菜单键弹出关于对话框" />
12
13  </LinearLayout>
```

在上面的代码中只定义了一个标记页面的 TextView 控件。

（2）修改 src/com.wyl.example/MainActivity.java 文件，代码如下：

```
01  //定义 MainActivity 继承自 Activity
02  public class MainActivity extends Activity {
03      /** Called when the activity is first created. */
04
05      @Override
06      public void onCreate(Bundle savedInstanceState) {
07          super.onCreate(savedInstanceState);
08          // 设置页面的布局文件
09          setContentView(R.layout.activity_main);
10      }
11
```

```
12    //在此方法中创建dialog
13    protected void creatdialog() {
14        //初始化AlertDialog构建器对象
15        AlertDialog.Builder builder = new Builder(MainActivity.this);
16        //设置dialog的信息
17        builder.setMessage("本程序是V1.0版本, 如有问题请及时提出! ");
18        //定义标题TextView
19        TextView tv = new TextView(MainActivity.this);
20        tv.setGravity(android.view.Gravity.CENTER_HORIZONTAL);
21        tv.setText("关于");
22        tv.setTextSize(30);
23        //创建对话框并且显示
24        AlertDialog a = builder.create();
25        a.setCustomTitle(tv);
26        a.show();
27    }
28
29    //在Activity中的键盘监听回调事件
30    public boolean onKeyDown(int keyCode, KeyEvent event) {
31        //当按下物理的menu键的时候触发事件
32        if (keyCode == KeyEvent.KEYCODE_MENU && event.getRepeatCount() == 0) {
33            creatdialog();
34        }
35        return false;
36    }
37 }
```

在上面代码中第 13 行定义 creatdialog 方法创建 AlertDialog, 在第 15～26 行根据 AlertDialog.Builder 创建 AlertDialog, 然后设置标题, 设置提示信息, 并且通过 setCustomTitle 方法设置了自定义的标题视图。在第 30 行监听键盘的按键回调是方法, 当用户按下菜单键的时候弹出对话框。

4. 实例扩展

Android 中 AlertDialog 可以进行自定义, 而且方式有很多, 如下所示。
- setCustomTitle 方法: 自定义 title 的布局形式。
- setView 方法: 自定义整个 AlertDialog 的布局形式。
- setIcon 方法: 设置对话框的图标。

大家可以灵活运用这些方法来构造属于自己的 AlertDialog。

范例 049 电话服务评价对话框

1. 实例简介

我们在使用 Android 应用的时候, 也经常遇到需要用户交互的对话框, 如评价调查、省份选择和班级选择。这些需求要求只有当用户单击某按钮的时候弹出, 然后根据用户的选择进行下一步处理。本实例就带领大家来使用 Android 中的自定义 Dialog 来完成电话服务评价对话框。

2. 运行效果

该实例运行效果如图 3.49 所示。

图 3.49　电话服务评价对话框

3. 实例程序讲解

在上面例子的效果中，当用户单击手机物理菜单键时弹出关于对话框，显示程序版本信息。想要实现我们上例的效果，步骤如下所示。

（1）修改 res/layout/activity_main.xml 文件，代码如下：

```xml
01 <?xml version="1.0" encoding="UTF-8"?>
02 <!--定义基础的 LinearLayout 布局   -->
03 <LinearLayout xmlns:android="http://schemas.android.com/apk/res/android"
04     android:layout_width="fill_parent"
05     android:layout_height="fill_parent" >
06
07 <!-- 定义 TextView 控件 -->
08 <TextView
09     android:layout_width="fill_parent"
10     android:layout_height="fill_parent"
11     android:text="单击物理返回键弹出评价对话框" />
12
13 </LinearLayout>
```

在上面的代码中只定义了一个标记页面的 TextView 控件。

（2）修改 src/com.wyl.example/MainActivity.java 文件，代码如下：

```java
01 //定义 MainActivity 继承自 Activity
02 public class MainActivity extends Activity {
03 private TextView Tv;
04
05 /** Called when the activity is first created. */
06
07 @Override
08 public void onCreate(Bundle savedInstanceState) {
```

```
09        super.onCreate(savedInstanceState);
10        // 设置页面的布局文件
11        setContentView(R.layout.activity_main);
12        Tv = (TextView)findViewById(R.id.Tv);
13    }
14
15    // 在此方法中创建 dialog
16    protected void creatdialog() {
17        //构造 dialog 对象
18        Dialog dialog = new AlertDialog.Builder(this)
19                //设置对话框的标题和图标
20                .setIcon(android.R.drawable.btn_dialog).setTitle("评价对话框")
21                //设置对话框的内容
22                .setMessage("请对我的服务进行评价: ")
23                //设置对话框的按钮
24                .setPositiveButton("很好", new OnClickListener() {
25
26                    @Override
27                    public void onClick(DialogInterface dialog, int which) {
28                        // TODO Auto-generated method stub
29                        Tv.setText("很好");
30                    }
31                }).setNegativeButton("一般", new OnClickListener() {
32
33                    @Override
34                    public void onClick(DialogInterface dialog, int which) {
35                        // TODO Auto-generated method stub
36                        Tv.setText("一般");
37                    }
38                }).setNeutralButton("有待改进", new OnClickListener() {
39
40                    @Override
41                    public void onClick(DialogInterface dialog, int which) {
42                        // TODO Auto-generated method stub
43                        Tv.setText("有待改进");
44                    }
45                }).create();
46        //显示对话框
47        dialog.show();
48    }
49
50    // 在 Activity 中的键盘监听回调事件
51    public boolean onKeyDown(int keyCode, KeyEvent event) {
52        // 当按下物理的 menu 键的时候触发事件
53        if (keyCode == KeyEvent.KEYCODE_BACK && event.getRepeatCount() == 0) {
54            creatdialog();
55        }
56        return false;
57    }
58 }
```

在上面代码中第 16~47 行定义了 Dialog 对象,然后设置它的标题,设置它的图标,设置它的提示信息,设置三个评价按钮。在第 51 行设置的键盘监听器,当用户按下返回键时调用 creatdialog 方法。

4．实例扩展

在 Android 中 Dialog 对话框可以添加多个按钮来给用户进行操作选择，但是如果涉及到很多选项的话，建议大家用 ListView 或者 Spinner 进行选择，否则你的 Dialog 会看起来很复杂，用户也会比较反感。

范例 050 数据加载成功的提示

1．实例简介

我们在使用 Android 应用的时候，经常会遇到这样一种提示，它对于用户来说需要知道，但是它的优先级又不像 Dialog 那么高，需要用户选择。例如，短信发送成功和图片下载成功等。这就需要介绍 Android 中的另一种提示方式了——Toast。本实例就带领大家来使用 Android 中的 Toast 来完成数据加载成功的提示。

2．运行效果

该实例运行效果如图 3.50 所示。

3．实例程序讲解

在上面例子的效果中，程序运行起来就在进行模拟

图 3.50 数据加载成功的提示

数据加载，等待几秒后加载完毕显示 Toast 提示信息，想要实现上例效果，步骤如下所示。

（1）修改 res/layout/activity_main.xml 文件，代码如下：

```xml
01  <?xml version="1.0" encoding="UTF-8"?>
02  <!--定义基础的 LinearLayout 布局   -->
03  <LinearLayout xmlns:android="http://schemas.android.com/apk/res/android"
04      android:layout_width="fill_parent"
05      android:layout_height="fill_parent" >
06
07  <!-- 定义 TextView 控件 -->
08  <TextView
09      android:id="@+id/Tv"
10          android:layout_width="fill_parent"
11          android:layout_height="fill_parent"
12          android:text="等待数据加载......" />
13
14  </LinearLayout>
```

在上面的代码中只定义了一个标记页面的 TextView 控件。

（2）修改 src/com.wyl.example/MainActivity.java 文件，代码如下：

```java
01  //定义 MainActivity 继承自 Activity
02  public class MainActivity extends Activity {
03      //定义 TextView 控件
04      private TextView Tv;
05      //定义 thread 返回的 id 表示
```

```
06  private final int HANDLER_TEST = 0;
07
08      //定义Handler对象
09      private Handler h = new Handler(){
10          //实现当handler接收到message信息的回调函数
11          @Override
12          public void handleMessage(Message msg) {
13              // TODO Auto-generated method stub
14              super.handleMessage(msg);
15              //判断message中的what字段的值
16              switch (msg.what) {
17              case HANDLER_TEST:
18                  Toast.makeText(MainActivity.this, "数据加载完毕", Toast.
                    LENGTH_SHORT).show();
19                  Tv.setText("数据加载完毕");
20                  break;
21              }
22          }
23      };
24
25  /** Called when the activity is first created. */
26
27      @Override
28      public void onCreate(Bundle savedInstanceState) {
29          super.onCreate(savedInstanceState);
30          // 设置页面的布局文件
31          setContentView(R.layout.activity_main);
32          //得到textview控件
33          Tv = (TextView)findViewById(R.id.Tv);
34          //定一个thread类的对象
35          new Thread(){
36
37              /* (non-Javadoc)
38               * @see java.lang.Thread#run()
39               */
40              //线程的run方法
41              @Override
42              public void run() {
43                  // TODO Auto-generated method stub
44                  super.run();
45                  //模拟数据加载的时间
46                  try {
47                      sleep(5000);
48                  } catch (InterruptedException e) {
49                      // TODO Auto-generated catch block
50                      e.printStackTrace();
51                  }
52                  //构造message对象
53                  Message msg = new Message();
54                  msg.what = HANDLER_TEST;
55                  //给handler发送message信息对象
56                  h.sendMessage(msg);
57              }
58          //启动线程
59          }.start();
60      }
61  }
```

在上面代码中第 9~23 行定义了 Handler 对象，用来接收其他线程发送到主线程的

message 信息。在第 35 行定义了线程，在此线程的 run 方法中模拟数据加载时间，这里线程暂停 5 秒，然后发送 handler 信息给你主线程。主线程接收到发送的信息后，通过 Toast 来显示数据加载完毕的信息。

4．实例扩展

Toast 是 Android 中独有的提示方式，其最大的特点就是它的出现不会得到程序的焦点，你可以在不影响正常操作的情况下看到 Toast 提示，而且提示的时间有限，显示固定时间后自动消失。所以使用 Toast 显示那些用户关注的优先级不是很高的信息，会增加软件的交互性。

范例 051　网络图片加载成功的提示

1．实例简介

我们在使用 Android 应用的时候，同样 Toast 最多的应用就是获取网络的数据，因为网络的数据获取不是立刻的，而是需要一定的时间的，而且用户希望当程序加载完网络数据后自动显示。例如，图片下载成功和网络数据请求等。本实例就带领大家来使用 Android 中的 Toast 来完成网络图片加载成功的提示。

2．运行效果

该实例运行效果如图 3.51 所示。

3．实例程序讲解

在上面例子的效果中，程序运行起来就会看到一个 ImageView 显示默认图片，等待图片加载完毕后在 ImageView 显示图片，并且显示 Toast 提示信息，想要实现上例效果，步骤如下所示。

图 3.51　网络图片加载成功的提示

（1）修改 res/layout/activity_main.xml 文件，代码如下：

```xml
01 <?xml version="1.0" encoding="UTF-8"?>
02 <!-- 定义基础的 LinearLayout 布局 -->
03 <LinearLayout xmlns:android="http://schemas.android.com/apk/res/android"
04     android:layout_width="fill_parent"
05     android:layout_height="fill_parent" >
06
07     <!-- 定义 ImageView 控件 -->
08     <ImageView
09         android:id="@+id/Iv"
10         android:layout_width="fill_parent"
11         android:layout_height="fill_parent"
12         android:src="@drawable/ic_launcher" />
13
14 </LinearLayout>
```

在上面的代码中定义了 ImageView 控件，设置 id 方便在 java 文件中获取，设置了默认的图片。

（2）修改 src/com.wyl.example/MainActivity.java 文件，代码如下：

```
001  //定义MainActivity继承自Activity
002  public class MainActivity extends Activity {
003      //定义TextView控件
004      private ImageView Iv;
005      //定义thread返回的id表示
006      private final int SUCCESS = 0;
007      private final int FAILED = 1;
008
009      //网络获取图片的地址
010      private final String url =
         "http://www.baidu.com/img/shouye_b5486898c692066bd2cbaeda86d74448.gif";
011      //网络获取图片的bitmap对象
012      private Bitmap bit = null;
013
014      //定义Handler对象
015      private Handler h = new Handler(){
016          //实现当handler接收到message信息的回调函数
017          @Override
018          public void handleMessage(Message msg) {
019              // TODO Auto-generated method stub
020              super.handleMessage(msg);
021              //判断message中的what字段的值
022              switch (msg.what) {
023              case SUCCESS:
024                  //加载图片成功的分支
025                  Toast.makeText(MainActivity.this, "图片加载成功！",
                         Toast.LENGTH_SHORT).show();
026                  Iv.setImageBitmap(bit);
027                  break;
028              case FAILED:
029                  //加载图片失败的分支
030                  Toast.makeText(MainActivity.this, "数据加载失败！",
                         Toast.LENGTH_SHORT).show();
031                  break;
032              }
033          }
034      };
035
036      /** Called when the activity is first created. */
037      @Override
038      public void onCreate(Bundle savedInstanceState) {
039          super.onCreate(savedInstanceState);
040          // 设置页面的布局文件
041          setContentView(R.layout.activity_main);
042          //得到ImageView控件
043          Iv = (ImageView)findViewById(R.id.Iv);
044
045          //判断网络是否通畅
046          if (isOpenNetwork()) {
047              //定一个thread类的对象
048              new Thread(){
049
050                  /* (non-Javadoc)
```

```
051                  * @see java.lang.Thread#run()
052                  */
053              //线程的run方法
054              @Override
055              public void run() {
056                  // TODO Auto-generated method stub
057                  super.run();
058                  //请求网络图片
059                  try {
060                      bit = getRemoteImage(new URL(url));
061                  } catch (MalformedURLException e) {
062                      // TODO Auto-generated catch block
063                      e.printStackTrace();
064                  }
065                  //根据加载的网络图片,返回不同的message对象
066                  if (bit != null) {
067                      //构造message对象
068                      Message msg = new Message();
069                      msg.what = SUCCESS;
070                      //给handler发送message信息对象
071                      h.sendMessage(msg);
072                  }
073                  else{
074                      //构造message对象
075                      Message msg = new Message();
076                      msg.what = FAILED;
077                      //给handler发送message信息对象
078                      h.sendMessage(msg);
079                  }
080              }
081          //启动线程
082          }.start();
083      }
084      else{
085          //网络不通畅时提示信息
086          Toast.makeText(MainActivity.this, "网络不通!",
                  Toast.LENGTH_SHORT).show();
087      }
088  }
089
090  //根据传入的url对象,请求网络图片
091  public Bitmap getRemoteImage(final URL aURL) {
092      try {
093          //建立url连接
094          final URLConnection conn = aURL.openConnection();
095          conn.connect();
096          //从url连接中读取图片流
097          final BufferedInputStream bis = new BufferedInputStream(
098                  conn.getInputStream());
099          //从图片流中得到bitmap图片
100          final Bitmap bm = BitmapFactory.decodeStream(bis);
101          //关闭图片流
102          bis.close();
103          return bm;
104
105      } catch (IOException e) {
106          Log.d("DEBUGTAG", "Oh noooz an error...");
107      }
108
```

```
109              return null;
110          }
111
112      //判断网络状态是否正常
113      private boolean isOpenNetwork() {
114          //得到系统的网络连接服务
115          ConnectivityManager connManager =
                 (ConnectivityManager)getSystemService(Context.CONNECTIVITY_SERVICE);
116          //判断是否网络可连接
117          if(connManager.getActiveNetworkInfo() != null) {
118              return connManager.getActiveNetworkInfo().isAvailable();
119          }
120          return false;
121      }
122  }
```

在上面代码中第 15～34 行定义了 Handler 对象，用来接收其他线程发送到主线程的 message 信息，当接收到成功的标记信息后显示图片，并提示图片载入成功，否则提示图片载入失败。在第 48 行定义了一个线程，在此线程的 run 方法中通过 91 行自定义的方法加载网络图片，等待网络图片加载完成后发送 handler 信息给你主线程。主线程接收到发送的信息后，通过 Toast 来显示数据加载完毕的信息。在第 113 行可以检查网络是否畅通的状态。由于此实例需要连接网络，所以需要在 Manifest 文件中添加网络访问权限如下：

```
<uses-permission android:name="android.permission.INTERNET"/>
<uses-permission android:name="android.permission.ACCESS_NETWORK_STATE"/>
```

4．实例扩展

Toast 也可以进行自定义，其中主要涉及到的方法如下所示。
❑ setMargin 方法：设置 View 的边距。
❑ setGravity 方法：设置 Toast 的对齐方式。
❑ setView 方法：设置 Toast 的布局视图。
通过如上方法大家可以定义出属于自己的 Toast 提示了。

范例 052　模拟收到短信的状态栏提示

1．实例简介

在 Android 中还有一种提示方式非常常见，就是状态栏的提醒。例如，当用户手机收到短信，当用户手机进入无线网覆盖的区域，当手机连接电脑时等。这就需要我们使用到 Android 中的另一个类——Notification 类。本实例就带领大家来使用 Android 中的 Notification 来完成模拟收到短信的状态栏提示。

2．运行效果

该实例运行效果如图 3.52 所示。

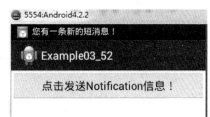

图 3.52　模拟收到短信的状态栏提示

3. 实例程序讲解

在上面例子的效果中,程序运行起来就会看到一个 Button 按钮,单击后发送了一条 Notification 提示信息,想要实现上例效果,步骤如下所示。

(1)修改 res/layout/activity_main.xml 文件,代码如下:

```xml
01 <?xml version="1.0" encoding="UTF-8"?>
02 <!-- 定义基础的 LinearLayout 布局 -->
03 <LinearLayout xmlns:android="http://schemas.android.com/apk/res/android"
04     android:layout_width="fill_parent"
05     android:layout_height="fill_parent" >
06
07     <!-- 定义 Button 控件 -->
08     <Button
09         android:id="@+id/Btn"
10         android:layout_width="fill_parent"
11         android:layout_height="wrap_content"
12         android:text="单击发送 Notification 信息!" />
13
14 </LinearLayout>
```

在上面的代码中定义了 Button 控件,设置 id 方便在 java 文件中获取,设置了 Button 的显示文字。

(2)修改 src/com.wyl.example/MainActivity.java 文件,代码如下:

```java
01 //定义 MainActivity 继承自 Activity
02 public class MainActivity extends Activity {
03 // 定义 Button 控件
04 private Button Btn;
05
06 /** Called when the activity is first created. */
07 @Override
08 public void onCreate(Bundle savedInstanceState) {
09     super.onCreate(savedInstanceState);
10     // 设置页面的布局文件
11     setContentView(R.layout.activity_main);
12     // 得到 Button 控件
13     Btn = (Button) findViewById(R.id.Btn);
14     // 设置 Button 的单击监听器
15     Btn.setOnClickListener(new OnClickListener() {
16
17         @Override
18         public void onClick(View v) {
19             // TODO Auto-generated method stub
20             // 发送 Notification 通知
21             sendNotificaction();
22         }
23     });
24 }
25
26 // 发送一个 Notification 通知
27 private void sendNotificaction() {
28     //得到系统的 Notification 服务对象
29     NotificationManager manager = (NotificationManager) this
```

```
30              .getSystemService(Context.NOTIFICATION_SERVICE);
31      // 创建一个 Notification 对象
32      Notification notification = new Notification();
33      // 设置显示 Notification 对象的图标
34      notification.icon = R.drawable.ic_launcher;
35      // 设置显示 Notification 对象的内容
36      notification.tickerText = "您有一条新的短消息!";
37
38      // 设置显示 Notification 对象的声音
39      notification.defaults = Notification.DEFAULT_SOUND;
40      // 置显示 Notification 对象的声音模式
41      notification.audioStreamType = android.media.AudioManager.
        ADJUST_LOWER;
42
43      //定义单击 Notification 的事件 Intent
44      Intent intent = new Intent(this, MainActivity.class);
45      PendingIntent pendingIntent = PendingIntent.getActivity(this, 0,
46          intent, PendingIntent.FLAG_ONE_SHOT);
47      // 单击状态栏的图标出现的提示信息设置
48      notification.setLatestEventInfo(this, "短消息内容", "我是一个短消息,
        愚人节快乐!",
49          pendingIntent);
50      //发送 Notification 消息
51      manager.notify(1, notification);
52  }
53  }
```

在上面代码中第 13 行拿到 Button 对象。在第 15 行给 Button 对象设置监听器,当单击按钮时调用 sendNotification 方法。在第 27~52 行实现发送 Notification,首先得到 NotificationManager 对象,然后设置 Notification 的图标,设置文字内容,设置提示声音,设置单击文字后的 Intent。然后发送消息。

4. 实例扩展

Notification 是系统级的提示,所以在 Android 中需要通过系统的 service 进行控制,我们这里通过 SystemService 得到了 NotificationManager 对象,然后才可以对 Notification 进行管理。

范例 053　模拟数据下载的状态栏提示

1. 实例简介

在 Android 中状态栏的提醒还有一中最常见的用法就是数据下载。例如,当用户要下载某个文件的时候,或者用户听音乐的时候等。这就需要我们去自定义 Android 中的 Notification。本实例就带领大家来自定义 Android 中的 Notification 来完成模拟数据下载的状态栏提示。

2. 运行效果

该实例运行效果如图 3.53 所示。

图 3.53　模拟数据下载的状态栏提示

3．实例程序讲解

在上面例子的效果中，程序运行起来就会看到两个 Button 按钮，单击第一个按钮后发送模拟文件下载的一条 Notification 提示信息，单击第二个按钮，清除此下载提示，想要实现上例效果，步骤如下所示。

（1）修改 res/layout/activity_main.xml 文件，代码如下：

```xml
01 <?xml version="1.0" encoding="utf-8"?>
02 <!-- 定义基础的 LinearLayout 布局 -->
03 <LinearLayout xmlns:android="http://schemas.android.com/apk/res/android"
04     android:layout_width="fill_parent"
05     android:layout_height="fill_parent"
06     android:orientation="vertical" >
07 <!-- 定义发送 notification 的按钮 -->
08     <Button
09         android:id="@+id/BtnSend"
10         android:layout_width="fill_parent"
11         android:layout_height="wrap_content"
12         android:text="发送下载 Notification" />
13 <!-- 定义取消 notification 的按钮 -->
14     <Button
15         android:id="@+id/BtnClean"
16         android:layout_width="fill_parent"
17         android:layout_height="wrap_content"
18         android:text="清除 Notification" />
19
20 </LinearLayout>
```

在上面的代码中定义了两个 Button 控件，设置 id 方便在 java 文件中获取，设置了 Button 的显示文字。

（2）新建 res/layout/layout_notification.xml 文件，代码如下：

```xml
01 <?xml version="1.0" encoding="utf-8"?>
02 <!-- 定义 Notification 的布局 -->
03 <LinearLayout xmlns:android="http://schemas.android.com/apk/res/android"
04     android:layout_width="wrap_content"
05     android:layout_height="wrap_content"
06     android:orientation="vertical" >
07 <!-- 定义通知布局的文本框 -->
08     <TextView
09         android:id="@+id/Tv"
10         android:layout_width="wrap_content"
11         android:layout_height="fill_parent"
```

```
12          android:text="下载中"
13          android:textColor="@android:color/white"
14          android:textSize="20sp" />
15  <!-- 定义下载进度 progressbar 控件 -->
16      <ProgressBar
17          android:id="@+id/Pb"
18          style="?android:attr/progressBarStyleHorizontal"
19          android:layout_width="260dip"
20          android:layout_height="wrap_content"
21          android:layout_gravity="center_vertical" />
22
23  </LinearLayout>
```

在上面的代码中定义 Notification 的显示效果，其中包含一个 TextView 和一个 ProgressBar，分别设置 id，方法在 java 文件中获取到此控件。

（3）修改 src/com.wyl.example/MainActivity.java 文件，代码如下：

```
01  //定义 MainActivity 继承自 Activity
01  //定义 MainActivity 继承自 Activity
02  public class MainActivity extends Activity {
03      //定义 Notification 的 id
04      private int notification_id = 1;
05      //定义 notificationManage 的对象
06      private NotificationManager nm;
07      //定义主线程的 Handler
08      private Handler handler = new Handler();
09      //定义 Notification 对象
10      private Notification notification;
11      //记录进度条的进度
12      private int count = 0;
13      //记录是否进度条取消
14      private Boolean isclean = false;
15
16      @Override
17      public void onCreate(Bundle savedInstanceState) {
18          super.onCreate(savedInstanceState);
19          //设置当前页面的布局
20          setContentView(R.layout.activity_main);
21
22          //得到页面中的按钮对象，并设置监听器
23          Button BtnSend = (Button) findViewById(R.id.BtnSend);
24          BtnSend.setOnClickListener(mylis);
25          Button BtnClean = (Button) findViewById(R.id.BtnClean);
26          BtnClean.setOnClickListener(mylis);
27
28          // 得到 NotificationManager 的服务对象
29          nm = (NotificationManager) getSystemService(NOTIFICATION_SERVICE);
30          //初始化 notification 对象
31          notification = new Notification(R.drawable.ic_launcher, "开始下载",
32              System.currentTimeMillis());
33          //得到 Notification 的视图对象
34          notification.contentView = new RemoteViews(getPackageName(),
35              R.layout.layout_notification);
36          // 设置视图中的 ProgressBar 对象
37          notification.contentView.setProgressBar(R.id.Pb, 100, 0, false);
38          // 定义单击通知的事件
39          Intent notificationIntent = new Intent(this, MainActivity.class);
```

```
40      PendingIntent contentIntent = PendingIntent.getActivity(this, 0,
41              notificationIntent, 0);
42      notification.contentIntent = contentIntent;
43  }
44  //自定义按钮单击监听器
45  OnClickListener mylis = new OnClickListener() {
46
47      @Override
48      public void onClick(View v) {
49          // TODO Auto-generated method stub
50          switch (v.getId()) {
51          case R.id.BtnClean:
52              // 取消 notification
53              nm.cancel(notification_id);
54              isclean = true;
55              break;
56          case R.id.BtnSend:
57              // 显示 notification
58              showNotification();
59              handler.post(run);
60              break;
61          default:
62              break;
63          }
64      }
65
66  };
67  //定义 Runnable 对象进行进度更新
68  Runnable run = new Runnable() {
69
70      @Override
71      public void run() {
72          // TODO Auto-generated method stub
73          //判断通知是否被取消
74          if (!isclean) {
75              //如果没有取消就进行进度的更新
76              count++;
77              notification.contentView.setProgressBar(R.id.Pb, 100, count,
78                      false);
79              // 更新 notification,就是更新进度条
80              showNotification();
81              // 200毫秒 count 加 1
82              if (count < 100)
83                  handler.postDelayed(run, 200);
84          }
85      }
86  };
87  //显示 notification
88  public void showNotification() {
89      nm.notify(notification_id, notification);
90  }
91  }
53  }
```

在上面代码中第 23~26 行拿到 Button 对象,并且设置相应的监听器。在第 29~42 行定义了 Notification 对象并且设置图标,设置自定义布局,设置单击事件。在第 53 行实现取消 Notification,在第 58 行实现发送 Notification。第 68 行定义了 Runable 对象,用来更新 Notification 中的进度条。这样就可以实现模拟短信接收的效果了。

4．实例扩展

自定义 Notification 有很多种方式，主要调用 Notification.contentView 对象来进行自定义，只要把你的布局生成 View，传递给 contentView 对象，那么 Notification 对象的显示就改变了。

3.4 小　　结

在本章节中主要介绍了 Android 中基本控件的使用及各种高级控件的使用，其中基本控件的使用是 Android 开发的基础，希望各位读者一定要掌握。高级控件的使用是 Android 中应用开发的难点，当然也是程序出彩的地方，因为这些高级组件可以根据应用的需要进行自定义修改，这让你的程序个性化。当然本章的实例大多数是静态的页面效果，下一章我们会讲述如何让你的程序能够与用户进行交互，也就是如何让你的程序动起来。

第 4 章　让你的程序和用户说话

上一章了解了 Android 应用最基本的用户界面的开发。但是我们现在使用的应用程序不但能够给用户提供界面，而且用户还可以对程序进行各种各样的操作，如键盘操作和屏幕滑动操作，而且随着现在智能手机硬件的功能逐渐提升，手机与用户的交互形式也越来越多。例如，摇一摇功能、手机的重力感应功能和手机的平衡感应功能，这些功能在良好的应用界面的基础上给用户带来了更加好的使用体验。所以在我们掌握了用户界面开发后，一定要掌握的就是 Android 应用如何与用户进行交互了。

Android 系统中提供了两种应用与用户进行交互的方式：

第一种是通过回调函数的形式，也就是实现某些固定的函数，然后当某个时间触发后，回调函数自动调用。这种方法使用起来比较简单，但是使用的范围有限。

第二种是通过监听器的方式，通过这种方式给你要接受用户操作的控件设置监听器，然后当用户操作此控件的时候，就把用户的操作事件传递给设置的监听器做处理。这样可以使事件的接受者和处理者分开，而且监听器的种类也很多，基本我们常见的事件类型都可以接受。

本章主要通过各种实例来介绍 Android 中常见的事件监听器，而且对于多线程处理也进行了实例讲解。希望读者阅读完本章内容后，可以根据自己的需求独立完成和用户进行各种交互的界面的开发，并且可以适当的通过多线程操作来完成一些相对复杂的界面交互。

4.1　Android 中基于回调函数的事件处理

范例 054　Activity 的声明周期回调

1. 实例简介

在上一章中我们主要讲解的是界面开发，而且基本上都是在一个 Activity 中通过界面的修改或布局的修改来完成一个令人耳目一新的界面，在其中我们的布局基本都是通过 setContentView 方法将一个 xml 布局设置给当前的 Activity。但是这些内容只是使用到了 Activity 的一个回调函数就是 onCreate。那么此函数在何时调用呢？Activity 是否还有其他的回调函数呢？我们通过本实例带领大家一起来看一下 Activity 的生命周期都有哪些回调函数。

2. 运行效果

该实例运行效果如图 4.1 所示。

图 4.1　Activity 的生命周期

3．实例程序讲解

想要实现如上效果，首先修改的地方在建立的工程下的 res/layout/activity_main.xml 文件，代码如下：

```xml
01 <?xml version="1.0" encoding="utf-8"?>
02 <!-- 定义基础的 LinearLayout 布局 -->
03 <LinearLayout xmlns:android="http://schemas.android.com/apk/res/android"
04     android:layout_width="fill_parent"
05     android:layout_height="fill_parent"
06     android:orientation="vertical" >
07     <!-- 定义 TextView 文本标签 -->
08     <TextView
09         android:layout_width="fill_parent"
10         android:layout_height="wrap_content"
11         android:text="Activity生命周期演示，请查看 logcat" />
12
13 </LinearLayout>
```

这是我们的 Activity 的布局文件，其中第 8～11 行构造了一个 TextView 控件，在此 TextView 中显示了一个文本提示效果。

在 src/com.wyl.example/MainActivity.java 代码中实现 Activity 的生命周期回调方法 onCreate、onDestroy、onPause、onRestart、onResume、onStart 和 onStop。代码如下：

```java
01 //定义MainActivity继承自Activity
02 public class MainActivity extends Activity {
03 public static final String TAG = "Activity Life Cycle";
04
05 @Override
06 public void onCreate(Bundle savedInstanceState) {
07     //当创建此Activity的时候回调
08     super.onCreate(savedInstanceState);
09     setContentView(R.layout.activity_main);
10     Log.e(TAG, "onCreate");
11 }
12
```

```
13    @Override
14    protected void onDestroy() {
15        //当销毁此 Activity 的时候回调
16        super.onDestroy();
17        Log.e(TAG, "onDestroy");
18    }
19
20    @Override
21    protected void onPause() {
22        //当暂停此 Activity 的时候回调
23        super.onPause();
24        Log.e(TAG, "onPause");
25    }
26
27    @Override
28    protected void onRestart() {
29        //当重新开始此 Activity 的时候回调
30        super.onRestart();
31        Log.e(TAG, "onRestart");
32    }
33
34    @Override
35    protected void onResume() {
36        //当显示展示此 Activity 的界面的时候回调
37        super.onResume();
38        Log.e(TAG, "onResume");
39    }
40
41    @Override
42    protected void onStart() {
43        //当使用此 Activity 可以接受用户操作的时候回调
44        super.onStart();
45        Log.e(TAG, "onStart");
46    }
47
48    @Override
49    protected void onStop() {
50        //当停止此 Activity 的时候回调
51        super.onStop();
52        Log.e(TAG, "onStop");
53    }
54 }
```

如上面中代码的第 6 行，实现了 onCreate 方法，第 14 行实现了 onDestroy 方法，第 21 行实现了 onPause 方法，第 28 行实现了 onRestart 方法，第 35 行实现了 onResume 方法，第 42 行实现了 onStart 方法，第 49 行实现了 onStop 方法，这些方法都是 Activity 的生命周期方法，执行的时机如下所示。

- onCreate：当 Activity 创建的时候，回调此方法，一般在此方法中写 Activity 的初始化内容。
- onResume：当 Activity 显示给用户的时候回调此方法，一般在此方法中设置 Activity 显示的内容。
- onStart：当 Activity 可以接受用户的操作的时候回调此方法，一般在此方法中设置监听器事件。
- onPause：当 Activity 从 onStart 状态转变成不可接受用户操作的时候回调此方法，

一般在此方法中得到用户输入的数据。
- onStop：当 Activity 的界面不可被用户看到的时候调用，一般在此方法中回收此 Activity 中的控件的内存。
- onRestart：当 Activity 的界面从 onStop 到 onStart 方法的过程中，会回调此方法，一般回复之前保存的数据。
- onDestroy：当 Activity 销毁的时候回调此方法，一般销毁在 onCreate 创建的对象的内存。

4．实例扩展

扩展 1：在 Activity 的回调函数中还有两个回调函数我们经常会用到，如下所示。
- onSaveInstanceState()：当 Activity 关闭的时候调用，一般在此方法中进行页面中用户输入数据的保存。
- onRestoreInstanceState()：当 Activity 在此启动的时候调用，一般在此方法中恢复之前关闭前保存的用户输入的数据。

这两个方法在早期的 API 中使用的比较多，现在的应用一般也不再使用了。

扩展 2：在此实例中我们用到了今后程序调试的最常见的一种方法就是打印 Log。对于 Android 中的 Log，要在你的程序中使用的话，一定要导入 android.util.Log 包，在此类中可以实现 Log 的打印。Log 在 Android 中根据其严重级别分为如下所示。
- Log.e：Error 级别，就是错误级别，一般会中止程序运行，是最严重的 Log 级别。
- Log.w：Warrgning 级别，就是警告级别，一般不会中止程序，但是可能会影响程序的执行结果。
- Log.d：Debug 级别，就是调试级别，一般不会中止程序，一般是程序员为了调试程序而打印的 log。
- Log.i：Info 级别，就是信息界级别，不会中止程序，一般是系统中执行操作的信息提示。
- Log.v：Verbose 级别，就是可见级别，一般是最低的信息提示。

在这里说明一下，在我程序的调试过程中为了能在 LogCat 中清楚的看到错误，所以我在本书写作的过程中有 Log 提示的话一般都是用 Log.e 级别的了。当程序调试完毕后，请将调试 Log 取消。

范例 055　用户名长度检测效果

1．实例简介

在我们使用应用的过程中，经常会使用到用户登录的功能，在登录的时候输入用户名，一般要满足一定的要求。例如，用户名的长度要在 5～13 个字符之间，不得包含特殊字符，或者密码的输入框中，一定是要字母和数字的组合，不能全是字母和数字等。一般遇到这样的功能我们都在用户名输入框的后面加一个用户名检测的按钮，当用户单击此按钮的时候就检测输入框的内容是否满足输入要求,根据判断的结果给出用户提示是合法和不合法，本例子就带领大家来实现一个用户名合法性检测的实例。

2. 运行效果

该实例运行效果如图 4.2 所示。

图 4.2 输入用户名后，单击按钮检测用户名的合法性

3. 实例程序讲解

想要实现本实例效果首先修改 res/layout/activity_main.xml 文件，代码如下：

```xml
01 <?xml version="1.0" encoding="utf-8"?>
02 <!-- 定义基础的 LinearLayout 布局 -->
03 <LinearLayout xmlns:android="http://schemas.android.com/apk/res/android"
04     android:layout_width="fill_parent"
05     android:layout_height="fill_parent"
06     android:orientation="vertical" >
07     <!-- 定义 EditText 文本输入框 -->
08     <EditText
09         android:id="@+id/Et"
10         android:layout_width="fill_parent"
11         android:layout_height="wrap_content"
12         android:hint="请输入用户名: "/>
13 
14     <!-- 定义 Button 按钮屏幕区域 -->
15     <Button
16         android:layout_width="fill_parent"
17         android:layout_height="wrap_content"
18         android:onClick="myclick"
19         android:text="单击我,检测用户名合法性! "/>
20 
21     <!-- 定义 TextView 文本标签 -->
22     <TextView
23         android:id="@+id/Tv"
24         android:layout_width="fill_parent"
25         android:layout_height="fill_parent"/>
26 
27 </LinearLayout>
```

这是 Activity 的布局文件。在其中第 15～19 行设置了一个 Button 按钮控件，其中添加了一个 onClick 属性，此属性代表当用户单击此 Button 时的回调方法名字。

然后修改 src/com.wyl.example/MainActivity.java 文件，代码如下：

```
01  //定义MainActivity继承自Activity
02  public class MainActivity extends Activity {
03      //定义TextView对象
04      private TextView Tv;
05      //定义EditText对象
06      private EditText Et;
07
08      @Override
09      public void onCreate(Bundle savedInstanceState) {
10          //当创建此Activity的时候回调
11          super.onCreate(savedInstanceState);
            //设置当前页面的布局xml
12          setContentView(R.layout.activity_main);
            //得到当前Activity中的控件对象
13          findView();
14      }
15
16      private void findView() {
17          // 得到当前布局的控件对象
18          Tv = (TextView)findViewById(R.id.Tv);
19          Et = (EditText)findViewById(R.id.Et);
20      }
21
22      //在xml中绑定的单击调用函数
23      public void myclick(View v){
24          //得到用户输入的用户名，得到长度
25          int len = Et.getText().toString().length();
26          //根据输入的用户名的长度，做出对应的提示
27          if (len > 5 && len < 9) {
                //满足条件显示合法用户
28              Tv.setText("用户名合法");
29          }else{
                //用户名长度不合法
30              Tv.setText("用户名长度非法");
31          }
32      }
33  }
```

此文件是 Activity 的代码文件，其中第 16~20 行得到了布局中的 TextView 对象和 EditText 对象。在第 22~32 行实现了当按钮单击后的回调函数。在此函数中得到 EditText 的文字长度，然后根据用户输入的字符串的长度设置 TextView 的对应信息。

4．实例扩展

在此实例中实现了 Button 按钮的单击回调事件，这里需要注意的一点就是回调函数的格式是固定的，返回值为 void，权限为 public，参数为 View，函数名根据要求定义即可。注意回调函数的格式不同的话，回调函数是无法调用的。

范例 056 打字游戏实现

1．实例简介

在 Android 中我们经常会遇到按键操作的情况。例如，拨电话的按钮负责调出电话的

拨打界面，挂断电话按钮负责挂断电话，按下电源键的时候锁定手机屏幕等。想要实现这些操作就需要当用户按下按钮的时候我们的程序能够得到用户的按钮事件。在本例中我们就利用 Android 的键盘按键操作的回调函数来实现一个打字游戏的界面效果。

2．运行效果

该实例运行效果如图 4.3 所示。

3．实例程序讲解

图 4.3　打字游戏效果

想要实现本例效果，首先要修改 res/layout/activity_main.xml 文件，代码如下：

```
01  <?xml version="1.0" encoding="utf-8"?>
01  <?xml version="1.0" encoding="utf-8"?>
02  <!-- 定义基础的 LinearLayout 布局 -->
03  <LinearLayout xmlns:android="http://schemas.android.com/apk/res/android"
04      android:layout_width="fill_parent"
05      android:layout_height="fill_parent"
06      android:orientation="vertical" >
07
08      <!-- 定义 TextView 文本标签 -->
09      <TextView
10          android:id="@+id/Tv"
11          android:layout_width="fill_parent"
12          android:layout_height="fill_parent"
13          android:text="i have a dream that one day this nation
14          will rise up and live out the true meaning of its creed ,
15          we hold these truths to be self-evident, that all men
16          are created equal."
17          />
18
19  </LinearLayout>
```

这是 Activity 的布局文件，在其中第 8～17 行定义了一个 TextView，在此 TextView 中显示了打字游戏需要用户依次输入的打字内容。

然后修改 src/com.wyl.example/MainActivity.java 文件，代码如下：

```
01  //定义 MainActivity 继承自 Activity
02  public class MainActivity extends Activity {
03    //定义 TextView 对象
04    private TextView Tv;
05    //用户输入的字母个数
06    private int count = 1;
07
08    @Override
09    public void onCreate(Bundle savedInstanceState) {
10        //当创建此 Activity 的时候回调
11        super.onCreate(savedInstanceState);
12        //设置当前页面的布局视图为 activity_main
12        setContentView(R.layout. activity_main);
13        findView();
14    }
15
```

```java
16
17  @Override
18  public boolean onKeyDown(int keyCode, KeyEvent event) {
19      //得到用户所按下的键
20      char ch = Character.toLowerCase(event.getDisplayLabel());
21      //判断用户按下的字母是那个按钮
22      if (ch == Tv.getText().charAt(count-1)
23              && event.getAction() == KeyEvent.ACTION_DOWN) {
24          //文本内容
25          SpannableString ss = new SpannableString(Tv.getText().toString());
26
27          //设置单个的字符颜色
28          ss.setSpan(new ForegroundColorSpan(Color.RED),
29              0, count,Spanned.SPAN_EXCLUSIVE_EXCLUSIVE);
30          ss.setSpan(new StyleSpan(Typeface.BOLD_ITALIC),
31              0, count,Spanned.SPAN_EXCLUSIVE_EXCLUSIVE);
            //记录用户输入的字符的个数
32          count++;
            //在 TextView 中显示修饰以后的字符串
33          Tv.setText(ss);
34      }
35
36      return super.onKeyDown(keyCode, event);
37  }
38
39  private void findView() {
40      // 得到当前布局的控件对象
41      Tv = (TextView)findViewById(R.id.Tv);
42  }
43  }
```

此文件是 Activity 的代码文件，第 39～42 行得到了布局中的 TextView 对象。第 17～37 行实现了当用户单击手机上的按键的回调函数。在此函数中得到了用户所按下的键，然后得到 TextView 中已经显示为红色字体的字母个数，如果相同，那么就说明用户输入正确了，这样就可以将用户输入的字母变成红色了，依次类推直到 TextView 中的文字全部变成红色。

4．实例扩展

在此实例中我们实现了用户打字的效果，现在基本的打字软件都是这种效果了，在此实例的基础之上，只要加上计时功能，我们的打字游戏就可以完成了。当用户第一次输入一个字母的时候开始计时，当最后一个字母变成红色时结束计时，当用户输入完全部的文字后，显示用户总共花费的打字时间，这样一个打字游戏的雏形就完成了，剩下的功能根据你的需要自己进行改造吧。

范例 057　长按播放 TextView 动画

1．实例简介

在我们 Android 应用中为了给用户更好的使用体验，可以加入一些动画，具体动画的内容我们会在今后的章节进行讲解。触发动画的形式多种多样，例如：按返回键退出应用时，会弹出确认动画；按某个按钮的时候，会弹出选中动画等。本例就带领大家实现一个长按键盘上的某个按键播放 TextView 动画的效果。

2．运行效果

该实例运行效果如图 4.4 所示。

3．实例程序讲解

想要实现本例效果首先定义动画的 xml 文件，创建 res/anim/anim.xml 文件，代码如下：

图 4.4　长按 S 键播放动画

```xml
01 <?xml version="1.0" encoding="utf-8"?>
02 <set xmlns:android="http://schemas.android.com/apk/res/android" >
03     <!-- 定义放大缩小的动画 xml -->
04     <scale
05         android:duration="3000"
06         android:fromXScale="0.0"
07         android:fromYScale="0.0"
08         android:interpolator="@android:anim/decelerate_interpolator"
09         android:repeatCount="1"
10         android:startOffset="0"
11         android:toXScale="1.5"
12         android:toYScale="1.5" />
13 
14 </set>
```

在此文件中定义了 TextView 的放大缩小动画。

然后修改 res/layout/activity_main.xml 文件，代码如下：

```xml
01 <?xml version="1.0" encoding="utf-8"?>
02 <!-- 定义基础的 LinearLayout 布局 -->
03 <LinearLayout xmlns:android="http://schemas.android.com/apk/res/android"
04     android:layout_width="fill_parent"
05     android:layout_height="fill_parent"
06     android:orientation="vertical" >
07 
08     <!-- 定义 TextView 文本标签 -->
09     <TextView
10         android:id="@+id/Tv"
11         android:layout_width="fill_parent"
12         android:layout_height="fill_parent"
13         android:text="长按 S 键播放 Tv 的动画"
14         />
15 
16 </LinearLayout>
```

这是 Activity 的布局文件，其中在第 8～14 行定义了会播放动画的 TextView，在其 text 属性中显示了播放动画的提示。

然后修改 src/com.wyl.example/MainActivity.java 文件，代码如下：

```java
01 //定义 MainActivity 继承自 Activity
02 public class MainActivity extends Activity {
03     //定义 TextView 对象
04     private TextView Tv;
```

```
05
06   @Override
07   public void onCreate(Bundle savedInstanceState) {
08       //当创建此 Activity 的时候回调
09       super.onCreate(savedInstanceState);
         //设置当前页面的布局视图为 activity_main
10       setContentView(R.layout.activity_main);
         //得到当前页面中的控件对象
11       findView();
12   }
13
14
15   //按键按下的回调方法
16   @Override
17   public boolean onKeyDown(int keyCode, KeyEvent event) {
18       //要开始事件的追踪器
19       event.startTracking();
20       //返回 true,代表需要继续处理此事件
21       return true;
22   }
23
24
25   //当用户长按键盘上某个按键的时候自动调用
26   @Override
27   public boolean onKeyLongPress(int keyCode, KeyEvent event) {
28       //得到用户长按的键
29       char ch = Character.toLowerCase(event.getDisplayLabel());
30
31       //当用户长按的键是 s 时执行
32       if ('s'== ch) {
33           //通过 AnimationUtils 读取动画 xml
34           Animation scale=AnimationUtils.loadAnimation(MainActivity.this,
             R.anim.anim);
             //设置 TextView 开始动画
35
36           Tv.startAnimation(scale);
37       }
38
39       return super.onKeyLongPress(keyCode, event);
40   }
41
42
43   private void findView() {
44       //得到当前布局的控件对象
45       Tv = (TextView)findViewById(R.id.Tv);
46   }
47   }
```

此文件是当前 Activity 的代码文件,在代码的第 43~46 行通过 findViewById 得到 TextView 的对象。第 16~22 行实现了 onKeyDown 方法,在此方法中没有实现具体的代码内容,只是 return true。在第 27~40 行实现了 Activity 的 onKeyLongPress 函数,当用户长按某个按键的时候调用,在此方法中也可以得到用户长按的按键,然后判断是否为要求的按键,如果是要求的按键就播放 TextView 的动画,否则不播放。

4.实例扩展

在此实例中实现了 onKeyDown 方法,但是没有写具体实现代码,这里强调一下,大

家在实现 onKeyLongPress 方法的时候一定要实现 onKeyDown 方法，而且返回值为 true，否则 onKeyLongPress 方法不会被调用。

范例 058 按钮的快捷键

1．实例简介

在我们使用 Android 应用的时候，有时候希望能够快速的进行一些复杂的操作。例如，一键拨号和一键清理手机内存等功能。本实例就带领大家一起来做一个按钮的快捷键的实现。

2．运行效果

该实例运行效果如图 4.5 所示。

3．实例程序讲解

在如上效果中，单击按钮和单击 A 键会触发相同的单击事件，这样也就实现了按钮的快捷键的功能。

图 4.5 按钮的快捷键

想要实现如上功能，首先要修改 res/layout/activity_main.xml 文件，代码如下：

```xml
01 <?xml version="1.0" encoding="utf-8"?>
02 <!-- 定义基础的 LinearLayout 布局 -->
03 <LinearLayout xmlns:android="http://schemas.android.com/apk/res/android"
04     android:layout_width="fill_parent"
05     android:layout_height="fill_parent"
06     android:orientation="vertical" >
07
08     <!-- 定义 TextView 文本标签 -->
09     <TextView
10         android:layout_width="fill_parent"
11         android:layout_height="wrap_content"
12         android:text="单击按钮或单击 A 键都可触发 Button 的单击事件"
13         />
14
15     <!-- 定义 Button 控件 -->
16     <Button
17         android:id="@+id/Btn"
18         android:layout_width="fill_parent"
19         android:layout_height="wrap_content"
20         android:onClick="myclick"
21         android:text="单击(A)"
22         />
23
24 </LinearLayout>
```

这是 Activity 的布局文件，第 15～22 行在当前布局中添加了一个 Button 控件代表我们要操作的按钮，在此按钮上设置了 onClick 属性，也就是当我们单击此 Button 时会调用 Activity 中的 myclik 方法。

然后修改 src/com.wyl.example/MainActivity.java 文件，代码如下：

```java
01  //定义MainActivity继承自Activity
02  public class MainActivity extends Activity {
03      //定义Button对象
04      private Button Btn;
05  
06      @Override
07      public void onCreate(Bundle savedInstanceState) {
08          //当创建此Activity的时候回调
09          super.onCreate(savedInstanceState);
10          setContentView(R.layout.activity_main);
11          findView();
12      }
13  
14  
15      //按键按下的回调方法
16      @Override
17      public boolean onKeyDown(int keyCode, KeyEvent event) {
18          //得到用户的按键
19          char ch = Character.toLowerCase(event.getDisplayLabel());
20  
21          //当用户长按的键是s时执行
22          if ('a'== ch) {
23              myclick(Btn);            //调用点击时间处理方法myclick
24          }
25          return super.onKeyDown(keyCode, event);
26      }
27  
28      //在xml中绑定的单击调用函数
29      public void myclick(View v){
         //显示Toast的提示信息
30          Toast.makeText(MainActivity.this, "按钮被单击了...", Toast.LENGTH_SHORT).show();
31      }
32  
33  
34      private void findView() {
35          // 得到当前布局的控件对象
36          Btn = (Button)findViewById(R.id.Btn);
37      }
38  }
```

在如上的代码中,第36行得到了布局中的Button按钮,在第28～31行实现此Button按钮的单击处理事件,在第15～26行实现了键盘按键的监听事件,在onKeyDown方法中实现了得到用户的每一次按键的消息,然后判断其是否为我们按钮的快捷键,如果用户按下了按钮的快捷键,那么同样也会调用Button的按键处理函数myclick,通过这种方法就可以实现,不论用户是按下了按钮的快捷键,还是单击按钮,都可以显示相同的操作了。

4. 实例扩展

在此实例中实现了按钮的快捷键的功能,当然同样的原理也可以实现某些功能的快捷键。例如,当我按下数字2的时候,就给某个固定的人打电话。这些功能和我们实例的实现原理是完全相同的,至于如何给某个号码打电话,我们会在后面的章节进行讲解。

范例 059　屏幕单击测试器

1．实例简介

随着手机越来越智能，所以用户与手机交互的手段也越来越多样化，其中最主要的一种形式就是用手指触摸屏幕实现的单击和双击操作，现在也有很多应用基于屏幕触摸操作来吸引用户。例如，打地鼠游戏、屏幕单击的虚拟键盘和短信的屏幕单击输入等。在 Android 手机中越来越多的手机已经转向了使用屏幕操作手机一切功能的方向上来。那么我们本实例就带领大家一起来完成一个屏幕单击测试器，看一下在 Android 中如何处理用户的屏幕单击操作吧。

2．运行效果

该实例运行效果如图 4.6 所示。

图 4.6　屏幕单击测试器

3．实例程序讲解

在上例效果中，当用户单击屏幕时，屏幕上的文字就会改变，改变成你单击位置的屏幕坐标。要想实现这样的效果，首先要修改 res/layout/activity_main.xml 文件，代码如下：

```
01  <?xml version="1.0" encoding="utf-8"?>
02  <!-- 定义基础的 LinearLayout 布局 -->
03  <LinearLayout xmlns:android="http://schemas.android.com/apk/res/android"
04      android:layout_width="fill_parent"
05      android:layout_height="fill_parent"
06      android:orientation="vertical" >
07
08      <!-- 定义 TextView 文本标签 -->
09      <TextView
10          android:id="@+id/Tv"
11          android:layout_width="fill_parent"
12          android:layout_height="wrap_content"
13          android:text="单击屏幕得到相对屏幕的位置"
14          />
15
16  </LinearLayout>
```

这是 Activity 的布局文件，其中第 8～14 行在当前布局中添加了一个 TextView 控件用来显示我们单击的屏幕的位置。

然后修改 src/com.wyl.example/MainActivity.java 文件，通过 onTouchEvent 回调方法得到用户单击屏幕的事件，并得到用户触摸屏幕的位置，然后修改 TextView 的显示内容。主要代码如下：

```
01  //定义 MainActivity 继承自 Activity
02  public class MainActivity extends Activity {
03      //定义 TextView 对象
04      private TextView Tv;
05
```

```
06    @Override
07    public void onCreate(Bundle savedInstanceState) {
08        //当创建此 Activity 的时候回调
09        super.onCreate(savedInstanceState);
          //设置当前页面的布局视图为 activity_main
10        setContentView(R.layout.activity_main);
          //得到当前 Activity 中的控件对象
11        findView();
12    }
13
14    @Override
15    public boolean onTouchEvent(MotionEvent event) {
16        //当按下屏幕的时候，获取单击位置的 x, y
17        if (MotionEvent.ACTION_DOWN == event.getAction()) {
              //得到点击的 x 点坐标
18            float x = event.getX();
              //得到点击的 y 点坐标
19            float y = event.getY();
              //在 TextView 中显示用户点击的 x, y 坐标
20
21            Tv.setText("您单击的位置是：\nx:"+x+"\n y:"+y);
22        }
23
24        return super.onTouchEvent(event);
25    }
26
27    private void findView() {
28        // 得到当前布局的控件对象
29        Tv = (TextView)findViewById(R.id.Tv);
30    }
31 }
```

在此代码中第 29 行通过 findViewById 得到 TextView 对象。在第 14~25 行实现了 Activity 的回调函数 onTouchEvent，在此方法中有一个 event 参数，代表用户的操作事件，通过此对象的 getAction 方法可以得到此用户的操作类型，当用户是触摸屏幕的时候，通过 getX 和 getY 得到用户触摸单击的位置，然后修改 TextView 的内容为屏幕单击的位置信息。这样就实现了本例的效果。

4．实例扩展

对于 MotionEvent 有很多种事件类型，这里使用的是 ACTION_DOWN 类型，也就是当用户按下屏幕时所触发的事件，当然除了此事件外，常见的用户事件类型如下所示。

- ACTION_DOWN：用户按下屏幕的事件。
- ACTION_MOVE：用户滑动的时间。
- ACTION_UP：用户手指从按下状态抬起屏幕的时间。

对于事件来说我们可以通过 getX 和 getY 得到用户触摸单击的位置，当然常见的事件函数还有如下几个。

- getAction 方法：得到操作事件的类型。
- getDownTime 方法：得到用户按下的时间。
- getEventTime 方法：得到用户操作的时间。
- getPressure 方法：得到用户的触摸压力值。

大家可以灵活使用这些事件的类型和事件常用的方法来构造属于自己的触摸事件操作。

范例 060 Activity 内容加载完毕提示

1．实例简介

在之前的例子里我们看到了 Activity 有它自己的生命周期，当到达某一个生命周期的时候，会调用不同的回调方法，但是在生命周期中又无法得到 Activity 需要的内容加载完毕的事件。在我们平时使用应用的时候，有时需要得到本 Activity 数据加载完毕的消息提示。本实例就带领大家一起来做一个当 Activity 需要的数据内容加载完毕的提示信息。

2．运行效果

该实例运行效果如图 4.7 所示。

图 4.7 Activity 内容加载完毕提示

3．实例程序讲解

要想实现这样的效果，首先修改当前页面的布局文件 res/layout/activity_main.xml，代码如下：

```
01  <?xml version="1.0" encoding="utf-8"?>
02  <!-- 定义基础的 LinearLayout 布局 -->
03  <LinearLayout xmlns:android="http://schemas.android.com/apk/res/android"
04      android:layout_width="fill_parent"
05      android:layout_height="fill_parent"
06      android:orientation="vertical" >
07
08      <!-- 定义 TextView 文本标签 -->
09      <TextView
10          android:layout_width="fill_parent"
11          android:layout_height="wrap_content"
12          android:text="页面加载完毕后显示 Toast 提示框"
13          />
14
15  </LinearLayout>
```

在这个布局中的第 9～13 行都定义了 TextView 控件设置了提示文本内容。

然后再修改 src/com.wyl.example/MainActivity.java 文件，代码如下：

```
01  //定义 MainActivity 继承自 Activity
02  public class MainActivity extends Activity {
03
04      @Override
05      public void onCreate(Bundle savedInstanceState) {
06          //当创建此 Activity 的时候回调
07          super.onCreate(savedInstanceState);
08          //设置当前页面的布局视图为 activity_main
08          setContentView(R.layout.activity_main);
09      }
10
11      //当 Activity 的焦点改变的时候自动回调此函数
```

```
12    @Override
13    public void onWindowFocusChanged(boolean hasFocus) {
14        // TODO Auto-generated method stub
15        super.onWindowFocusChanged(hasFocus);
16        //如果是得到焦点
17        if(hasFocus)
18        {
19            //构造 Toast 对象，设置显示的内容
20            Toast t = Toast.makeText(MainActivity.this,
21                "页面加载完毕",Toast.LENGTH_SHORT);
22            //设置 Toast 对象的对齐方式
23            t.setGravity(Gravity.CENTER, 0, 0);
24            //显示 Toast 对象
25            t.show();
26        }
27    }
28    }
```

在此 Java 文件中实现了 onWindowFocusChanged 方法，此方法在 Activity 的焦点改变的时候自动回调，刚好在 Activity 的内容加载完毕后 Activity 得到焦点，会第一次调用此方法，这样就可以实现 Activity 的内容加载完毕后的事件监听效果了。

4．实例扩展

对于 Android 中的 Toast 对象，我们之前简单的介绍过一点。这里我们又用到了 Toast，它主要使用在得不到焦点的情况下，给用户某些提示信息，当然 Toast 也有对齐的方式的，如下所示。

- Gravity.CENTER：居中对齐。
- Grivaty.LEFT：左对齐。
- Grivaty.RIGHT：右对齐。
- Grivaty.TOP：向上对齐。
- Grivaty.END：向下对齐。

大家可以根据自己的需要进行定义自己的 Toast 样式。

范例 061　横竖界面自动切换

1．实例简介

在我们使用 Android 的一些应用的时候经常会见到这样一种功能，就是当用户垂直拿手机的时候界面为纵向显示效果，而把屏幕横过来，界面效果自动切换。例如，手机的拨号界面、手机的短信发送界面和视频的播放界面等。本实例就带领大家一起来做一个随着手机摆放方向不同，界面自动进行切换的应用。

2．运行效果

该实例运行效果如图 4.8 所示。

图 4.8　横竖界面自动切换

3．实例程序讲解

在实例效果中当你的手机纵向放置的时候显示纵向的布局效果，当把你的手机横向放置的时候，本应用界面就会自动切换成横向的布局效果。首先修改 res/layout/activity_main.xml 文件，代码如下：

```
01  <?xml version="1.0" encoding="utf-8"?>
02  <!-- 定义基础的 LinearLayout 布局 -->
03  <LinearLayout xmlns:android="http://schemas.android.com/apk/res/android"
04      android:layout_width="fill_parent"
05      android:layout_height="fill_parent"
06      android:orientation="vertical" >
07  
08      <!-- 定义 TextView 文本标签 -->
09      <TextView
10          android:id="@+id/Tv"
11          android:layout_width="fill_parent"
12          android:layout_height="wrap_content"
13          android:text="我是纵向布局"
14          />
15  
16  </LinearLayout>
```

在上面代码的第 8～14 行定义了 TextView 控件，此控件仅仅是为了显示当前的布局标示。

然后创建 res/layout/ activity_main_horizontal.xml 文件，代码如下：

```
01  <?xml version="1.0" encoding="utf-8"?>
02  <!-- 定义基础的 LinearLayout 布局 -->
03  <LinearLayout xmlns:android="http://schemas.android.com/apk/res/android"
04      android:layout_width="fill_parent"
05      android:layout_height="fill_parent"
06      android:orientation="vertical" >
07  
08      <!-- 定义 TextView 文本标签 -->
09      <TextView
10          android:id="@+id/Tv"
11          android:layout_width="fill_parent"
12          android:layout_height="wrap_content"
13          android:text="我是横向布局"
14          />
15  
16  </LinearLayout>
```

在上面代码的第 8～14 行定义了 TextView 控件，此控件仅仅是为了显示当前的布局标示。

然后修改 src/com.wyl.example/MainActivity.java 文件，代码如下：

```java
01  //定义 MainActivity 继承自 Activity
02  public class MainActivity extends Activity {
03
04      @Override
05      public void onCreate(Bundle savedInstanceState) {
06          //当创建此 Activity 的时候回调
07          super.onCreate(savedInstanceState);
            //设置当前页面的布局视图为 activity_main
08          setContentView(R.layout.activity_main);
09      }
10
11      /**
12       * 屏幕旋转时调用此方法
13       */
14      @Override
15      public void onConfigurationChanged(Configuration newConfig) {
16          super.onConfigurationChanged(newConfig);
17          //newConfig.orientation 获得当前屏幕状态是横向或者竖向
18          //Configuration.ORIENTATION_PORTRAIT 表示竖向
19          //Configuration.ORIENTATION_LANDSCAPE 表示横屏
20          //通过当前切换后的屏幕的方向，设置不同的显示视图
21          if(newConfig.orientation==Configuration.ORIENTATION_PORTRAIT){
                //当竖屏幕的时候提示显示竖屏幕
22              Toast.makeText(MainActivity.this, "现在是竖屏", Toast.LENGTH_SHORT).show();
                //设置当前页面的布局视图为 activity_main
23              setContentView(R.layout.activity_main);
24          }else if(newConfig.orientation==Configuration.ORIENTATION_LANDSCAPE){
                //当竖屏幕的时候提示显示竖屏幕
25              Toast.makeText(MainActivity.this, "现在是横屏", Toast.LENGTH_SHORT).show();
                //设置当前页面的布局视图为 activity_main_horizontal
26              setContentView(R.layout.activity_main_horizontal);
27          }
28
29      }
30  }
```

在此 Activity 的代码中实现了 onConfigurationChanged 方法，当我们的手机屏幕方向改变的时候，会自动调用此方法，所以在方法中通过 newConfig 的 getorientation 方法得到当前的屏幕方向，然后做相应的视图修改就可以了。

4．实例扩展

本实例可以在真实手机上进行测试，也可以在 AVD 上进行测试，当在 AVD 上进行测试的时候，按下 Ctrl+F11 可以进行手机屏幕方向的更改。

范例 062 动态添加联系人列表

1．实例简介

在 Android 中也会遇到这样一种情况，就是一个界面的布局是不固定的，当单击某个

按钮或进行某个操作的时候去修改当前的页面布局。例如，发短信的时候，可以依次添加多个联系人，发邮件的时候也可以，删除列表中的数据的时候也可以连续选中多个进行删除等。这些就要求当我们接收到用户某个操作的时候，我们能够动态的修改页面的布局。本实例就带领大家一起来实现一个动态添加联系人的实例。

2．运行效果

该实例运行效果如图 4.9 所示。

图 4.9　动态添加联系人输入框

3．实例程序讲解

在图 4.9 效果中每当用户单击添加联系人按钮，下方就会多出一个联系人的输入框。要想实现这样的效果，首先修改当前页面的布局文件 res/layout/activity_main.xml，代码如下：

```xml
01  <?xml version="1.0" encoding="utf-8"?>
02  <!-- 定义基础的 LinearLayout 布局 -->
03  <LinearLayout xmlns:android="http://schemas.android.com/apk/res/android"
04      android:id="@+id/Ll"
05      android:layout_width="fill_parent"
06      android:layout_height="fill_parent"
07      android:orientation="vertical" >
08
09      <!-- 定义 Button 对象 -->
10      <Button
11          android:id="@+id/Btn"
12          android:layout_width="fill_parent"
13          android:layout_height="wrap_content"
14          android:text="单击动态添加联系人"
15          />
16
17  </LinearLayout>
18  </LinearLayout>
```

此文件是当前 Activity 的布局文件，在上面代码的第 9～15 行定义了一个 Button 控件，并设置了 id 方便我们下面在 Activity 中取得此对象。

然后再修改 src/com.wyl.example/MainActivity.java 文件，代码如下：

```java
01  //定义 MainActivity 继承自 Activity
02  public class MainActivity extends Activity {
03      //定义 TextView 对象
04      private Button Btn;
05      //定义 LinearLayout 线性布局对象
06      private LinearLayout Ll;
07      //定义 EditText 的数量
08      private int count = 0;
09
10      @Override
11      public void onCreate(Bundle savedInstanceState) {
12          //当创建此 Activity 的时候回调
13          super.onCreate(savedInstanceState);
14          //设置当前页面的布局视图为 activity_main
```

```
14          setContentView(R.layout.activity_main);
            //得到当前页面布局中的视图控件
15          findView();
            //设置视图控件的监听器
16          setListener();
17      }
18
19      private void setListener() {
20          //设置 Button 的监听器
21          Btn.setOnClickListener(new OnClickListener() {
22              @Override
23              public void onClick(View v) {
24                  //添加联系人
25                  addEditText();
26              }
27          });
28      }
29
30      //当 Activity 调用 setContentView 或者 addContentView 时回调
31      @Override
32      public void onContentChanged() {
33          super.onContentChanged();
34          //显示已有的联系人的数量
35          Toast.makeText(MainActivity.this,
36              "已经添加了"+count+"个联系人!",
37              Toast.LENGTH_SHORT).show();
38          count++;
39      }
40
41      private void addEditText(){
42          //初始化一个 EditText 对象,设置一个默认文字内容,hint 值
43          EditText e = new EditText(MainActivity.this);
44          e.setHint("请输入第"+count+"个联系人的信息!");
45
46          //将建立好的 EditText 对象加入 Linearlayout 布局中
47          Ll.addView(e);
48          //设置当前页面的布局是 LinearLayout 对象
49          setContentView(Ll);
50      }
51
52      private void findView() {
53          //得到当前布局的控件对象
54          Btn = (Button)findViewById(R.id.Btn);
55          Ll = (LinearLayout)findViewById(R.id.Ll);
56      }
57  }
```

在如上的代码中第 52～56 行得到了布局中的 Button 对象和整体布局的 LinearLayout 对象方便对布局进行修改。在第 19～28 行定义了给 Button 对象设置了单击事件的监听器并且调用第 41～50 行定义的 addEditText 方法去在 LinearLayout 中添加控件。当 Activity 调用 setContentView 或 addContentView 方法的时候就会修改当前 Activity 的布局,这时候 onContentChanged 方法会被自动回调,在此方法中实现了添加联系人功能,并且增加我们联系人的计数变量。

4．实例扩展

在 Activity 中通过 setContentView 和 addContentView 都可以修改 Activity，但是他们的修改是有区别的，setContentView 是给当前的布局设置一个新的视图，或者布局 id，是用新的布局替换旧的布局；而对于 addContentView 方法来说是在已有的布局上添加一层，新的布局和旧的布局同时存在，而且新的布局覆盖旧的布局。所以大家在使用过程中加以区分，根据自己的情况选择修改布局的方法。

4.2　Android 中基于监听器的事件处理

范例 063　宝宝看图识字软件

1．实例简介

现在 Android 中的幼儿教育的软件越来越多了，其实实现的原理很简单，就是当用户单击某张图片或某种颜色的时候，展示给用户他所单击的位置对应的汉字。当然这类软件也有很多了，例如，看图识颜色、看图识动物和看图识蔬菜等等。本实例就带领大家一起来做一个幼儿教育的简单软件——看图识字软件。

2．运行效果

该实例运行效果如图 4.10 所示。

3．实例程序讲解

在图 4.10 效果中单击三种颜色，或者单击下面三种动物后，上面的文本标签就切换成了你选中的图案的描述。

图 4.10　宝宝看图识字软件

例如，你选中红色后，就会提示你选择了红色；选择了老虎后，会提示你选择了老虎。在本程序中用到了三张资源的图片 elephant.png、rat.png 和 tiger.png，我已经放到了工程的/src/drawable 目录下了。有了这些资源图片，我们就可以开始写代码了。要想实现如图 4.10 所示的效果，首先修改当前页面的布局文件 res/layout/activity_main.xml，在其中添加一个基础的 LinearLayout 布局，然后通过 TextView 和 LinearLayout 布局组合成我们如图 4.10 的样子。代码如下：

```
01  <?xml version="1.0" encoding="utf-8"?>
02  <!-- 定义基础的 LinearLayout 布局 -->
03  <LinearLayout xmlns:android="http://schemas.android.com/apk/res/android"
04      android:layout_width="fill_parent"
05      android:layout_height="fill_parent"
06      android:orientation="vertical" >
07
08      <!-- 定义 TextView 对象 -->
09      <TextView
10          android:id="@+id/Tv"
11          android:layout_width="fill_parent"
```

```xml
12          android:layout_height="wrap_content"
13          android:textSize="20dip"
14          android:text="请选出如下图形: "
15          />
16      <LinearLayout
17          android:layout_width="fill_parent"
18          android:layout_height="wrap_content"
19          android:orientation="horizontal"
20          >
21              <!-- 定义红色的 TextView 对象 -->
22              <TextView
23                  android:id="@+id/TvRed"
24                  android:layout_width="fill_parent"
25                  android:layout_height="100dip"
26                  android:background="@color/red"
27                  android:layout_weight="1"
28                  />
29              <!-- 定义绿色的 TextView 对象 -->
30              <TextView
31                  android:id="@+id/TvGreen"
32                  android:layout_width="fill_parent"
33                  android:layout_height="100dip"
34                  android:background="@color/green"
35                  android:layout_weight="1"
36                  />
37              <!-- 定义蓝色的 TextView 对象 -->
38              <TextView
39                  android:id="@+id/TvBlue"
40                  android:layout_width="fill_parent"
41                  android:layout_height="100dip"
42                  android:background="@color/blue"
43                  android:layout_weight="1"
44                  />
45      </LinearLayout>
46      <LinearLayout
47          android:layout_width="fill_parent"
48          android:layout_height="wrap_content"
49          android:orientation="horizontal"
50          >
51              <!-- 定义老鼠的 TextView 对象 -->
52              <TextView
53                  android:id="@+id/TvRat"
54                  android:layout_width="fill_parent"
55                  android:layout_height="100dip"
56                  android:background="@drawable/rat"
57                  android:layout_weight="1"
58                  />
59              <!-- 定义大象的 TextView 对象 -->
60              <TextView
61                  android:id="@+id/TvElephant"
62                  android:layout_width="fill_parent"
63                  android:layout_height="100dip"
64                  android:background="@drawable/elephant"
65                  android:layout_weight="1"
66                  />
67              <!-- 定义老虎的 TextView 对象 -->
68              <TextView
69                  android:id="@+id/TvTiger"
70                  android:layout_width="fill_parent"
71                  android:layout_height="100dip"
```

```
72                    android:background="@drawable/tiger"
73                    android:layout_weight="1"
74            />
75        </LinearLayout>
76   </LinearLayout>
```

在上面代码的第 22、30、38、52、60 和 68 行分别定义了六个 TextView 控件,这六个 TextView 分别代表红色、绿色、蓝色、老虎、老鼠和大象,然后分别设置对应的 id,方便我们下面在 Activity 中去取得这些控件的对应对象。

然后修改 src/com.wyl.example/MainActivity.java 文件,在其中得到对应的 TextView 控件,并且设置相应的单击监听器,然后根据用户的单击控件提示不同的单击信息,具体代码如下:

```
01   //定义 MainActivity 继承自 Activity
02   public class MainActivity extends Activity {
03       //定义 TextView 对象
04       private TextView Tv;
05       //定义红色的 TextView 对象
06       private TextView TvRed;
07       //定义蓝色的 TextView 对象
08       private TextView TvBlue;
09       //定义绿色的 TextView 对象
10       private TextView TvGreen;
11       //定义老虎的 TextView 对象
12       private TextView TvTiger;
13       //定义老鼠的 TextView 对象
14       private TextView TvRat;
15       //定义大象的 TextView 对象
16       private TextView TvElephant;
17
18       @Override
19       public void onCreate(Bundle savedInstanceState) {
20           //当创建此 Activity 的时候回调
21           super.onCreate(savedInstanceState);
22           setContentView(R.layout.activity_main);
23           findView();
24           setListener();
25       }
26
27       private void setListener() {
28           //设置 TextView 的监听器
29           TvBlue.setOnClickListener(new OnClickListener() {
30               @Override
31               public void onClick(View v) {
32                   //设置对应的文字
33                   Tv.setText("您选择了蓝色!");
34               }
35           });
36           //设置 TextView 的监听器
37           TvRed.setOnClickListener(new OnClickListener() {
38               @Override
39               public void onClick(View v) {
40                   //设置对应的文字
41                   Tv.setText("您选择了红色!");
42               }
43           });
```

```
44          //设置TextView的监听器
45          TvGreen.setOnClickListener(new OnClickListener() {
46              @Override
47              public void onClick(View v) {
48                  //设置对应的文字
49                  Tv.setText("您选择了绿色!");
50              }
51          });
52          //设置TextView的监听器
53          TvTiger.setOnClickListener(new OnClickListener() {
54              @Override
55              public void onClick(View v) {
56                  //设置对应的文字
57                  Tv.setText("您选择了老虎!");
58              }
59          });
60          //设置TextView的监听器
61          TvElephant.setOnClickListener(new OnClickListener() {
62              @Override
63              public void onClick(View v) {
64                  //设置对应的文字
65                  Tv.setText("您选择了大象!");
66              }
67          });
68          //设置TextView的监听器
69          TvRat.setOnClickListener(new OnClickListener() {
70              @Override
71              public void onClick(View v) {
72                  //设置对应的文字
73                  Tv.setText("您选择了老鼠!");
74              }
75          });
76      }
77
78      private void findView() {
79          //得到当前布局的控件对象
80          Tv = (TextView)findViewById(R.id.Tv);
81          TvBlue = (TextView)findViewById(R.id.TvBlue);
82          TvRed = (TextView)findViewById(R.id.TvRed);
83          TvGreen = (TextView)findViewById(R.id.TvGreen);
84          TvElephant = (TextView)findViewById(R.id.TvElephant);
85          TvTiger = (TextView)findViewById(R.id.TvTiger);
86          TvRat = (TextView)findViewById(R.id.TvRat);
87      }
88  }
```

在如上代码中的第78~87行得到了对应的七个TextView对象，分别代表红色、蓝色、绿色、老虎、大象、老鼠和结果TextView。在第27~76行分别设置了如上六个TextView的单击事件，根据不同的控件的单击，在结果TextView中显示不同的结果信息。这样就能够实现类似看图识字的效果了。

4．实例扩展

对于监听器来说，本例都是采用匿名内部类的对象的方式来实现的，其实如果在一个Activity中有多个onClickListener对象的话，可以自己实现这个类的对象，然后把所有的控件的onClickListener监听器都设置为同一个监听器，然后在onClick方法的View参数中去判断到底单击的是哪个 View。示例代码如下所示：

```
01    OnClickListener myonclicklistener = new OnClickListener() {
02        @Override
03        public void onClick(View v) {
04            switch (v.getId()) {
05            case R.id.Tv:
06                //当用户点击 Tv 控件的时候的处理逻辑
07                break;
08            case R.id.TvTiger:
09                //当用户点击老虎 Tv 控件的时候的处理逻辑
10                break;
11                //当用户点击其他 Tv 控件的时候的处理逻辑
12                ...
13            default:
14                break;
15            }
16        }
17    };
```

范例 064　控件的拖动效果

1．实例简介

在 Android 中可以要求我们实现的控件按照我们的要求来移动，包括可以用手指进行控件的拖动。例如，对于有些改版的 Android 系统来说你想要操作桌面上的应用程序的话，你只要用手指把这个应用的图标拖动到指定区域就可以删除此应用了。本实例就带领大家一起来做一个用手指拖动控件的应用。

2．运行效果

该实例运行效果如图 4.11 所示。

3．实例程序讲解

在图 4.11 效果中我们可以通过手指将上面方块中的内容拖动出来，任意移动，当把上面方块的内容拖动至下面方块上方的时候，松手后两个方块的颜色会进行交换。要想实现这样的效果，步骤如下所示。

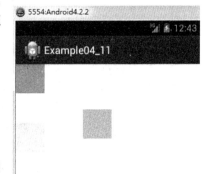

图 4.11　控件的拖动效果

（1）在此实例中用到了两个自定义的 View 控件，一个为被拖动控件，一个为被覆盖控件，首先创建 src/com.wyl.example/AreaOne.java 文件，代码如下：

```
01    //定义一个 View 类的子类，作为被拖动 View
02    public class AreaOne extends View {
03        //当前 View 的构造方法
04        public AreaOne(Context context) {
05            super(context);        //当前 View 的构造方法
06        }
07        //当前 View 的带属性参数的构造方法
08        public AreaOne(Context context, AttributeSet attrs) {
```

```
09         super(context, attrs);  //当前 View 的构造方法
10     }
11
12 }
```

创建 src/com.wyl.example/AreaTwo.java 文件，代码如下：

```
01 //定义一个 View 类的子类，作为放置拖动 View 的 View
02 public class AreaTwo extends View {
03     //当前 View 的构造方法
04     public AreaTwo(Context context) {
05         super(context);
06     }
07     //当前 View 的带属性参数的构造方法
08     public AreaTwo(Context context, AttributeSet attrs) {
09         super(context, attrs);
10     }
11
12 }
```

（2）定义界面的布局，修改 res/layout/activity_main.xml 布局文件，代码如下：

```
01 <?xml version="1.0" encoding="utf-8"?>
02 <!-- 定义基础的 LinearLayout 布局 -->
03 <LinearLayout xmlns:android = "http://schemas.android.com/apk/res/android"
04     android:orientation = "vertical"
05     android:layout_width = "fill_parent"
06     android:layout_height = "fill_parent"
07     >
08     <!-- 定义 com.wyl.example.AreaOne 对象 -->
09     <com.wyl.example.AreaOne
10         android:id = "@+id/ViewOne"
11         android:layout_width = "50dip"
12         android:layout_height = "50dip"
13         android:background = "#FFFF00"
14         />
15     <!-- 定义空白区域 -->
16     <TextView
17         android:layout_width = "50dip"
18         android:layout_height = "50dip"
19         />
20     <!-- 定义 com.wyl.example.AreaTwo 对象 -->
21     <com.wyl.example.AreaTwo
22         android:id = "@+id/ViewTwo"
23         android:layout_width = "50dip"
24         android:layout_height = "50dip"
25         android:background = "#00FF00"
26         />
27
28 </LinearLayout>
```

其中第 8~14 行定义了 AreOne 的布局，在第 20~26 行定义了自定义的 AreaTwo 的布局。

（3）在 Java 文件中获取相应的控件对象，然后设置 setOnLongClickListener 监听器，并且设置了对象的 setOnDragListener 监听拖动的事件。修改 src/com.wyl.example/MainActivity.java 文件，代码如下：

```
001 //定义 MainActivity 继承自 Activity
```

```java
002 public class MainActivity extends Activity {
003     public View ViewOne;
004     public View ViewTwo;
005
006     @Override
007     public void onCreate(Bundle savedInstanceState) {
008         super.onCreate(savedInstanceState);
009         // 设置当前页面的布局
010         setContentView(R.layout.activity_main);
011         //得到当前视图中的试图对象
012         findView();
            //设置当前页面中视图的监听器
013         setListener();
014     }
015
016     private void setListener() {
017         //设置 ViewOne 和 ViewTwo 的监听器
018         ViewOne.setOnLongClickListener(new OnLongClickListener() {
019             public boolean onLongClick(View view) {
020                 // 长按 AreaOne 后开始拖动
021                 ViewOne.startDrag(null, new DragShadowBuilder(view),
022                         (Object) view, 0);
023                 return true;
024             }
025         });
026
027         ViewTwo.setOnDragListener(new OnDragListener() {
028             //开始拖动时的回调事件
029             public boolean onDrag(View mDropView, DragEvent event) {
030                 boolean result = false;
031
032                 switch (event.getAction()) {
033                 //开始拖动的 action
034                 case DragEvent.ACTION_DRAG_STARTED: {
035
036                     Log.e("OnDragListener", "View 开始被拖动!");
037                     /**
038                      *在拖动开始时,只有返回true,后面的动作(ACTION_DRAG_
039                      *ENTERED,ACTION_DRAG_LOCATION,ACTION_DROP)才会被执行!
040                      */
041                     result = true;
042
043                     break;
044                 }
045                 //被拖动的 View 进入当前 View 时被调用
046                 case DragEvent.ACTION_DRAG_ENTERED: {
047                     Log.e("OnDragListener", "被拖动的 View 进入当前 View!");
048
049                     break;
050                 }
051                 //被拖动的 View 进入当前 View 后,位置改变时被回调
052                 case DragEvent.ACTION_DRAG_LOCATION: {
053                     Log.e("OnDragListener", "被拖动的 View 进入当前 View
                             后,位置发生改变!");
054
055                     break;
056                 }
057                 //拖动的 View 被放入当前 View 时被回调
```

```
058                    case DragEvent.ACTION_DROP: {
059
060                        Log.e("OnDragListener","拖动的View被放入当前View!");
061
062                        /**
063                         * 松开拖动时,交换View的背景
064                         */
065                        View mDragView = (View) event.getLocalState();
                            //得到拖动视图的背景
066                        Drawable mDragViewBackgroud = mDragView.getBackground();
                            //设置拖动的View的背景
067                        mDragView.setBackgroundDrawable(mDropView.
                            getBackground());
068                        mDropView.setBackgroundDrawable(mDragViewBackgroud);
069
070                        break;
071                    }
072                    //拖动结束时被回调
073                    case DragEvent.ACTION_DRAG_ENDED: {
074                        Log.e("OnDragListener", "拖动结束!");
075                        //拖动结束时调用的分支
076                        break;
077                    }
078                    //拖动完成时被回调
079                    case DragEvent.ACTION_DRAG_EXITED: {
080                        Log.e("OnDragListener", "拖动退出!");
081                        //拖动完成时走入的分支
082                        break;
083                    }
084
085                    default: {
086
087                        break;
088                    }
089                }
090                return result;
091            }
092        });
093    }
094
095    //得到布局中的控件对象
096    private void findView() {
097        ViewOne = findViewById(R.id.ViewOne);
098        ViewTwo = findViewById(R.id.ViewTwo);
099    }
100 }
```

在上面代码的第 95~99 行获得自定义 View 控件。在第 18 行为 AreaOne 控件设置长按监听器。在第 27 行为 AreaTwo 设置了拖动监听器,并且实现了 onDrag 回调方法,在此方法中实现了拖动控件和移动覆盖的逻辑。

4. 实例扩展

在 Android 中所有的 View 都可以进行自定义,所以可以实现一些自由度比较大的功能,大家在遇到某些界面需要设计和实现的时候,一定要多多关注 Android 官方的 API,多动脑筋找出用户需求的效果的解决方案。

范例 065 Email 格式的检测

1. 实例简介

在我们使用 Android 应用的时候最常用的一个功能就是登录注册，其中免不了要使用邮箱的输入。但是对于邮箱来说又有着相对复杂的规则，所以我们经常会看到这样的效果，在邮箱的输入框后面有邮箱格式检测的图片，当用户输入的邮箱为合理邮箱时显示正确的图案，否则显示错误的图案。本实例就带领大家一起来做一个 Email 格式的检测程序，随着用户的输入去检测输入的数据是否符合邮箱的规则然后显示正确的提示图案。

2. 运行效果

该实例运行效果如图 4.12 所示。

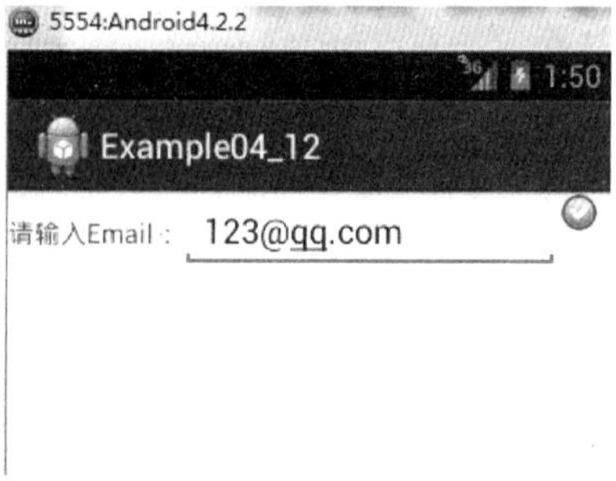

图 4.12 Email 格式的检测效果

3. 实例程序讲解

在实现本例效果的过程中需要两张图片 right.png 和 wrong.png，我已经提前复制到了 /res/drawable 目录下。然后修改 res/layout/activity_main.xml 文件，代码如下：

```xml
01 <?xml version="1.0" encoding="utf-8"?>
02 <!-- 定义基础的 LinearLayout 布局 -->
03 <LinearLayout xmlns:android="http://schemas.android.com/apk/res/android"
04     android:layout_width="fill_parent"
05     android:layout_height="fill_parent"
06     android:orientation="horizontal" >
07     <!-- 定义 Email 的提示文本标签 -->
08     <TextView
09         android:id="@+id/showInfo"
10         android:layout_width="wrap_content"
11         android:layout_height="wrap_content"
12         android:text="请输入 Email: " />
13     <!-- 定义 Email 的文本输入框 -->
14     <EditText
15         android:id="@+id/inputInfo"
16         android:layout_width="200dip"
17         android:layout_height="wrap_content"
18         android:selectAllOnFocus="true"/>
19     <!-- 定义 Email 的校验的图片 -->
20     <ImageView
21         android:id="@+id/showImg"
22         android:layout_width="wrap_content"
23         android:layout_height="wrap_content"/>
24
25 </LinearLayout>
```

这是当前 Activity 的布局文件，在如上代码的第 8 行建立了一个 TextView 节点，在第 14 行建立了一个 Email 的输入框，在第 20 行建立了一个 ImageView 节点来代表输入检测的结果图片。

然后修改 src/com.wyl.example/MainActivity.java 文件，代码如下：

```
01 //定义MainActivity继承自Activity
02 public class MainActivity extends Activity {
03 //输入框,email的输入框
04     private EditText Tvinput = null;
05 //显示正确与错误的图片控件
06     private ImageView Iv = null;
07
08     @Override
09     public void onCreate(Bundle savedInstanceState) {
10         super.onCreate(savedInstanceState);
            //设置当前页面的布局视图为activity_main
11         setContentView(R.layout.activity_main);
            //得到当前页面中的视图控件对象
12         findView();
            //设置视图对象的监听器
13         setListener();
14     }
15
16 private void setListener() {
17     //为输入框组件绑定键盘事件
18     Tvinput.setOnKeyListener(new View.OnKeyListener() {
19
20         @Override
21         public boolean onKey(View v, int keyCode, KeyEvent event) {
22             switch (event.getAction()) {         //得到操作类型
23             case KeyEvent.ACTION_DOWN:           //键盘按下
24                 String inputString = Tvinput.getText().toString();
                                                    //获得输入框的内容
25                 if (inputString.matches("\\w+@\\w+\\.\\w+")) {
                                                    //验证通过
26                     Iv.setImageResource(R.drawable.right);
                                                    //设置为正确的图片
27                 }else {
28                     Iv.setImageResource(R.drawable.wrong);
                                                    //设置为错误的图片
29                 }
30                 break;
31             case KeyEvent.ACTION_UP:             //键盘弹起
32
33                 break;
34             }
35             return false;
36         }
37     });
38 }
39
40 private void findView() {
41     // 获得输入框控件
42     Tvinput = (EditText) findViewById(R.id.inputInfo);
43     // 获得图片控件
44     Iv = (ImageView) findViewById(R.id.showImg);
45 }
46 }
```

在此代码中的第 40～45 行得到了布局文件中的 EditText 对象和 ImageView 对象，在代码的第 18 行给 EditText 设置了 OnKeyListener，当在 EditText 中每次输入一个字母都会回调此 onKey 方法，然后将得到输入的 Email 字符串进行正则表达式匹配，如果匹配成功，显示正确匹配的图片，否则显示不满足 Email 规则的提示图片。

4．实例扩展

在 OnKeyListener 中也可以得到用户操作的事件的对象，然后根据事件的对象的 action 类型来判断用户的操作是否处理。

范例 066　隐藏导航栏

1．实例简介

在 Android 3.0 以后，Google 提出了要加强 Android 手机的虚拟按键技术，弱化物理按键，所以基本从 Android 3.0 版本后，大部分的手机都采用虚拟键盘的形式来显示菜单键、返回键和 Home 键，我们把这些按钮显示的位置叫做导航栏。但是针对一些应用的场合，导航栏却给我们带来了不大不小的麻烦。例如，当视频播放的时候我们希望导航栏隐藏，当我们操作应用的时候我们希望导航栏显示。本实例就带领大家一起来做一个控制导航栏显示和隐藏的效果。

2．运行效果

该实例运行效果如图 4.13 所示。

3．实例程序讲解

在图 4.13 效果中，当用户单击屏幕的时候，虚拟导航栏隐藏，再次单击屏幕时，虚拟导航栏显示。要想实现这样的效果，需要修改 res/layout/activity_main.xml 文件，代码如下：

图 4.13　隐藏导航栏

```
01  <?xml version="1.0" encoding="utf-8"?>
02  <!-- 定义基础的 LinearLayout 布局 -->
03  <LinearLayout xmlns:android="http://schemas.android.com/apk/res/android"
04      android:layout_width="fill_parent"
05      android:layout_height="fill_parent"
06      android:orientation="horizontal" >
07      <!-- 定义 TextView 的文本标签 -->
08      <TextView
09          android:id="@+id/Tv"
10          android:layout_width="fill_parent"
11          android:layout_height="fill_parent"
12          android:text="单击屏幕，隐藏或显示导航栏！" />
13
14  </LinearLayout>
```

在如上代码的第 8 行定义了一个 TextView，此控件仅仅显示一个文本标签提示的作用。

然后修改 src/com.wyl.example/MainActivity.java 文件，代码如下：

```java
01  //定义 MainActivity 继承自 Activity
02  public class MainActivity extends Activity {
03  //获得系统的 TextView 文本标签对象
04      private TextView Tv;
05      //获得当前页面的根 View 对象
06  private View rootView;
07
08  @Override
09      public void onCreate(Bundle savedInstanceState) {
10          super.onCreate(savedInstanceState);
            //设置当前页面的布局视图为 activity_main
11          setContentView(R.layout.activity_main);
            //得到当前页面的视图控件
12          findView();
13          //设置视图控件的监听器
14          setListener();
15      }
16
17  private void setListener() {
18      //设置 Textview 的单击监听器
19      Tv.setOnClickListener(new OnClickListener() {
20          @Override
21          public void onClick(View paramView) {
22              //得到当前根 View 的显示状态
23              int i = rootView.getSystemUiVisibility();
24              Log.e("setOnClickListener", i+"");
25
26              //如果当前导航栏显示，就设置为隐藏
27              if (i == View.SYSTEM_UI_FLAG_VISIBLE) {
                    //设置当前根视图的显示形式
28                  rootView.setSystemUiVisibility(
29                      View.SYSTEM_UI_FLAG_HIDE_NAVIGATION);
30              }
31          }
32      });
33
34      rootView.setOnSystemUiVisibilityChangeListener(
35              new OnSystemUiVisibilityChangeListener() {
36          //当状态栏的状态改变的时候，回调此方法
37          @Override
38          public void onSystemUiVisibilityChange(int visibility) {
39              // TODO Auto-generated method stub
40              Log.e("onSystemUiVisibilityChange", visibility+"");
41
42              //根据当前 View 的显示状态进行不同的提示
43              if (visibility == View.VISIBLE) {
44                  Toast.makeText(MainActivity.this,
45                      "显示虚拟导航按钮！",
46                      Toast.LENGTH_SHORT).show();
47              }
48              else{
49                  Toast.makeText(MainActivity.this,
50                      "隐藏虚拟导航按钮！",
51                      Toast.LENGTH_SHORT).show();
52              }
53          }
54      });
```

```
55  }
56
57  private void findView() {
58      //得到当前窗口的根 View
59      rootView=getWindow().getDecorView();
60      //设置根 Viewd 的导航栏为隐藏
61      rootView.setSystemUiVisibility(
62              View.SYSTEM_UI_FLAG_HIDE_NAVIGATION);
63      //得到当前页面的 TextView 对象
64      Tv = (TextView)findViewById(R.id.Tv);
65  }
66  }
```

在如上代码中第 58~64 行得到了当前窗口的根 View,然后设置了当前系统 UI 的显示形式为隐藏导航栏,但是当用户单击屏幕的时候导航栏会自动的显示出来,这时候我们通过第 19 行设置 TextView 的单击事件来实现切换系统 UI 的显示模式,这样就可以实现控制导航栏显示与否了。

4．实例扩展

对于本实例来说在 AVD 上无法看出效果,因为我们的 AVD 都是没有虚拟导航栏的,大家可以在一些真机上进行测试。例如,三星 Nexus S、三星 Galaxy Nexus 或任何具有虚拟按键导航栏的手机上进行测试效果。

范例 067 屏幕多点触摸测试器

1．实例简介

随着智能手机的逐步发展,人们对于智能手机的要求越来越高,现在我们对于手机的操作已经全部都通过手指的触摸屏幕来实现了,当然在其中不单纯是单点触摸,用户还有一些更加复杂的手指触摸动作。例如,滑动和多点触摸。本实例就带领大家一起开发一个 Android 的多点触摸测试应用。

2．运行效果

该实例运行效果如图 4.14 所示。

3．实例程序讲解

在图 4.14 的效果中,当用户通过多根手指触摸屏幕时会显示以手指触摸屏幕的中心点为圆心的一个实心圆形,而且在屏幕的上方显示每根手指的触摸的屏幕位置。要想实现本实例这样的效

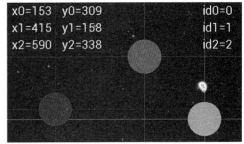

图 4.14 屏幕多点触摸测试器

果,首先需要定义一个属于我们自己的 View,新建 src/com.wyl.example/MyView.java 文件,代码如下:

```
001  //自定义View继承自surfaceView类,并且实现了surfaceholder类中的callback接口
002  public class MyView extends SurfaceView implements SurfaceHolder.Callback {
003      //最多的触摸点数量
```

```
004     private static final int MAX_TOUCHPOINTS = 10;
005     //提示的问题
006     private static final String START_TEXT = "请单点或多点触摸屏幕进行测试";
007     //文字画笔
008     private Paint textPaint = new Paint();
009     //圆形画笔
010     private Paint touchPaints[] = new Paint[MAX_TOUCHPOINTS];
011     //对应每一个圆形画笔的颜色
012     private int colors[] = new int[MAX_TOUCHPOINTS];
013
014     //记录屏幕的宽度和高度
015     private int width, height;
016     //放大的倍数
017     private float scale = 1.0f;
018
019     public MyView(Context context) {
020         super(context);
021         //得到当前的view的surfaceholder对象
022         SurfaceHolder holder = getHolder();
023         //设置当前holder的回调方法
024         holder.addCallback(this);
025         // 确保我们的View能获得输入焦点
026         setFocusable(true);
027         // 确保能接收到触屏事件
028         setFocusableInTouchMode(true);
029         //初始化颜色值
030         init();
031     }
032
033     private void init() {
034         // 初始化文字笔的颜色
035         textPaint.setColor(Color.WHITE);
036         //定义十种按键的颜色值
037         colors[0] = Color.BLUE;
038         colors[1] = Color.RED;
039         colors[2] = Color.GREEN;
040         colors[3] = Color.YELLOW;
041         colors[4] = Color.CYAN;
042         colors[5] = Color.MAGENTA;
043         colors[6] = Color.DKGRAY;
044         colors[7] = Color.WHITE;
045         colors[8] = Color.LTGRAY;
046         colors[9] = Color.GRAY;
047         //分别初始化每个手指的颜色值的笔
048         for (int i = 0; i < MAX_TOUCHPOINTS; i++) {
049             touchPaints[i] = new Paint();
050             touchPaints[i].setColor(colors[i]);
051         }
052     }
053
054     /*
055      * 处理触屏事件
056      */
057     @Override
058     public boolean onTouchEvent(MotionEvent event) {
059         // 获得屏幕触点数量
060         int pointerCount = event.getPointerCount();
061         if (pointerCount > MAX_TOUCHPOINTS) {
```

```
062             pointerCount = MAX_TOUCHPOINTS;
063         }
064         // 锁定 Canvas，开始进行相应的界面处理
065         Canvas c = getHolder().lockCanvas();
066         if (c != null) {
067             //定义 canvas 的背景颜色值为黑色
068             c.drawColor(Color.BLACK);
069             if (event.getAction() == MotionEvent.ACTION_UP) {
070                 // 当手离开屏幕时，清屏
071             } else {
072                 //先在屏幕上画一个十字，横向贯穿屏幕，纵向贯穿屏幕
073                 for (int i = 0; i < pointerCount; i++) {
074                     // 获取一个触点的坐标，然后开始绘制
075                     int id = event.getPointerId(i);
076                     int x = (int) event.getX(i);
077                     int y = (int) event.getY(i);
078                     drawCrosshairsAndText(x, y,
079                         touchPaints[id], i, id, c);
080                 }
081                 //使用不同的颜色在每个手指的位置画圆
082                 for (int i = 0; i < pointerCount; i++) {
083                     int id = event.getPointerId(i);
084                     int x = (int) event.getX(i);
085                     int y = (int) event.getY(i);
086                     drawCircle(x, y, touchPaints[id], c);
087                 }
088             }
089             // 画完后，解锁显示
090             getHolder().unlockCanvasAndPost(c);
091         }
092         return true;
093     }
094
095     /**
096      * 画交叉十字线及坐标信息
097      *
098      * @param x: 线的 x 坐标
099      * @param y: 线的 y 坐标
100      * @param paint: 线的颜色
101      * @param ptr: 第几个点
102      * @param id: id 值
103      * @param c: 画布
104      */
105     private void drawCrosshairsAndText(int x, int y, Paint paint, int ptr,
106         int id, Canvas c) {
107         //在（0, y）和（width, y）这两个点上画直线
108         c.drawLine(0, y, width, y, paint);
109         //在（x, 0）和（x, height）这两个点上画直线
110         c.drawLine(x, 0, x, height, paint);
111         //定义文字的大小
112         int textY = (int) ((15 + 20 * ptr) * scale);
113         //画出 x 的值
114         c.drawText("x" + ptr + "=" + x,
115             10 * scale, textY, textPaint);
116         //画出 y 的值
117         c.drawText("y" + ptr + "=" + y,
118             70 * scale, textY, textPaint);
```

```java
119         //画出id的值
120         c.drawText("id" + ptr + "=" + id,
121             width - 55 * scale, textY, textPaint);
122     }
123
124     /**
125      * 画手指单击的实心圆
126      *
127      * @param x: 实心圆的x值
128      * @param y: 实心圆的y值
129      * @param paint: 实心圆的画笔
130      * @param c: 在这个画布上画
131      */
132     private void drawCircle(int x, int y, Paint paint, Canvas c) {
133         //在canvas上画圆
134         c.drawCircle(x, y, 20 * scale, paint);
135     }
136
137     /*
138      * 进入程序时背景画成黑色,然后把"START_TEXT"写到屏幕
139      */
140     public void surfaceChanged(SurfaceHolder holder, int format, int width,
141             int height) {
142         //得到屏幕的宽度
143         this.width = width;
144         //得到屏幕的高度
145         this.height = height;
146         //得到屏幕的放大比例
147         if (width > height) {
148             this.scale = width / 480f;
149         } else {
150             this.scale = height / 480f;
151         }
152         //通过放大比例计算出字体大小
153         textPaint.setTextSize(14 * scale);
154         //得到当前View的holder对象
155         Canvas c = getHolder().lockCanvas();
156         //设置背景为黑色
157         if (c != null) {
158             // 背景黑色
159             c.drawColor(Color.BLACK);
160             //在屏幕中间画上提示语
161             float tWidth = textPaint.measureText(START_TEXT);
162             c.drawText(START_TEXT,
163                 width / 2 - tWidth / 2, height / 2,
164                 textPaint);
165             //解锁显示
166             getHolder().unlockCanvasAndPost(c);
167         }
168     }
169
170     public void surfaceCreated(SurfaceHolder holder) {
171     }
172
173     public void surfaceDestroyed(SurfaceHolder holder) {
174     }
175 }
```

在此 Activity 文件中第 19 行实现了 MyView 的构造方法，其中初始化了当前 View 的一系列的初始值，在第 33 行定义 init 函数用来初始化画实心圆的笔的颜色。在第 58 行实现了当前 View 的触摸事件监听器，在此监听器中得到手指触摸屏幕的数量并且开始画当前页面，包括先画十字、再画实心圆和再画左上角每个手指的位置。在第 140 行得到屏幕的宽度、高度来计算和画笔的字体的大小。

然后修改我们 Activity 的代码文件 src/com.wyl.example/MainActivity.java，代码如下：

```
01  //定义 MainActivity 继承自 Activity
02  public class MainActivity extends Activity {
03  @Override
04  public void onCreate(Bundle savedInstanceState) {
05      super.onCreate(savedInstanceState);
06      // 隐藏标题栏
07      requestWindowFeature(Window.FEATURE_NO_TITLE);
08      // 设置成全屏
09      getWindow().setFlags(WindowManager.LayoutParams.FLAG_FULLSCREEN,
10              WindowManager.LayoutParams.FLAG_FULLSCREEN);
11      // 设置为上面的 MyView 为当前的页面对象
12      setContentView(new MyView(this));
13  }
14  }
```

在如上代码的第 12 行新建了一个 MyView 类的对象，并且设置当前页面展示 MyView 类的对象。

4．实例扩展

对于 Android 来说它是一个非常灵活的框架，大家可以通过自定义的 View 来实现一些意想不到的功能。

范例 068 图片的平移、缩放和旋转

1．实例简介

人们对于 Android 手机的多点触摸的使用要求越来越高，多点触摸最经典的代表应用就是图片的查看，当用户打开一张图片时，图片的大小为原始大小，用户可以通过单指点中图片，然后进行拖动平移，也可以进行双指同时向外滑动的操作来放大和缩小图片，甚至可以两指点中图片后按照一定角度进行顺时针和逆时针的旋转。本实例就带领大家一起来通过手指的单点和多点触摸操作实现一个图片平移、放大、缩小和旋转的效果。

2．运行效果

该实例运行效果如图 4.15 所示。

3．实例程序讲解

本实例的效果就是用户可以通过单指点中图片，然

图 4.15 图片的平移、缩放和旋转

后进行拖动平移，也可以进行双指同时向外滑动的操作，可以放大和缩小图片，甚至可以两指点中图片后按照一定角度进行顺时针和逆时针的旋转。要想实现这样的效果，首先需要自定义文件 src/com.wyl.example/MyView.java 文件，代码如下：

```java
001  //自定义 MyView 类继承自 ImageView
002  public class MyView extends ImageView {
003      private float x_down = 0;
004      private float y_down = 0;
005      //起始点的坐标
006      private PointF start = new PointF();
007      //中心点的坐标
008      private PointF mid = new PointF();
009      private float oldDist = 1f;                     //原始距离
010      private float oldRotation = 0;                  //旋转角度
011      private Matrix matrix = new Matrix();           //矩阵对象
012      private Matrix matrix1 = new Matrix();
013      private Matrix savedMatrix = new Matrix();
014
015      private static final int NONE = 0;
016      private static final int DRAG = 1;
017      private static final int ZOOM = 2;
018      private int mode = NONE;
019
020      private boolean matrixCheck = false;
021
022      //记录当前屏幕的宽度
023      private int widthScreen;
024      //记录当前屏幕的高度
025      private int heightScreen;
026
027      //在页面中显示的 Bitmap 图片
028      private Bitmap kenan;
029
030      public MyView(Activity activity) {
031          super(activity);
032          //通过 Bitampfactory 读取 drawable 目录下的 kenan 资源
033          kenan = BitmapFactory.
034              decodeResource(getResources(), R.drawable.kenan);
035
036          //定义图片一个显示矩阵
037          DisplayMetrics dm = new DisplayMetrics();
038          //得到当前屏幕的显示矩阵存入 dm 变量
039          activity.getWindowManager().
040              getDefaultDisplay().getMetrics(dm);
041          //通过显示矩阵得到当前屏幕的宽度和高度的像素值
042          widthScreen = dm.widthPixels;
043          heightScreen = dm.heightPixels;
044
045          matrix = new Matrix();
046      }
047
048      //显示 view 的时候回调 onDraw
049      protected void onDraw(Canvas canvas) {
050          //首先保存当前页面已有的图像
051          canvas.save();
052          //按照当前的矩阵绘制 kenan 图片
053          canvas.drawBitmap(kenan, matrix, null);
```

```
054         //画图板恢复
055         canvas.restore();
056     }
057
058     //当用户触摸此视图的时候回调此方法
059     public boolean onTouchEvent(MotionEvent event) {
060         //得到touch的事件类型
061         switch (event.getAction() & MotionEvent.ACTION_MASK) {
062         case MotionEvent.ACTION_DOWN:
063             //当按下屏幕时,记录当前的状态为拖动
064             mode = DRAG;
065             //记录xy坐标
066             x_down = event.getX();
067             y_down = event.getY();
068             //保存当前的矩阵
069             savedMatrix.set(matrix);
070             break;
071         case MotionEvent.ACTION_POINTER_DOWN:
072             //多个手指触摸的状态
073             mode = ZOOM;
074             //记录之前的两手指间距
075             oldDist = spacing(event);
076             //记录之前的角度
077             oldRotation = rotation(event);
078             //保存当前的图片矩阵
079             savedMatrix.set(matrix);
080             //得到旋转的中心点
081             midPoint(mid, event);
082             break;
083         case MotionEvent.ACTION_MOVE:
084             //当手指移动时的状态
085             if (mode == ZOOM) {
086                 //缩放并且平移
087                 matrix1.set(savedMatrix);
088                 //得到旋转的角度
089                 float rotation =
090                         rotation(event) - oldRotation;
091                 //得到距离
092                 float newDist = spacing(event);
093                 //得到放大倍数
094                 float scale = newDist / oldDist;
095                 //缩放倍数
096                 matrix1.postScale(scale, scale, mid.x, mid.y);
097                 //得到旋转角度
098                 matrix1.postRotate(rotation, mid.x, mid.y);
099                 //得到图片是否出边界
100                 matrixCheck = matrixCheck();
101                 if (matrixCheck == false) {
102                     matrix.set(matrix1);
103                     invalidate();
104                 }
105             } else if (mode == DRAG) {
106                 //平行移动
107                 matrix1.set(savedMatrix);
108                 matrix1.postTranslate(event.getX() - x_down
109                         , event.getY() - y_down);         //平移
110                 matrixCheck = matrixCheck();
111                 matrixCheck = matrixCheck();
```

```
112                if (matrixCheck == false) {
113                    matrix.set(matrix1);
114                    invalidate();
115                }
116            }
117            break;
118        case MotionEvent.ACTION_UP:
119        case MotionEvent.ACTION_POINTER_UP:
120            mode = NONE;
121            break;
122        }
123        return true;
124    }
125
126    //对图片的矩阵进行检测
127    private boolean matrixCheck() {
128        float[] f = new float[9];
129        matrix1.getValues(f);
130        // 图片 4 个顶点的坐标
131        float x1 = f[0] * 0 + f[1] * 0 + f[2];
132        float y1 = f[3] * 0 + f[4] * 0 + f[5];
133        float x2 = f[0] * kenan.getWidth()
134                + f[1] * 0 + f[2];
135        float y2 = f[3] * kenan.getWidth()
136                + f[4] * 0 + f[5];
137        float x3 = f[0] * 0 + f[1] *
138                kenan.getHeight() + f[2];
139        float y3 = f[3] * 0 + f[4] *
140                kenan.getHeight() + f[5];
141        float x4 = f[0] * kenan.getWidth() +
142                f[1] * kenan.getHeight() + f[2];
143        float y4 = f[3] * kenan.getWidth() +
144                f[4] * kenan.getHeight() + f[5];
145        // 图片现宽度
146        double width = Math.sqrt((x1 - x2) *
147                (x1 - x2) + (y1 - y2) * (y1 - y2));
148        // 缩放比率判断
149        if (width < widthScreen / 3 || width > widthScreen * 3) {
150            return true;
151        }
152        // 出界判断
153        if ((x1 < widthScreen / 3 && x2 < widthScreen / 3
154                && x3 < widthScreen / 3
155                && x4 < widthScreen / 3)
156                || (x1 > widthScreen * 2 / 3
157                        && x2 > widthScreen * 2 / 3
158                        && x3 > widthScreen * 2 / 3
159                        && x4 > widthScreen * 2 / 3)
160                || (y1 < heightScreen / 3
161                        && y2 < heightScreen / 3
162                        && y3 < heightScreen / 3
163                        && y4 < heightScreen / 3)
164                || (y1 > heightScreen * 2 / 3
165                        && y2 > heightScreen * 2 / 3
166                        && y3 > heightScreen * 2 / 3
167                        && y4 > heightScreen * 2 / 3)) {
168            return true;
169        }
170        return false;
```

```
171     }
172
173     // 触碰两点间距离
174     private float spacing(MotionEvent event) {
175         //通过三角函数得到两点间的距离
176         float x = event.getX(0) - event.getX(1);
177         float y = event.getY(0) - event.getY(1);
178         return FloatMath.sqrt(x * x + y * y);
179     }
180
181     // 取手势中心点
182     private void midPoint(PointF point, MotionEvent event) {
183         //得到手势中心点的位置
184         float x = event.getX(0) + event.getX(1);
185         float y = event.getY(0) + event.getY(1);
186         point.set(x / 2, y / 2);
187     }
188
189     // 取旋转角度
190     private float rotation(MotionEvent event) {
191         //得到两个手指间的旋转角度
192         double delta_x = (event.getX(0) - event.getX(1));
193         double delta_y = (event.getY(0) - event.getY(1));
194         double radians = Math.atan2(delta_y, delta_x);
195         return (float) Math.toDegrees(radians);
196     }
197 }
```

这个类是我们本例子的核心类，此类继承自 ImageView 类，在如上代码的第 30 行初始化本类的对象，包含得到本 View 需要显示的图片，得到屏幕的宽度高度，然后实例化了一个 Matrix 矩阵。在第 49 行实现了 View 的 onDraw 方法，当本 View 进行显示的时候自动回到此方法，在此方法中保存当前画布，然后绘制旋转后的图片，然后恢复画布。在第 59 行实现了本 View 的 onTouchEvent 方法，我们本例子的所有操作都在此方法中实现，在第 62 行获取事件的类型为手指按在屏幕上的处理工作。在第 71 行是当有多个手指按在屏幕上的操作。在第 83 行实现了当手指滑动时的操作处理，这里又分为两种情况就是之前是单指还是双指操作，单指的话为拖动，双指的话为缩放或旋转。在 174 行实现通过事件得到两个手指间距离。在第 182 行实现得到两指触摸的中心点，在第 190 行实现通过两个手指触摸的位置获得事件的旋转角度的方法。

然后修改 src/com.wyl.example/MainActivity.java 文件，代码如下：

```
01 //定义 MainActivity 继承自 Activity
02 public class MainActivity extends Activity {
03     @Override
04     public void onCreate(Bundle savedInstanceState) {
05         super.onCreate(savedInstanceState);
06
07         //定义自定义 View 的对象
08         MyView myview = new MyView(this);
09         //设置当前页面的视图为自定义的 myview
10         setContentView(myview);
11     }
12 }
```

在此文件中的主要操作是定义一个之前的 MyView 对象，然后通过 setContentView 方法设置当前页面的布局为此 View。在此代码的第 8 行定义了一个 MyView 类的对象，在第 10 行将此对象设置给了当前的 Activity 的布局。

4．实例扩展

由于本实例需要用到手机的多点触摸的功能，而对于模拟器来说它不支持多点触摸，所以本例中所有多点触摸的效果在 AVD 中是无法看到的，只有在真实支持多点触摸的手机中才能看到效果。

范例 069　图片浏览器滑动切换图片

1．实例简介

在 Android 中关于触摸屏幕的操作不单有拖动、双指放大缩小和旋转。还有一种最常见的操作就是滑动操作，此操作在我们平常也经常用到。例如，对于手机的浏览器向左滑动可以返回上一个页面，向右滑动可以得到下一个页面，还有现在的图片浏览器基本也都是用滑动来切换图片了。本实例就带领大家一起完成一个滑动切换图片的效果。

2．运行效果

该实例运行效果如图 4.16 所示。

3．实例程序讲解

在图 4.16 效果中通过单指在屏幕滑动可以实现图片的切换效果。想要实现此效果，首先需要修改 res/layout/activity_main.xml 文件，代码如下：

图 4.16　滑动切换图片

```
01  <?xml version="1.0" encoding="utf-8"?>
02  <!-- 定义基础的 LinearLayout 布局 -->
03  <LinearLayout xmlns:android="http://schemas.android.com/apk/res/android"
04      android:layout_width="fill_parent"
05      android:layout_height="fill_parent"
06      android:orientation="horizontal" >
07      <!-- 定义 ImageView 控件 -->
08      <ImageView
09          android:id="@+id/Iv"
10          android:layout_width="fill_parent"
11          android:layout_height="fill_parent"
12          android:src="@drawable/k1" />
13  
14  </LinearLayout>
```

此文件为当前 Activity 的布局文件，其中的第 8~12 行定义了一个 ImageView，并设置了 id，方便我们在 Activity 中获取此对象。

然后再修改 src/com.wyl.example/MainActivity.java 文件，代码如下：

```
01  //定义 MainActivity 继承自 Activity
```

```java
02  public class MainActivity extends Activity {
03      // 定义保存 ImageView 的对象
04      private ImageView Iv;
05      //定义手势检测器对象
06      private GestureDetector gestureDetector;
07      //定义图片的资源数组
08      private int[] ResId = new int[]{
09          R.drawable.k1,
10          R.drawable.k2,
11          R.drawable.k3
12      };
13      //定义当前显示的图片的下标
14      private int count = 0;
15
16      @Override
17      public void onCreate(Bundle savedInstanceState) {
18          super.onCreate(savedInstanceState);
19          //设置当前页面的布局视图为 activity_main
20          setContentView(R.layout.activity_main);
            //得到当前页面中的视图控件
21          findView();
            //设置当前视图中的监听器事件
22          setListener();
23      }
24
25      private void setListener() {
26          //设置手势监听器的处理效果由 onGestureListener 来处理
27          gestureDetector = new GestureDetector(MainActivity.this,
28                  onGestureListener);
29      }
30
31      @Override
32      public boolean onTouchEvent(MotionEvent event) {
33          //当前 Activity 被触摸时回调
34          return gestureDetector.onTouchEvent(event);
35      }
36
37      private void findView() {
38          //得到当前页面的 imageview 控件
39          Iv = (ImageView) findViewById(R.id.Iv);
40      }
41      //定义了 GestureDetector 的手势识别监听器
42      private GestureDetector.OnGestureListener onGestureListener
43              = new GestureDetector.SimpleOnGestureListener() {
44          //当识别的手势是滑动手势时回调 onFinger 方法
45          @Override
46          public boolean onFling(MotionEvent e1, MotionEvent e2, float velocityX,
47                  float velocityY) {
48              //得到滑动手势的起始和结束点的 x，y 坐标，并进行计算
49              float x = e2.getX() - e1.getX();
50              float y = e2.getY() - e1.getY();
51
52              //通过计算结果判断用户是向左滑动或者向右滑动
53              if (x > 0) {
54                  count++;
55                  count %= 3;
56              } else if (x < 0) {
57                  count--;
58                  count = (count + 3) % 3;
```

```
59          }
60          //切换 imageview 的图片
61          changeImg();
62          return true;
63      }
64  };
65
66  public void changeImg() {
67      //设置当前位置的图片资源
68      Iv.setImageResource(ResId[count]);
69  }
70  }
```

在代码的第 37～40 行通过 findViewById 方法得到了 ImageView 对象。在第 25～29 行设置了一个手势识别器。在代码的第 32～35 行实现了当前 Activity 的 onTouchEvent 方法，然后把所有的 Touch 事件都转交给了前面的手势识别器来进行处理。在第 41～64 行自定义了一个手势识别器，其中实现了 onFing 方法，当有滑动事件的时候自动回调此方法，在此回调方法中拿到两点之间的位置，然后通过两点位置的差别，判断此滑动是向左滑动，还是向右滑动，然后调用在第 66～69 行定义的切换图片的方法进行图片的切换。

4．实例扩展

本例切换的图片都存储在本地，在一些应用中经常用到的图片都是从网络获取的，方法类似，然后在其中加上网络获取图片的代码就可以了。

范例 070　简易画板

1．实例简介

对于屏幕触摸滑动操作的应用在 Android 中确实很多，其中比较经典的应用就是画板应用。大家经常会启动一个画板应用，然后通过手指在上面进行绘制，或者大家平时发送短信也是在屏幕写出你想要写的字即可。本实例就带领大家一起来实现一个简单的画板效果。

2．运行效果

该实例运行效果如图 4.17 所示。

3．实例程序讲解

在图 4.17 效果中打开应用可以看到一个画板，大家可以通过手指在此画板上进行滑动，这时你会看到画板上会把你的手指当做一根笔，手指划过的地方都会有笔迹的存在。要想实现这样的效果，首先需要自定义 src/com.wyl.example/MyView.java 文件，代码如下：

图 4.17　简易画板

```
01  //自定义 MyView 类继承自 ImageView
02  public class MyView extends View {
```

```java
03    //上次触屏的位置
04    private int mLastX, mLastY;
05    //当前触屏的位置
06    private int mCurrX, mCurrY;
07    //保存每次绘画的结果
08    private Bitmap mBitmap;
09    //绘图的笔
10    private Paint mPaint;
11
12    //构造函数
13    public MyView(Context context) {
14        super(context);
15        //初始化画笔
16        mPaint = new Paint();
17        mPaint.setStrokeWidth(6);
18    }
19
20    //当前view显示的时候自动回调ondraw方法
21    @Override
22    protected void onDraw(Canvas canvas) {
23        super.onDraw(canvas);
24        //得到当前view的宽度和高度
25        int width = getWidth();
26        int height = getHeight();
27
28        //如果bitmap为空的话，就初始化bitmap
29        if (mBitmap == null) {
30            mBitmap = Bitmap.createBitmap(width, height,
31                    Bitmap.Config.ARGB_8888);
32        }
33
34        //将之前的bitmap的结果画到当前的页面上
35        Canvas tmpCanvas = new Canvas(mBitmap);
36        //在当前的页面上划线
37        tmpCanvas.drawLine(mLastX, mLastY, mCurrX, mCurrY, mPaint);
38
39        //再把Bitmap画到canvas上
40        canvas.drawBitmap(mBitmap, 0, 0, mPaint);
41    }
42
43    //当用户触摸此view时自动回调
44    @Override
45    public boolean onTouchEvent(MotionEvent event) {
46        //记录当前的x,y坐标
47        mLastX = mCurrX;
48        mLastY = mCurrY;
49        //获取当前单击的位置
50        mCurrX = (int) event.getX();
51        mCurrY = (int) event.getY();
52
53        switch (event.getAction()) {
54        case MotionEvent.ACTION_DOWN:
55            mLastX = mCurrX;
56            mLastY = mCurrY;
57            break;
58        default:
```

```
59          break;
60      }
61      //重绘view
62      invalidate();
63
64      return true; // 必须返回true
65  }
66 }
```

如上代码是自定义了一个 View 类，在第 13 行实现了此 View 类的构造方法，初始化了画笔，设置了画笔的粗细。在第 22 行实现了 View 类的绘制函数 onDraw 方法。在第 45 行实现了此 View 类的触摸事件监听的方法，在此方法中得到触摸的位置和触摸的类型，然后重回当前 View。

然后在修改 src/com.wyl.example/MainActivity.java 文件，代码如下：

```
01 //定义MainActivity继承自Activity
02 public class MainActivity extends Activity {
03
04 @Override
05 public void onCreate(Bundle savedInstanceState) {
06     super.onCreate(savedInstanceState);
07     //定义自定义View的对象
08     MyView myview = new MyView(MainActivity.this);
09     //将当前Activity的布局设置为自定义的View
10     setContentView(myview);
11 }
12 }
```

在代码的第 8～10 行定义了一个 MyView 类的对象，然后将其设置给当前的 Activity 作为页面的显示布局。

4．实例扩展

本例讲述完毕后大家可以对于本例进行扩展的演练了。例如，本例的画笔颜色都是黑色的，是否可以实现功能，让用户进行颜色的选择呢，画笔的粗细是否可以选择呢等等，这些功能大家都可以进行完善了。

范例 071　登录和注册页面的 ViewFlipper 效果

1．实例简介

在开发 Android 应用的时候，经常会因为手机的屏幕有限，而且设计一些不同于 PC 软件的技巧。例如，在 PC 软件上会有下拉菜单，而在 Android 手机上有 Menu；在 PC 上会用标签页来显示两个并列的页面，而在 Android 手机中会用 Tabhost 等等。在这些为手机特定的功能效果中，其中滑屏切换页面的效果也很常见。本实例就带领大家一起使用 ViewFlippe，来实现一个用户登录和注册页滑动切换的效果。

2．运行效果

该实例运行效果如图 4.18 所示。

Android 开发范例实战宝典

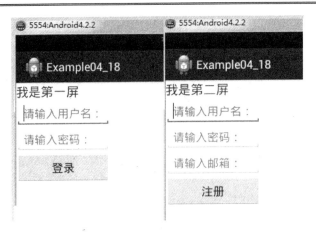

图 4.18　滑屏切换登录和注册页面的 ViewFlipper 效果

3．实例程序讲解

在图 4.18 效果中主要用一个 ViewFlipper 控件并且给此控件设置了屏幕滑动的监听器。要想实现这样的效果，首先需要修改 res/layout/activity_main.xml 文件，代码如下：

```xml
01 <?xml version="1.0" encoding="utf-8"?>
02 <!-- 定义 Linearlayout 基本布局 -->
03 <LinearLayout xmlns:android="http://schemas.android.com/apk/res/android"
04     android:layout_width="fill_parent"
05     android:layout_height="fill_parent"
06     android:orientation="horizontal" >
07
08     <!-- 定义 ViewFlipper 视图切换器 -->
09     <ViewFlipper
10         android:id="@+id/ViewFlipper01"
11         android:layout_width="wrap_content"
12         android:layout_height="wrap_content" >
13     <!-- 视图切换器的第一个界面 -->
14         <LinearLayout
15             android:id="@+id/LinearLayout01"
16             android:layout_width="wrap_content"
17             android:layout_height="wrap_content"
18             android:orientation="vertical" >
19         <!-- 提示文本标签 -->
20         <TextView
21             android:layout_width="match_parent"
22             android:layout_height="wrap_content"
23             android:textSize="20dip"
24             android:text="我是第一屏" />
25         <!-- 登录界面的用户输入框 -->
26         <EditText
27             android:layout_width="match_parent"
28             android:layout_height="wrap_content"
29             android:hint="请输入用户名："
30             />
31         <!-- 登录界面的密码输入框 -->
32         <EditText
33             android:layout_width="match_parent"
34             android:layout_height="wrap_content"
35             android:hint="请输入密码："
```

```
36          />
37       <!-- 登录界面的登录按钮 -->
38       <Button
39          android:layout_width="match_parent"
40          android:layout_height="wrap_content"
41          android:text="登录"
42          />
43
44       </LinearLayout>
45
46       <!-- 视图切换器的第二个界面 -->
47       <LinearLayout
48          android:id="@+id/LinearLayout02"
49          android:layout_width="wrap_content"
50          android:layout_height="wrap_content"
51          android:orientation="vertical" >
52       <!-- 注册界面文本提示框 -->
53       <TextView
54          android:id="@+id/TextView02"
55          android:layout_width="match_parent"
56          android:layout_height="wrap_content"
57          android:textSize="20dip"
58          android:text="我是第二屏" />
59       <!-- 注册界面的用户名输入框 -->
60       <EditText
61          android:layout_width="match_parent"
62          android:layout_height="wrap_content"
63          android:hint="请输入用户名："
64          />
65       <!-- 注册界面的密码输入框 -->
66       <EditText
67          android:layout_width="match_parent"
68          android:layout_height="wrap_content"
69          android:hint="请输入密码："
70          />
71       <!-- 注册界面的邮箱输入框 -->
72       <EditText
73          android:layout_width="match_parent"
74          android:layout_height="wrap_content"
75          android:hint="请输入邮箱："
76          />
77       <!-- 注册界面的注册按钮 -->
78       <Button
79          android:layout_width="match_parent"
80          android:layout_height="wrap_content"
81          android:text="注册"
82          />
83       </LinearLayout>
84    </ViewFlipper>
85
86 </LinearLayout>
```

在如上代码中第 9～84 行定义了一个 ViewFlipper 控件，在此控件中定义了两个 LinearLayout，分别代表登录和注册页面。在第 14～44 行中定义了登录页面的布局，在第 47～83 行中定义了注册页面的布局。

然后修改 src/com.wyl.example/MainActivity.java 文件，代码如下：

```java
01  //定义MainActivity继承自Activity
02  public class MainActivity extends Activity implements OnGestureListener {
03      //定义页面中的ViewFilpper对象
04      private ViewFlipper flipper;
05      //定义手势识别监听器对象
06      private GestureDetector detector;
07
08      @Override
09      protected void onCreate(Bundle savedInstanceState) {
10          super.onCreate(savedInstanceState);
            //设置当前页面的布局视图为activity_main
11          setContentView(R.layout.activity_main);
12          //初始化手势识别监听器
13          detector = new GestureDetector(this);
14          //得到filpper对象
15          flipper = (ViewFlipper) this.findViewById(R.id.ViewFlipper01);
16      }
17
18      //当页面被触摸时自动回调,把拿到的事件给手势识别器进行处理
19      @Override
20      public boolean onTouchEvent(MotionEvent event) {
21          return this.detector.onTouchEvent(event);
22      }
23
24      //当页面被按下时自动回调
25      @Override
26      public boolean onDown(MotionEvent e) {
27          return false;
28      }
29
30      //当在页面中滑动时自动回调
31      @Override
32      public boolean onFling(MotionEvent e1, MotionEvent e2, float velocityX,
33              float velocityY) {
34          //得到滑动过程中的亮点的x和y坐标,进行判断
35          if (e1.getX() > e2.getX()) {
36              //当向左滑动时显示上一张
37              this.flipper.showNext();
38          } else if (e1.getX() < e2.getX()) {
39              //当向右滑动时显示下一张
40              this.flipper.showPrevious();
41          } else {
42              return false;
43          }
44          return true;
45      }
46
47      //当在界面长按时自动调用
48      @Override
49      public void onLongPress(MotionEvent e) {
50      }
51
52      //界面滚动的时候自动调用
53      @Override
54      public boolean onScroll(MotionEvent e1, MotionEvent e2, float distanceX,
55              float distanceY) {
56          return false;
57      }
58
```

```
59  //当轻击界面的时候自动调用
60  @Override
61  public boolean onSingleTapUp(MotionEvent e) {
62      return false;
63  }
64
65  //当界面被按压的时候自动调用
66  @Override
67  public void onShowPress(MotionEvent e) {
68      // TODO Auto-generated method stub
69
70  }
71 }
```

在代码第 13 行定义了一个 GestureDetector 对象，它用来识别我们常见的 Android 中的手势。在第 15 行通过 findViewById 得到了页面布局中的 ViewFlipper 对象。在第 20~22 行实现了 onTouchEvent 回调函数，并且把滑动事件拿到后传给 GestureDetector 对象来处理。在第 32~45 行实现了 GestureDetector 的 onFling 方法，在用户对屏幕进行滑动的时候调用，在此方法中通过滑动事件的起始和结束点的位置，确定滑动的方向，然后切换 ViewFlipper 显示的内容。

4．实例扩展

在 GestureDetector 类的识别中，不单能够识别 onFling 的滑动事件，还能够识别滚动事件，回调 onScroll 方法；还有屏幕的轻击时间，回调 onSingleTapUp 方法；还有屏幕被按住的事件，回调 onShowPress 方法。通过这些方法的组合可以组合出更多的手势识别。大家根据自己的需求进行组合改进即可。

范例 072　神庙逃亡的操作模拟效果

1．实例简介

在 Android 中手势识别被运用的越来越多，而且越来越有创意，最近流行一款跑酷游戏，叫做神庙逃亡，在此游戏中用户通过对屏幕的向上、向下、向左和向右的滑动来控制主人公躲避各种障碍。它的开发是采用国外比较著名的游戏引擎 unity3D 来实现的，我们这里不去讲解 unity3D，但是通过简单的手势识别也可以实现神庙逃亡的模拟操作。本实例就带领大家一起使用对于屏幕的事件的监听来实现一个神庙逃亡的操作模拟效果。

2．运行效果

该实例运行效果如图 4.19 所示。

3．实例程序讲解

在本例效果中，用户可以通过向上、向下、向左和

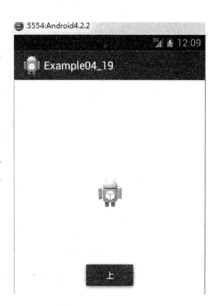

图 4.19　神庙逃亡的操作模拟效果

向右滑动来控制图片中的小人进行运动,并且在运动完毕后,图片会自动返回原始位置。
想要实现本例效果,首先要修改 res/layout/activity_main.xml 文件,代码如下:

```xml
01 <?xml version="1.0" encoding="utf-8"?>
02 <!-- 定义Linearlayout基本布局 -->
03 <LinearLayout xmlns:android="http://schemas.android.com/apk/res/android"
04     android:layout_width="fill_parent"
05     android:layout_height="fill_parent"
06     >
07     <!-- 定义显示人物图片的ImageView控件 -->
08     <ImageView
09         android:id="@+id/Iv"
10         android:layout_width="fill_parent"
11         android:layout_height="fill_parent"
12         android:src="@drawable/ic_launcher"
13         android:scaleType="center"
14         />
15
16 </LinearLayout>
```

在如上代码的第 8~14 行定义了一个显示代表人物的 ImageView 控件对象。

然后修改 src/com.wyl.example/MainActivity.java 文件,代码如下:

```java
001 //定义MainActivity继承自Activity
002 public class MainActivity extends Activity implements OnGestureListener {
003     //定义手势识别监听器对象
004     private GestureDetector detector;
005     //定义ImageView对象
006     private ImageView Iv;
007     //向上的动画
008     private Animation amUp;
009     //向下的动画
010     private Animation amDown;
011     //向左的动画
012     private Animation amLeft;
013     //向右的动画
014     private Animation amRight;
015
016     @Override
017     protected void onCreate(Bundle savedInstanceState) {
018         super.onCreate(savedInstanceState);
019         setContentView(R.layout.activity_main);
020         //初始化手势识别监听器
021         detector = new GestureDetector(this);
022         //得到ImageView对象
023         Iv = (ImageView)findViewById(R.id.Iv);
024         initAnimation();        //初始化对象的动画效果
025     }
026
027     private void initAnimation() {
028         int screenWidth = getWindowManager().getDefaultDisplay().getWidth();
029         int screenHeight = getWindowManager().getDefaultDisplay().getHeight();
030
031         Log.e("screenWidth",screenWidth+ "");
032         Log.e("screenHeight",screenHeight+ "");
033
034         //向上的动画设定
035         amUp = new TranslateAnimation(0, 0, 0, -screenHeight/2);
```

```
036        //动画开始到结束的执行时间(1000 = 1 秒)
037        amUp. setDuration (500 );
038        //重复次数为 1 次
039        amUp.setRepeatCount(1);
040        //设置回播
041        amUp.setRepeatMode(Animation.REVERSE);
042
043         //向下的动画设定
044        amDown = new TranslateAnimation(0, 0, 0, screenHeight/2);
045        //动画开始到结束的执行时间(1000 = 1 秒)
046        amDown. setDuration (500);
047        //重复次数为 1 次
048        amDown.setRepeatCount(1);
049        //设置回播
050        amDown.setRepeatMode(Animation.REVERSE);
051
052        //向左的动画设定
053        amLeft = new TranslateAnimation(0, -screenWidth/2, 0, 0);
054        //动画开始到结束的执行时间(1000 = 1 秒)
055        amLeft. setDuration (500 );
056        //重复次数为 1 次
057        amLeft.setRepeatCount(1);
058        //设置回播
059        amLeft.setRepeatMode(Animation.REVERSE);
060
061
062        //向右的动画设定
063        amRight = new TranslateAnimation(0, screenWidth/2, 0, 0);
064        //动画开始到结束的执行时间(1000 = 1 秒)
065        amRight. setDuration (500 );
066        //重复次数为 1 次
067        amRight.setRepeatCount(1);
068        //设置回播
069        amRight.setRepeatMode(Animation.REVERSE);
070    }
071
072    //当页面被触摸时自动回调,把拿到的事件给手势识别器进行处理
073    @Override
074    public boolean onTouchEvent(MotionEvent event) {
075        return this.detector.onTouchEvent(event);
076    }
077
078    //当页面被按下时自动回调
079    @Override
080    public boolean onDown(MotionEvent e) {
081        return false;
082    }
083
084    //当在页面中滑动时自动回调
085    @Override
086    public boolean onFling(MotionEvent e1, MotionEvent e2, float velocityX,
087            float velocityY) {
088        //得到滑动过程中的亮点的 x 和 y 坐标,进行判断
089        if (e1.getX() - e2.getX() > 100) {
            //判断是否用户向左滑动
090            Iv.startAnimation(amLeft);
091            Toast.makeText(MainActivity.this, "左", Toast.LENGTH_SHORT).show();
092            return true;
```

```
093            } else if (e1.getX() - e2.getX() < -100) {
094                //判断用户是否向右滑动
094                Iv.startAnimation(amRight);
095                Toast.makeText(MainActivity.this, "右", Toast.LENGTH_SHORT).show();
096                return true;
097            } else if (e1.getY() - e2.getY() > 100) {
                   //判断用户是否向上滑动
098                Iv.startAnimation(amUp);
099                Toast.makeText(MainActivity.this, "上", Toast.LENGTH_SHORT).show();
100                return true;
101            } else if (e1.getY() - e2.getY() < -100) {
                   //判断用户是否向右滑动
102                Iv.startAnimation(amDown);
103                Toast.makeText(MainActivity.this, "下", Toast.LENGTH_SHORT).show();
104                return true;
105            }
106            return true;
107        }
108
109        //当在界面长按时自动调用
110        @Override
111        public void onLongPress(MotionEvent e) {
112        }
113
114        //界面滚动的时候自动调用
115        @Override
116        public boolean onScroll(MotionEvent e1, MotionEvent e2, float distanceX,
117                float distanceY) {
118            return false;
119        }
120
121        //当轻击界面的时候自动调用
122        @Override
123        public boolean onSingleTapUp(MotionEvent e) {
124            return false;
125        }
126
127        //当界面被按压的时候自动调用
128        @Override
129        public void onShowPress(MotionEvent e) {
130            // TODO Auto-generated method stub
131
132        }
133   }
```

在如上代码的第 21 行定义了 GestureDetector 对象，用来识别用户的滑动事件。在第 27～70 行初始化了向上、向下、向左和向右的四个动画，方便在后面进行播放。在第 74～76 行实现了 onTouchEvent 方法，并且将拿到的滑动事件传给 detector 对象进行处理。在第 86～107 行实现了 onFling 方法，当用户滑动屏幕的时候回调此方法，在此方法中通过滑动前后手指的位置确定用户滑动的方向，然后分别给我们得到的 ImageView 对象播放不同的动画。通过这种方法就可以实现类似神庙逃亡中人物的效果了。

4．实例扩展

本实例仅仅实现了一个基本的滑动事件的监听和识别的效果，大家在今后设计应用或者游戏的时候可以适当考虑加上屏幕滑动的效果，不但可以提高用户对于软件的使用体验，还能够以更多的形式展现应用的功能。

范例 073　手势库的创建及手势识别

1．实例简介

在 Android 中我们使用手势,不单纯是实现滑动和单击双击,当然也可以自定义手势。例如,现在的手写输入法,就是根据用户输入的手势,然后跟库中的文字进行比较得到的文字。本实例就带领大家一起创建一个手势库,并且实现定义手势后,用户再次输入手势可以进行识别。

2．运行效果

该实例运行效果如图 4.20 所示。

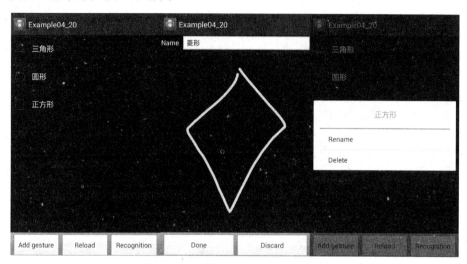

图 4.20　手势库的创建及手势识别

3．实例程序讲解

在本例的效果中有三个页面,如下所示。

手势列表页:在此页面中主要显示手势库中已有的手势,长按某一个手势可以重命名或者删除某一个手势,并且设置了添加手势、读取手势和识别手势三个按钮,添加手势和识别手势分别跳转到添加手势页面和识别手势页面。

添加手势页面:在此页面中输入手势的名字,并且画出手势的形状,单击 Done 按钮,手势就添加到手势库中了。

识别手势的页面:在此页面中滑动画出某一个手势,程序会自动进行识别,如果和手势库的某一个手势相似则显示对应手势的名称。

想要实现本例效果,步骤如下所示。

(1) 修改 res/layout/activity_main.xml 文件,代码如下:

```
01  <?xml version="1.0" encoding="utf-8"?>
02  <!-- 基本的 LinearLayout 基础布局 -->
```

```
03  <LinearLayout xmlns:android="http://schemas.android.com/apk/res/android"
04      android:layout_width="match_parent"
05      android:layout_height="match_parent"
06      android:background="@android:color/black"
07      android:orientation="vertical" >
08  <!-- 手势列表 -->
09      <ListView
10          android:id="@android:id/list"
11          android:layout_width="match_parent"
12          android:layout_height="0dip"
13          android:layout_weight="1.0" />
14  <!-- 空行文本标签 -->
15      <TextView
16          android:id="@android:id/empty"
17          android:layout_width="match_parent"
18          android:layout_height="0dip"
19          android:layout_weight="1.0"
20          android:gravity="center"
21          android:text="@string/gestures_loading"
22          android:textAppearance="?android:attr/textAppearanceMedium" />
23  <!-- 页面下方的三个并列的按钮 -->
24      <LinearLayout
25          style="@android:style/ButtonBar"
26          android:layout_width="match_parent"
27          android:layout_height="wrap_content"
28          android:orientation="horizontal" >
29      <!-- 定义添加手势的按钮 -->
30          <Button
31              android:id="@+id/addButton"
32              android:layout_width="0dip"
33              android:layout_height="wrap_content"
34              android:layout_weight="1"
35              android:enabled="false"
36              android:onClick="addGesture"
37              android:text="@string/button_add" />
38      <!-- 定义重新读取手势库的按钮-->
39          <Button
40              android:id="@+id/reloadButton"
41              android:layout_width="0dip"
42              android:layout_height="wrap_content"
43              android:layout_weight="1"
44              android:enabled="false"
45              android:onClick="reloadGestures"
46              android:text="@string/button_reload" />
47      <!-- 定义识别手势的按钮-->
48          <Button
49              android:id="@+id/recognitionButton"
50              android:layout_width="0dip"
51              android:layout_height="wrap_content"
52              android:layout_weight="1"
53              android:enabled="true"
54              android:onClick="recognitionGestures"
55              android:text="@string/button_ident" />
56      </LinearLayout>
57
58  </LinearLayout>
```

此代码是手势列表页面的布局文件，在第 9～13 行定义了一个 ListView 控件用来显示手势库中的手势。在第 15～22 行定义了一个 TextView 控件为了显示读取手势的提示文字。

在第 24～56 行定义了一个 LinearLayout，用来显示添加手势、读取手势和识别手势三个按钮。

（2）新建 res/layout/ dialog_rename.xml 文件，代码如下：

```xml
01  <?xml version="1.0" encoding="utf-8"?>
02  <!-- 定义基本的修改名字的 Dialog 布局 -LinearLayout 布局 -->
03  <LinearLayout xmlns:android="http://schemas.android.com/apk/res/android"
04      android:layout_width="match_parent"
05      android:layout_height="wrap_content"
06      android:padding="20dip"
07      android:orientation="vertical">
08  <!-- 定义一个 TextView 控件显示修改名字的文字提示 -->
09      <TextView
10          android:id="@+id/label"
11          android:layout_height="wrap_content"
12          android:layout_width="wrap_content"
13          android:text="@string/gestures_rename_label"
14          android:gravity="left"
15          android:textAppearance="?android:attr/textAppearanceMedium" />
16  <!-- 定义一个 EditText 控件定义用户修改后的名字输入框 -->
17      <EditText
18          android:id="@+id/name"
19          android:layout_height="wrap_content"
20          android:layout_width="match_parent"
21          android:scrollHorizontally="true"
22          android:autoText="false"
23          android:capitalize="none"
24          android:gravity="fill_horizontal"
25          android:textAppearance="?android:attr/textAppearanceMedium" />
26
27  </LinearLayout>
```

上面代码是修改手势名字的 Dialog 控件的布局文件，在其中定义了提示用户输入新的手势名字的 TextView 控件，而且定义了一个 EditText 控件，用来接受用户输入的手势库名字。

（3）新建 res/layout/gestures_item.xml 文件，代码如下：

```xml
01  <?xml version="1.0" encoding="utf-8"?>
02  <!-- 定义显示手势列表的 item 布局 -->
03  <TextView xmlns:android="http://schemas.android.com/apk/res/android"
04      android:id="@android:id/text1"
05      android:layout_width="match_parent"
06      android:layout_height="wrap_content"
07
08      android:gravity="center_vertical"
09      android:minHeight="?android:attr/listPreferredItemHeight"
10      android:textColor="@android:color/background_light"
11      android:drawablePadding="12dip"
12      android:paddingLeft="6dip"
13      android:paddingRight="6dip"
14
15      android:ellipsize="marquee"
16      android:singleLine="true"
17      android:textAppearance="?android:attr/textAppearanceLarge" />
```

在此文件中定义了手势库列表中显示的手势 item 的样式。

（4）新建 res/layout/activity_creategesture.xml 文件，代码如下：

```xml
01 <?xml version="1.0" encoding="utf-8"?>
02 <!-- 创建手势的基本布局 Linearlayout -->
03 <LinearLayout
04     xmlns:android="http://schemas.android.com/apk/res/android"
05     android:layout_width="match_parent"
06     android:layout_height="match_parent"
07     android:background="@android:color/black"
08     android:orientation="vertical">
09     <!-- 定义创建手势页面中的提示用户输入手势名称的控件 -->
10     <LinearLayout
11         android:layout_width="match_parent"
12         android:layout_height="wrap_content"
13         android:orientation="horizontal">
14
15         <TextView
16             android:layout_width="wrap_content"
17             android:layout_height="wrap_content"
18             android:layout_marginRight="6dip"
19             android:textColor="@android:color/background_light"
20             android:text="@string/prompt_gesture_name"
21             android:textAppearance="?android:attr/textAppearanceMedium" />
22
23         <EditText
24             android:id="@+id/gesture_name"
25             android:layout_width="0dip"
26             android:layout_weight="1.0"
27             android:textColor="@android:color/black"
28             android:layout_height="wrap_content"
29             android:maxLength="40"
30             android:singleLine="true" />
31
32     </LinearLayout>
33     <!-- 定义识别手势的控件 -->
34     <android.gesture.GestureOverlayView
35         android:id="@+id/gestures_overlay"
36         android:layout_width="match_parent"
37         android:layout_height="0dip"
38         android:layout_weight="1.0"
39         android:gestureStrokeType="multiple" />
40     <!-- 定义添加手势按钮和放弃添加按钮 -->
41     <LinearLayout
42         style="@android:style/ButtonBar"
43         android:layout_width="match_parent"
44         android:layout_height="wrap_content"
45         android:orientation="horizontal">
46         //定义完成按钮
47         <Button
48             android:id="@+id/done"
49             android:layout_width="0dip"
50             android:layout_height="wrap_content"
51             android:layout_weight="1"
52             android:enabled="false"
53             android:onClick="addGesture"
54             android:text="@string/button_done" />
55         //定义放弃按钮
56         <Button
57             android:layout_width="0dip"
58             android:layout_height="wrap_content"
59             android:layout_weight="1"
60
```

```
61          android:onClick="cancelGesture"
62          android:text="@string/button_discard" />
63
64     </LinearLayout>
65
66 </LinearLayout>
67
```

此文件是添加手势页面的布局文件,在第 10~32 行定义了一个 LinearLayout,其中包含一个 TextView 和一个 EditText 对象,用来提示用户输入需要添加的手势的名称。在第 34~39 行定义了一个手势识别的控件 GestureoverLayView,用户在此控件上滑动得到手势对象。在第 41~64 行定义了一个 LinearLayout 对象,用来显示确认添加按钮和取消添加按钮。

(5) 新建 res/layout/activity_gestureidentify.xml 文件,代码如下:

```
01 <?xml version="1.0" encoding="utf-8"?>
02 <!-- 定义手势识别的页面的布局 -->
03 <LinearLayout xmlns:android="http://schemas.android.com/apk/res/android"
04     android:layout_width="fill_parent"
05     android:layout_height="fill_parent"
06     android:background="@android:color/black"
07     android:orientation="vertical" >
08     <!-- 提示用户在下面的控件上绘制手势 -->
09     <TextView
10         android:layout_width="fill_parent"
11         android:layout_height="wrap_content"
12         android:textColor="@android:color/background_light"
13         android:text="请在下面绘制手势: " />
14     <!--手势识别控件
15     注意  android.gesture.要加,否则报错,估计是找不到包-->
16     <android.gesture.GestureOverlayView
17         android:id="@+id/gesture_overlay_view_test"
18         android:layout_width="match_parent"
19         android:layout_height="0dip"
20         android:layout_weight="1.0"
21         android:gestureStrokeType="multiple" />
22
23 </LinearLayout>
```

此页面是手势识别的页面布局,在第 9~13 行定义了一个提示用户输入的文本标签,在第 14~21 行定义了能够得到用户手势的 GestureOverlayView 控件。

(6) 创建 src/com.wyl.example/CreateGestureActivity 文件,代码如下:

```
001 //创建手势的 Activity
002 public class CreateGestureActivity extends Activity {
003     private static final float LENGTH_THRESHOLD = 120.0f;
004     //定义手势对象
005     private Gesture mGesture;
006     //完成的按钮视图
007     private View mDoneButton;
008
009     @Override
010     protected void onCreate(Bundle savedInstanceState) {
011         super.onCreate(savedInstanceState);
012
013         setContentView(R.layout.activity_creategesture);
```

```
014         //得到对应的按钮控件
015         mDoneButton = findViewById(R.id.done);
016
017         //得到GestureOverlayView对象
018         GestureOverlayView overlay =
019             (GestureOverlayView) findViewById(R.id.gestures_overlay);
020         //添加手势识别监听器
021         overlay.addOnGestureListener(new GesturesProcessor());
022     }
023
024     //当时Activity暂停的时候回调
025     @Override
026     protected void onSaveInstanceState(Bundle outState) {
027         super.onSaveInstanceState(outState);
028         //暂停手势
029         if (mGesture != null) {
030             outState.putParcelable("gesture", mGesture);
031         }
032     }
033
034     //当Activity恢复的时候回调
035     @Override
036     protected void onRestoreInstanceState(Bundle savedInstanceState) {
037         super.onRestoreInstanceState(savedInstanceState);
038         //得到之前保存的参数
039         mGesture = savedInstanceState.getParcelable("gesture");
040         if (mGesture != null) {
041             final GestureOverlayView overlay =
042                 (GestureOverlayView)
043                     findViewById(R.id.gestures_overlay);
044             //执行手势视图的设置
045             overlay.post(new Runnable() {
046                 public void run() {
047                     overlay.setGesture(mGesture);
048                 }
049             });
050             //设置完成按钮为可单击
051             mDoneButton.setEnabled(true);
052         }
053     }
054
055     //添加手势按钮的onclick方法
056     @SuppressWarnings({"UnusedDeclaration"})
057     public void addGesture(View v) {
058         if (mGesture != null) {
059             //得到手势的名字
060             final TextView input =
061                 (TextView) findViewById(R.id.gesture_name);
062             final CharSequence name = input.getText();
063             if (name.length() == 0) {
064                 input.setError(getString(R.string.error_missing_name));
065                 return;
066             }
067             //得到手势库
068             final GestureLibrary store = MainActivity.getStore();
069             //添加新的手势
070             store.addGesture(name.toString(), mGesture);
071             //保存手势
072             store.save();
```

```
073            setResult(RESULT_OK);
074
075            final String path =
076                new File(Environment.getExternalStorageDirectory(),
077                    "gestures").getAbsolutePath();
078            Toast.makeText(this,
079                getString(R.string.save_success, path),
080                Toast.LENGTH_LONG).show();
081        } else {
082            setResult(RESULT_CANCELED);
083        }
084
085        finish();
086    }
087
088
089    //取消手势的创建的方法
090    @SuppressWarnings({"UnusedDeclaration"})
091    public void cancelGesture(View v) {
092        setResult(RESULT_CANCELED);
093        finish();
094    }
095
096    //手势监听器
097    private class GesturesProcessor
098        implements GestureOverlayView.OnGestureListener {
099        public void onGestureStarted(GestureOverlayView overlay,
100            MotionEvent event) {
101            mDoneButton.setEnabled(false);        //设置按钮不可单击
102            mGesture = null;
103        }
104
105        public void onGesture(GestureOverlayView overlay,
106            MotionEvent event) {
107        }
108
109        //得到手势,添加到手势库
110        public void onGestureEnded(GestureOverlayView overlay,
111            MotionEvent event) {
112            mGesture = overlay.getGesture();
113            if (mGesture.getLength() < LENGTH_THRESHOLD) {
114                overlay.clear(false);
115            }
116            mDoneButton.setEnabled(true);
117        }
118
119        public void onGestureCancelled(GestureOverlayView overlay,
120            MotionEvent event) {
121        }
122    }
123 }
```

此文件是创建用户手势的页面,在此页面中第 15~21 行得到对应的按钮和 GestureOverlayView 对象,并设置了手势监听器对象。在第 57~88 行实现了添加手势的功能,首先判断手势名称是否为空,如果不为空的话,得到手势库对象,然后进行手势的添加。

(7) 创建 src/com.wyl.example/ RecognitionGestureActivity 文件,代码如下:

```
01  //识别手势的Activity
02  public class RecognitionGestureActivity extends Activity {
03      //定义手势库对象
04      GestureLibrary mGestureLib;
05
06      /** Called when the activity is first created. */
07      @Override
08      public void onCreate(Bundle savedInstanceState) {
09          super.onCreate(savedInstanceState);
10          setContentView(R.layout.activity_gestureidentify);
11
12          //手势画板
13          GestureOverlayView gestures = (GestureOverlayView) findViewById
            (R.id.gesture_overlay_view_test);
14          //手势识别的监听器
15          gestures.addOnGesturePerformedListener(new GestureOverlayView.
            OnGesturePerformedListener() {
16
17              @Override
18              public void onGesturePerformed(GestureOverlayView overlay,
19                      Gesture gesture) {
20                  //从手势库中查询匹配的内容，匹配的结果可能包括多个相似的结果，匹配度高的结果放在最前面
21                  ArrayList<Prediction> predictions = mGestureLib
22                          .recognize(gesture);
23                  if (predictions.size() > 0) {
24                      Prediction prediction = (Prediction) predictions.get(0);
25                      //匹配的手势
26                      if (prediction.score > 1.0) {
27                          Toast.makeText(RecognitionGestureActivity.this,
28                                  prediction.name, Toast.LENGTH_SHORT).show();
29                      }
30                  }
31              }
32          });
33
34      mGestureLib = MainActivity.getStore();
35      }
36  }
```

此页面实现了手势识别的主要功能，在第 13 行得到了手势获取的控件。然后给这个控件添加了手势识别的监听器。在此监听器中将得到的用户输入的手势和手势库中的手势进行比较，会得到手势和手势库中手势的匹配得分，当得分大于 1 的时候说明具有一定的相似度。最后得到手势的名字进行显示。

（8）修改 src/com.wyl.example/MainActivity.java 文件，代码如下：

```
001  //定义MainActivity继承自Activity
002  public class MainActivity extends ListActivity {
003      //定义状态值
004      private static final int STATUS_SUCCESS = 0;
005      private static final int STATUS_CANCELLED = 1;
006      private static final int STATUS_NO_STORAGE = 2;
007      private static final int STATUS_NOT_LOADED = 3;
008
009      //定义上下文菜单的id
010      private static final int MENU_ID_RENAME = 1;
011      private static final int MENU_ID_REMOVE = 2;
012
```

```
013    //dialog的重命名id
014    private static final int DIALOG_RENAME_GESTURE = 1;
015
016    //添加新的手势的标示
017    private static final int REQUEST_NEW_GESTURE = 1;
018
019    //获取的手势的id字段
020    private static final String GESTURES_INFO_ID = "gestures.info_id";
021
022    //手势文件的存取位置
023    private final File mStoreFile = new File(
024            Environment.getExternalStorageDirectory(), "gestures");
025
026    //定义了手势的比较器
027    private final Comparator<NamedGesture> mSorter = new Comparator
       <NamedGesture>() {
028        public int compare(NamedGesture object1, NamedGesture object2) {
029            return object1.name.compareTo(object2.name);
030        }
031    };
032    //手势库兑现
033    private static GestureLibrary sStore;
034
035    //手势适配器
036    private GesturesAdapter mAdapter;
037    private GesturesLoadTask mTask;
038    private TextView mEmpty;
039
040    private Dialog mRenameDialog;
041    private EditText mInput;
042    private NamedGesture mCurrentRenameGesture;
043
044    @Override
045    protected void onCreate(Bundle savedInstanceState) {
046        super.onCreate(savedInstanceState);
047
048        setContentView(R.layout.activity_main);
049
050        //初始化手势适配器
051        mAdapter = new GesturesAdapter(this);
052        setListAdapter(mAdapter);
053
054        //判断手势库是否为空
055        if (sStore == null) {
056            sStore = GestureLibraries.fromFile(mStoreFile);
057        }
058        mEmpty = (TextView) findViewById(android.R.id.empty);
059        loadGestures();            //读取手势
060
061        //注册上下文菜单
062        registerForContextMenu(getListView());
063    }
064    //其他文件得到手势库
065    static GestureLibrary getStore() {
066        return sStore;
067    }
068
069    //重新读取手势库按钮的回调方法
070    @SuppressWarnings({ "UnusedDeclaration" })
```

```java
071    public void reloadGestures(View v) {
072        loadGestures();
073    }
074
075    //添加手势按钮的回调方法
076    @SuppressWarnings({ "UnusedDeclaration" })
077    public void addGesture(View v) {
078        Intent intent = new Intent(this, CreateGestureActivity.class);
079        startActivityForResult(intent, REQUEST_NEW_GESTURE);
080    }
081
082    //识别手势的回调方法
083    public void recognitionGestures(View v) {
084        Intent intent = new Intent(this, RecognitionGestureActivity.class);
085        startActivity(intent);
086    }
087
088    //取得返回的数据的回调方法
089    @Override
090    protected void onActivityResult(int requestCode, int resultCode, Intent data) {
091        super.onActivityResult(requestCode, resultCode, data);
092
093        if (resultCode == RESULT_OK) {
094            switch (requestCode) {
095            case REQUEST_NEW_GESTURE:
096                loadGestures();
097                break;
098            }
099        }
100    }
101
102    //通过异步任务读取手势
103    private void loadGestures() {
104        if (mTask != null
105                && mTask.getStatus() != GesturesLoadTask.Status.FINISHED) {
106            mTask.cancel(true);
107        }
108        mTask = (GesturesLoadTask) new GesturesLoadTask().execute();
109    }
110
111    //当前 Activity 结束的时候停止异步任务
112    @Override
113    protected void onDestroy() {
114        super.onDestroy();
115        //结束手势监听的事件
116        if (mTask != null
117                && mTask.getStatus() != GesturesLoadTask.Status.FINISHED) {
118            mTask.cancel(true);
119            mTask = null;
120        }
121        //清理对话框
122        cleanupRenameDialog();
123    }
124
125    //检测手势库的数据是否为空
126    private void checkForEmpty() {
127        if (mAdapter.getCount() == 0) {
128            mEmpty.setText(R.string.gestures_empty);
129        }
```

```java
130        }
131
132        //离开Activity时保存数据信息
133        @Override
134        protected void onSaveInstanceState(Bundle outState) {
135            super.onSaveInstanceState(outState);
136            //保存手势状态
137            if (mCurrentRenameGesture != null) {
138                outState.putLong(GESTURES_INFO_ID,
139                    mCurrentRenameGesture.gesture.getID());
140            }
141        }
142
143        //恢复Activity时恢复保存的数据信息
144        @Override
145        protected void onRestoreInstanceState(Bundle state) {
146            super.onRestoreInstanceState(state);
147            //得到之前保存的手势信息
148            long id = state.getLong(GESTURES_INFO_ID, -1);
149            if (id != -1) {
150                final Set<String> entries = sStore.getGestureEntries();
151                out: for (String name : entries) {           //找到手势对象
152                    for (Gesture gesture : sStore.getGestures(name)) {
153                        if (gesture.getID() == id) {           //获取手势信息
154                            mCurrentRenameGesture = new NamedGesture();
155                            mCurrentRenameGesture.name = name;
156                            mCurrentRenameGesture.gesture = gesture;
157                            break out;
158                        }
159                    }
160                }
161            }
162        }
163
164        //构造上下文菜单
165        @Override
166        public void onCreateContextMenu(ContextMenu menu, View v,
167                ContextMenu.ContextMenuInfo menuInfo) {
168            //创建菜单
169            super.onCreateContextMenu(menu, v, menuInfo);
170            //显示菜单的Adapter效果
171            AdapterView.AdapterContextMenuInfo info = (AdapterView.
                   AdapterContextMenuInfo) menuInfo;
172            menu.setHeaderTitle(((TextView) info.targetView).getText());
173            //添加菜单项
174            menu.add(0, MENU_ID_RENAME, 0, R.string.gestures_rename);
175            menu.add(0, MENU_ID_REMOVE, 0, R.string.gestures_delete);
176        }
177
178        //当菜单被单击时的回调方法
179        @Override
180        public boolean onContextItemSelected(MenuItem item) {
181            final AdapterView.AdapterContextMenuInfo menuInfo = (AdapterView.
                   AdapterContextMenuInfo) item
182                   .getMenuInfo();
            //得到用户的手势视图
183            final NamedGesture gesture = (NamedGesture) menuInfo.targetView
184                   .getTag();
            //根据菜单的选项id,重命名或者删除手势
```

```java
186         switch (item.getItemId()) {
187         case MENU_ID_RENAME:
188             renameGesture(gesture);
189             return true;
190         case MENU_ID_REMOVE:
191             deleteGesture(gesture);
192             return true;
193         }
194
195         return super.onContextItemSelected(item);
196     }
197
198     //重新命名手势
199     private void renameGesture(NamedGesture gesture) {
200         mCurrentRenameGesture = gesture;
201         showDialog(DIALOG_RENAME_GESTURE);
202     }
203
204     //创建Dialog
205     @Override
206     protected Dialog onCreateDialog(int id) {
            //重命名对话框
207         if (id == DIALOG_RENAME_GESTURE) {
208             return createRenameDialog();
209         }
210         return super.onCreateDialog(id);
211     }
212
213     //准备创建Dialog
214     @Override
215     protected void onPrepareDialog(int id, Dialog dialog) {
216         super.onPrepareDialog(id, dialog);
217         if (id == DIALOG_RENAME_GESTURE) {
218             mInput.setText(mCurrentRenameGesture.name);
219         }
220     }
221
222     //创建重命名对话框
223     private Dialog createRenameDialog() {
            //读取自定义的xml布局,得到view
224
225         final View layout = View.inflate(this, R.layout.dialog_rename, null);
            //得到布局中的相应控件
226
227         mInput = (EditText) layout.findViewById(R.id.name);
228         ((TextView) layout.findViewById(R.id.label))
229                 .setText(R.string.gestures_rename_label);
230
231         //实例化AlertDialog.Builder对象来创建AlertDialog
232         AlertDialog.Builder builder = new AlertDialog.Builder(this);
233         builder.setIcon(0);
234         builder.setTitle(getString(R.string.gestures_rename_title));
235         builder.setCancelable(true);
236         //设置按钮的取消监听器
237         builder.setOnCancelListener(new Dialog.OnCancelListener() {
238             public void onCancel(DialogInterface dialog) {
239                 cleanupRenameDialog();
240             }
241         });
242         //设置按钮取消按钮事件
243         builder.setNegativeButton(getString(R.string.cancel_action),
244                 new Dialog.OnClickListener() {
```

```java
245                    public void onClick(DialogInterface dialog, int which) {
246                        cleanupRenameDialog();
247                    }
248                });
249         //设置按钮的重命名按钮的事件监听器
250         builder.setPositiveButton(getString(R.string.rename_action),
251                 new Dialog.OnClickListener() {
252                    public void onClick(DialogInterface dialog, int which) {
253                        changeGestureName();
254                    }
255                });
256         builder.setView(layout);
257         return builder.create();
258     }
259
260     //修改手势的名字
261     private void changeGestureName() {
262         final String name = mInput.getText().toString();
263         if (!TextUtils.isEmpty(name)) {
264             final NamedGesture renameGesture = mCurrentRenameGesture;
265             final GesturesAdapter adapter = mAdapter;
266             final int count = adapter.getCount();
267
268             // 改变手势的名称
269             // 改变手势的 id 等信息
270             for (int i = 0; i < count; i++) {
271                 final NamedGesture gesture = adapter.getItem(i);
272                 if (gesture.gesture.getID() == renameGesture.gesture.getID()) {
273                     sStore.removeGesture(gesture.name, gesture.gesture);
274                     gesture.name = mInput.getText().toString();
275                     sStore.addGesture(gesture.name, gesture.gesture);
276                     break;
277                 }
278             }
279             //通知 adapter，数据已经修改
280             adapter.notifyDataSetChanged();
281         }
282         mCurrentRenameGesture = null;
283     }
284
285     //清理重命名对话框
286     private void cleanupRenameDialog() {
287         if (mRenameDialog != null) {
288             mRenameDialog.dismiss();
289             mRenameDialog = null;
290         }
291         mCurrentRenameGesture = null;
292     }
293
294     //删除手势
295     private void deleteGesture(NamedGesture gesture) {
296         sStore.removeGesture(gesture.name, gesture.gesture);
297         sStore.save();
298         //删除手势对象
299         final GesturesAdapter adapter = mAdapter;
300         adapter.setNotifyOnChange(false);
301         adapter.remove(gesture);
302         adapter.sort(mSorter);
303         checkForEmpty();
304         adapter.notifyDataSetChanged();
```

```java
305         //提示手势删除成功
306         Toast.makeText(this, R.string.gestures_delete_success,
307                 Toast.LENGTH_SHORT).show();
308     }
309
310     //手势读取的异步任务
311     private class GesturesLoadTask extends
312             AsyncTask<Void, NamedGesture, Integer> {
313         private int mThumbnailSize;
314         private int mThumbnailInset;
315         private int mPathColor;
316
317         @Override
318         protected void onPreExecute() {
319             super.onPreExecute();
320             //获取系统资源
321             final Resources resources = getResources();
322             mPathColor = resources.getColor(R.color.gesture_color);
323             //得到距离常量
324             mThumbnailInset = (int) resources
325                     .getDimension(R.dimen.gesture_thumbnail_inset);
326             mThumbnailSize = (int) resources
327                     .getDimension(R.dimen.gesture_thumbnail_size);
328             //设置对应按钮的 enable 属性
329             findViewById(R.id.addButton).setEnabled(false);
330             findViewById(R.id.reloadButton).setEnabled(false);
331
332             mAdapter.setNotifyOnChange(false);
333             mAdapter.clear();
334         }
335
336         //后台执行的内容
337         @Override
338         protected Integer doInBackground(Void... params) {
339             //判断是否被取消
340             if (isCancelled())
341                 return STATUS_CANCELLED;
342             if (!Environment.MEDIA_MOUNTED.equals(Environment
343                     .getExternalStorageState())) {
344                 return STATUS_NO_STORAGE;
345             }
346
347             //读取手势库
348             final GestureLibrary store = sStore;
349             //判断手势是否读取成功
350             if (store.load()) {
351                 for (String name : store.getGestureEntries()) {
352                     if (isCancelled())
353                         break;
354                     //循环得到手势的对象
355                     for (Gesture gesture : store.getGestures(name)) {
356                         final Bitmap bitmap = gesture.toBitmap
                                (mThumbnailSize, mThumbnailSize,
357                                 mThumbnailInset, mPathColor);
358                         final NamedGesture namedGesture = new NamedGesture();
359                         namedGesture.gesture = gesture;
360                         namedGesture.name = name;
361                         //设置手势对应的视图效果
362                         mAdapter.addBitmap(namedGesture.gesture.
```

```
                        getID(), bitmap);
363                     publishProgress(namedGesture);
364             }
365         }
366
367         return STATUS_SUCCESS;
368     }
369
370     return STATUS_NOT_LOADED;
371 }
372
373 //更新读取进度
374 @Override
375 protected void onProgressUpdate(NamedGesture... values) {
376     super.onProgressUpdate(values);
377     //定义手势 adapter
378     final GesturesAdapter adapter = mAdapter;
379     adapter.setNotifyOnChange(false);
380     //添加手势
381     for (NamedGesture gesture : values) {
382         adapter.add(gesture);
383     }
384     //手势排序
385     adapter.sort(mSorter);
386     adapter.notifyDataSetChanged();
387 }
388
389 //执行异步任务前的回调函数
390 @Override
391 protected void onPostExecute(Integer result) {
392     super.onPostExecute(result);
393     //异步任务的状态成功
394     if (result == STATUS_NO_STORAGE) {
395         getListView().setVisibility(View.GONE);
396         mEmpty.setVisibility(View.VISIBLE);
397         mEmpty.setText(getString(R.string.gestures_error_loading,
398                 mStoreFile.getAbsolutePath()));
399     } else {    //异步任务完成，修改按钮的属性
400         findViewById(R.id.addButton).setEnabled(true);
401         findViewById(R.id.reloadButton).setEnabled(true);
402         checkForEmpty();
403     }
404 }
405 }
406 //等译命名手势类
407 static class NamedGesture {
408     String name;
409     Gesture gesture;
410 }
411
412 //自定义的手势适配器
413 private class GesturesAdapter extends ArrayAdapter<NamedGesture> {
414     private final LayoutInflater mInflater;
415     private final Map<Long, Drawable> mThumbnails = Collections
416             .synchronizedMap(new HashMap<Long, Drawable>());
417     //定义手势适配器
418     public GesturesAdapter(Context context) {
419         super(context, 0);
420         mInflater = (LayoutInflater) context
421                 .getSystemService(Context.LAYOUT_INFLATER_SERVICE);
```

```
422         }
423
424         //添加手势 bitmap
425         void addBitmap(Long id, Bitmap bitmap) {
426             mThumbnails.put(id, new BitmapDrawable(bitmap));
427         }
428
429         //得到每一个 item 的视图
430         @Override
431         public View getView(int position, View convertView, ViewGroup parent) {
432             if (convertView == null) {
433                 convertView = mInflater.inflate(R.layout.gestures_item, parent,
434                         false);
435             }
436             //通过点击位置得到手势对象
437             final NamedGesture gesture = getItem(position);
438             final TextView label = (TextView) convertView;
439             //设置对象的相关显示信息
440             label.setTag(gesture);
441             label.setText(gesture.name);
442             label.setCompoundDrawablesWithIntrinsicBounds(
443                     mThumbnails.get(gesture.gesture.getID()), null,
                        null, null);
444
445             return convertView;
446         }
447     }
448 }
```

这是当前实例的主页面,显示手势库中已有的手势列表,包括对于手势库列表中 item 的重命名和删除功能。在第 26~31 行定义了手势的比较器。在第 50~52 行定义手势列表的 adapter。在第 70~73 行实现了手势的读取功能。在第 76~80 行实现了从添加手势页面返回的结果值。在第 82~86 行实现了识别手势的页面返回值。在第 164~176 行定义了当单击某一个手势 item 的时候弹出的菜单。在第 198~202 行实现弹出重命名和删除 item 的对话框。在第 222~258 行定义了修改手势库中手势名字的对话框。在第 260~283 行定义了删除手势库中手势的对话框。在第 331 行定义了手势读取的异步任务。在第 413 行定义了手势显示的 Adapter。

4．实例扩展

本实例是 Google 自带的一个实例,之所以在这里讲解这个实例,一方面希望大家能够多多阅读 Google 自带的实例代码,另一方面很多朋友在手势库的添加过程中遇到了很多问题,在网上给我进行了很多的留言,借此机会来给大家一个相对完整的手势库的创建和识别的例子。

范例 074 滑动切换 Activity 的背景效果

1．实例简介

在 Android 中很多应用都已经添加了手势识别的操作,为了方便用户对于某些常用功能的操作,例如,浏览器的上一页、下一页和滑动切换当前页面的背景。本实例就带领大家一起来完成一个滑动切换 Activity 背景效果。

2．运行效果

该实例运行效果如图 4.21 所示。

3．实例程序讲解

本实例实现了一个简单的登录页面，它特殊的地方就在于当你水平滑动屏幕的时候，此 Activity 的背景可以进行改变。想要实现这样的效果首先要修改 res/layout/activity_main.xml 文件，代码如下：

图 4.21 滑动切换 Activity 背景效果

```xml
01  <?xml version="1.0" encoding="utf-8"?>
02  <!-- 定义基础的 LinearLayout 布局 -->
03  <LinearLayout xmlns:android="http://schemas.android.com/apk/res/android"
04      android:id="@+id/Ll"
05      android:layout_width="fill_parent"
06      android:layout_height="fill_parent"
07      android:background="@drawable/k1"
08      android:orientation="vertical" >
09
10      <!-- 定义 EditText 输入框 -->
11      <EditText
12          android:layout_width="fill_parent"
13          android:layout_height="wrap_content"
14          android:hint="请输入用户名： " />
15      <!-- 定义 EditText 输入框 -->
16      <EditText
17          android:layout_width="fill_parent"
18          android:layout_height="wrap_content"
19          android:hint="请输入密码： " />
20      <!-- 定义登录 Button -->
21      <Button
22          android:layout_width="fill_parent"
23          android:layout_height="wrap_content"
24          android:text="登录"
25          />
26  </LinearLayout>
```

在此布局定义了我们比较常见的一个登录框。

然后修改 res/layout/activity_main.xml 文件，代码如下：

```java
01  //定义 MainActivity 继承自 Activity
02  public class MainActivity extends Activity {
03      // 定义保存 ImageView 的对象
04      private LinearLayout Ll;
05      //定义手势检测器对象
06      private GestureDetector gestureDetector;
07      //定义图片的资源数组
08      private int[] ResId = new int[]{
09          R.drawable.k1,
10          R.drawable.k2,
11          R.drawable.k3
12      };
13      //定义当前显示的图片的下标
```

```java
14  private int count = 0;
15
16  @Override
17  public void onCreate(Bundle savedInstanceState) {
18      super.onCreate(savedInstanceState);
19      //设置当前页面的布局视图为 activity_main
20      setContentView(R.layout.activity_main);
        //得到当前页面中的控件对象
21      findView();
22      setListener();         //设置监听器
23  }
24
25  private void setListener() {
26      //设置手势监听器的处理效果由 onGestureListener 来处理
27      gestureDetector = new GestureDetector(MainActivity.this,
28              onGestureListener);
29  }
30
31  @Override
32  public boolean onTouchEvent(MotionEvent event) {
33      //当前 Activity 被触摸时回调
34      return gestureDetector.onTouchEvent(event);
35  }
36
37  private void findView() {
38      //得到当前页面的 imageview 控件
39      Ll = (LinearLayout) findViewById(R.id.Ll);
40  }
41  //定义了 GestureDetector 的手势识别监听器
42  private GestureDetector.OnGestureListener onGestureListener =
                  new GestureDetector.SimpleOnGestureListener() {
43          //当识别的手势是滑动手势时回调 onFinger 方法
44          @Override
45          public boolean onFling(MotionEvent e1, MotionEvent e2, float velocityX,
46                  float velocityY) {
47              //得到滑动手势的起始和结束点的 x,y 坐标,并进行计算
48              float x = e2.getX() - e1.getX();
49              float y = e2.getY() - e1.getY();
50
51              //通过计算结果判断用户是向左滑动或者向右滑动
52              if (x > 0) {
53                  count++;              //计数变量++
54                  count %= 3;
55              } else if (x < 0) {
56                  count--;              //计数变量--
57                  count = (count + 3) % 3;
58              }
59              //切换 imageview 的图片
60              changeImg();
61              return true;
62          }
63  };
64
65  public void changeImg() {
66      //设置当前位置的图片资源
67      Ll.setBackgroundResource(ResId[count]);
68  }
69  }
```

在如上代码中第 6 行定义了一个 GestureDetector 对象,用来识别手势,在此文件的第 31～35 行重写了 onTouchEvent 方法,用来处理用户对于屏幕的滑动事件,在第 42 行定义了一个手势监听器的对象,其中实现了 onFling 方法,当用户滑动屏幕的时候此方法就会回调。

4．实例扩展

其实 Android 中手势滑动的效果应用的场景很多很多,因此今后的手机基本上都会以屏幕的触摸和各种手势来操纵。

范例 075　按钮控制小人儿移动

1．实例简介

在我们常见的 Android 游戏中,触摸屏幕控制人物的移动是其中的一种方式,还有一种方式就是单击某些按钮时,游戏人物位置的移动。本实例就带领大家通过 Android 中的按钮来实现控制游戏中的人物移动的实例。

2．运行效果

该实例运行效果如图 4.22 所示。

3．实例程序讲解

图 4.22　通过按钮来控制小人儿的移动

在上面的例子效果中通过单击对应的向上、向下、向左和向右的按钮就可以控制图片的移动。想要实现如上效果。首先要新建 src/com.wyl.example/MyViewGroup.java 文件,代码如下:

```
01  //自定义 MyViewGroup 继承自 ViewGroup
02  public class MyViewGroup extends LinearLayout {
03      //定义滚动器对象
04      private Scroller mScroller;
05      //定义每次移动的位置间隔
06      private int space = 10;
07      //自定义布局的构造方法
08      public MyViewGroup(Context context) {
09          super(context);
10          initCustomLinearLayout(context);
11      }
12      //自定义布局的带参数的构造方法
13      public MyViewGroup(Context context, AttributeSet attrs) {
14          super(context, attrs);
15          initCustomLinearLayout(context);
16      }
17      //自定义布局的向右滚动方法
18      public void scrollToRight() {
19          //以提供的起始点和将要滑动的距离开始滚动。滚动会使用默认值 250ms 作为持续时间
20          mScroller.startScroll(mScroller.getCurrX(),mScroller.getCurrY(),
```

```
21              -space,0, 250);
22          //重绘视图
23          invalidate();
24      }
    //自定义布局的向上滚动方法
25  public void scrollToUp() {
26      //以提供的起始点和将要滑动的距离开始滚动。滚动会使用默认值250ms作为持续时间
27      mScroller.startScroll(mScroller.getCurrX(),mScroller.getCurrY(),
28              0, space, 250);
29      //重绘视图
30      invalidate();
31  }
    //自定义布局的向下滚动方法
32  public void scrollToDown() {
33      //以提供的起始点和将要滑动的距离开始滚动。滚动会使用默认值250ms作为持续时间
34      mScroller.startScroll(mScroller.getCurrX(),mScroller.getCurrY(),
35              0, -space, 250);
36      //重绘视图
37      invalidate();
38  }
    //自定义布局的向左滚动方法
39
40  public void scrollToLeft() {
41      //以提供的起始点和将要滑动的距离开始滚动。滚动会使用默认值250ms作为持续时间
42      mScroller.startScroll(mScroller.getCurrX(),mScroller.getCurrY(),
43              space, 0,250);
44      //重绘视图
45      invalidate();
46  }
47  @Override
48  public void onLayout(boolean changed,int left,int top,int right, int bottom) {
49      super.onLayout(changed, left, top, right, bottom);
50  }
51
52  @Override
53  public void computeScroll() {
54      //判断滚动是否结束
55      if (!mScroller.isFinished()) {
56          // computeScrollOffset() 返回false表示滚动结束
57          if (mScroller.computeScrollOffset()) {
58              //设置滚动到何处
59              scrollTo(mScroller.getCurrX(), mScroller.getCurrY());
60
61              //让系统重绘视图
62              invalidate();
63          }
64      }
65
66  }
    //初始化滚动其对象
67
68  private void initCustomLinearLayout(Context context) {
69      mScroller = new Scroller(context);
70  }
71 }
```

在此文件中定义了一个自己的 LinearLayout 类。在如上代码中第 2 行此类继承自 LinearLayout 类，在第 8 和 13 行定义了两个构造方法。在第 18 行、第 25 行、第 32 行和第 40 行定义了当用户单击按钮后的移动图片的事件。在第 53~64 行实现了计算滚动是否

结束的方法。

然后修改 res/layout/activity_main.xml 文件，代码如下：

```xml
01  <?xml version="1.0" encoding="utf-8"?>
02  <LinearLayout xmlns:android="http://schemas.android.com/apk/res/android"
03      android:id="@+id/layout"
04      android:layout_width="fill_parent"
05      android:layout_height="fill_parent"
06      android:orientation="vertical"
07      android:background="@color/skyblue">
08
09      <!-- 定义自定义的视图组 -->
10      <com.wyl.example.MyViewGroup
11          android:id="@+id/mygroupview"
12          android:layout_width="fill_parent"
13          android:layout_height="fill_parent"
14          android:layout_weight="1" />
15
16      <!-- 实现左右移动的按钮 -->
17      <LinearLayout
18          android:layout_width="wrap_content"
19          android:layout_height="wrap_content"
20          android:orientation="horizontal"
21          android:layout_gravity="center_horizontal">
22
23          <Button
24              android:id="@+id/BtnLeft"
25              android:layout_width="wrap_content"
26              android:layout_height="wrap_content"
27              android:text="@string/left_button" />
28
29          <Button
30              android:id="@+id/BtnRight"
31              android:layout_width="wrap_content"
32              android:layout_height="wrap_content"
33              android:text="@string/right_button" />
34      </LinearLayout>
35      <!-- 实现上下移动的按钮 -->
36      <LinearLayout
37          android:layout_width="wrap_content"
38          android:layout_height="wrap_content"
39          android:orientation="horizontal"
40          android:layout_gravity="center_horizontal">
41          <Button
42              android:id="@+id/BtnUp"
43              android:layout_width="wrap_content"
44              android:layout_height="wrap_content"
45              android:text="@string/up_button" />
46          <Button
47              android:id="@+id/BtnDown"
48              android:layout_width="wrap_content"
49              android:layout_height="wrap_content"
50              android:text="@string/down_button" />
51      </LinearLayout>
52
53  </LinearLayout>
```

此文件是本实例的主页面的布局，在此布局中，第 10～14 行定义了自定义的 ViewGroup 对象。然后分别定义了向上、向下、向左和向右移动的按钮控件。

然后修改 src/com.wyl.example/MainActivity.java 文件，代码如下：

```
01  //定义MainActivity继承自Activity
02  public class MainActivity extends Activity {
03      //定义上下左右移动的四个按钮对象
04      private Button rightButton;
05      private Button leftButton;
06      private Button downButton;
07      private Button upButton;
08      //定义视图组
09      private MyViewGroup myviewgroup;
10  
11      @Override
12      public void onCreate(Bundle savedInstanceState) {
13          super.onCreate(savedInstanceState);
14          setContentView(R.layout.activity_main);
15          WindowManager wm = (WindowManager) this.getSystemService
            (Context.WINDOW_SERVICE);
16          //屏幕宽度
17          int width = wm.getDefaultDisplay().getWidth();
18          //屏幕高度
19          int height = wm.getDefaultDisplay().getHeight();
20  
21  
22          //获取自定义ViewGroup
23          myviewgroup = (MyViewGroup) findViewById(R.id.mygroupview);
24  
25          //定义移动的Textview对象
26          TextView textView = new TextView(this);
27          //设置TextView的文字、背景及控件宽度高度
28          textView.setWidth(80);
29          textView.setHeight(80);
30          textView.setLayoutParams(new LayoutParams(80, 80));
31          textView.setBackgroundResource(R.drawable.ic_launcher);
32          //添加到自定义视图组中
33          myviewgroup.setGravity(Gravity.CENTER);
34          myviewgroup.addView(textView);
35          //得到页面中的控件对象
36          findView();
37          setListener();
38      }
39  
40      private void setListener() {
41          //设置上下左右按钮的监听器
42          rightButton.setOnClickListener(new OnClickListener() {
43  
44              @Override
45              public void onClick(View v) {
46                  myviewgroup.scrollToRight();
47              }
48          });
49          leftButton.setOnClickListener(new OnClickListener() {
50  
51              @Override
52              public void onClick(View v) {
53                  myviewgroup.scrollToLeft();
54              }
55          });
56          downButton.setOnClickListener(new OnClickListener() {
```

```
57
58          @Override
59          public void onClick(View v) {
60              myviewgroup.scrollToDown();
61          }
62      });
63      upButton.setOnClickListener(new OnClickListener() {
64
65          @Override
66          public void onClick(View v) {
67              myviewgroup.scrollToUp();
68          }
69      });
70  }
71
72  private void findView() {
73      //获得向右按钮
74      rightButton = (Button) findViewById(R.id.BtnRight);
75      //获得向左按钮
76      leftButton = (Button) findViewById(R.id.BtnLeft);
77      //获得向下按钮
78      downButton = (Button) findViewById(R.id.BtnDown);
79      //获得向上按钮
80      upButton = (Button) findViewById(R.id.BtnUp);
81  }
82  }
```

此代码是当前 Activity 的代码，在其中的 onCreate 方法中初始化了页面中的控件。在第 40 行设置了四个按钮的监听器，分别调用 myviewgroup 的 scrollToRight、scrollToLeft、scrollToDown 和 scrollToUp 方法来移动图中的 ImageView。这样就能够实现如图 4.22 实例的效果了。

4．实例扩展

本实例实现了通过按钮控制图片移动的效果，在我们经常玩儿的 RPG 游戏中经常会看到这样的效果。而且对于我们经常玩儿的华容道和拖箱子，这样的游戏也要使用这种方式来识别用户的操作动作，大家根据自己的需求进行设计就可以了。

4.3　Android 中多线程处理

范例 076　异步请求广告图片

1．实例简介

我们在使用 Android 应用的时候，经常会看到一些网络图片，它们不是存储在你的手机上的，而是从网络上下载下来的，我们这次的内容不是网络下载，而是在下载过程中有可能会需要一定的时间。例如，如果是大图片的话，时间比较长；如果用手机 2/3G 网络的话，比 WIFI 的时间长等。这样我们就需要了解 Android 中的异步加载机制。本实例就带领大家通过 Android 中的 Handler 来完成一个常见的网络广告图片加载的实例。

2. 运行效果

该实例运行效果如图 4.23 所示。

3. 实例程序讲解

上面的例子效果中我们在程序运行开始的时候,只看到了三个空白的 imageView,如果网络条件通畅的话,在一段时间后会自动加载网络上我指定的几张图片。想要实现这个效果,首先要修改 res/layout/activity_main.xml 文件,代码如下:

图 4.23 网络广告图片加载

```xml
01 <?xml version="1.0" encoding="utf-8"?>
02 <!-- 定义页面的基本布局,滚动布局 -->
03 <ScrollView xmlns:android="http://schemas.android.com/apk/res/android"
04     android:layout_width="fill_parent"
05     android:layout_height="fill_parent" >
06     <!-- 在滚动布局中加入线性布局 -->
07     <LinearLayout
08         android:id="@+id/layout"
09         android:layout_width="fill_parent"
10         android:layout_height="fill_parent"
11         android:layout_gravity="center_horizontal"
12         android:background="@color/skyblue"
13         android:orientation="vertical" >
14
15     <!-- 加入图片控件 1 -->
16         <ImageView
17             android:id="@+id/Iv1"
18             android:layout_width="wrap_content"
19             android:layout_height="wrap_content" />
20     <!-- 加入图片控件 2 -->
21         <ImageView
22             android:id="@+id/Iv2"
23             android:layout_width="wrap_content"
24             android:layout_height="wrap_content" />
25     <!-- 加入图片控件 3 -->
26         <ImageView
27             android:id="@+id/Iv3"
28             android:layout_width="wrap_content"
29             android:layout_height="wrap_content" />
30     </LinearLayout>
31
32 </ScrollView>
```

在如上代码中第 7~30 行定义了一个 LinearLayout,在此 LinearLayout 中定义了三个 ImageView,并且分别设置了对应的 id,方法在后面的 Activity 中得到对应的对象,并且给此 LinearLayout 设置的 orientation 为 vertical,代表垂直排列这些 ImageView。

然后新建 src/com.wyl.example/ ImageHandle.java 文件,代码如下:

```java
01 //定义 ImageHandle 继承自 Handler 类
02 public class ImageHandle extends Handler{
03
04     //定义一个 Activity 的上下文对象
```

```
05  Activity context;
06
07  //定义了 ImageHandle 的构造函数,传入上下文对象
08  public ImageHandle(Activity context){
09      this.context=context;
10  }
11
12  //接收 thread 发送过来的 message 信息
13  @Override
14  public void handleMessage(Message msg) {
15      super.handleMessage(msg);
16      //从 message 信息中得到从网络请求下来的图片信息,并且设置到 imageview 对象上
17      ImageView img=(ImageView)context.findViewById(msg.arg1);
18      img.setImageDrawable((Drawable)msg.obj);
19  }
20
21  //读取网络图片的函数,第一个参数为网络图片的 url,第二个图片为需要设置的图片控件的 id
22  public void loadImg(final String imgUrl,final int viewId){
23      //建立一个线程
24      Thread thread=new Thread(){
25          @Override
26          public void run() {
27              try {
28                  //读取网络上的图片
29                  Drawable drawable=
                        Drawable.createFromStream(new URL(imgUrl).
                        openStream(), "img.png");
30                  //定义 Message 对象
31                  Message msg=ImageHandle.this.obtainMessage();
32                  //设置 Message 对象的参数
33                  msg.arg1=viewId;
34                  msg.obj=drawable;
35                  //调用 handle 的 sendMessage 函数
36                  ImageHandle.this.sendMessage(msg);
37              } catch (Exception e) {
38                  e.printStackTrace();
39              }
40          }
41      };
42      //启动线程
43      thread.start();
44      thread=null;
45  }
46  }
```

在此 Java 文件中定义了一个类,继承自 Handler,在第 8 行定义了此类的构造方法,并且实现了 Handler 类的 handleMessage 方法,用来接收传入的消息,在此传入消息中包括了网络上加载下来的图片资源。在第 22~45 行,实现了如何从网络上获取图片的方法,这里通过 Drawable 类的 createFromStream 方法得到 imgUrl 地址对应的图片对象,并且发送 Message 消息给当前的 handler 对象。

然后修改 src/com.wyl.example/MainActivity.java 文件,代码如下:

```
01  //定义 MainActivity 继承自 Activity
02  public class MainActivity extends Activity {
03  @Override
04  public void onCreate(Bundle savedInstanceState) {
05      super.onCreate(savedInstanceState);
```

```
06          setContentView(R.layout.activity_main);
07          //开启一个加载图片的handler,设置Iv1的图片
08          ImageHandle imgHandle1 =new ImageHandle(this);
09          imgHandle1.loadImg("http://www.baidu.com/img/baidu_logo.gif",R.id.Iv1);
10
11          //开启一个加载图片的handler,设置Iv2的图片
12          ImageHandle imgHandle2 =new ImageHandle(this);
13          imgHandle2.loadImg(
                    "http://a2.att.hudong.com/10/96/300000931099127952960461732
                    .jpg", R.id.Iv2);
14
15          //开启一个加载图片的handler,设置Iv3的图片
16          ImageHandle imgHandle3 =new ImageHandle(this);
17          imgHandle3.loadImg(
                    "http://www.chenguangblog.com/wp-content/uploads/2011/07/
                    sbaidu.jpg", R.id.Iv3);
18     }
19  }
```

在此代码中第 8~9 行、第 12~13 行和第 16~17 行，分别调用了三次 ImageHander 去加载三张网络的图片，然后设置到三个 ImageView 的 id 对应的控件中。

4．实例扩展

本实例只实现了一种最简单的异步图片加载，当然要了解异步加载的原理，首先大家需要了解 JavaSE 中的多线程的内容，还有什么是 Android 中的 UI 线程，以及在除了 UI 线程以外的线程无法更新 UI 的问题，这时候你才能真正了解异步任务的意义所在。

范例 077 本地三国演义文本的异步加载

1．实例简介

我们在使用 Android 应用的时候，不单是网络应用会很耗时，有时候对于一些本地应用也会消耗很多的时间。例如，本地文件的压缩、解压缩和本地数据库的大量操作，本地大文件的读取操作等。这样就需要我们大家在做本地操作的时候，如果此操作需要花费大量的时间，那么也应该加上异步操作。本实例就带领大家通过 Android 中读取本地文件，完成一个本地三国演义文本的异步加载。

2．运行效果

该实例运行效果如图 4.24 所示。

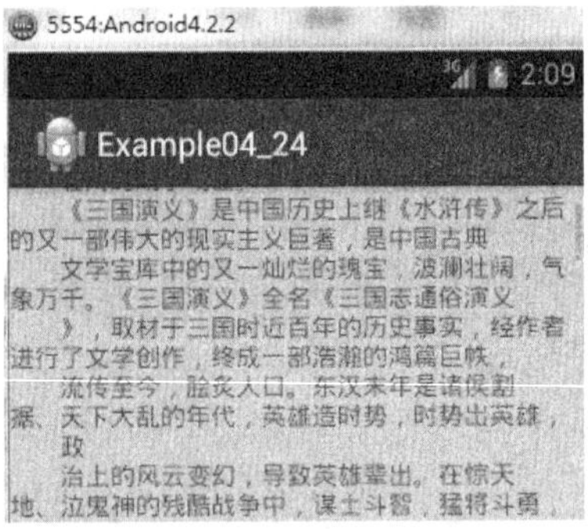

3．实例程序讲解

在上面的例子效果中，首先我需要用到另一个外部的大文本文件，这里提前在 AVD 的 sdcard 根目录下放置了一个 sanguo.txt，大约几百 K 的大小，这样的话当我运行此实例的时候会发现我们看到了一个 TextView 显示文件中的内容，但是并不是一次性全部显示出来的，因为如果这样的话，你的应用程序启动起来会很慢，我们的程序效果是文本内容分阶段加载，通过我们应用右侧的滚动条

图 4.24 本地三国演义文本的异步加载

的状态也可以看到。想要实现我们这个实例，首先要修改 res/layout/activity_main.xml 文件，代码如下：

```xml
01  <?xml version="1.0" encoding="utf-8"?>
02  <!-- 定义页面的基本布局，滚动布局 -->
03  <ScrollView xmlns:android="http://schemas.android.com/apk/res/android"
04      android:layout_width="fill_parent"
05      android:layout_height="fill_parent" >
06
07      <!-- 在滚动布局中加入线性布局 -->
08      <LinearLayout
09          android:id="@+id/layout"
10          android:layout_width="fill_parent"
11          android:layout_height="fill_parent"
12          android:layout_gravity="center_horizontal"
13          android:background="@color/skyblue"
14          android:orientation="vertical" >
15      <!-- 定义显示小说文本的 TextView -->
16      <TextView
17          android:id="@+id/Tv"
18          android:layout_width="fill_parent"
19          android:layout_height="wrap_content" />
20      </LinearLayout>
21
22  </ScrollView>
```

在如上代码中第 8～20 行定义了一个 LinearLayout——线性布局。在此线性布局中定义了一个 TextView 对象，并且设置了 id 属性，方便我们在 Activity 中得到此控件。

然后新建 src/com.wyl.example/ FileRead.java 文件，代码如下：

```java
01  //定义文件读取类
02  public class FileRead {
03      //标记文件是否读取完毕
04      public boolean readfinish = false;
05      //记录已经读取的字符串列表
06      public List<String> list = null;
07
08      //定义了读取文件的线程
09      public class ReadFileThread extends Thread {
10
11          //在此线程中异步读取文件
12          public void run() {
13              //初始化 ArrayList 数组
14              list = new ArrayList<String>(100);
15              //将已有的数组数据清除
16              list.clear();
17              //表示还没有读取完毕
18              readfinish = false;
19              //读取 sdcard 下面的 sanguo.txt
20              try {
21                  //定义随机读取的文件对象
22                  RandomAccessFile raf = new RandomAccessFile("/sdcard/sanguo.
23                  txt","r");
24                  //当文件没有读取到文件结束的时候，一直读取，并且加入 arraylist 中
25                  while (raf.getFilePointer() < raf.length()) {
26                      //把文件中读取的字符串数据进行字符集转换
```

```
27                      list.add(new String(raf.readLine().getBytes("iso8859-1"),
                            "utf-8"));
28                  }
29
30              } catch (Exception e1) {
31                  // TODO Auto-generated catch block
32                  e1.printStackTrace();
33              }
34              //标记文件读取完毕
35              readfinish = true;
36          }
37      };
38  }
```

此类继承自 Thread 类，并且在第 12～36 行实现了父类的 run 方法，在此方法中定义了一个 String 数组，然后通过 RandomAccess 类，读取本地的 sanguo.txt 文件，然后通过循环依次读取文件内容。最后放入 List 中。

然后修改 src/com.wyl.example/MainActivity.java 文件，代码如下：

```
01  //定义 MainActivity 继承自 Activity
02  public class MainActivity extends Activity {
03      //定义一个自定义的 FileRead 类的
04      FileRead fr = null;
05      //定义接收子线程发送的 message
06      Handler mHandler = null;
07
08      int curi = 0;
09      //更新 UI 的线程
10      Runnable updateui = null;
11      //接收已经加载完毕的字符串数组
12      String[] tmp = null;
13      String s = "";
14      TextView tv = null;
15
16      //定义数据监听线程,当已经读取的字符串的数量超过 10 条时更新 UI 内容
17      class ReadListener extends Thread {
18          public void run() {
19              int i = 0, newi = 0;
20              //判断 fileread 是否读取完毕
21              while (!fr.readfinish) {
22                  //得到已经读取的数据的条数
23                  newi = fr.list.size();
24                  //已经读取的数据的条数大于 10 条时,进行 UI 界面的更行
25                  if ((newi - i) > 10)
26                  {
27                      i = newi;
28                      tmp = (String[]) fr.list.toArray(new String[fr.list.
                            size()]);
29                      //发送更新的信息给 handler
30                      mHandler.post(updateui);
31                  }
32              }
33              //当数据读取完毕后,发送更新请求
34              tmp = (String[]) fr.list.toArray(new String[fr.list.size()]);
35              mHandler.post(updateui);
36          }
37      };
```

```
38
39    @Override
40    public void onCreate(Bundle savedInstanceState) {
41        super.onCreate(savedInstanceState);
42        setContentView(R.layout.activity_main);
43        //得到页面中的 TextView 控件
44        tv = (TextView) findViewById(R.id.Tv);
45        //得到 FileRead 类的对象
46        fr = new FileRead();
47        ReadFileThread readThread = fr.new ReadFileThread();
48        updateui = new Runnable()// 更新 UI 的线程
49        {
50            @Override
51            public void run() {
52                // TODO Auto-generated method stub
53
54                int i = 0;
55                //得到字符串数组合并为一个字符串
56                for (i = curi; i < tmp.length; i++) {
57                    s += tmp[i] + "\n";
58                }
59                //设置 TextView 的内容为合并后的字符串
60                tv.setText(s);
61                curi = i;
62            }
63        };
64        //开启异步读取线程
65        readThread.start();
66        ReadListener updateThread = new ReadListener();
67        mHandler = new Handler();
68        updateThread.start();
69    }
70    }
```

在此代码中定义了当前实例的唯一一个 Activity，在第 4 行定义了一个 FileRead 对象，方便我们来读取文件，在第 6 行定义了一个当前线程的 Handler 对象，在第 10 行定义了 Runnable 对象。在第 17～37 行定义了读取文件的线程，时刻检查 FileRead 对象是否读取完毕，并且发送读取进度给当前 Activity 的 Handler 对象。在第 44 行得到当前布局中的 TextView 对象，在第 48～63 行用来更新 TextView 的显示文字。在第 68 行启动更新线程。这样就可以时刻根据文件的读写速度更新 TextView 的显示状态了。

4．实例扩展

本文实例只是演示了在 Android 中读取本地大文件的时候需要花费大量的时间。当然真实项目中经常用到的文本或者内容都是从网络获取的，在后面的内容中我们会学习如何获取网络中的数据。

范例 078　应用程序的启动动画

1．实例简介

我们经常会在使用 Android 应用的时候，看到这样一种效果，就是在程序启动的时候显示程序的 Logo 图片，然后显示指定的时间后，程序自动跳转到应用首页。例如，常见的手机 QQ 软件、淘宝客户端和 QQ 网购客户端等基本都有这种程序启动的动画。这个动画是自动完成的，所以我们之前学过的内容还无法解决。本实例就带领大家使用 Android 中的 Handler 的 postDelayed 来完成一个软件启动的 Logo 展示页面。

2. 运行效果

该实例运行效果如图 4.25 所示。

图 4.25 应用程序的启动 Logo 界面

3. 实例程序讲解

在上面的例子效果中可以看到，应用程序首先显示一个 Activity，其中显示了程序的 Logo 图片，当等待指定的时间后，自动跳转到程序的主页面。对于我们这个实例，首先要修改 res/layout/activity_main.xml 文件，代码如下：

```
01  <?xml version="1.0" encoding="utf-8"?>
02  <!-- 在滚动布局中加入线性布局 -->
03  <LinearLayout xmlns:android="http://schemas.android.com/apk/res/android"
04      android:id="@+id/layout"
05      android:layout_width="fill_parent"
06      android:layout_height="fill_parent"
07      android:layout_gravity="center_horizontal"
08      android:orientation="vertical" >
09      <!-- 启动 Logo 的 ImageView 控件 -->
10      <ImageView
11          android:layout_width="fill_parent"
12          android:layout_height="fill_parent"
13          android:scaleType="fitXY"
14          android:src="@drawable/baidu" />
15
16  </LinearLayout>
```

在如上代码中第 10～14 行定义了一个 ImageView 控件。这个控件的 src 为我之前放到工程的 drawable 目录中的 baidu.jpg，也就是我们的模拟 Logo 图片。

然后新建 res/layout/ activity_next.xml 文件，代码如下：

```
01  <?xml version="1.0" encoding="utf-8"?>
02  <!-- 在滚动布局中加入线性布局 -->
```

```
03  <LinearLayout xmlns:android="http://schemas.android.com/apk/res/android"
04      android:id="@+id/layout"
05      android:layout_width="fill_parent"
06      android:layout_height="fill_parent"
07      android:layout_gravity="center_horizontal"
08      android:orientation="vertical" >
09      <!-- 定义基本的文字标签  -->
10      <TextView
11          android:layout_width="fill_parent"
12          android:layout_height="wrap_content"
13          android:text="I am NextActivity!" />
14
15  </LinearLayout>
```

在如上代码中第 10~13 行定义了一个 TextView 控件。这个控件用来标记当前为新的 Activity 页面。

然后新建 src/com.wyl.example/ NextActivity.java 文件，代码如下：

```
1  //定义 MainActivity 继承自 Activity
2  public class NextActivity extends Activity {
3
4    @Override
5    public void onCreate(Bundle savedInstanceState) {
6        super.onCreate(savedInstanceState);
7        setContentView(R.layout.activity_next);
8    }
9  }
```

在此 Activity 中加载 activity_next.xml 布局，代表程序的主页面。

然后修改 src/com.wyl.example/MainActivity.java 文件，代码如下：

```
01  //定义 MainActivity 继承自 Activity
02  public class MainActivity extends Activity {
03
04  @Override
05  public void onCreate(Bundle savedInstanceState) {
06      super.onCreate(savedInstanceState);
07      setContentView(R.layout.activity_main);
08
09      new Handler().postDelayed(new Runnable(){
10          // 为了减少代码使用匿名 Handler 创建一个延时的调用
11                  public void run() {
12                      Intent i = new Intent(MainActivity.this, NextActivity.
                        class);
13                      //通过 Intent 打开最终真正的主界面 Main 这个 Activity
14                      MainActivity.this.startActivity(i);//启动 Main 界面
15                      MainActivity.this.finish();     //关闭自己这个开场屏
16                  }
17          }, 3000);    //logo 图片暂停 3 秒钟
18  }
19  }
```

在此代码中的第 7 行加载当前页面的布局为 activity_main.xml。在第 9~17 行，定义了一个 Handler，然后调用了 postDelayed 方法，延迟发送一个 Runnable 对象，这里延迟时间为 3 秒，也就是当用户打开此页面后，此页面显示三秒后，自动执行此 Runnable 对象。在此对象的 run 方法中通过 Intent 跳转到 NextActivity 页面，并且把当前的 Activity 结束。

4．实例扩展

在我们本例中应用到了 Activity 中的 finish 方法，由于在 Android 中 Activity 的存储方式是以栈的方式存储的，所以正常情况下新打开一个界面都会进行压栈的操作，当用户单击 Android 手机的返回键时，会调用出栈操作。但是当你不希望用户再返回到某一个 Activity 时，可以对当前的 Activity 调用 finish 方法，这样此 Activity 就从栈中清除了，用户按返回键会返回到在上层的界面。

范例 079 NBA 球星信息介绍的网格视图

1．实例简介

我们在使用 Android 应用的时候，经常会看到有网格列表的视图，第 3 章讲过此视图为 GridView，但是其中的 item 中都有图片，而且都是从网络上获取得到的。例如，QQ 软件的好友列表页面，新浪微博的微博列表页面等。这就要求我们在实现 GridView 的过程中每个 item 中都要有一个异步请求和请求网络中的图片。本实例就带领大家使用 Android 中的 GridView+异步请求图片，来完成一个 NBA 球星信息介绍的网格视图效果。

2．运行效果

该实例运行效果如图 4.26 所示。

3．实例程序讲解

在上面的例子效果中可以看到，在程序中以双列的方式显示了 NBA 球星的列表记录，而且每个记录都有球星的照片和姓名，其中的照片都是从网络异步获取下来的。对于我们这个实例，首先要修改 res/layout/activity_main.xml 文件，代码如下：

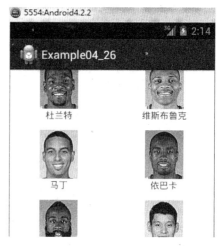

图 4.26 NBA 球星信息介绍的网格视图

```xml
01 <?xml version="1.0" encoding="utf-8"?>
02 <!-- 定义基本的相对布局 -->
03 <RelativeLayout xmlns:android="http://schemas.android.com/apk/res/android"
04     android:layout_width="fill_parent"
05     android:layout_height="wrap_content" >
06     <!-- 定义表格布局 GridView -->
07     <GridView
08         android:id="@+id/Gv"
09         android:layout_width="fill_parent"
10         android:layout_height="fill_parent"
11         android:columnWidth="70dp"
12         android:gravity="center"
13         android:horizontalSpacing="10dp"
14         android:numColumns="2"
15         android:stretchMode="columnWidth"
16         android:verticalSpacing="10dp" />
```

```
17
18 </RelativeLayout>
```

在如上代码中第 7～16 行定义了一个 GridView 控件，在此控件中设置了对齐方式 gravity 为 center 居中，定义了 numColumns 为 2，代表一共有两列，并且设置了 id，方便我们在 Activity 中得到此 GridView 控件。

然后新建 res/layout/ griditem.xml 文件，代码如下：

```xml
01 <?xml version="1.0" encoding="utf-8"?>
02 <!-- 定义 GridView 的 item 的布局文件 -->
03 <RelativeLayout xmlns:android="http://schemas.android.com/apk/res/android"
04     android:layout_width="fill_parent"
05     android:layout_height="wrap_content"
06     android:paddingBottom="4dip" >
07     <!-- item 的图片控件 -->
08     <ImageView
09         android:id="@+id/ItemIv"
10         android:layout_width="wrap_content"
11         android:layout_height="wrap_content"
12         android:layout_centerHorizontal="true" >
13     </ImageView>
14     <!-- item 的文字控件 -->
15     <TextView
16         android:id="@+id/ItemTv"
17         android:layout_width="wrap_content"
18         android:layout_height="wrap_content"
19         android:layout_below="@+id/ItemIv"
20         android:layout_centerHorizontal="true" >
21     </TextView>
22
23 </RelativeLayout>
```

此文件为 GridView 控件中每个 item 的布局视图，基本布局为 RelativeLayout 布局，其中第 8～13 行定义了一个 ImageView，代表球星的头像。在其中第 15～21 行，定义了一个 TextView，代表球星简介。

然后定义 src/com.wyl.example/ViewCache.java 文件，代码如下：

```java
01 //定义视图的占位符类
02 public class ViewCache
03 {
04     //定义基本的视图对象
05     private View baseView;
06     //定义显示的文字标签对象
07     private TextView textView;
08     //定义显示的图片对象
09     private ImageView imageView;
10
11     public ViewCache(View baseView)
12     {
13         this.baseView = baseView;        .//ViewCache 的构造方法
14     }
15
16     //得到文字标签对象
17     public TextView getTextView()
18     {
19         if (textView == null)
20         {    //得到对应 TextView 对象
```

```
21            textView = (TextView) baseView.findViewById(R.id.ItemTv);
22        }
23        return textView;
24    }
25    //得到图片控件对象
26    public ImageView getImageView()
27    {
28        if (imageView == null)
29        {   //得到ImageView对象
30            imageView = (ImageView) baseView.findViewById(R.id.ItemIv);
31        }
32        return imageView;
33    }
34 }
```

此文件定义了一个占位符的实体类,代表每一个 GridView 中的一个 item 的占位符,其中每一个对象都包括了一个 TextView 对象和一个 Image 对象。在我们后面的 Adapter 类中会用到此类。

然后定义 src/com.wyl.example/GirdItem.java 文件,代码如下:

```
01 //定义为 GridView 的一个 item 对象的实体类
02 public class GirdItem {
03    //item 图片地址
04    private String imageUrl;
05    //item 的文字标签
06    private String text;
07
08    //构造函数
09    public GirdItem(String imageUrl, String text) {
10        this.imageUrl = imageUrl;
11        this.text = text;
12    }
13
14    //得到图片地址
15    public String getImageUrl() {
16        return imageUrl;
17    }
18    //得到文字标签值
19    public String getText() {
20        return text;
21    }
22 }
```

此代码定义了一个实体类,这里叫做 GirdItem 类,其在本工程中代表球员信息的实体类,其中包括了网络图片地址 imageUrl 和球员信息 text 两个属性。

然后定义 src/com.wyl.example/ AsyncImageLoader.java 文件,代码如下:

```
01 //定义异步加载图片类
02 public class AsyncImageLoader
03 {
04    //定义异步加载图片的缓存哈希图
05    private HashMap<String, SoftReference<Drawable>> imageCache;
06
07    public AsyncImageLoader()
08    {
09        imageCache = new HashMap<String, SoftReference<Drawable>>();
10    }
```

```
11  //读取图片的方法
12  public Drawable loadDrawable(final String imageUrl,
13          final ImageCallback imageCallback)
14  {
15      //如果在图片缓存中有此图片,则直接得到此图片,然后返回
16      if (imageCache.containsKey(imageUrl))
17      {
18          SoftReference<Drawable> softReference=imageCache.get(imageUrl);
19          Drawable drawable = softReference.get();
20          if (drawable != null)
21          {
22              return drawable;
23          }
24      }
25      //定义 handler 对象来发送 message
26      final Handler handler = new Handler()
27      {
28          public void handleMessage(Message message)
29          {
30              imageCallback.imageLoaded((Drawable) message.obj, imageUrl);
31          }
32      };
33      //定义一个新的线程用来读取网络上的图片地址
34      new Thread()
35      {
36          @Override
37          public void run()
38          {
39              //读取图片地址 umageUrl
40              Drawable drawable = loadImageFromUrl(imageUrl);
41              //将读取到的地址添加到缓存列表中
42              imageCache.put(imageUrl, new SoftReference<Drawable>
                    (drawable));
43              //定义 message 对象,设置内容为加载到的图片对象
44              Message message = handler.obtainMessage(0, drawable);
45              //发送图片
46              handler.sendMessage(message);
47          }
48      }.start();
49      return null;
50  }
51
52  //读取从图片地址读取图片
53  public static Drawable loadImageFromUrl(String url)
54  {
55      //定义 url 对象,及 inputstream 对象
56      URL m;
57      InputStream i = null;
58      try
59      {
60          //通过图片的 url 得到图片的 inputstream 对象
61          m = new URL(url);
62          i = (InputStream) m.getContent();
63      }
64      catch (MalformedURLException e1)
65      {
66          e1.printStackTrace();
67      }
68      catch (IOException e)
```

```
69      {
70          e.printStackTrace();
71      }
72      //通过 inputstream 对象得到图片的 drawable 对象
73      Drawable d = Drawable.createFromStream(i, "src");
74      return d;
75  }
76
77  //图片获取到后的接口
78  public interface ImageCallback
79  {
80      public void imageLoaded(Drawable imageDrawable, String imageUrl);
81  }
82 }
```

此类定义了一个异步图片加载类，其中第 5 行定义了一个 HashMap，用来缓存加载过的图片对象。在第 12～50 行，定义了 loadDrawable 方法，通过传入的 imageUrl 获取网络上的图片，在第 16 行，首先要判断此图片资源是否之前加载过，如果之前有加载过的话，就直接采用缓存的图片资源对象。在第 26～32 行定义了一个 Handler 来接收消息。在第 34～48 行定义了一个线程去加载网络上的图片，加载完毕后发消息给 handler 对象。在第 53～75 行实现了如何根据图片的 url 从网络上获取图片的方法。

然后定义 src/com.wyl.example/GridViewAdapter.java 文件，代码如下：

```
01  //自定义 Adatper，继承自 ArrayAdapter
02  public class GridViewAdapter extends ArrayAdapter<GirdItem> {
03      //定义 GridView 对象
04      private GridView gridView;
05      //定义异步图片加载对象
06      private AsyncImageLoader asyncImageLoader;
07
08      public GridViewAdapter(Activity activity,
09              List<GirdItem> imageAndTexts, GridView gridView) {
10          super(activity, 0, imageAndTexts);
11          //初始化数据
12          this.gridView = gridView;
13          asyncImageLoader = new AsyncImageLoader();
14      }
15
16      public View getView(int position, View convertView, ViewGroup parent) {
17          Activity activity = (Activity) getContext();
18
19          //定义 GridView 的 item 布局，读取 griditem.xml 布局
20          View rowView = convertView;
21          ViewCache viewCache;
22          if (rowView == null) {
23              //读取 griditem.xml 为 item 的布局
24              LayoutInflater inflater = activity.getLayoutInflater();
25              rowView = inflater.inflate(R.layout.griditem, null);
26              //通过当前的布局视图，初始化缓存视图
27              viewCache = new ViewCache(rowView);
28              //设置当前视图的标签
29              rowView.setTag(viewCache);
30          } else {
31              viewCache = (ViewCache) rowView.getTag();
32          }
33          //得到需要显示的对象
```

```
34        GirdItem imageAndText = getItem(position);
35
36        //读取对象的相应内容值
37        String imageUrl = imageAndText.getImageUrl();
38        ImageView imageView = viewCache.getImageView();
39        imageView.setTag(imageUrl);
40        //读取缓存图片
41        Drawable cachedImage = asyncImageLoader.loadDrawable(imageUrl,
42                new com.wyl.example.AsyncImageLoader.ImageCallback() {
43                    public void imageLoaded(Drawable imageDrawable,
44                         String imageUrl) {
45                        ImageView imageViewByTag = (ImageView) gridView
46                                .findViewWithTag(imageUrl);
47                        if (imageViewByTag != null) {
48                            imageViewByTag.setImageDrawable(imageDrawable);
49                        }
50                    }
51                });
52        //如果缓存图片为空,则显示默认图片
53        if (cachedImage == null) {
54            imageView.setImageResource(R.drawable.ic_launcher);
55        } else {
56            imageView.setImageDrawable(cachedImage);
57        }
58        //显示文字
59        TextView textView = viewCache.getTextView();
60        textView.setText(imageAndText.getText());
61        return rowView;
62    }
63 }
```

在此类中自定义了一个 Adapter,在其中的第 6 行定义了一个异步请求的任务。在第 8~14 行实现了此 Adapter 的数据初始化。在第 16~62 行实现了 GridView 的每个 item 的布局显示。在第 20~21 行定义了一个 view 的占位符类 ViewCache,定义了一个 View,代表当前 item 的 View。第 22 行判断如果当前的视图为空的话,就去加载 griditem 布局构成新的视图。否则直接加载缓存的视图。在第 33 行得到本 item 需要显示的数据对象,其中此对象中的文字属性直接显示,如是网络图片地址,就开启一个异步任务进行下载,加载完成后显示下载的图片。

然后定义 src/com.wyl.example/MainActivity.java 文件,代码如下:

```
01 //定义了本实例的主要Activity
02 public class MainActivity extends Activity {
03
04    @Override
05    public void onCreate(Bundle savedInstanceState) {
06        super.onCreate(savedInstanceState);
07        setContentView(R.layout.activity_main);
08        //得到GridView控件
09        GridView gridView = (GridView) findViewById(R.id.Gv);
10        //定义数据源list
11        List<GirdItem> list = new ArrayList<GirdItem>();
12        //添加数据
13        list.add(new GirdItem(
14            "http://www.sinaimg.cn/ty/nba/players/2008/4244.jpg",
15            "杜兰特"));
16        list.add(new GirdItem(
```

```
17                "http://www.sinaimg.cn/ty/nba/players/2008/4390.jpg",
18                "维斯布鲁克"));
19      list.add(new GirdItem(
20                "http://www.sinaimg.cn/ty/nba/players/2008/3843.jpg",
21                "马丁"));
22      list.add(new  GirdItem("http://www.sinaimg.cn/ty/nba/players/2008/
        4486.jpg",
23                "依巴卡"));
24      list.add(new GirdItem(
25                "http://www.sinaimg.cn/ty/nba/players/2008/4563.jpg",
26                "哈登"));
27      list.add(new GirdItem(
28                "http://www.sinaimg.cn/ty/nba/players/2008/4795.jpg",
29                "JimLin"));
30      list.add(new GirdItem(
31                "http://www.sinaimg.cn/ty/nba/players/2008/4920.jpg",
32                "帕森斯"));
33      //设置 GridView 的数据源
34      gridView.setAdapter(new GridViewAdapter(this, list, gridView));
35  }
36  }
```

此文件为当前实例的主要 Activity 文件，在此文件中得到了 GridView 控件，然后根据自定义的数据 list 创建 Adapter，最后 GridView 就显示了 List 中的数据内容。

4．实例扩展

本实例中的效果在各大购物应用客户端中也很常见，但是其中样式各不相同，有些是只有一列的，有些是两列的，有些是三列的，这些需求根据项目的要求大家可以自行修改 GridView 布局中的属性字段即可，异步加载的内容是相同的。

范例 080 NBA 球星信息介绍的列表视图

1．实例简介

我们在使用 Android 应用的时候，不单会看到有网格列表的视图，而且经常会遇到列表视图和网格视图相互切换的效果，其中的列表视图的 item 中也有图片，而且都是从网络上获取得到的。例如，QQ 软件的好友列表页面和新浪微博的微博列表页面等。这就要求我们在实现 ListView 的过程中每个 item 中都要有一个异步请求，请求网络中的图片。本实例就带领大家使用 Android 中的 ListView+异步请求图片，来完成一个 NBA 球星信息介绍的列表视图效果。

2．运行效果

该实例运行效果如图 4.27 所示。

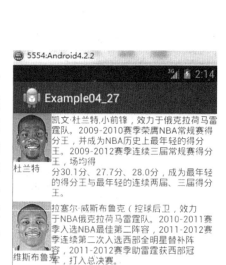

图 4.27 NBA 球星信息介绍的列表视图

3. 实例程序讲解

在上面的例子效果中可以看到，在程序中以列表的方式显示了 NBA 球星的列表记录，而且每个记录都有球星的照片、姓名和简介，其中的照片都是从网络异步获取下来的。对于我们这个实例，首先要修改 res/layout/activity_main.xml 文件，代码如下：

```xml
01  <?xml version="1.0" encoding="utf-8"?>
02  <!-- 定义基本的相对布局 -->
03  <RelativeLayout xmlns:android="http://schemas.android.com/apk/res/android"
04      android:layout_width="fill_parent"
05      android:layout_height="wrap_content" >
06      <!-- 定义表格布局 ListView -->
07      <ListView
08          android:id="@+id/Lv"
09          android:layout_width="fill_parent"
10          android:layout_height="fill_parent"
11          />
12  
13  </RelativeLayout>
```

在如上代码中第 6~11 行定义了一个 ListView 控件，在此控件中设置了 id，方便我们在 Activity 中得到此 ListView 控件。

然后新建 res/layout/listviewitem.xml 文件，代码如下：

```xml
01  <?xml version="1.0" encoding="utf-8"?>
02  <!-- 定义 GridView 的 item 的布局文件 -->
03  <RelativeLayout xmlns:android="http://schemas.android.com/apk/res/android"
04      android:layout_width="fill_parent"
05      android:layout_height="wrap_content"
06      android:paddingBottom="4dip" >
07      <!-- item 的图片控件 -->
08      <ImageView
09          android:id="@+id/ItemIv"
10          android:layout_width="wrap_content"
11          android:layout_height="wrap_content"
12          android:layout_alignParentLeft="true" >
13      </ImageView>
14      <!-- item 的姓名文字控件 -->
15      <TextView
16          android:id="@+id/ItemTvName"
17          android:layout_width="wrap_content"
18          android:layout_height="wrap_content"
19          android:layout_alignParentLeft="true"
20          android:layout_below="@+id/ItemIv" >
21      </TextView>
22  
23      <!-- item 的简介文字控件 -->
24      <TextView
25          android:id="@+id/ItemTvInfo"
26          android:layout_width="wrap_content"
27          android:layout_height="wrap_content"
28          android:layout_toRightOf="@+id/ItemIv" >
29      </TextView>
30  
31  </RelativeLayout>
```

此文件为 ListView 控件中每个 item 的布局视图，基本布局为 RelativeLayout 布局，其中第 8~13 行定义了一个 ImageView,代表球星的头像。其中第 15~21 行,定义了一个 TextView,代表球星的姓名。其中第 24~29 行,定义了一个 TextView,代表球星的简介。

然后定义 src/com.wyl.example/ViewCache.java 文件,代码如下:

```
01  //定义视图的占位符类
02  public class ViewCache
03  {
04      //定义基本的视图对象
05      private View baseView;
06      //定义显示的姓名文字标签对象
07      private TextView textViewname;
08
09      //定义显示的简介文字标签对象
10      private TextView textViewinfo;
11      //定义显示的图片对象
12      private ImageView imageView;
13
14      public ViewCache(View baseView)
15      {
16          this.baseView = baseView;
17      }
18
19      //得到姓名文字标签对象
20      public TextView getnameTextView()
21      {
22          if (textViewname == null)
23          {   //得到姓名文字标签控件
24              textViewname = (TextView) baseView.findViewById(R.id.ItemTvName);
25          }
26          return textViewname;
27      }
28
29      //得到简介文字标签对象
30      public TextView getinfoTextView()
31      {
32          if (textViewinfo == null)
33          {   //得到球员简介文字标签控件
34              textViewinfo = (TextView) baseView.findViewById(R.id.ItemTvInfo);
35          }
36          return textViewinfo;
37      }
38      //得到图片控件对象
39      public ImageView getImageView()
40      {
41          if (imageView == null)
42          {{  //得到球员头像图片控件
43              imageView = (ImageView) baseView.findViewById(R.id.ItemIv);
44          }
45          return imageView;
46      }
47  }
```

此文件定义了一个占位符的实体类,代表每一个 ListView 中的一个 item 的占位符,其中每一个对象都包括了两个 TextView 对象和一个 Image 对象。在我们后面的 Adapter 类中会用到此类。

然后定义 src/com.wyl.example/ListItem.java 文件,代码如下:

```
01 //定义为 ListView 的一个 item 对象的实体类
02 public class ListItem {
03 //item 图片地址
04 private String imageUrl;
05 //item 的姓名文字标签文字
06 private String name;
07 //item 的简介文字标签文字
08 private String info;
09
10 //构造函数
11 public ListItem(String imageUrl, String namestr,String infostr) {
12     this.imageUrl = imageUrl;
13     this.name = namestr;
14     this.info = infostr;
15 }
16
17 //得到图片地址
18 public String getImageUrl() {
19     return imageUrl;
20 }
21
22 //得到简介文字标签值
23 public String getinfo() {
24     return info;
25 }
26 //得到姓名文字标签值
27 public String getName() {
28     return name;
29 }
30 }
```

此代码定义了一个实体类，这里叫做 ListItem 类，其在本工程中代表球员信息的实体类，其中包括了网络图片地址 imageUrl、球员姓名 name 和球员信息 info 三个属性。

然后定义 src/com.wyl.example/ AsyncImageLoader.java 文件，代码如下：

```
01 //定义异步加载图片类
02 public class AsyncImageLoader
03 {
04 //定义异步加载图片的缓存哈希图
05 private HashMap<String, SoftReference<Drawable>> imageCache;
06
07 public AsyncImageLoader()
08 {
09     imageCache = new HashMap<String, SoftReference<Drawable>>();
10 }
11 //读取图片的方法
12 public Drawable loadDrawable(final String imageUrl,
13         final ImageCallback imageCallback)
14 {
15     //如果在图片缓存中有此图片，则直接得到此图片，然后返回
16     if (imageCache.containsKey(imageUrl))
17     {
18         SoftReference<Drawable> softReference = imageCache.get(imageUrl);
19         Drawable drawable = softReference.get();
20         if (drawable != null)
21         {
22             return drawable;
23         }
```

```java
        }
        //定义handler对象来发送message
        final Handler handler = new Handler()
        {
            public void handleMessage(Message message)
            {
                imageCallback.imageLoaded((Drawable) message.obj, imageUrl);
            }
        };
        //定义一个新的线程用来读取网络上的图片地址
        new Thread()
        {
            @Override
            public void run()
            {
                //读取图片地址 umageUrl
                Drawable drawable = loadImageFromUrl(imageUrl);
                //将读取到的地址添加到缓存列表中
                imageCache.put(imageUrl, new SoftReference<Drawable>(drawable));
                //定义message对象，设置内容为加载到的图片对象
                Message message = handler.obtainMessage(0, drawable);
                //发送图片
                handler.sendMessage(message);
            }
        }.start();
        return null;
    }

    //从图片地址读取图片
    public static Drawable loadImageFromUrl(String url)
    {
        //定义url对象，及inputstream对象
        URL m;
        InputStream i = null;
        try
        {
            //通过图片的url得到图片的InputStream对象
            m = new URL(url);
            i = (InputStream) m.getContent();
        }
        catch (MalformedURLException e1)
        {
            e1.printStackTrace();
        }
        catch (IOException e)
        {
            e.printStackTrace();
        }
        //通过InputStream对象得到图片的Drawable对象
        Drawable d = Drawable.createFromStream(i, "src");
        return d;
    }

    //图片获取到后的接口
    public interface ImageCallback
    {
        public void imageLoaded(Drawable imageDrawable, String imageUrl);
    }
}
```

此类定义了一个异步图片加载类，其中第 5 行定义了一个 HashMap，用来缓存加载过的图片对象。在第 12～50 行，定义了 loadDrawable 方法，通过传入的 imageUrl，获取网络上的图片，在第 16 行，首先要判断此图片资源是否之前加载过，如果之前有加载过的话，就直接采用缓存的图片资源对象。在第 26～32 行定义了一个 Handler 来接收消息。在第 34～48 行定义了一个线程去加载网络上的图片，加载完毕后发消息给 handler 对象。在第 53～75 行实现了如何根据图片的 url 从网络上获取图片的方法。

然后定义 src/com.wyl.example/ListViewAdapter.java 文件，代码如下：

```
01  //自定义 Adatper,继承自 ArrayAdapter
02  public class ListViewAdapter extends ArrayAdapter<ListItem> {
03      //
04      private ListView listview;
05      //定义异步图片加载对象
06      private AsyncImageLoader asyncImageLoader;
07
08      public ListViewAdapter(Activity activity,
09              List<ListItem> imageAndTexts, ListView listView) {
10          super(activity, 0, imageAndTexts);
11          //初始化数据
12          this.listview = listView;
13          asyncImageLoader = new AsyncImageLoader();
14      }
15
16      public View getView(int position, View convertView, ViewGroup parent) {
17          Activity activity = (Activity) getContext();
18
19          //定义 GridView 的 item 布局,读取 griditem.xml 布局
20          View rowView = convertView;
21          ViewCache viewCache;
22          if (rowView == null) {
23              //读取 griditem.xml 为 item 的布局
24              LayoutInflater inflater = activity.getLayoutInflater();
25              rowView = inflater.inflate(R.layout.listviewitem, null);
26              //通过当前的布局视图,初始化缓存视图
27              viewCache = new ViewCache(rowView);
28              //设置当前视图的标签
29              rowView.setTag(viewCache);
30          } else {
31              viewCache = (ViewCache) rowView.getTag();
32          }
33          //得到需要显示的对象
34          ListItem imageAndText = getItem(position);
35
36          //读取对象的相应内容值
37          String imageUrl = imageAndText.getImageUrl();
38          ImageView imageView = viewCache.getImageView();
39          imageView.setTag(imageUrl);
40          //读取缓存图片
41          Drawable cachedImage = asyncImageLoader.loadDrawable(imageUrl,
42                  new com.wyl.example.AsyncImageLoader.ImageCallback() {
43                      public void imageLoaded(Drawable imageDrawable,
44                              String imageUrl) {
                            //得到对应 ImageView 对象
45                          ImageView imageViewByTag = (ImageView) listview
46                                  .findViewWithTag(imageUrl);
```

```
47                    //设置图片控件显示的图像
                      if (imageViewByTag != null) {
48                        imageViewByTag.setImageDrawable(imageDrawable);
49                    }
50                }
51            });
52        //如果缓存图片为空,则显示默认图片
53        if (cachedImage == null) {
54            imageView.setImageResource(R.drawable.ic_launcher);
55        } else {
56            imageView.setImageDrawable(cachedImage);
57        }
58        //显示姓名文字
59        TextView textViewname = viewCache.getnameTextView();
60        textViewname.setText(imageAndText.getName());
61        //显示简介文字
62        TextView textViewinfo = viewCache.getinfoTextView();
63        textViewinfo.setText(imageAndText.getinfo());
64        return rowView;
65    }
66 }
```

在此类中自定义了一个 Adapter,在其中的第 6 行定义了一个异步请求的任务。在第 8~14 行实现了此 Adapter 的数据初始化。在第 16~65 行实现了 ListView 的每个 item 的布局显示。在第 20~21 行定义了一个 view 的占位符类 ViewCache,定义了一个 View,代表当前 item 的 View。第 22 行判断如果当前的视图为空的话,就去加载 griditem 布局构成新的视图。否则直接加载缓存的视图。在第 33 行得到本 item 需要显示的数据对象,其中此对象中的文字属性直接显示,如是网络图片地址,就开启一个异步任务进行下载,加载完成后显示下载的图片。

然后定义 src/com.wyl.example/MainActivity.java 文件,代码如下:

```
01 //定义了本实例的主要 Activity
02 public class MainActivity extends Activity {
03
04 @Override
05 public void onCreate(Bundle savedInstanceState) {
06     super.onCreate(savedInstanceState);
07     setContentView(R.layout.activity_main);
08     //得到 GridView 控件
09     ListView listview = (ListView) findViewById(R.id.Lv);
10     //定义数据源 list
11     List<ListItem> list = new ArrayList<ListItem>();
12     //添加数据
13     list.add(new ListItem(
14         "http://www.sinaimg.cn/ty/nba/players/2008/4244.jpg",
15         "杜兰特","凯文·杜兰特..."));
16     list.add(new ListItem(
17         "http://www.sinaimg.cn/ty/nba/players/2008/4390.jpg",
18         "维斯布鲁克","拉塞尔·威斯布鲁克..."));
19     list.add(new ListItem(
20         "http://www.sinaimg.cn/ty/nba/players/2008/3843.jpg",
21         "马丁","凯文·马丁..."));
22     list.add(new ListItem("http://www.sinaimg.cn/ty/nba/players/2008/
23         4486.jpg","依巴卡","司职大前锋..."));
24     list.add(new ListItem(
```

```
25                "http://www.sinaimg.cn/ty/nba/players/2008/4563.jpg",
26                "哈登","詹姆斯•哈登..."));
27      list.add(new ListItem(
28                "http://www.sinaimg.cn/ty/nba/players/2008/4795.jpg",
29                "JimLin","林书豪（控球后卫，...")));
30      list.add(new ListItem(
31                "http://www.sinaimg.cn/ty/nba/players/2008/4920.jpg",
32                "帕森斯","钱德勒..."));
33      //设置ListView的数据源
34      listview.setAdapter(new ListViewAdapter(this, list, listview));
35   }
36  }
```

此文件为当前实例的主要 Activity 文件，在此文件中得到了 ListView 控件，然后根据自定义的数据 list 创建 Adapter，最后 ListView 就显示了 List 中的数据内容。

4．实例扩展

现在常见的应用程序客户端的软件层出不穷，但是对于列表的展示一直是主要的方式，当然每个应用也都在努力的去改善，使得自己的应用与其他应用有个性设计存在，通过异步请求可以实现个性化的列表加载效果。

范例 081　文件下载

1．实例简介

我们在使用 Android 应用的时候，经常会遇到需要下载应用、下载文件、下载音乐和下载图片的功能。例如，QQ 应用的自动更新功能和安卓市场的应用搜索下载功能等。这就要求我们要在程序中实现文件下载的功能，而且文件下载的时间一般取决于手机的性能和网络的带宽，所以我们需要通过多线程来实现。本实例就带领大家使用 Android 中的多线程来完成一个文件下载器程序。

2．运行效果

该实例运行效果如图 4.28 所示。

3．实例程序讲解

在上面的例子效果中上方的 EditText 中输入需要下载的文件的地址，然后单击下面的"开始下载"按钮，然后在下方的进度条就会显示开始下载并且时刻显示下载进度。对于我们这个实例，首先修改 res/layout/activity_main.xml 的代码如下：

图 4.28　文件下载器程序

```
01  <?xml version="1.0" encoding="utf-8"?>
02  <!-- 定义基本的Linearlayout布局 -->
03  <LinearLayout xmlns:android="http://schemas.android.com/apk/res/android"
04      android:layout_width="fill_parent"
05      android:layout_height="fill_parent"
06      android:orientation="vertical" >
```

```xml
07      <!-- 定义文本提示标签-->
08      <TextView
09          android:id="@+id/Tv"
10          android:layout_width="fill_parent"
11          android:layout_height="wrap_content"
12          android:text="请输入下载地址,然后单击下载按钮: " />
13      <!-- 定义用户输入的下载地址的控件-->
14      <EditText
15          android:id="@+id/Et"
16          android:layout_width="fill_parent"
17          android:layout_height="wrap_content"
18          android:text="http://p1.s.hjfile.cn/thread/201207/20120718111554562
            _336_o.jpg"
19          />
20      <!-- 定义用户单击下载的按钮-->
21      <Button
22          android:id="@+id/Btn"
23          android:layout_width="fill_parent"
24          android:layout_height="wrap_content"
25          android:text="单击开始下载"
26          />
27      <!-- 定义下载过程中显示的进度条对象-->
28      <ProgressBar
29          android:id="@+id/Pb"
30          style="?android:attr/progressBarStyleHorizontal"
31          android:layout_width="fill_parent"
32          android:layout_height="wrap_content"
33          android:max="100" />
34
35  </LinearLayout>
```

本代码是当前 Activity 的布局文件,在其中定义了基本布局为 LinearLayout 布局,在此布局中加入图片地址输入框 EditText 和一个 Button 按钮,单击后开始下载,一个 ProgressBar,设置最大值。而且给此三个控件都设置了 id,方便我们在 Activity 中获取。

然后修改 src/com.wyl.example/MainActivity.java 文件,代码如下:

```java
001 //定义了本实例的主要Activity
002 public class MainActivity extends Activity {
003     //定义进度条对象
004     private ProgressBar pb;
005     //定义文本提示标签,单击按钮后显示下载进度
006     private TextView tv;
007     //定义按钮对象
008     private Button btn;
009     //定义输入的下载文件地址的输入框
010     private EditText et;
011     //定义需要下载的文件的大小变量
012     private int fileSize;
013     //定义已经下载了的文件的大小
014     private int downLoadFileSize;
015     //定义下载的文件名对象
016     private String filename;
017     //定义界面的处理其他线程message的handler对象
018     private Handler handler = new Handler() {
019         //当接收到其他线程的message消息后自动回调此函数
020         @Override
021         public void handleMessage(Message msg) {
```

```
022            //如果没有发送消息的线程还没有中止的话，进行处理
023            if (!Thread.currentThread().isInterrupted()) {
024                switch (msg.what) {
025                //发送消息0，代表设置progressBar的总长度
026                case 0:
027                    pb.setMax(fileSize);
028                //发送消息1，代表文件正在下载中
029                case 1:
030                    //设置下载进度为当前已经下载的大小
031                    pb.setProgress(downLoadFileSize);
032                    //通过当前下载的文件的大小和文件的总大小得到下载的文件的进度
033                    int result = downLoadFileSize * 100 / fileSize;
034                    tv.setText(result + "%");
035                    break;
036                //发送消息2，代表文件下载完成
037                case 2:
038                    Toast.makeText(MainActivity.this,
039                        "文件下载完成，请在sdcard目录下查看下载文件",
040                        1).show();
041                    break;
042                //发送消息-1，代表文件下载出错
043                case -1:
044                    String error = msg.getData().getString("error");
045                    Toast.makeText(MainActivity.this, error, 1).show();
046                    break;
047                }
048            }
049            super.handleMessage(msg);
050        }
051    };
052
053
054
055    @Override
056    public void onCreate(Bundle savedInstanceState) {
057        super.onCreate(savedInstanceState);
058        setContentView(R.layout.activity_main);
059
060        //得到对应的控件对象
061        findView();
062        //设置控件的监听器
063        setListener();
064    }
065
066    private void setListener() {
067        //设置btn按钮的单击事件
068        btn.setOnClickListener(new OnClickListener() {
069            //当按钮被单击时自动回调onclick方法
070            @Override
071            public void onClick(View v) {
072                //定义一个线程开始文件下载工作
073                new Thread() {
074                    public void run() {
075                        try {
076                            //得到Et输入框中的需要下载的文件的地址
077                            String str = et.getText().toString();
078                            //调用downfile函数下载，下载后保存在sdcard目录下
079                            down_file(
080                                str,
```

```
081                                "/sdcard/");
082                    } catch (ClientProtocolException e) {
083                        // TODO Auto-generated catch block
084                        e.printStackTrace();
085                    } catch (IOException e) {
086                        // TODO Auto-generated catch block
087                        e.printStackTrace();
088                    }
089                }
090            }.start();
091        }
092    });
093 }
094
095 private void findView() {
096     //得到布局中的文件
097     pb = (ProgressBar) findViewById(R.id.Pb);
098     tv = (TextView) findViewById(R.id.Tv);
099     btn = (Button)findViewById(R.id.Btn);
100     et = (EditText)findViewById(R.id.Et);
101 }
102
103 public void down_file(String url, String path) throws IOException {
104     //通过文件下载地址得到需要下载的文件名
105     filename = url.substring(url.lastIndexOf("/") + 1);
106     //定义 URL 对象
107     URL myURL = new URL(url);
108     //得到 URLConnection 对象
109     URLConnection conn = myURL.openConnection();
110     //连接此 URL
111     conn.connect();
112     //得到 URL 中的文件流
113     InputStream is = conn.getInputStream();
114     //得到文件的大小
115     fileSize = conn.getContentLength();         //根据响应获取文件大小
116     //如果文件大小小于 0,文件失败
117     if (fileSize <= 0)
118         throw new RuntimeException("无法获知文件大小 ");
119     //如果得到的文件流为空,文件为空
120     if (is == null)
121         throw new RuntimeException("stream is null");
122
123     //定义本地文件的读取流对象
124     FileOutputStream fos = new FileOutputStream(path + filename);
125     //定义文件读取的字节数组
126     byte buf[] = new byte[1024];
127     downLoadFileSize = 0;
128     //发送文件开始下载的消息
129     sendMsg(0);
130     //通过循环得到文件流的对象,然后依次写入本地文件
131     do {
132         //循环读取
133         int numread = is.read(buf);
134         //当读取到文件结束后,退出循环
135         if (numread == -1) {
136             break;
137         }
138         fos.write(buf, 0, numread);
```

```
139                downLoadFileSize += numread;
140                //发送正在下载的消息，更新进度条
141                sendMsg(1);
142            } while (true);
143
144            //发送文件下载完成的消息
145            sendMsg(2);
146            //关闭文件流
147            try {
148                is.close();
149            } catch (Exception ex) {
150                Log.e("tag", "error: " + ex.getMessage(), ex);
151            }
152
153        }
154
155        //从子线程发送 message 给主线程
156        private void sendMsg(int flag) {
157            Message msg = new Message();
158            msg.what = flag;
159            handler.sendMessage(msg);
160        }
161  }
```

本代码是当前 Activity 的源代码，其中第 18~51 行定义了一个 Handler 对象，用来接收和处理子线程传入的消息，其中首先通过 isInterrupted 方法判断当前线程是否中断，然后根据接收到的消息标示进行不同的动作，如果为 0，代表设置 progressBar 的总大小；如果为 1，代表传入了下载的进度；如果为 2，代表文件下载完成，请到 sdcard 下查找；如果返回的是-1，代表下载出错。在第 95~101 行得到了布局中的所有控件。在第 66~93 行为控件设置了监听器，当用户单击按钮的时候，开启线程去下载文件到 sdcard。在第 103~153 行实现了如何通过一个文件的 url 下载此文件保存到 sdcard 的步骤，首先通过 url 得到 URL 对象，然后通过 URLConnection 的 getinputStream 方法得到网络中的资源流。得到文件的大小后发送消息 0，下载的过程中发送消息 1。直到所有的文件都被下载完毕，返回消息 2。

4. 实例扩展

对于文件的下载来说有很多种方法，但是我们的异步请求的实现是不变的，只要将耗时的操作放到子线程中去做，然后再通过发送 handler 消息，就可以修改 UI 线程中的布局控件了。

范例 082　中断文件下载

1. 实例简介

我们在使用 Android 应用的时候，经常会遇到需要下载应用，而且在某些条件下可能希望下载到一半的时候中断。例如，下载音乐的时候发现资源很慢，下载过程中发现自己下载错误等。这就要求我们要在程序中实现文件下载过程中的中断功能，而且文件下载的过程是在子线程中进行的，如何来停止正在进行的子线程操作呢。本实例就带领大家完成

可以中断的文件下载程序。

2. 运行效果

该实例运行效果如图 4.29 所示。

3. 实例程序讲解

在上面的例子效果中，上方的 EditText 中输入需要下载的文件的地址，然后单击下面的"开始下载"按钮，然后在下方的进度条就会显示开始下载并且时刻显示下载进度，单击"停止下载"按钮，下载终止。对于我们这个实例，首先修改 res/layout/activity_main.xml 文件，代码如下：

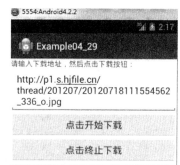

图 4.29 中断文件下载器程序

```xml
01 <?xml version="1.0" encoding="utf-8"?>
02 <!-- 定义基本的 Linearlayout 布局 -->
03 <LinearLayout xmlns:android="http://schemas.android.com/apk/res/android"
04     android:layout_width="fill_parent"
05     android:layout_height="fill_parent"
06     android:orientation="vertical" >
07     <!-- 定义文本提示标签-->
08     <TextView
09         android:id="@+id/Tv"
10         android:layout_width="fill_parent"
11         android:layout_height="wrap_content"
12         android:text="请输入下载地址，然后单击下载按钮： " />
13     <!-- 定义用户输入的下载地址的控件-->
14     <EditText
15         android:id="@+id/Et"
16         android:layout_width="fill_parent"
17         android:layout_height="wrap_content"
18         android:text="http://p1.s.hjfile.cn/thread/201207/20120718111554562_336_o.jpg"
19         />
20     <!-- 定义用户单击下载的按钮-->
21     <Button
22         android:id="@+id/BtnStart"
23         android:layout_width="fill_parent"
24         android:layout_height="wrap_content"
25         android:text="单击开始下载"
26         />
27     <!-- 定义用户终止下载的按钮-->
28     <Button
29         android:id="@+id/BtnStop"
30         android:layout_width="fill_parent"
31         android:layout_height="wrap_content"
32         android:text="单击终止下载"
33         />
34     <!-- 定义下载过程中显示的进度条对象-->
35     <ProgressBar
36         android:id="@+id/Pb"
37         style="?android:attr/progressBarStyleHorizontal"
38         android:layout_width="fill_parent"
39         android:layout_height="wrap_content"
```

```
40            android:max="100" />
41
42    </LinearLayout>
```

本代码是当前 Activity 的布局文件,其中定义了基本布局为 LinearLayout 布局,在此布局中加入图片地址输入框 EditText、两个 Button 按钮、单击后开始下载和单击后停止下载和一个 ProgressBar,设置最大值。而且给此四个控件都设置了 id,方便我们在 Activity 中获取。

然后修改 src/com.wyl.example/MainActivity.java 文件,代码如下:

```
001  //定义了本实例的主要 Activity
002  public class MainActivity extends Activity {
003      //定义进度条对象
004      private ProgressBar pb;
005      //定义文本提示标签,单击按钮后显示下载进度
006      private TextView tv;
007      //定义开始下载按钮对象
008      private Button btnstart;
009      //定义终止下载按钮对象
010      private Button btnstop;
011      //定义输入的下载文件地址的输入框
012      private EditText et;
013      //定义需要下载的文件的大小变量
014      private int fileSize;
015      //定义已经下载了的文件的大小
016      private int downLoadFileSize;
017      //定义下载的文件名对象
018      private String filename;
019      //定义本节目中的线程对象
020      private Thread thread;
021
022      //定义界面的处理其他线程 message 的 handler 对象
023      private Handler handler = new Handler() {
024          //当接收到其他线程的 message 消息后自动回调此函数
025          @Override
026          public void handleMessage(Message msg) {
027              //如果没有发送消息的线程还没有中止的话,进行处理
028              if (!Thread.currentThread().isInterrupted()) {
029                  switch (msg.what) {
030                      //发送消息 0,代表设置 progressBar 的总长度
031                      case 0:
032                          pb.setMax(fileSize);
033                      //发送消息 1,代表文件正在下载中
034                      case 1:
035                          //设置下载进度为当前已经下载的大小
036                          pb.setProgress(downLoadFileSize);
037                          //通过当前下载的文件的大小和文件的总大小得到下载的文件的进度
038                          int result = downLoadFileSize * 100 / fileSize;
039                          tv.setText(result + "%");
040                          break;
041                      //发送消息 2,代表文件下载完成
042                      case 2:
043                          Toast.makeText(MainActivity.this,
044                              "文件下载完成,请在 sdcard 目录下查看下载文件",
045                              1).show();
046                          break;
```

```
047            //发送消息3，下载终止
048            case 3:
049                Toast.makeText(MainActivity.this,
050                    "文件下载终止",
051                    1).show();
052                pb.setProgress(0);
053                tv.setText("0%");
054                break;
055            //发送消息-1,代表文件下载出错
056            case -1:
057                String error = msg.getData().getString("error");
058                Toast.makeText(MainActivity.this, error, 1).show();
059                break;
060            }
061        }
062        super.handleMessage(msg);
063    }
064 };
065
066 @Override
067 public void onCreate(Bundle savedInstanceState) {
068     super.onCreate(savedInstanceState);
069     setContentView(R.layout.activity_main);
070
071     //得到对应的控件对象
072     findView();
073     //设置控件的监听器
074     setListener();
075 }
076
077 private void setListener() {
078     //设置btnstart按钮的单击事件
079     btnstart.setOnClickListener(new OnClickListener() {
080         //当按钮被单击时自动回调onclick方法
081         @Override
082         public void onClick(View v) {
083             //定义一个线程开始文件下载工作
084             thread = new Thread() {
085                 public void run() {
086                     try {
087                         //得到Et输入框中需要下载的文件的地址
088                         String str = et.getText().toString();
089                         //调用downfile函数下载,下载后保存在sdcard目录下
090                         down_file(
091                             str,
092                             "/sdcard/");
093                     } catch (ClientProtocolException e) {
094                         // TODO Auto-generated catch block
095                         e.printStackTrace();
096                     } catch (IOException e) {
097                         // TODO Auto-generated catch block
098                         e.printStackTrace();
099                     }
100                 }
101             };
102             thread.start();
103         }
104     });
105
```

```java
106         //设置btnstop按钮的单击事件
107         btnstop.setOnClickListener(new OnClickListener() {
108             //当按钮被单击时自动回调onClick方法
109             @Override
110             public void onClick(View v) {
111                 //终止线程
112                 thread.interrupt();
113             }
114         });
115     }
116
117     private void findView() {
118         //得到布局中的文件
119         pb = (ProgressBar) findViewById(R.id.Pb);
120         tv = (TextView) findViewById(R.id.Tv);
121         btnstart = (Button)findViewById(R.id.BtnStart);
122         btnstop = (Button)findViewById(R.id.BtnStop);
123         et = (EditText)findViewById(R.id.Et);
124     }
125
126     public void down_file(String url, String path) throws IOException {
127         //通过文件下载地址得到需要下载的文件名
128         filename = url.substring(url.lastIndexOf("/") + 1);
129         //定义URL对象
130         URL myURL = new URL(url);
131         //得到URLConnection对象
132         URLConnection conn = myURL.openConnection();
133         //连接此URL
134         conn.connect();
135         //得到URL中的文件流
136         InputStream is = conn.getInputStream();
137         //得到文件的大小
138         fileSize = conn.getContentLength();// 根据响应获取文件大小
139         //如果文件大小小于0,文件失败
140         if (fileSize <= 0)
141             throw new RuntimeException("无法获知文件大小 ");
142         //如果得到的文件流为空,文件为空
143         if (is == null)
144             throw new RuntimeException("stream is null");
145
146         //定义本地文件的读取流对象
147         FileOutputStream fos = new FileOutputStream(path + filename);
148         //定义文件读取的字节数组
149         byte buf[] = new byte[1024];
150         downLoadFileSize = 0;
151         //发送文件开始下载的消息
152         sendMsg(0);
153         //通过循环得到文件流的对象,然后依次写入本地文件
154         do {
155             // 循环读取
156             int numread = is.read(buf);
157
158             if (Thread.currentThread().isInterrupted()) {
159                 sendMsg(3);
160                 //关闭文件流
161                 try {
162                     is.close();
163                 } catch (Exception ex) {
```

```
164                     Log.e("tag", "error: " + ex.getMessage(), ex);
165                     }
166                     return;
167                 }
168                 //当读取到文件结束后，退出循环
169                 if (numread == -1) {
170                     break;
171                 }
172                 fos.write(buf, 0, numread);
173                 downLoadFileSize += numread;
174                 //发送正在下载的消息，更新进度条
175                 sendMsg(1);
176             } while (true);
177
178             // 发送文件下载完成的消息
179             sendMsg(2);
180             //关闭文件流
181             try {
182                 is.close();
183             } catch (Exception ex) {
184                 Log.e("tag", "error: " + ex.getMessage(), ex);
185             }
186
187         }
188
189         //从子线程发送 message 给主线程
190         private void sendMsg(int flag) {
191             Message msg = new Message();
192             msg.what = flag;
193             handler.sendMessage(msg);
194         }
195
196 }
```

本代码是当前 Activity 的源代码，其中第 23～64 行定义了一个 Handler 对象，用来接收和处理子线程传入的消息，其中首先通过 isInterrupted 方法判断当前线程是否中断，然后根据接收到的消息标示进行不同的动作，如果为 0，代表设置 progressBar 的总大小；如果为 1，代表传入了下载的进度；如果为 2，代表文件下载完成，请到 sdcard 下查找；如果返回的是–1，代表下载出错。在第 117～124 行得到了布局中的所有控件。在第 77～104 行为控件设置了监听器，当用户单击按钮的时候，开启线程去下载文件到 sdcard，当用户单击停止下载的按钮时，调用线程的 interrupt 方法来终止线程。在第 126～187 行实现了如何通过一个文件的 url 下载此文件保存到 sdcard 的步骤，首先通过 url 得到 URL 对象，然后通过 URLConnection 的 getinputStream 方法得到网络中的资源流。得到文件的大小后发送消息 0，下载的过程中发送消息 1。直到所有的文件都被下载完毕，返回消息 2。

4．实例扩展

对于子线程的终止来说，其实是没有特别好的办法直接终止的，本实例采用的思路就是在 handler 接收到子线程的信息时判断线程是否终止,但是真实的执行过程中有可能你终止了某个线程，某个线程还会再执行一段时间，这就需要大家要在线程执行的过程中尽量多的去判断当前线程是否已经终止。

范例 083　线程间通讯

1．实例简介

我们在使用 Android 应用的时候，如果遇到需要耗时的工作的时候，要使用多线程来解决，多线程中最主要的一个问题就是线程之间如何通讯。例如，我开启一个线程实现下载功能，那么如何通知主线程下载进度，我开启了一个线程进行文件的压缩，那么如何通知用户已经解压完毕呢等。这就要求大家一定要了解线程之间通讯的基本原理就是 handler+message 机制。本实例就带领大家一起实现一个最基本的 handler+message 机制的实例。

2．运行效果

该实例运行效果如图 4.30 所示。

3．实例程序讲解

在上面例子的效果中，我们通过主线程接受来自子线程的消息，并且打印在 TextView 中。想要实现我们上例的效果，首先要修改 res/layout/activity_main.xml 文件，代码如下：

图 4.30　线程间通讯

```
01 <?xml version="1.0" encoding="utf-8"?>
02 <!-- 定义基本的 Linearlayout 布局 -->
03 <LinearLayout xmlns:android="http://schemas.android.com/apk/res/android"
04     android:layout_width="fill_parent"
05     android:layout_height="fill_parent"
06     android:orientation="vertical" >
07     <!-- 定义文本提示标签-->
08     <TextView
09         android:id="@+id/Tv"
10         android:layout_width="fill_parent"
11         android:layout_height="wrap_content"
12         android:text="接受来自子线程的消息如下：" />
13
14 </LinearLayout>
```

在如上代码中第 8～12 行定义了一个 TextView 控件，并且设置其 id，方便我们在 Activity 中获取此控件，然后显示子线程传递的消息。

然后修改 src/com.wyl.example/MainActivity.java 文件，代码如下：

```
01 //定义了本实例的主要 Activity
02 public class MainActivity extends Activity {
03     //定义文本标签对象
04     private TextView tv;
05     //定义接受其他线程数据的 handler 对象
06     private Handler handler = new Handler(){
07         @Override
08         public void handleMessage(Message msg) {
09             // TODO Auto-generated method stub
10             super.handleMessage(msg);
```

```java
11              //接收到其他线程的消息后，设置文本标签对象的值
12              String t = tv.getText().toString();
13              tv.setText(t+"\n"+msg.obj);
14          }
15      };
16
17
18      @Override
19      public void onCreate(Bundle savedInstanceState) {
20          super.onCreate(savedInstanceState);
21
22          setContentView(R.layout.activity_main);
23          //得到布局文件中的TextView对象
24          tv = (TextView)findViewById(R.id.Tv);
25          //定义一个Thread，用来开启另一个线程
26          new Thread(){
27
28              @Override
29              public void run() {
30                  // TODO Auto-generated method stub
31                  super.run();
32                  //定义计数变量，这里循环20次
33                  int i = 0;
34                  //用来得到当前的系统时间
35                  String time;
36                  //定义日期格式
37                  SimpleDateFormat df = new SimpleDateFormat("yyyy-MM-dd HH:mm:ss");
                    //设置日期格式
38                  //在子线程中得到系统的时间，然后发送消息给主线程
39                  while(i++ < 20){
40                      //得到系统的时间
41                      time = df.format(new Date());
42                      Log.e("Thread---The Thread id is :"
43                          ,Thread.currentThread().getId()+"");
44                      Log.e("Thread---Time:",time);
45                      //定义message对象，并且设置message对象信息为上面得到的系统时间
46                      Message message=new Message();
47                      message.obj = time;
48                      //发送信息给主线程
49                      handler.sendMessage(message);
50                      try {
51                          sleep(2000);
52                      } catch (InterruptedException e) {
53                          // TODO Auto-generated catch block
54                          e.printStackTrace();
55                      }
56                  }
57              }
58          //启动子线程
59          }.start();
60      }
61  }
```

在上面代码中第6～15行定义了当前页面的Handler对象，并且实现了handleMessage方法，此方法当有其他线程发送消息到当前线程的时候会自动调用，在handleMessage方法中实现了得到消息内容，并且显示在TextView中。在第26～59行，定义了一个线程，

在其 run 方法中实现了循环 20 次，然后每隔 2 秒发送一次线程中得到的日期给主线程。这样当主线程收到 message 信息后，就会显示在 TextView 中了。

4．实例扩展

本实例讲述的是 Android 中多线程的基本原理，当然，也可以在本实例中加入多个子线程，分别发送消息给主线程，这就是实现一个及时通讯软件的最基本的原理了。

范例 084　本地图片加载速度测试器

1．实例简介

对于 Android 手机来说，市面上有很多种类，而且每款手机各不相同，对于手机的性能评价有一项指标就是读取本地文件的速度。例如，手机读取本地图片的速度是多少。这就要求有一些可以测试手机读取本地文件的应用。本实例就带领大家使用 Android 中的多线程来完成一个测试手机读取本地图片速度的应用程序。

2．运行效果

该实例运行效果如图 4.31 所示。

3．实例程序讲解

在上面例子的效果中，当我们单击开始加载图片，

图 4.31　本地图片加载速度测试器

这时候就会开启一个线程加载 100 张图片，然后计算平均加载一张图片所需要的时间。在其中主要用到的就是子线程加载图片，以及子线程中传递加载时间给主线程。想要实现我们上例的效果，首先要修改 res/layout/activity_main.xml 文件，代码如下：

```
01  <?xml version="1.0" encoding="utf-8"?>
02  <!-- 定义基本的 Linearlayout 布局 -->
03  <LinearLayout xmlns:android="http://schemas.android.com/apk/res/android"
04      android:layout_width="fill_parent"
05      android:layout_height="fill_parent"
06      android:orientation="vertical" >
07      <!-- 定义文本提示标签-->
08      <ImageView
09          android:id="@+id/Iv"
10          android:layout_width="match_parent"
11          android:layout_height="wrap_content"
12          android:src="@drawable/baidu"/>
13
14      <!-- 定义文本提示标签-->
15      <Button
16          android:id="@+id/Btn"
17          android:layout_width="match_parent"
18          android:layout_height="wrap_content"
19          android:text="单击开始循环加载图片"/>
```

```
020
021         <!-- 定义文本提示标签-->
022         <TextView
023             android:id="@+id/Tv"
024             android:layout_width="match_parent"
025             android:layout_height="wrap_content"
026             android:text="接受来自子线程的消息如下：\n" />
027
028    </LinearLayout>
```

在如上代码中第 8～12 行定义了 ImageView 控件，显示加载的图片。在第 15～19 行定义了一个 Button 按钮，当用户单击此按钮的时候开始加载。在第 21～26 行定义了一个 TextView 控件，显示图片的加载张数及消耗的时间。并且设置各自的 id 属性，方便我们在 Activity 中获取此控件。

然后修改 src/com.wyl.example/MainActivity.java 文件，代码如下：

```
001  //定义了本实例的主要 Activity
002  public class MainActivity extends Activity {
003      //代表系统正在读取图片
004      public final static int LOAD_PROGRESS = 0;
005      //代表系统读取 100 张图片完成
006      public final static int LOAD_COMPLETE = 1;
007      //定义的 Button 对象
008      Button btn = null;
009      //定义 TextView 对象
010      TextView tv = null;
011      //开始加载时的时间
012      Long mLoadStatr = 0L;
013      //结束加载时的时间
014      Long mLoadEnd = 0L;
015      // 接收传递过来的信息
016      Handler handler = new Handler() {
017          @Override
018          public void handleMessage(Message msg) {
019              switch (msg.what) {
020              //当系统在读取图片时，显示读取的进度
021              case LOAD_PROGRESS:
022                  tv.setText(tv.getText()+"时间："
                          +msg.arg2+"\t 当前读取到第" + msg.arg1 + "张图片\n");
023                  break;
024              //当读取完毕时，显示读取图片的平均时间
025              case LOAD_COMPLETE:
026                  tv.setText("读取结束一共耗时"
                          + msg.arg1 + "毫秒\t 平均读取时间:"+msg.arg2+"毫秒\n");
027                  break;
028              }
029              super.handleMessage(msg);
030          }
031      };
032
033      @Override
034      protected void onCreate(Bundle savedInstanceState) {
035          super.onCreate(savedInstanceState);
036          //设置当前 Activity 的页面布局
037          setContentView(R.layout.activity_main);
038
```

```
039            findView();
040            setListener();
041        }
042
043        private void setListener() {
044            //设置btn的单击监听器
045            btn.setOnClickListener(new OnClickListener() {
046                @Override
047                public void onClick(View arg0) {
048                    // 开始读取图片
049                    LoadImage();
050                }
051            });
052        }
053
054        private void findView() {
055            //得到布局中的控件对象
056            btn = (Button) findViewById(R.id.Btn);
057            tv = (TextView) findViewById(R.id.Tv);
058            tv.setText("单击按钮开始测试加载图片的速度：\n");
059        }
060
061        //定义子线程，循环加载图片
062        public void LoadImage() {
063            new Thread() {
064                @Override
065                public void run() {
066                    //得到加载图片开始的时间
067                    mLoadStatr = System.currentTimeMillis();
068
069                    for (int i = 0; i < 100; i++) {
070                        // 这里循环加载图片100遍
071                        ReadPic(MainActivity.this, R.drawable.baidu);
072
073                        // 每读取完一张图片发送消息给handler
074                        Message msg = new Message();
075                        msg.what = LOAD_PROGRESS;
076                        msg.arg1 = i + 1;
077                        msg.arg2 = (int) System.currentTimeMillis();
078                        handler.sendMessage(msg);
079                    }
080
081                    // 得到加载图片结束的时间
082                    mLoadEnd = System.currentTimeMillis();
083
084                    // 100张图片加载完成
085                    Message msg = new Message();
086                    msg.what = LOAD_COMPLETE;
087                    msg.arg1 = (int) (mLoadEnd - mLoadStatr);
088                    msg.arg2 = msg.arg1/100;
089                    handler.sendMessage(msg);
090                }
091            }.start();
092
093        }
094
095        public Bitmap ReadPic(Context context, int resId) {
096            //得到图片工程的选项
097            BitmapFactory.Options opt = new BitmapFactory.Options();
```

```
098          opt.inPreferredConfig = Bitmap.Config.RGB_565;
099          opt.inPurgeable = true;
100          opt.inInputShareable = true;
101          // 获取资源图片
102          InputStream is = context.getResources().openRawResource(resId);
103          return BitmapFactory.decodeStream(is, null, opt);
104      }
105 }
```

在上面代码中第 16～31 行定义了一个 Handler 对象，其中实现了 handleMessage 方法，在其收到消息后，如果是 LOAD_PROGRESS，说明图片在加载中；如果消息为 LOAD_COMPLETE，代表图片已经加载完毕。在第 54～59 行得到控件对象，在 43～52 行设置当按钮单击后，开始加载图片。在第 61～93 行启动线程，循环加载 100 次图片，并且在加载过程中发送消息给主线程。在第 95～104 行定义了如何加载一张本地图片代码。

4．实例扩展

其实本实例仅仅讲述了在 Android 中如何测试加载图片的速度。大家可以扩展，完成测试 Android 中加载本地文本的测试，测试 Android 中加载本地音乐的测试器等。原理与本实例相同。

范例 085 Surface 的读写刷新

1．实例简介

我们在使用 Android 应用的时候，也经常看到有一些动画效果。例如，程序中的某些启动展开动画、画板应用的图形描绘和游戏中的动画等功能。本实例就带领大家使用 Android 中的 Surface+多线程来完成一个动画播放的效果。

2．运行效果

该实例运行效果如图 4.32 所示。

3．实例程序讲解

在上面例子的效果中，可以通过单击单线程读写图片按钮，通过一个线程完成读写工作，单击开启双线程完成工作，那么就开启两个线程：一个复责读图片，一个负责写图片。

图 4.32 Surface 的读写刷新

想要实现我们上例的效果，首先修改 res/layout/activity_main.xml 文件，代码如下：

```
01 <?xml version="1.0" encoding="utf-8"?>
02 <!-- 定义当前布局的基本 LinearLayout -->
03 <LinearLayout xmlns:android="http://schemas.android.com/apk/res/android"
04     android:layout_width="fill_parent"
05     android:layout_height="fill_parent"
06     android:orientation="vertical" >
07     <!-- 定义页面上部的水平的两个按钮的 LinearLayout -->
08     <LinearLayout
09         android:layout_width="match_parent"
```

```
10          android:layout_height="wrap_content"
11          android:background="@android:color/black" >
12       <!-- 打开单个独立线程加载图片的按钮 -->
13       <Button
14          android:id="@+id/Btn1"
15          android:layout_width="wrap_content"
16          android:layout_height="wrap_content"
17          android:background="@android:color/white"
18          android:layout_weight="1"
19          android:text="单个独立线程" >
20       </Button>
21       <!-- 打开两个独立线程加载图片的按钮 -->
22       <Button
23          android:id="@+id/Btn2"
24          android:layout_width="wrap_content"
25          android:layout_height="wrap_content"
26          android:background="@android:color/white"
27          android:layout_weight="1"
28          android:text="两个独立线程" >
29       </Button>
30    </LinearLayout>
31    <!-- 图片显示的SurfaceView控件 -->
32    <SurfaceView
33       android:id="@+id/Sv"
34       android:layout_width="fill_parent"
35       android:layout_height="fill_parent" >
36    </SurfaceView>
37
38 </LinearLayout>
```

在如上代码中第 13～29 行定义了两个按钮，分别代表一个独立线程读写 Surface 和两个线程分别读写图片。在第 32～36 行定义了一个 Surface 控件用来绘制用户加载的图片。

然后修改 src/com.wyl.example/MainActivity.java 文件，代码如下：

```
001 //定义了本实例的主要Activity
002 public class MainActivity extends Activity {
003    //定义单个线程加载图片的按钮
004    Button btnSingleThread;
005    //定义两个线程加载图片的按钮
006    Button btnDoubleThread;
007    //定义SurfaceView对象
008    SurfaceView sfv;
009    //定义SurfaceHolder对象
010    SurfaceHolder sfh;
011    //得到资源中的所有图片资源id
012    ArrayList<Integer> imgList = new ArrayList<Integer>();
013
014    int imgWidth, imgHeight;
015    //需要绘制的Bitmap对象
016    Bitmap bitmap;
017
018    @Override
019    public void onCreate(Bundle savedInstanceState) {
020       super.onCreate(savedInstanceState);
021       setContentView(R.layout.activity_main);
022
023       findView();
024       setListener();
```

```
025         }
026
027     private void setListener() {
028         //设置按钮的单击监听器
029         btnSingleThread.setOnClickListener(new myOnClickListener());
030         btnDoubleThread.setOnClickListener(new myOnClickListener());
031         // 自动运行 surfaceCreated 以及 surfaceChanged
032         sfh.addCallback(new MyCallBack());
033     }
034
035     private void findView() {
036         //得到布局中的控件对象
037         btnSingleThread = (Button) this.findViewById(R.id.Btn1);
038         btnDoubleThread = (Button) this.findViewById(R.id.Btn2);
039         sfv = (SurfaceView) this.findViewById(R.id.Sv);
040         sfh = sfv.getHolder();
041     }
042
043     //自定义 myOnClickListener 类实现 OnClickListener 接口
044     class myOnClickListener implements View.OnClickListener {
045         //当单击按钮时自动回调 onClick 方法
046         @Override
047         public void onClick(View v) {
048             //当单击 btnSingleThread 时,开启一条读取线程
049             if (v == btnSingleThread) {
050                 new Load_DrawImage(0, 0).start();   //开一条线程读取并绘图
051             } else if (v == btnDoubleThread) {
052                 new LoadImage().start();              //开一条线程读取
053                 new DrawImage(imgWidth + 10, 0).start();//开一条线程绘图
054             }
055         }
056     }
057     //自定义 MyCallBack 类实现了 SurfaceHolder.Callback 接口
058     class MyCallBack implements SurfaceHolder.Callback {
059         //当 surface 改变后自动回调 surfaceChanged 方法
060         @Override
061         public void surfaceChanged(SurfaceHolder holder, int format, int width,
062                 int height) {
063             Log.i("Surface:", "Change");
064
065         }
066         //当 surface 创建时自动回调 surfaceCreated 方法
067         @Override
068         public void surfaceCreated(SurfaceHolder holder) {
069             Log.i("Surface:", "Create");
070
071             // 用反射机制来获取资源中的图片 ID 和尺寸
072             Field[] fields = R.drawable.class.getDeclaredFields();
073             for (Field field : fields) {
074                 // 除了 ic_launcher 之外的图片
075                 if (!"ic_launcher".equals(field.getName()))
076                 {
077                     int index = 0;
078                     try {
079                         index = field.getInt(R.drawable.class);
080                     } catch (IllegalArgumentException e) {
081                         // TODO Auto-generated catch block
082                         e.printStackTrace();
```

```java
083                } catch (IllegalAccessException e) {
084                    // TODO Auto-generated catch block
085                    e.printStackTrace();
086                }
087                // 保存图片ID
088                imgList.add(index);
089            }
090        }
091        // 取得图像大小
092        Bitmap bmImg = BitmapFactory.decodeResource(getResources(),
093                imgList.get(0));
094        imgWidth = bmImg.getWidth();
095        imgHeight = bmImg.getHeight();
096    }
097    //当surfaceview销毁的时候回调此方法
098    @Override
099    public void surfaceDestroyed(SurfaceHolder holder) {
100        Log.i("Surface:", "Destroy");
101    }
102
103 }
104
105 //自定义线程，读取并显示图片
106 class Load_DrawImage extends Thread {
107     int x, y;
108     int imgIndex = 0;
109
110     //构造函数初始化x，y的值
111     public Load_DrawImage(int x, int y) {
112         this.x = x;
113         this.y = y;
114     }
115
116     public void run() {
117         while (true) {
118             //通过surfaceholder锁定canvas，得到canvas对象
119             Canvas c = sfh.lockCanvas(new Rect(this.x, this.y, this.x
120                     + imgWidth, this.y + imgHeight));
121             //根据资源id读取对应的资源
122             Bitmap bmImg = BitmapFactory.decodeResource(getResources(),
123                     imgList.get(imgIndex));
124             //在canvas的指定位置上画bmimg对象
125             c.drawBitmap(bmImg, this.x, this.y, new Paint());
126             //循环资源id
127             imgIndex++;
128             if (imgIndex == imgList.size())
129                 imgIndex = 0;
130             //屏幕解锁更新屏幕内容
131             sfh.unlockCanvasAndPost(c);            // 更新屏幕显示内容
132         }
133     }
134 };
135
136 //自定义Thread类，只负责绘图
137 class DrawImage extends Thread {
138     int x, y;
139     //构造函数初始化x，y的值
140     public DrawImage(int x, int y) {
141         this.x = x;
```

```
142                this.y = y;
143            }
144
145            public void run() {
146                while (true) {
147                    // 如果图像加载成功
148                    if (bitmap != null) {
149                        //锁定 canvas
150                        Canvas c = sfh.lockCanvas(new Rect(this.x, this.y,
151                            this.x+ imgWidth, this.y + imgHeight));
152                        //在 canvas 指定的位置，绘制 btimap 对象
153                        c.drawBitmap(bitmap, this.x, this.y, new Paint());
154                        //解锁 canvas,更新显示内容
155                        sfh.unlockCanvasAndPost(c);       // 更新屏幕显示内容
156                    }
157                }
158            }
159        };
160
161        //自定义 Thread 类，只负责读取图片
162        class LoadImage extends Thread {
163            int imgIndex = 0;
164
165            public void run() {
166                while (true) {
167                    //读取对应的图片资源
168                    bitmap = BitmapFactory.decodeResource(getResources(),
169                        imgList.get(imgIndex));
170                    //循环加载资源图片
171                    imgIndex++;
172                    if (imgIndex == imgList.size())
173                        imgIndex = 0;
174                }
175            }
176        };
177    }
```

在上面代码中第 35～41 行得到了页面中的控件对象，在第 27～33 行设置了控件的单击事件。在第 43～56 行自定义了一个 OnClickListener 类，然后实现其 onClick 方法，分别开启线程去加载图片。在第 58～103 行定义了 Surface 的 Callback 接口。然后在第 106 行、第 137 行和第 162 行分别定义了三个线程来实现图片的读写功能。

4．实例扩展

Android 中主要的展示类有两个，一个是 View，一个是 Surface，此两个类主要区别在于 View 是内部控件的父类，而 Surface 类主要擅长图片的显示及内容的多次清除和展示。

范例 086　按两次物理返回键退出程序

1．实例简介

我们在使用 Android 应用的时候，在退出程序时一般程序都会有所提示，防止用户误操作退出。例如，当我们单击物理返回键退出程序时可能会弹出 AlertDialog，让我们确认

退出程序，或者现在比较流行的一种方式是当你第一次按下物理的返回键时，提醒用户在规定时间内再按一次返回键就退出程序。这就要求当用户第一次按下返回键的时候进行计时，如果规定时间内用户再次按下返回键，那就退出程序，否则就不退出。本实例就带领大家使用 Android 中的 handler 完成一个用户连续按下两次物理返回键然后就退出程序的效果。

2．运行效果

该实例运行效果如图 4.33 所示。

3．实例程序讲解

在上面例子的效果中，当用户第一次按下返回键的时候，通过 toast 提示用户想要退出程序的话再次按下返回键，如果在规定时间内用户再次按下了返回键，那么就退出程序，否则不退出程序。想要实现我们上例的效果，首先要修改 res/layout/activity_main.xml 文件，代码如下：

图 4.33　按两次物理返回键退出程序

```xml
01 <?xml version="1.0" encoding="utf-8"?>
02 <!-- 定义当前布局的基本 LinearLayout -->
03 <LinearLayout xmlns:android="http://schemas.android.com/apk/res/android"
04     android:layout_width="fill_parent"
05     android:layout_height="fill_parent"
06     android:orientation="vertical" >
07
08     <!-- 定义页面文字标签 -->
09     <TextView
10         android:layout_width="fill_parent"
11         android:layout_height="fill_parent"
12         android:text="在 2 秒内单击两次返回键，可退出程序....." />
13
14 </LinearLayout>
```

在如上代码中第 9~12 行定义了 TextView 控件，用来提醒用户连续单击两次物理返回键退出程序。

然后修改 src/com.wyl.example/MainActivity.java 文件，代码如下：

```java
01 //定义了本实例的主要 Activity
02 public class MainActivity extends Activity {
03     //定义是否退出程序的标记
04     private boolean isExit=false;
05     //定义接收用户发送信息的 handler
06     private Handler mHandler = new Handler(){
07         @Override
08         public void handleMessage(Message msg) {
09             super.handleMessage(msg);
10             //标记用户不退出状态
11             isExit=false;
```

```
12      }
13    };
14
15    @Override
16    public void onCreate(Bundle savedInstanceState) {
17        super.onCreate(savedInstanceState);
18        setContentView(R.layout.activity_main);
19
20    }
21    //监听手机的物理按键单击事件
22    @Override
23    public boolean onKeyDown(int keyCode, KeyEvent event) {
24        //判断用户是否单击的是返回键
25        if(keyCode == KeyEvent.KEYCODE_BACK){
26            //如果 isExit 标记为 false，提示用户再次按键
27            if(!isExit){
28                isExit=true;
29                Toast.makeText(getApplicationContext(),
                        "再按一次退出程序", Toast.LENGTH_SHORT).show();
30                //如果用户没有在 2 秒内再次按返回键的话，就发送消息标记用户为不退出状态
31                mHandler.sendEmptyMessageDelayed(0, 2000);
32            }
33            //如果 isExit 标记为 true，退出程序
34            else{
35                //退出程序
36                finish();
37                System.exit(0);
38            }
39        }
40        return false;
41    }
42 }
```

在上面代码中第 6～13 行定义了一个 Handler 对象，用来接收子线程发送的消息。在第 23～41 行实现了 Activity 的回调函数 okKeyDown，用来监听用户对于物理键的操作，当用户第一次按下返回键时，修改标记为 true，然后发送一个延迟消息，延迟 2 秒发送，就是在 2 秒内，如果用户再次按下返回键的话，就退出程序了。

4．实例扩展

其实想要实现这个效果有很多种方法，当然本质的原理都是相似的，就是在指定时间内看用户是否连续单击两次物理键，当然大家也可以使用 Java 中的 Timer 来实现此实例。

范例 087　线程嵌套

1．实例简介

我们在使用 Android 应用的时候，经常会用到多线程的功能，但是在一些复杂应用中还会发生一种这样的情况，就是在子线程中又可以生成子线程，这样的话我们的程序如何来管理子线程和子线程的线程呢？这就要求大家要对 Java 中的线程有一个完整的认识。本实例就带领大家实现一个在子线程中再创建子线程的实例。

2. 运行效果

该实例运行效果如图 4.34 所示。

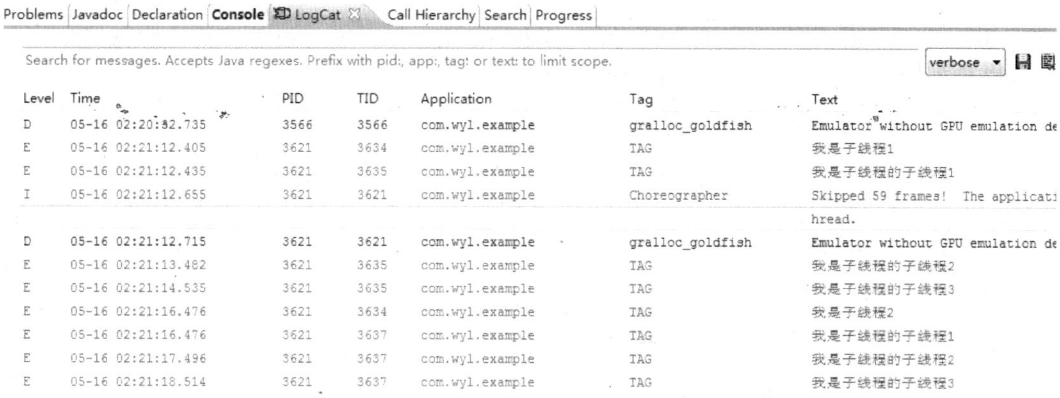

```
Level  Time              PID    TID    Application       Tag                Text
D      05-16 02:20:52.735  3566   3566   com.wyl.example   gralloc_goldfish   Emulator without GPU emulation de
E      05-16 02:21:12.405  3621   3621   com.wyl.example   TAG                我是子线程1
E      05-16 02:21:12.435  3621   3635   com.wyl.example   TAG                我是子线程的子线程1
I      05-16 02:21:12.655  3621   3621   com.wyl.example   Choreographer      Skipped 59 frames! The applicati
                                                                              hread.
D      05-16 02:21:12.715  3621   3621   com.wyl.example   gralloc_goldfish   Emulator without GPU emulation de
E      05-16 02:21:13.482  3621   3635   com.wyl.example   TAG                我是子线程的子线程2
E      05-16 02:21:14.535  3621   3635   com.wyl.example   TAG                我是子线程的子线程3
E      05-16 02:21:16.476  3621   3634   com.wyl.example   TAG                我是子线程2
E      05-16 02:21:16.476  3621   3637   com.wyl.example   TAG                我是子线程的子线程1
E      05-16 02:21:17.496  3621   3637   com.wyl.example   TAG                我是子线程的子线程2
E      05-16 02:21:18.514  3621   3637   com.wyl.example   TAG                我是子线程的子线程3
```

图 4.34　子线程的嵌套

3. 实例程序讲解

在上面例子的效果中，我们可以通过查看 Eclipse 的 Logcat 窗口来查看线程的创建情况。想要实现本例的效果，首先修改 res/layout/activity_main.xml 文件，代码如下：

```xml
01  <?xml version="1.0" encoding="utf-8"?>
02  <!-- 定义当前布局的基本 LinearLayout -->
03  <LinearLayout xmlns:android="http://schemas.android.com/apk/res/android"
04      android:layout_width="fill_parent"
05      android:layout_height="fill_parent"
06      android:orientation="vertical" >
07
08      <!-- 定义页面文字标签 -->
09      <TextView
10          android:layout_width="fill_parent"
11          android:layout_height="fill_parent"
12          android:text="请查看 LogCat 中的子线程创建情况....." />
13
14  </LinearLayout>
```

在如上代码中第 9～12 行定义了 TextView 控件用来提醒用户查看 LogCat 窗口。

然后修改 src/com.wyl.example/MainActivity.java 文件，代码如下：

```java
01  //定义了本实例的主要 Activity
02  public class MainActivity extends Activity{
03      //定义主线程的 handler 对象
04      private Handler mainhandler;
05      //定义子线程对象
06      private Thread thread;
07      // 开辟子线程
08      private Thread th;
09      //是否继续让子线程继续创建子线程的标志
10      private boolean FLAG_RUN = true;
11      //子线程的计数变量
```

```java
12  private int i = 0;
13  //子线程的子线程的计数变量
14  private int j = 0;
15
16  @Override
17  public void onCreate(Bundle savedInstanceState) {
18      super.onCreate(savedInstanceState);
19      setContentView(R.layout.activity_main);
20      //初始化主线程的 Handler 对象
21      mainhandler = new Handler(cb);
22      //使用 Runnable 对象初始化子线程对象
23      thread = new Thread(r);
24      //启动子线程对象
25      thread.start();
26  }
27  //自定义 Runnable 对象,实现相应的回调接口
28  Runnable r = new Runnable(){
29      @Override
30      public void run() {
31          //根据 FLAG_RUN 标记判断是否继续创建子线程
32          while (FLAG_RUN) {
33              i++;
34              Log.e("TAG","我是子线程" + i);
35              //子线程再次创建子线程
36              childThread();
37              try {
38                  thread.sleep(4000);
39                  //发送 message 给主线程的 handler
40                  mainhandler.obtainMessage(1).sendToTarget();
41              } catch (InterruptedException e) {
42                  e.printStackTrace();
43              }
44          }
45      }
46  };
47  //自定义 Callback 对象,实现相应的回调接口
48  Callback cb = new Callback(){
49      //当 handler 接收到 message 消息后回调此函数
50      @Override
51      public boolean handleMessage(Message msg) {
52          if (msg.what == 1) {
53              //如果已经创建了 2 个子线程,那么就不再创建
54              if (i < 2) {
55                  Toast.makeText(MainActivity.this,
                          "子线程"+i+"创建成功", Toast.LENGTH_SHORT).show();
56                  FLAG_RUN = true;
57              } else {
58                  FLAG_RUN = false;
59              }
60          }
61          return true;
62      }
63  };
64
65  //定义方法,实现在子线程中再次创建子线程
66  private void childThread() {
67      j = 0;
```

```
68          //定义子线程创建的子线程
69          th = new Thread() {
70              @Override
71              public void run() {
72                  while (true) {
73                      //如果创建了 3 个，就结束，否则继续创建
74                      if (j >= 3) {
75                          break;
76                      } else {
77                          j++;
78                          Log.e("TAG","我是子线程的子线程" + j);
79                          try {
80                              th.sleep(1000);
81                          } catch (InterruptedException e) {
82                              e.printStackTrace();
83                          }
84                      }
85
86                  }
87              }
88          };
89          //线程启动
90          th.start();
91      }
92  }
```

在上面代码中第 21~25 行定义了一个主线程的 Handler 对象，然后实例化了一个 Thread 对象，然后启动了此对象。第 28~46 行定义了一个 Runnable 对象，用来给主线程发送消息。在第 47~63 行实现了一个 CallBack 接口用来实现收到消息后的处理过程。在第 66~91 行实现了 childTrhead 方法，用来创建子线程，在子线程中每隔 1 秒创建一个线程。

4．实例扩展

在我们这个实例中通过模拟子线程创建的方式，希望各位读者能够明白在 Android 中线程之间是没有区别的，不论你是由谁创建的，他们都可以独立的运行，而且可以相互传递消息。

范例 088 异步任务加载网络图片

1．实例简介

我们在前几个应用中使用了 Handler+Message 的方式来给大家呈现多线程的问题，主要原因是 Android 的线程安全，也就是只有在主线程中才可以修改程序的 UI，但是这种方法有时候用起来比较复杂，我们要建立一系列的线程而且让它们与主线程进行消息传递。例如，新闻的图片列表和广告列表等，这样的话在我们的程序中就会有多个 Thread 的创建操作，而且我们还要维护他们之间的关系。Android 中通过另一种简单的方式可以实现异步操作，那就是异步任务。本实例就带领大家使用 Android 中的异步任务来完成一个网络图片加载的效果。

2. 运行效果

该实例运行效果如图 4.35 所示。

图 4.35　异步任务加载网络图片

3. 实例程序讲解

在本例子的效果中，首先看到了图片的默认加载图片，当图片从网络加载完成后，显示为网络加载的图片。想要实现我们上例的效果，首先修改 res/layout/activity_main.xml 文件，代码如下：

```xml
01  <?xml version="1.0" encoding="utf-8"?>
02  <!-- 定义当前布局的基本 LinearLayout -->
03  <LinearLayout xmlns:android="http://schemas.android.com/apk/res/android"
04      android:layout_width="fill_parent"
05      android:layout_height="fill_parent"
06      android:orientation="vertical" >
07      <!-- 定义异步任务加载图片的 imageview -->
08      <ImageView
09          android:id="@+id/Iv"
10          android:layout_width="match_parent"
11          android:layout_height="match_parent"
12          android:layout_gravity="center"
13          android:src="@drawable/load" />
14
15  </LinearLayout>
```

此文件为本例 Activity 的布局文件，相对比较简单，就是在基本的线性布局中放入了需要展示加载图片的 ImageView 控件。

然后创建 src/com.wyl.example/ ImageAsynTask.java 文件，代码如下：

```java
01  //自定义 ImageAsynTask 类，继承自异步任务类
02  public class ImageAsynTask extends AsyncTask<Void, Void, Drawable> {
03      //用来存储加载图片成功后展示图片的 ImageView 对象
04      private ImageView m;
05      //记录需要加载的网络图片的地址
06      private String imageurl;
07  
08      //构造函数，传入 imageview 对象和网络图片地址
09      public ImageAsynTask(ImageView i,String t){
10          m = i;
11          imageurl = t;
12      }
13  
14      //后台执行耗时的操作
15      @Override
16      protected Drawable doInBackground(Void... params) {
17          return loadImages(imageurl);
18      }
19  
20      //执行后台操作完毕后自动回调此函数
21      @Override
22      protected void onPostExecute(Drawable result) {
23          super.onPostExecute(result);
24          //如果图片加载
25          if (null != result) {
26              m.setImageDrawable(result);
27          }
28          else{
29              m.setImageResource(R.drawable.failed);
30          }
31      }
32  
33      //执行后台操作前回调的函数
34      @Override
35      protected void onPreExecute() {
36          super.onPreExecute();
37      }
38  
39      //网路图片的下载方法
40      public Drawable loadImages(String url) {
41          try {
42              //根据传入的图片 url 地址，得到 Drawable 对象
43              return Drawable.createFromStream(
44                      (InputStream) (new URL(url)).openStream(), "test");
45          } catch (IOException e) {
46              e.printStackTrace();
47          }
48          return null;
49      }
50  }
```

此文件主要定义了异步图片加载的类，在此文件中的第 9～12 行实现了构造方法，初始化了 ImageView 和需要加载的图片地址 imageurl 的值。在第 16～18 行实现了异步任务

的接口方法 doInBackground 方法，代表在异步任务类中需要执行的耗时的操作。在第 22～31 行实现了异步任务的 onPostExecute 方法，在异步任务的耗时工作完毕后会自动将结果传递给此方法进行处理，这里是将异步任务中加载的图片设置给对应的 ImageView 控件。在第 40～49 行实现了从网络上加载图片的具体代码。

然后修改 src/com.wyl.example/MainActivity.java 文件，代码如下：

```
01  //定义了本实例的主要 Activity
02  public class MainActivity extends Activity {
03      //定义需要展示图片的 ImageView 对象
04      private ImageView mImage;
05
06      @Override
07      protected void onCreate(Bundle savedInstanceState) {
08          super.onCreate(savedInstanceState);
09          setContentView(R.layout.activity_main);
10          //得到布局中的对象
11          mImage = (ImageView) findViewById(R.id.Iv);
12          //新建一个异步任务，开始加载网络图片
13          new ImageAsynTask(mImage, "http://img1.3lian.com/img2011/07/20/05.jpg")
14                  .execute();
15      }
16  }
```

在上面代码中第 11 行得到了页面中的 ImageView 控件对象，在第 13 行中启动了异步任务，去加载网络图片，并且把之前得到的 ImageView 控件传入此方法。

4．实例扩展

AsynTask 异步任务在 Android 中使用的方式很多，其效果能够和 Handler+Message 的效果一致，而且异步任务还有一些特殊的用法，我们在后面的实例中会进行讲解。

范例 089　网站源代码查看器

1．实例简介

我们平时常见的 Android 客户端大部分都作为数据的呈现，其真实数据一般都是从网络获取的。例如，QQ 网购客户端和淘宝客户端等。这就要求在 Android 的手机上可以获取到网络的数据，而且最好是异步的。本实例就带领大家实现一个网络中网站源代码的查看器效果。

2．运行效果

该实例运行效果如图 4.36 所示。

3．实例程序讲解

在上面例子的效果中，我们能看到一个输入框，一个按钮，还有一个网络源代码的查看的 TextView。

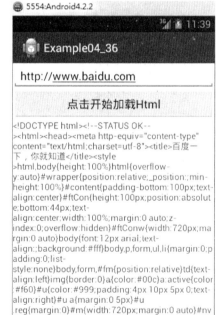

图 4.36　网站源代码查看器

当用户在输入框中，输入想要查看的网址，然后单击"开始加载"按钮，在结果的 TextView 中就会加载指定网站的源代码了。想要实现我们上例的效果，首先修改 res/layout/activity_main.xml 文件，代码如下：

```xml
01  <?xml version="1.0" encoding="utf-8"?>
02  <!-- 定义当前布局的基本 LinearLayout -->
03  <LinearLayout xmlns:android="http://schemas.android.com/apk/res/android"
04      android:layout_width="fill_parent"
05      android:layout_height="fill_parent"
06      android:orientation="vertical" >
07      <!-- 定义获得用户需要加载的网址的输入框 -->
08      <EditText
09          android:id="@+id/Et"
10          android:layout_width="match_parent"
11          android:layout_height="wrap_content"
12          android:text="http://www.baidu.com"
13          />
14      <!-- 定义用户单击按钮控件 -->
15      <Button
16          android:id="@+id/Btn"
17          android:layout_width="match_parent"
18          android:layout_height="wrap_content"
19          android:text="单击开始加载 Html"
20          />
21
22      <ScrollView
23          android:layout_width="match_parent"
24          android:layout_height="match_parent"
25          >
26          <!-- 定义异步任务加载 Html 源代码的 TextView -->
27          <TextView
28              android:id="@+id/Tv"
29              android:layout_width="match_parent"
30              android:layout_height="match_parent"/>
31
32      </ScrollView>
33  </LinearLayout>
```

此文件为 Activity 的布局文件，其中以线性布局为基本布局，其中包括了一个 EditText、一个 Button 和一个 TextView。

新建 src/com.wyl.example/ HtmlAsynTask.java 文件，代码如下：

```java
01  //自定义 HtmlAsynTask 类，继承自异步任务类
02  public class HtmlAsynTask extends AsyncTask<Void, Void, String> {
03      //用来存储加载网址内容成功后展示网络源代码的 TextView 对象
04      private TextView m;
05      //记录需要加载的网络地址
06      private String url;
07
08      //构造函数，传入 TextView 对象和网络地址
09      public HtmlAsynTask(TextView i,String t){
10          m = i;
11          url = t;
12      }
13
14      //后台执行耗时的操作
15      @Override
```

```
16    protected String doInBackground(Void... params) {
17        return requestByHttpGet(url);
18    }
19
20    //执行后台操作完毕后自动回调此函数
21    @Override
22    protected void onPostExecute(String result) {
23        super.onPostExecute(result);
24        //如果加载得到的网络地址的源代码不为空的话,设置 textview 对象的值
25        if (null != result) {
26            m.setText(result);
27        }
28        //如果加载的内容为空的话,显示加载失败
29        else{
30            m.setText("加载失败");
31        }
32    }
33
34    //执行后台操作前回调的函数
35    @Override
36    protected void onPreExecute() {
37        super.onPreExecute();
38    }
39
40    //通过加载 url 的网络内容
41    public String requestByHttpGet(String url) {
42        //新建 HttpGet 对象
43        HttpGet httpGet = new HttpGet(url);
44        //定义 HttpClient 对象
45        HttpClient httpClient = new DefaultHttpClient();
46        //定义 HttpResponse 实例
47        HttpResponse httpResp;
48        try {
49            httpResp = httpClient.execute(httpGet);
50            // 判断是否请求成功
51            if (httpResp.getStatusLine().getStatusCode() == 200) {
52                // 获取返回的数据
53                String result = EntityUtils.toString(httpResp.getEntity(), "UTF-8");
54                Log.e("TAG", "HttpGet 方式请求成功,返回数据如下: ");
55                return result;
56            } else {
57                Log.e("TAG", "HttpGet 方式请求失败");
58            }
59        } catch (ClientProtocolException e) {
60            // TODO Auto-generated catch block
61            e.printStackTrace();
62        } catch (IOException e) {
63            // TODO Auto-generated catch block
64            e.printStackTrace();
65        }
66        return null;
67    }
68 }
```

此文件为异步加载网站源代码的类的实现,在此类中定义了记录网址的变量 url 和需要展现结果的 TextView 对象。在第 16~17 行定义在异步任务中操作的耗时工作,这里加载网站的源代码。在第 22~32 行实现了 onPostExecute 方法,将加载下来的网络源代码

设置给 TextView 控件。在第 40～67 行实现了根据网址获得源代码的步骤。

然后修改 src/com.wyl.example/MainActivity.java 文件，代码如下：

```
01  //定义了本实例的主要Activity
02  public class MainActivity extends Activity{
03      //定义布局中的TextView控件
04      private TextView tv;
05      //定义布局中的EditText控件
06      private EditText et;
07      //定义布局中的Button控件
08      private Button btn;
09
10      @Override
11      protected void onCreate (Bundle savedInstanceState) {
12          super.onCreate(savedInstanceState);
13          setContentView(R.layout.activity_main);
14          findView();
15          setListener();
16      }
17
18  private void setListener() {
19      //设置btn的单击监听器
20      btn.setOnClickListener(new OnClickListener() {
21          @Override
22          public void onClick(View v) {
23              //初始textview的内容
24              tv.setText("Html 加载中.........");
25              //获取EditText的用户输入值
26              String str = et.getText().toString();
27              //启动异步任务加载用户输入的url中的网络html
28              new HtmlAsynTask(tv, str).execute();
29          }
30      });
31  }
32
33  private void findView() {
34      //得到布局中的TextView的对象
35      tv = (TextView) findViewById(R.id.Tv);
36      //得到布局中的EditText的对象
37      et = (EditText)findViewById(R.id.Et);
38      //得到布局中的EditText的对象
39      btn = (Button)findViewById(R.id.Btn);
40  }
41  }
```

在上面代码中第 33～40 行得到了布局中的控件对象，在第 18～31 行实现了 btn 按钮的单击监听器，在用户单击此按钮的时候启动异步加载 Html 源代码的异步任务。这样就可以实现本例效果了。

4．实例扩展

在我们这个实例中需要获得网站的源代码，所以需要获得 Android 系统应用的读取网络的权限 <uses-permission android:name="android.permission.INTERNET"/>，要在你的 manifest 文件中加入此权限你的应用程序才可以访问网络。

范例 090　终止异步任务操作

1．实例简介

使用 Handler+Message 的方式来实现异步任务最大的麻烦之处就在于操作的停止，对于异步任务来说实现的相对比较简单了。本实例就带领大家来实现一个终止异步任务效果。

2．运行效果

该实例运行效果如图 4.37 所示。

3．实例程序讲解

在上面例子的效果中，我们能看到一个输入框、两个按钮、一个开始加载、一个停止加载和一个网络源代码的查看的 TextView。当用户在输入框中，输入想要查看的网址，然后单击"开始加载"按钮，在加载出来之前单击停止加载，就停止了异步任务。在结果的 TextView 中就会加载指定网站的源代码了。想要

图 4.37　终止异步任务

实现我们上例的效果，首先修改 res/layout/activity_main.xml 文件，代码如下：

```
01  <?xml version="1.0" encoding="utf-8"?>
02  <!-- 定义当前布局的基本 LinearLayout -->
03  <LinearLayout xmlns:android="http://schemas.android.com/apk/res/android"
04      android:layout_width="fill_parent"
05      android:layout_height="fill_parent"
06      android:orientation="vertical" >
07      <!-- 定义获得用户需要加载的网址的输入框 -->
08      <EditText
09          android:id="@+id/Et"
10          android:layout_width="match_parent"
11          android:layout_height="wrap_content"
12          android:text="http://www.baidu.com"
13          />
14      <!-- 定义用户单击开始加载按钮控件 -->
15      <Button
16          android:id="@+id/BtnStart"
17          android:layout_width="match_parent"
18          android:layout_height="wrap_content"
19          android:text="单击开始加载 Html"
20          />
21      <!-- 定义用户单击终止加载按钮控件 -->
22      <Button
23          android:id="@+id/BtnStop"
24          android:layout_width="match_parent"
25          android:layout_height="wrap_content"
26          android:text="单击终止加载 Html"
27          />
28
29      <ScrollView
30          android:layout_width="match_parent"
```

```
31            android:layout_height="match_parent"
32            >
33            <!-- 定义异步任务加载 Html 源代码的 TextView -->
34        <TextView
35            android:id="@+id/Tv"
36            android:layout_width="match_parent"
37            android:layout_height="match_parent"/>
38
39        </ScrollView>
40  </LinearLayout>
```

此文件为 Activity 的布局文件，以线性布局为基本布局，其中包括了一个 EditText、两个 Button、一个开始按钮、一个终止按钮和一个 TextView。

新建 src/com.wyl.example/ HtmlAsynTask.java 文件，代码如下：

```
01  //自定义 HtmlAsynTask 类，继承自异步任务类
02  public class HtmlAsynTask extends AsyncTask<Void, Void, String> {
03      //用来存储加载网址内容成功后展示网络源代码的 TextView 对象
04      private TextView m;
05      //记录需要加载的网络地址
06      private String url;
07
08      //构造函数，传入 TextView 对象和网络地址
09      public HtmlAsynTask(TextView i,String t){
10          m = i;
11          url = t;
12      }
13
14      //后台执行耗时的操作
15      @Override
16      protected String doInBackground(Void... params) {
17          return requestByHttpGet(url);
18      }
19
20      //执行后台操作完毕后自动回调此函数
21      @Override
22      protected void onPostExecute(String result) {
23          super.onPostExecute(result);
24          Log.e("isCancelled()", isCancelled()+"");
25          //如果用户取消此异步任务
26          if (isCancelled()) {
27              m.setText("取消加载");
28          }
29          //如果加载得到的网络地址的源代码不为空的话，设置 textview 对象的值
30          else if (null != result) {
31              m.setText(result);
32          }
33          //如果加载的内容为空的话，显示加载失败
34          else{
35              m.setText("加载失败");
36          }
37  }
38
39      //执行后台操作前回调的函数
40      @Override
41      protected void onPreExecute() {
42          super.onPreExecute();
43      }
```

```
44
45    //通过加载url的网络内容
46    public String requestByHttpGet(String url) {
47        //新建HttpGet对象
48        HttpGet httpGet = new HttpGet(url);
49        //定义HttpClient对象
50        HttpClient httpClient = new DefaultHttpClient();
51        //定义HttpResponse实例
52        HttpResponse httpResp;
53        try {
54            httpResp = httpClient.execute(httpGet);
55            // 判断是否请求成功
56            if (httpResp.getStatusLine().getStatusCode() == 200) {
57                // 获取返回的数据
58                String result = EntityUtils.toString(httpResp.getEntity(),
                   "UTF-8");
59                Log.e("TAG", "HttpGet方式请求成功,返回数据如下: ");
60                return result;
61            } else {
62                Log.e("TAG", "HttpGet方式请求失败");
63            }
64        } catch (ClientProtocolException e) {
65            // TODO Auto-generated catch block
66            e.printStackTrace();
67        } catch (IOException e) {
68            // TODO Auto-generated catch block
69            e.printStackTrace();
70        }
71        return null;
72    }
73  }
```

此文件为异步加载网站源代码的类的实现,在此类中定义了记录网址的变量 url 和需要展现结果的 TextView 对象。本实例和上一个实例最大的区别在于它在执行结果的时候判断此异步任务是否被取消了,如果被取消则不再进行设置操作了。

然后修改 src/com.wyl.example/MainActivity.java 文件,代码如下:

```
01  //定义了本实例的主要Activity
02  public class MainActivity extends Activity{
03  //定义布局中的TextView控件
04      private TextView tv;
05      //定义布局中的EditText控件
06      private EditText et;
07      //定义布局中的开始加载Button控件
08  private Button btnstart;
09  //定义布局中的终止加载Button控件
10  private Button btnstop;
11  //定义异步请求html代码的异步任务
12  protected HtmlAsynTask hat;
13
14      @Override
15      protected void onCreate (Bundle savedInstanceState) {
16          super.onCreate(savedInstanceState);
17          setContentView(R.layout.activity_main);
18          findView();
19          setListener();
20      }
```

```
21
22  private void setListener() {
23      //设置 btn 的单击监听器
24      btnstart.setOnClickListener(new OnClickListener() {
25          @Override
26          public void onClick(View v) {
27              //初始 textview 的内容
28              tv.setText("Html 加载中.........");
29              //获取 EditText 的用户输入值
30              String str = et.getText().toString();
31              //启动异步任务加载用户输入的 url 中的网络 html
32              hat = new HtmlAsynTask(tv, str);
33              hat.execute();
34          }
35      });
36      btnstop.setOnClickListener(new OnClickListener() {
37          @Override
38          public void onClick(View v) {
39              //终止异步任务
40              hat.cancel(true);
41          }
42      });
43  }
44
45  private void findView() {
46      //得到布局中的 TextView 的对象
47      tv = (TextView) findViewById(R.id.Tv);
48      //得到布局中的 EditText 的对象
49      et = (EditText)findViewById(R.id.Et);
50      //得到布局中的开始加载的 Button 的对象
51      btnstart = (Button)findViewById(R.id.BtnStart);
52      //得到布局中的开始加载的 Button 的对象
53      btnstop = (Button)findViewById(R.id.BtnStop);
54  }
55  }
```

在上面代码中第 45～54 行得到了布局中的控件对象，在第 22～43 行实现了 btn 按钮的单击监听器。在用户单击"开始按钮"的时候启动异步加载 Html 源代码的异步任务，单击终止任务的时候终止异步任务。这样就可以实现本例效果了。

4．实例扩展

在异步任务的终止实现的时候，也会出现很多问题。例如，终止后异步任务仍然在执行，只有在 doInBackground 结束后才会知道终止，所以就要求大家在执行异步任务的过程中要时刻判断当前异步任务是否终止，然后做出相应的处理。

范例 091　异步任务进度展示

1．实例简介

我们在使用 Android 应用的时候，异步任务有时候也需要给主线程显示任务进行的进度。例如，文件下载进度和文件上传进度等。本实例就带领大家制作一个显示异步任务进度的效果。

2. 运行效果

该实例运行效果如图 4.38 所示。

3. 实例程序讲解

在上面例子的效果中，单击按钮开启异步任务，同时在下面的进度条中展现了异步任务的进度。想要实现我们上例的效果，首先要修改 res/layout/activity_main.xml 文件，代码如下：

图 4.38 异步任务进度展示

```xml
01 <?xml version="1.0" encoding="utf-8"?>
02 <!-- 定义当前布局的基本 LinearLayout -->
03 <LinearLayout xmlns:android="http://schemas.android.com/apk/res/android"
04     android:layout_width="fill_parent"
05     android:layout_height="fill_parent"
06     android:orientation="vertical" >
07     <!-- 定义用户单击开始加载按钮控件 -->
08     <Button
09         android:id="@+id/Btn"
10         android:layout_width="match_parent"
11         android:layout_height="wrap_content"
12         android:text="单击开始异步任务加载"
13         />
14     <ProgressBar
15         android:id="@+id/Pb"
16         android:layout_width="match_parent"
17         android:layout_height="wrap_content"
18         style="?android:attr/progressBarStyleHorizontal"
19         android:layout_gravity="center_vertical"
20         android:max="100" />
21 
22 </LinearLayout>
```

此文件为当前实例的页面布局文件。在如上代码中定义了一个 Button 控件用来开启异步任务，定义了一个 ProgressBar 控件，用来显示异步任务的进度。

然后新建 src/com.wyl.example/HtmlAsynTask.java 文件，代码如下：

```java
01 //自定义 HtmlAsynTask 类，继承自异步任务类
02 public class HtmlAsynTask extends AsyncTask<Void, Void, Integer> {
03 //用来显示异步任务的 ProgressBar 对象
04 private ProgressBar m;
05 
06 //构造函数，传入 ProgressBar 对象
07 public HtmlAsynTask(ProgressBar i) {
08     m = i;
09 }
10 
11 //后台执行耗时的操作
12 @Override
13 protected Integer doInBackground(Void... params) {
14     //模拟后台异步任务的加载进度
15     int i = 0;
16     while (i < 100) {
17         i+=10;
18         //更新 pb 的进度条
19         publishProgress(i);
```

```
20          try {
21              Thread.sleep(1000);
22          } catch (InterruptedException e) {
23          }
24      }
25      return null;
26  }
27
28  //执行后台操作完毕后自动回调此函数
29  @Override
30  protected void onPostExecute(Integer result) {
31      super.onPostExecute(result);
32
33  }
34
35  //执行后台操作前回调的函数
36  @Override
37  protected void onPreExecute() {
38      super.onPreExecute();
39  }
40
41  protected void publishProgress(int i) {
42      m.setProgress(i);
43  }
44  }
```

本文件定义了一部人物类，在此类中模拟异步任务的耗时操作，并且发送进度给主线程的进度条。在第 11～26 行模拟的后台的耗时操作，并随时发送进度。在第 41～43 行实现了设置 ProgressBar 的进度值。

然后修改 src/com.wyl.example/MainActivity.java 文件，代码如下：

```
01  //定义了本实例的主要 Activity
02  public class MainActivity extends Activity{
03      //定义布局中的开始加载 Button 控件
04  private Button btn;
05  //定义布局中的 ProgressBar 控件
06  private ProgressBar pb;
07
08      @Override
09      protected void onCreate (Bundle savedInstanceState) {
10          super.onCreate(savedInstanceState);
            //设置当前页面的布局视图为 activity_main
11          setContentView(R.layout.activity_main);
12          findView();           //得到当前视图中的控件对象
13          setListener();        //设置控件的监听器
14      }
15
16  private void setListener() {
17      //设置 btn 的单击监听器
18      btn.setOnClickListener(new OnClickListener() {
19          @Override
20          public void onClick(View v) {
21              //启动异步任务
22              new HtmlAsynTask(pb).execute();
23          }
24      });
25  }
26
```

```
27    private void findView() {
28        //得到布局中的开始加载的 Button 的对象
29        btn = (Button)findViewById(R.id.Btn);
30        //得到布局中的进度条对象
31        pb = (ProgressBar)findViewById(R.id.Pb);
32    }
33  }
```

在上面代码中第 27~32 行得到了布局中的按钮和进度条控件，在代码的第 16~25 行实现了单击按钮启动异步任务的代码。

4. 实例扩展

异步任务我们就做这几个实例，但是对于异步任务的用法还有很多，希望大家多多参考 Android 的官方开发者文档。

4.4 小　　结

在本章节中主要介绍了 Android 中的两种事件处理的方式，第一种是基于回调函数的事件处理方式，第二种是基于监听器的事件处理方式。还有 Android 中的多线程处理，也包括两种方式，一种是 Handler 的方法，另一种是异步任务的方式。本章的内容在实际项目开发使用的地方很多，用户与程序交互的方式也是吸引用户的主要手段之一，当然与用户交互的方式还有很多。例如，陀螺仪的应用和方向传感器的应用等，这些内容我们会在后面的章节继续讲解。下一章我们会讲述如何在 Android 内部组件之间进行交互。

第 5 章　Android 程序内部的信息传递者

上一章了解了 Android 中应用与用户进行交互的两种基本形式，一种是通过回调函数，一种是通过设置监听器方式，还有 Android 常见的多线程的处理 Handler 和异步任务。有了上一章的内容讲解，大家基本上可以创建一个简单的 Android 应用了，其中包括一个 Activity 以及在此 Activity 中根据用户的操作做出相应的功能相应。但是这对于一个完整的应用来说还是完全不够的，因为一个应用程序一般不会只有一个 Activity，而且我们经常看到的一种情况是在我们的应用程序中还可以调用其他 Android 系统的应用功能。例如，单击我们程序中的某个按钮可以直接拨打电话，或者可以直接发送短信。这也是 Android 系统开放性的体现，也是相对于 IOS 系统的优势所在。那么本章就给大家介绍 Android 中信息的传递者 Intent 的使用，通过 Intent 大家可以方便的和程序内部组件以及外部组件之间进行交互，实现各种功能。

Android 系统中的 Intent 主要分为两种：

第一种是调用系统的 Intent，打开系统的某些功能模块。这种用法使用起来比较简单，而且可以方便用户快速地打开 Android 系统中已完成的功能模块。

第二种是调用自定义 Intent，这种方法使用的情况也比较多，包括启动 Activity、启动 Broadcact 和启动 Service 等。Intent 是 Android 系统中四大组建之间传递消息的唯一手段，并且可以通过 Intent 携带一定的数据。

本章主要通过各种实例来介绍 Android 中常用的 Intent 的使用方法。希望读者阅读完本章内容后，在自己的应用程序中灵活运用 Intent，为自己的应用增加一些亮点功能。

5.1　Android 中系统 Intent 的使用

范例 092　Google 搜索内容

1．实例简介

在一些应用中当用户执行某些操作的时候，可以将选中的文字进行 Google 搜索，得到 Google 搜索的结果在浏览器中显示。例如，在我们查看某些新闻的时候，新闻中有一个新名词，我们希望搜索一下，这时候只要单击此名词，其 Google 搜索结果就会在浏览器中显示了。通过本实例带领大家一起来制作一个 Google 快捷搜索的应用。

2．运行效果

该实例运行效果如图 5.1 所示。

图 5.1　Google 搜索内容

3．实例程序讲解

在上例中，用户在输入框中输入希望搜索的关键字，然后单击搜索按钮，最后用户就看到在 Google 中搜索关键字的结果。想要实现如上效果，首先修改我们建立的工程下的 res/layout/activity_main.xml 文件，代码如下：

```xml
01  <?xml version="1.0" encoding="utf-8"?>
02  <!-- 定义当前布局的基本 LinearLayout -->
03  <LinearLayout xmlns:android="http://schemas.android.com/apk/res/android"
04      android:layout_width="fill_parent"
05      android:layout_height="fill_parent"
06      android:orientation="vertical" >
07
08      <!-- 定义用户输入搜索关键字的输入控件 -->
09      <EditText
10          android:id="@+id/Et"
11          android:layout_width="match_parent"
12          android:layout_height="wrap_content"
13          android:hint="请输入要搜索的关键字"
14          />
15
16      <!-- 定义用户开始搜索按钮控件 -->
17      <Button
18          android:id="@+id/Btn"
19          android:layout_width="match_parent"
20          android:layout_height="wrap_content"
21          android:text="单击开始 Google 查询"
22          />
23  </LinearLayout>
```

这是我们的 Activity 的布局文件，其中第 9～14 行定义了一个 EditText 控件，用来接收用户输入的搜索关键字。在第 17～22 行，定义了一个 Button 控件，当用户单击此控件的时候打开浏览器搜索文本框的关键字。

然后修改 src/com.wyl.example/MainActivity.java 文件，代码如下：

```java
01  //定义了本实例的主要 Activity
02  public class MainActivity extends Activity{
03      //定义布局中的开始跳转 Button 控件
04      private Button btn;
```

```
05      //定义布局中的输入查找关键字控件
06      private EditText Et;
07
08          @Override
09          protected void onCreate (Bundle savedInstanceState) {
10              super.onCreate(savedInstanceState);
11              //设置当前 Activity 的布局文件为 activity_main
12              setContentView(R.layout.activity_main);
13              //得到浏览器中的控件对象
14              findView();
15              //设置对象的监听器
16              setListener();
17          }
18
19      private void setListener() {
20          //设置 btn 的单击监听器
21          btn.setOnClickListener(new OnClickListener() {
22              @Override
23              public void onClick(View v) {
24                  //定义 Intent 对象
25                  Intent intent = new Intent();
26                  //设置 Intent 对象的 action 为 ACTION_WEB_SEARCH,代表通过 Google
                        浏览器搜索
27                  intent.setAction(Intent.ACTION_WEB_SEARCH);
28                  //设置 Intent 对象的附加内容为 SearchManager.QUERY,代表搜索关键字
29                  //然后得到用户输入在输入框中的文字传入
30                  intent.putExtra(SearchManager.QUERY,Et.getText().toString());
31                  //启动 Activity
32                  startActivity(intent);
33              }
34          });
35      }
36
37      private void findView() {
38          //得到布局中的开始加载 Button 的对象
39          btn = (Button)findViewById(R.id.Btn);
40          //得到布局中的开始加载 EditText 的对象
41          Et = (EditText)findViewById(R.id.Et);
42      }
43  }
```

如上面中代码的第 4 行定义了 Button 对象,在第 6 行定义了 EditText 对象。在第 14 行调用自定义方法 findView 来得到布局中的所有的控件,实现代码在第 37~42 行。在第 16 行设置得到控件的监听器对象,实现代码在第 19~35 行,其中给 btn 对象设置了单击监听器,当用户单击此按钮时新建 intent,然后设置 intent 的 action 属性为 Intent.ACTION_WEB_SEARCH,附带的参数为 SearchManager.QUERY,也就是搜索的关键字,这样当调用启动 startactivity 的时候系统就打开了此 Intent 对应的功能,也就是浏览器去搜索对应关键字了。

4. 实例扩展

由于此实例需要使用 Google 浏览器进行搜索,所以一定要保证你的手机是可以联网的而且你的手机需要有 Chrome 浏览器,否则效果无法显示。

范例 093　打开浏览器浏览网页

1．实例简介

在我们使用应用的过程中，经常会使用到用户单击某按钮跳转到某个网页的效果。例如，用户单击关于按钮，打开此应用的官方网站，或者单击反馈信息按钮，显示了提交用户意见的网页。一般遇到这样的功能，我们一般是当用户进行某个操作的时候。例如，单击某个按钮的时候，发送 Intent 给本地浏览器显示对应的网页，本例子就带领大家来实现一个通过用户的操作打开浏览器固定网页的实例。

2．运行效果

该实例运行效果如图 5.2 所示。

图 5.2　打开浏览器浏览网页

3．实例程序讲解

在本实例中，提供一个用户希望打开的网站的地址的输入框，然后用户单击打开按钮，界面跳转浏览器打开相应的网址显示。想要实现本实例效果，首先修改 res/layout/activity_main.xml 文件，代码如下：

```
01  <?xml version="1.0" encoding="utf-8"?>
02  <!-- 定义当前布局的基本 LinearLayout -->
03  <LinearLayout xmlns:android="http://schemas.android.com/apk/res/android"
04      android:layout_width="fill_parent"
05      android:layout_height="fill_parent"
06      android:orientation="vertical" >
07
08      <!-- 定义用户输入 URL 网址的输入控件 -->
09      <EditText
10          android:id="@+id/Et"
11          android:layout_width="match_parent"
12          android:layout_height="wrap_content"
13          android:hint="请输入要打开的网址"
14          />
```

```
15
16          <!-- 定义用户打开浏览器按钮控件 -->
17          <Button
18              android:id="@+id/Btn"
19              android:layout_width="match_parent"
20              android:layout_height="wrap_content"
21              android:text="单击打开浏览器显示网址"
22              />
23  </LinearLayout>
```

这是我们的 Activity 的布局文件。在其中第 8~14 行定义了一个 EditText 控件，用来接收用户输入的地址。在第 16~22 行定义了一个 Button 对象用来打开浏览器显示用户输入的网址。

然后修改 src/com.wyl.example/MainActivity.java 文件，代码如下：

```
01  //定义了本实例的主要 Activity
02  public class MainActivity extends Activity{
03      //定义布局中的开始跳转 Button 控件
04      private Button btn;
05      //定义布局中的输入需要打开的网站地址控件
06      private EditText Et;
07
08      @Override
09      protected void onCreate (Bundle savedInstanceState) {
10          super.onCreate(savedInstanceState);
11          //设置当前 Activity 的布局文件为 activity_main
12          setContentView(R.layout.activity_main);
13          //得到浏览器中的控件对象
14          findView();
15          //设置对象的监听器
16          setListener();
17      }
18
19  private void setListener() {
20      //设置 btn 的单击监听器
21      btn.setOnClickListener(new OnClickListener() {
22          @Override
23          public void onClick(View v) {
24              //得到用户输入的网站地址
25              String url = Et.getText().toString();
26              //当用户输入不为空时
27              if (!"".equals(url)) {
28                  //在用户输入的地址前加上 http://，一般用户输入网址的时候不会加
29                  //然后通过处理后的网址构成 Uri 对象
30                  Uri uri = Uri.parse("http://"+url);
31                  //定义 intent 对象,通过 Intent.ACTION_VIEW 来显示此 Uri 的内容
32                  Intent it = new Intent(Intent.ACTION_VIEW,uri);
33                  //启动 Activity
34                  startActivity(it);
35              }
36              else{
37                  //如果用户输入的 url 为空的话，使用 Toast 提示用户
38                  Toast.makeText(MainActivity.this,"请输入要跳转的网址...",
39                      Toast.LENGTH_SHORT).show();
```

```
40              }
41          }
42      });
43  }
44
45  private void findView() {
46      //得到布局中的开始加载的 Button 的对象
47      btn = (Button)findViewById(R.id.Btn);
48      //得到布局中的开始加载的 EditText 的对象
49      Et = (EditText)findViewById(R.id.Et);
50  }
51  }
```

此文件是 Activity 的代码文件，在其中第 4 行定义了一个 Button 对象，在第 6 行定义了一个 EditText 对象。在第 14 行的 findView 方法中得到了布局中的所有控件，具体实现在第 45～50 行。在第 16 行通过 setListener 方法设置的控件的监听器，具体方法在 19～43 行实现。其中当用户单击 Btn 按钮的时候首先获取用户在 EditText 中输入的文字内容，如果为空通过 Toast 提示用户再次输入，如果不为空则通过用户输入的网址初始化 Uri 对象，然后定义 Intent 对象设置 Action 为 Intent.ACTION_VIEW，然后显示对象为此 uri，然后调用 startActivity 方法打开对应 Activity。这样就可以显示浏览器并且打开用户输入的网址的内容了。

4．实例扩展

在此实例中当用户单击 Button 的时候，如果用户手机安装了多个浏览器，系统会自动提示用户从多个浏览器中选择一个打开此网址，因为对于一个网址来说有多个浏览器都是可以打开的，所以需要用户选择。

范例 094　电话拨号软件

1．实例简介

在我们使用应用的过程中，经常会使用到用户单击某按钮的时候直接拨打某个电话的效果。例如，在一些购物客户端中，当用户单击商家电话按钮，直接拨打了商家电话；当用户单击了电话投诉的按钮，直接拨打了商家指定的投诉电话。遇到这样的功能，我们一般是当用户进行某个操作的时候，例如，单击某个按钮的时候，发送 Intent 给本地的拨号界面进行拨号。本例子就带领大家来实现一个通过用户的操作拨打电话的实例。

2．运行效果

该实例运行效果如图 5.3 所示。

3．实例程序讲解

在本实例中，提供一个用户希望拨打的电话号码的输入框，然后用户单击拨打电话按钮，界面跳转到系统拨号界面直接拨打电话。想要实现本实例效果，首先修改 res/layout/activity_main.xml 文件，代码如下：

第 5 章 Android 程序内部的信息传递者

图 5.3 电话拨号软件

```
01  <?xml version="1.0" encoding="utf-8"?>
02  <!-- 定义当前布局的基本 LinearLayout -->
03  <LinearLayout xmlns:android="http://schemas.android.com/apk/res/android"
04      android:layout_width="fill_parent"
05      android:layout_height="fill_parent"
06      android:orientation="vertical" >
07
08      <!-- 定义用户输入电话号码的输入控件 -->
09      <EditText
10          android:id="@+id/Et"
11          android:layout_width="match_parent"
12          android:layout_height="wrap_content"
13          android:hint="请输入拨打的电话号码"
14          />
15
16      <!-- 定义用户开始拨号的按钮控件 -->
17      <Button
18          android:id="@+id/Btn"
19          android:layout_width="match_parent"
20          android:layout_height="wrap_content"
21          android:text="单击开始拨打电话"
22          />
23  </LinearLayout>
```

这是我们的 Activity 的布局文件。在其中第 8~14 行定义了一个 EditText 控件,用来接收用户输入的要拨打的电话号码。在第 16~22 行定义了一个 Button 对象,当用户单击的时候直接拨打电话。

然后修改 src/com.wyl.example/MainActivity.java 文件,代码如下:

```
01  //定义了本实例的主要 Activity
02  public class MainActivity extends Activity{
03      //定义布局中的开始跳转 Button 控件
04      private Button btn;
05      //定义布局中的输入需要拨打的电话号码控件
06      private EditText Et;
07
08      @Override
09      protected void onCreate (Bundle savedInstanceState) {
```

```java
10      super.onCreate(savedInstanceState);
11      //设置当前Activity的布局文件为activity_main
12      setContentView(R.layout.activity_main);
13      //得到浏览器中的控件对象
14      findView();
15      //设置对象的监听器
16      setListener();
17  }
18
19  private void setListener() {
20      //设置btn的单击监听器
21      btn.setOnClickListener(new OnClickListener() {
22          @Override
23          public void onClick(View v) {
24              //通过EditView的gettext方法得到用户输入的电话号码
25              String phonenum = Et.getText().toString();
26              //如果电话号码不为空的话
27              if (!"".equals(phonenum)) {
28                  //通过电话号码构成uri对象，需要tel:加上电话号码
29                  Uri uri = Uri.parse("tel:" + phonenum);
30                  //通过Intent.ACTION_CALL打开上述uri
31                  Intent it = new Intent(Intent.ACTION_CALL, uri);
32                  //通过intent打开activity
33                  startActivity(it);
34              }
35              else{
36                  //如果用户输入的电话号码为空，通过Toast提示用户输入
37                  Toast.makeText(MainActivity.this,"请输入您要拨打的电话号码...",
38                          Toast.LENGTH_SHORT).show();
39              }
40          }
41      });
42  }
43
44  private void findView() {
45      //得到布局中的开始加载的Button的对象
46      btn = (Button)findViewById(R.id.Btn);
47      //得到布局中的开始加载的EditText的对象
48      Et = (EditText)findViewById(R.id.Et);
49  }
50 }
```

此文件是Activity的代码文件，在其中第4行定义了一个Button对象，在第6行定义了一个EditText对象。在第14行的findView方法中得到了布局中的所有控件，具体实现在第44~49行。在第16行通过setListener方法设置控件的监听器，具体方法在19~42行实现。其中当用户单击Btn按钮的时候首先获取用户在EditText中输入的电话号码，如果为空通过Toast提示用户再次输入，如果不为空则通过用户输入的电话号码初始化Uri对象，然后定义Intent对象设置Action为Intent.ACTION_CALL，然后显示对象为此uri，最后调用startActivity方法打开对应Activity。这样就可以显示拨号界面直接拨打之前用户输入的电话号码了。

4. 实例扩展

在此实例中当用户单击Button的时候，如果用户手机安装了多个拨号软件，系统会自

动提示用户从多个拨号软件中选择一个来进行拨号,而且此程序要求获取拨打电话的权限,所以需要在 Manifest 文件中添加<uses-permission android:name="android.permission.CALL_PHONE" />,获取拨打电话的权限。

范例 095　分享短信

1．实例简介

在我们使用应用的过程中,经常会使用到用户单击某按钮的时候分享短信效果。例如,在一些社交客户端中,当用户单击通过短信分享此软件的下载地址的时候,或者在一些软件中可以通过短信反馈程序的问题等功能。遇到这样的功能,我们一般是当用户进行某个操作的时候,例如,单击某个按钮的时候,发送 Intent 给本地的短信发送界面进行短信发送,本例子就带领大家来实现一个通过用户的操作分享短信的实例。

2．运行效果

该实例运行效果如图 5.4 所示。

图 5.4　短信分享软件

3．实例程序讲解

在本实例中,提供一个用户希望分享的短信内容的输入框,然后用户单击"分享短信"按钮,界面跳转到系统发送短信界面,输入接收者电话号码后,可以发送短信。想要实现本实例效果,首先修改 res/layout/activity_main.xml 文件,代码如下:

```
01  <?xml version="1.0" encoding="utf-8"?>
02  <!-- 定义当前布局的基本 LinearLayout -->
03  <LinearLayout xmlns:android="http://schemas.android.com/apk/res/android"
04      android:layout_width="fill_parent"
05      android:layout_height="fill_parent"
```

```
06        android:orientation="vertical" >
07
08        <!-- 定义用户输入分享短信内容的输入控件 -->
09        <EditText
10            android:id="@+id/Et"
11            android:layout_width="match_parent"
12            android:layout_height="wrap_content"
13            android:hint="请输入要分享的短信内容"
14            />
15
16        <!-- 定义用户打开短信界面的按钮控件 -->
17        <Button
18            android:id="@+id/Btn"
19            android:layout_width="match_parent"
20            android:layout_height="wrap_content"
21            android:text="单击开始分享"
22            />
23    </LinearLayout>
```

这是我们的 Activity 的布局文件。在其中第 8~14 行定义了一个 EditText 控件，用来接收用户输入的要分享的短信内容。在第 16~22 行定义了一个 Button 对象，当用户单击的时候打开短信发送界面。

然后修改 src/com.wyl.example/MainActivity.java 文件，代码如下：

```
01 //定义了本实例的主要 Activity
02 public class MainActivity extends Activity{
03     //定义布局中的分享短信 Button 控件
04     private Button btn;
05     //定义布局中的输入需要分享短信内容的控件
06     private EditText Et;
07
08     @Override
09     protected void onCreate (Bundle savedInstanceState) {
10         super.onCreate(savedInstanceState);
11         //设置当前 Activity 的布局文件为 activity_main
12         setContentView(R.layout.activity_main);
13         //得到浏览器中的控件对象
14         findView();
15         //设置对象的监听器
16         setListener();
17     }
18
19 private void setListener() {
20     //设置 btn 的单击监听器
21     btn.setOnClickListener(new OnClickListener() {
22         @Override
23         public void onClick(View v) {
24             //获取用户输入的短信内容
25             String msg = Et.getText().toString();
26             //如果短信内容不为空，就发送 Intent
27             if (!"".equals(msg)) {
28                 //定义 Intent 对象，action 设置为 Intent.ACTION_VIEW
29                 Intent it = new Intent(Intent.ACTION_VIEW);
30                 //设置 sms_body 参数为接收到的短信内容
31                 it.putExtra("sms_body", msg);
32                 //设置 Intent 的类型为 vnd.android-dir/mms-sms
33                 //代表打开类型为短信打开
```

```
34                     it.setType("vnd.android-dir/mms-sms");
35                     //通过Intent打开activity
36                     startActivity(it);
37                 }
38                 else{
39                     //如果短信内容为空,通过Toast提示用户输入短信内容
40                     Toast.makeText(MainActivity.this,"请输入您要分享的短信
                       内容...",
41                             Toast.LENGTH_SHORT).show();
42                 }
43             }
44         });
45     }
46
47     private void findView() {
48         //得到布局中的开始加载的Button的对象
49         btn = (Button)findViewById(R.id.Btn);
50         //得到布局中的开始加载的EditText的对象
51         Et = (EditText)findViewById(R.id.Et);
52     }
53 }
```

此文件是 Activity 的代码文件,在其中第 4 行定义了一个 Button 对象,在第 6 行定义了一个 EditText 对象。在第 14 行的 findView 方法中得到了布局中的所有控件,具体实现在第 47～52 行。在第 16 行通过 setListener 方法设置的控件的监听器,具体方法在 19～45 行实现。其中当用户单击 Btn 按钮的时候首先获取用户在 EditText 中输入的分享短信的内容,如果为空通过 Toast 提示用户再次输入,如果不为空则定义 Intent 对象设置 Action 为 Intent.ACTION_VIEW,然后设置 Intent 的附加内容 sms_body。设置 Intent 的类型为 vnd.android-dir/mms-sms,代表通过短信的形式打开。然后调用 startActivity 方法打开对应 Activity。这样就可以显示发送短信界面输入需要发送的电话号码就可以发送短信了。

4. 实例扩展

在此实例中当用户单击 Button 的时候,如果用户手机安装了多个短信发送软件,系统会自动提示用户从多个短信软件中选择一个来进行短信发送。

范例 096 短信发送客户端

1. 实例简介

在我们使用应用的过程中,经常会使用到用户单击某按钮的时候直接发送短信效果,例如,在一些群发短信的客户端,电话号码发送的客户端。遇到这样的功能,我们一般是当用户进行某个操作的时候,例如,单击某个按钮的时候,通过 Intent 直接进行短信发送。本例子就带领大家来实现一个通过用户的操作直接发送短信的实例。

2. 运行效果

该实例运行效果如图 5.5 所示。

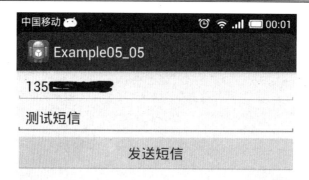

图 5.5　短信发送软件

3. 实例程序讲解

在本实例中，提供一个接收短信的号码输入框和一个短信内容的输入框，然后当用户单击发送短信按钮时，程序直接发送短信。想要实现本实例效果，首先修改 res/layout/activity_main.xml 文件，代码如下：

```xml
01 <?xml version="1.0" encoding="utf-8"?>
02 <!-- 定义当前布局的基本 LinearLayout -->
03 <LinearLayout xmlns:android="http://schemas.android.com/apk/res/android"
04     android:layout_width="fill_parent"
05     android:layout_height="fill_parent"
06     android:orientation="vertical" >
07
08     <!-- 定义用户输入接收短信的电话号码的输入控件 -->
09     <EditText
10         android:id="@+id/EtNum"
11         android:layout_width="match_parent"
12         android:layout_height="wrap_content"
13         android:hint="请输入要接收短信的电话号码"
14         />
15
16     <!-- 定义用户输入短信内容的输入控件 -->
17     <EditText
18         android:id="@+id/EtMsg"
19         android:layout_width="match_parent"
20         android:layout_height="wrap_content"
21         android:hint="请输入短信内容"
22         />
23
24     <!-- 定义用户单击发送短信的按钮控件 -->
25     <Button
26         android:id="@+id/Btn"
27         android:layout_width="match_parent"
28         android:layout_height="wrap_content"
29         android:text="发送短信"
30         />
31 </LinearLayout>
```

这是我们的 Activity 的布局文件。在其中第 8~14 行定义了一个 EditText 控件，用来接收用户输入的短信号码，在第 17~22 行定义另一个 EditText 控件，用来接收用户输入的需要发送短信的内容。在第 25~30 行定义了一个 Button 对象，当用户单击的时候直接发送短信。

然后修改 src/com.wyl.example/MainActivity.java 文件，代码如下：

```java
01  //定义了本实例的主要 Activity
02  public class MainActivity extends Activity {
03  //定义布局中的发送短信的 Button 控件
04  private Button btn;
05  //定义布局中的输入短信内容的控件
06  private EditText EtMsg;
07  //定义布局中的输入短信号码的控件
08  private EditText EtNum;
09
10  @Override
11  protected void onCreate(Bundle savedInstanceState) {
12      super.onCreate(savedInstanceState);
13      //设置当前 Activity 的布局文件为 activity_main
14      setContentView(R.layout.activity_main);
15      //得到浏览器中的控件对象
16      findView();
17      //设置对象的监听器
18      setListener();
19  }
20
21  private void setListener() {
22      //设置 btn 的单击监听器
23      btn.setOnClickListener(new OnClickListener() {
24          @Override
25          public void onClick(View v) {
26              //得到用户需要发送的短信的内容
27              String msg = EtMsg.getText().toString();
28              //得到用户需要接收短信的电话号码
29              String number = EtNum.getText().toString();
30              //当短信的接收电话号码和短信内容都不为空时进入分支
31              if (!"".equals(msg) && !"".equals(number)) {
32                  //得到系统的 SmsManger 对象
33                  SmsManager smsManager = SmsManager.getDefault();
34                  //初始化 PendingIntent
35                  PendingIntent sendIntent = PendingIntent.getBroadcast(
36                      MainActivity.this, 0, new Intent(), 0);
37                  //通过 smsManager 的 sendTextMessage 方法来发送到短信
38                  smsManager.sendTextMessage(number,null,msg,sendIntent,
39                      null);
40                  //通过 Toast 提示短信发送成功
41                  Toast.makeText(MainActivity.this, "发送成功",
42                      Toast.LENGTH_SHORT).show();
43              } else {
44                  //通过 Toast 提示用户短信号码或者短信内容为空
45                  Toast.makeText(MainActivity.this, "短信号码或短信内容为空...",
46                      Toast.LENGTH_SHORT).show();
47              }
48          }
49      });
```

```
50    }
51
52    private void findView() {
53        //得到布局中的开始加载的 Button 的对象
54        btn = (Button) findViewById(R.id.Btn);
55        //得到布局中的开始加载的 EditText 的对象
56        EtMsg = (EditText) findViewById(R.id.EtMsg);
57        //得到布局中的开始加载的 EditText 的对象
58        EtNum = (EditText) findViewById(R.id.EtNum);
59    }
60  }
```

此文件是 Activity 的代码文件,在其中第 4 行定义了一个 Button 对象,在第 6 行定义了短信内容的接收 EditText 对象,在第 8 行定义了短信号码的输入 EditText 对象。在第 16 行的 findView 方法中得到了布局中的所有控件,具体实现在第 52～59 行。在第 18 行通过 setListener 方法设置的控件的监听器,具体方法在 21～50 行实现。其中当用户单击 Btn 按钮的时候首先获取用户输入的短信号码和短信内容,如果两项中有一项为空,通过 Toast 提示用户再次输入,如果不为空则通过 SmsManager 得到系统短信的管理服务,然后初始化 PendingIntent 对象,调用 smsManager 对象的 sendTextMesage 方法来进行短信的发送,如果发送成功通过 Toast 进行提醒。这样就可以直接发送短信了。

4．实例扩展

在此实例中当用户单击 Button 的时候,会直接发送短信,其中使用到了 Android 系统的短信发送的权限,需要在 Manifest 文件中添加<uses-permission android:name="android.permission.SEND_SMS"/>权限。因为此例子需要真实的发送短信,而 AVD 毕竟是模拟的短信发送,所以一定要在真机下才可以看到短信发送的效果。

范例 097　彩信分享客户端

1．实例简介

在我们使用应用的过程中,经常会使用到用户单击某按钮的时候通过彩信分享的效果,例如:在一些社交应用的客户端,用户希望分享图片给朋友。遇到这样的功能,我们一般是当用户进行某个操作的时候,例如:单击某个按钮的时候,通过 Intent 打开彩信发送界面,本例子就带领大家来实现一个分享彩信的实例。

2．运行效果

该实例运行效果如图 5.6 所示。

3．实例程序讲解

在本实例中,提供一个接收彩信的号码输入框、一个彩信主题的输入框和一个彩信内容的输入框,然后当用户单击发送彩信按钮时,程序打开发送彩信的界面,并且附带系统

图 5.6　彩信分享软件

的一张图片。想要实现本实例效果，首先修改 res/layout/activity_ main.xml 文件，代码如下：

```xml
01 <?xml version="1.0" encoding="utf-8"?>
02 <!-- 定义当前布局的基本 LinearLayout -->
03 <LinearLayout xmlns:android="http://schemas.android.com/apk/res/android"
04     android:layout_width="fill_parent"
05     android:layout_height="fill_parent"
06     android:orientation="vertical" >
07
08     <!-- 定义用户输入接收彩信的电话号码的输入控件 -->
09     <EditText
10         android:id="@+id/EtNum"
11         android:layout_width="match_parent"
12         android:layout_height="wrap_content"
13         android:hint="请输入要接收彩信的电话号码"
14         />
15
16     <!-- 定义用户输入接收彩信主题的输入控件 -->
17     <EditText
18         android:id="@+id/EtSub"
19         android:layout_width="match_parent"
20         android:layout_height="wrap_content"
21         android:hint="请输入彩信的主题"
22         />
23
24     <!-- 定义用户输入彩信内容的输入控件 -->
25     <EditText
26         android:id="@+id/EtMsg"
27         android:layout_width="match_parent"
28         android:layout_height="wrap_content"
29         android:hint="请输入彩信内容"
30         />
31
32     <!-- 定义用户单击发送彩信的按钮控件 -->
33     <Button
34         android:id="@+id/Btn"
35         android:layout_width="match_parent"
36         android:layout_height="wrap_content"
37         android:text="发送彩信"
38         />
39 </LinearLayout>
```

这是我们的 Activity 的布局文件。在其中第 8～14 行定义了一个 EditText 控件，用来接收用户输入的接收彩信号码。在第 16～22 行定义另一个 EditText 控件，用来接收用户输入的需要发送彩信的主题。在第 24～30 行定义另一个 EditText 控件，用来接收用户输入的需要发送的彩信的内容。在第 32～38 行定义了一个 Button 对象，当用户单击的时候打开彩信发送界面。

然后修改 src/com.wyl.example/MainActivity.java 文件，代码如下：

```java
01 //定义了本实例的主要 Activity
02 public class MainActivity extends Activity {
03     //定义布局中的发送彩信的 Button 控件
04     private Button btn;
05     //定义布局中的输入彩信内容的控件
06     private EditText EtMsg;
07     //定义布局中的输入彩信号码的控件
```

```
08  private EditText EtNum;
09  //定义布局中彩信主题的输入控件
10  private EditText EtSub;
11
12  @Override
13  protected void onCreate(Bundle savedInstanceState) {
14      super.onCreate(savedInstanceState);
15      //设置当前Activity的布局文件为activity_main
16      setContentView(R.layout.activity_main);
17      //得到浏览器中的控件对象
18      findView();
19      //设置对象的监听器
20      setListener();
21  }
22
23  private void setListener() {
24      //设置btn的单击监听器
25      btn.setOnClickListener(new OnClickListener() {
26          @Override
27          public void onClick(View v) {
28              //得到用户需要发送的彩信内容
29              String msg = EtMsg.getText().toString();
30              //得到用户需要发送的彩信的号码
31              String number = EtNum.getText().toString();
32              //得到用户需要发送的彩信主题
33              String sub = EtSub.getText().toString();
34              //如果彩信号码,彩信主题,彩信内容都不为空则发送彩信
35              if (!"".equals(msg) && !"".equals(number)
36                      && !"".equals(sub)) {
37                  //定义了系统中的1号图片的uri
38                  Uri uri =Uri.parse("content://media/external/images/media/01");
39                  //定义Intent标记action为Intent.ACTION_SEND
40                  Intent intent = new Intent(Intent.ACTION_SEND);
41                  //设置Intent的flags为Intent.FLAG_ACTIVITY_NEW_TASK
42                  intent.addFlags(Intent.FLAG_ACTIVITY_NEW_TASK);
43                  //设置intent的附件参数为之前的uri图片
44                  intent.putExtra(Intent.EXTRA_STREAM,uri);
45                  //设置intent的主题内容
46                  intent.putExtra("subject", sub);
47                  //设置intent的接收者的电话号码
48                  intent.putExtra("address", number);
49                  //设置intent的彩信内容
50                  intent.putExtra("sms_body", msg);
51                  //设置intent的彩信附加文字
52                  intent.putExtra(Intent.EXTRA_TEXT, "it's EXTRA_TEXT");
53                  //设置彩信附件类型为image
54                  intent.setType("image/*");    //彩信附件类型
55                  //设置intent的彩信发送activity
56                  intent.setClassName("com.android.mms",
                            "com.android.mms.ui.ComposeMessageActivity");
57                  //启动activity
58                  startActivity(intent);
59              } else {
60                  //如果彩信信息为空,通过Toast提示用户输入
61                  Toast.makeText(MainActivity.this,"彩信号码或彩信内容为空...",
62                          Toast.LENGTH_SHORT).show();
63              }
```

```
64          }
65      });
66  }
67
68  private void findView() {
69      //得到布局中的开始加载的 Button 的对象
70      btn = (Button) findViewById(R.id.Btn);
71      //得到布局中的开始加载的 EditText 的对象
72      EtMsg = (EditText) findViewById(R.id.EtMsg);
73      //得到布局中的开始加载的 EditText 的对象
74      EtNum = (EditText) findViewById(R.id.EtNum);
75      //得到布局中的开始加载的 EditText 的对象
76      EtSub = (EditText) findViewById(R.id.EtSub);
77  }
78  }
```

此文件是 Activity 的代码文件，在其中第 4 行定义了一个 Button 对象，在第 6 行定义了彩信内容的接收 EditText 对象，在第 8 行定义了彩信号码的输入 EditText 对象，在第 10 行定义了彩信主题的输入 EditText 对象。在第 18 行的 findView 方法中得到了布局中的所有控件，具体实现在第 68～77 行。在第 20 行通过 setListener 方法设置的控件的监听器，具体方法在 23～66 行实现。其中当用户单击 Btn 按钮的时候首先获取用户输入的彩信号码、彩信内容和彩信主题，如果三项中有一项为空，通过 Toast 提示用户再次输入，如果不为空则获取用户系统的一张图片的 uri，然后定义 intent，设置 intent 的动作 action 为 Intent.ACTION_SEND，设置 flagellate 为 Intent.FLAG_ACTIVITY_NEW_TASK，设置彩信的主题、彩信的内容、彩信的接收号码、彩信的图片附件及彩信的附件类型，然后通过 startactivity 启动彩信发送界面，然后用户可以发送彩信。

4．实例扩展

在此实例中当用户单击 Button 的时候，会打开彩信发送界面，其中彩信的接收号码、主题、内容和附件都会包含在此彩信中。本实例尽量使用真机测试，因为此例子需要发送彩信，而 AVD 毕竟是模拟的彩信发送，所以一定要在真机下才可以看到彩信发送的效果。

范例 098　Email 发送客户端

1．实例简介

在我们使用应用的过程中，经常会使用到用户单击某按钮的时候发送邮件的效果，例如，在一些购物网站的客户端，用户希望通过邮箱分享软件信息给朋友。遇到这样的功能，我们一般是当用户进行某个操作的时候，例如：单击某个按钮的时候，通过 Intent 打开 Email 发送界面。本例子就带领大家来实现一个发送邮件的实例。

2．运行效果

该实例运行效果如图 5.7 所示。

图 5.7　Email 发送软件

3. 实例程序讲解

在本实例中，提供一个接收邮件的邮箱地址的输入框、一个邮件主题的输入框和一个邮件内容的输入框，然后当用户单击发送邮件按钮时，程序打开发送邮件的程序界面。想要实现本实例效果，首先修改 res/layout/activity_main.xml 文件，代码如下：

```xml
01  <?xml version="1.0" encoding="utf-8"?>
02  <!-- 定义当前布局的基本 LinearLayout -->
03  <LinearLayout xmlns:android="http://schemas.android.com/apk/res/android"
04      android:layout_width="fill_parent"
05      android:layout_height="fill_parent"
06      android:orientation="vertical" >
07
08      <!-- 定义用户输入接收邮件的邮箱的输入控件 -->
09      <EditText
10          android:id="@+id/EtEmail"
11          android:layout_width="match_parent"
12          android:layout_height="wrap_content"
13          android:hint="请输入要接收邮件的邮箱地址"
14          />
15
16      <!-- 定义用户输入接收邮件的主题的输入控件 -->
17      <EditText
18          android:id="@+id/EtSub"
19          android:layout_width="match_parent"
20          android:layout_height="wrap_content"
21          android:hint="请输入邮件的主题"
22          />
23
24      <!-- 定义用户输入邮件内容的输入控件 -->
25      <EditText
26          android:id="@+id/EtMsg"
27          android:layout_width="match_parent"
28          android:layout_height="wrap_content"
29          android:hint="请输入邮件的内容"
30          />
31
32      <!-- 定义用户单击发送邮件的按钮控件 -->
33      <Button
34          android:id="@+id/Btn"
35          android:layout_width="match_parent"
36          android:layout_height="wrap_content"
37          android:text="发送邮件"
38          />
39  </LinearLayout>
```

这是我们的 Activity 的布局文件。在其中第 8~14 行定义了一个 EditText 控件，用来接收用户输入的接收邮箱地址。在第 17~22 行定义另一个 EditText 控件，用来接收用户输入的需要发送邮件的主题。在第 24~30 行定义另一个 EditText 控件，用来接收用户输入的需要发送邮件的内容。在第 32~38 行定义了一个 Button 对象当用户单击的时候打开邮件发送界面。

然后修改 src/com.wyl.example/MainActivity.java 文件，代码如下：

```java
01  //定义了本实例的主要 Activity
02  public class MainActivity extends Activity {
```

```
03    //定义布局中的发送邮件的Button控件
04    private Button btn;
05    //定义布局中的输入邮件内容的控件
06    private EditText EtMsg;
07    //定义布局中的输入接收邮件地址的控件
08    private EditText EtEmail;
09    //定义布局中邮件的主题的输入控件
10    private EditText EtSub;
11
12    @Override
13    protected void onCreate(Bundle savedInstanceState) {
14        super.onCreate(savedInstanceState);
15        //设置当前Activity的布局文件为activity_main
16        setContentView(R.layout.activity_main);
17        //得到浏览器中的控件对象
18        findView();
19        //设置对象的监听器
20        setListener();
21    }
22
23    private void setListener() {
24        //设置btn的单击监听器
25        btn.setOnClickListener(new OnClickListener() {
26            @Override
27            public void onClick(View v) {
28                //得到用户输入的邮件内容
29                String msg = EtMsg.getText().toString();
30                //得到用户输入的邮件地址
31                String emailadd = EtEmail.getText().toString();
32                //得到用户输入的邮件主题
33                String sub = EtSub.getText().toString();
34                //判断邮件地址、邮件主题和邮件内容都不为空
35                if (!"".equals(msg) && !"".equals(emailadd)
36                        && !"".equals(sub)) {
37                    //定义Intent,并且设置action为Intent.ACTION_SEND
38                    Intent i = new Intent(Intent.ACTION_SEND);
39                    //设置邮件类型为文字类型
40                    i.setType("text/plain");              //模拟器请使用这行
41                    i.setType("message/rfc822") ;         //真机上使用这行
42                    //设置Email的接收地址
43                    i.putExtra(Intent.EXTRA_EMAIL, new String[]{emailadd});
44                    //设置Email的主题
45                    i.putExtra(Intent.EXTRA_SUBJECT,sub);
46                    //设置Email的内容
47                    i.putExtra(Intent.EXTRA_TEXT,msg);
48                    //启动activity选择一个邮件发送客户端
49                    startActivity(Intent.createChooser(i,"请选择发送邮件的
                        客户端"));
50                } else {
51                    //如果邮箱地址、邮箱主题和邮箱内容为空的话,通过Toast提示用户
52                    Toast.makeText(MainActivity.this, "接收邮件地址或邮件内
                        容为空...",
53                            Toast.LENGTH_SHORT).show();
54                }
55            }
```

```
56          });
57    }
58
59    private void findView() {
60        //得到布局中的开始加载的 Button 的对象
61        btn = (Button) findViewById(R.id.Btn);
62        //得到布局中的开始加载的 EditText 的对象
63        EtMsg = (EditText) findViewById(R.id.EtMsg);
64        //得到布局中的开始加载的 EditText 的对象
65        EtEmail = (EditText) findViewById(R.id.EtEmail);
66        //得到布局中的开始加载的 EditText 的对象
67        EtSub = (EditText) findViewById(R.id.EtSub);
68    }
69 }
```

此文件是 Activity 的代码文件，在其中第 4 行定义了一个 Button 对象，在第 6 行定义了邮件内容的接收 EditText 对象，在第 8 行定义了邮件地址的输入 EditText 对象，在第 10 行定义了邮件主题的输入 EditText 对象。在第 18 行的 findView 方法中得到了布局中的所有控件，具体实现在第 59～68 行。在第 20 行通过 setListener 方法设置控件的监听器，具体方法在 23～57 行实现。其中当用户单击 Btn 按钮的时候首先获取用户输入的邮箱地址、邮件内容和邮件主题，如果三项中有一项为空，通过 Toast 提示用户再次输入，如果不为空然后定义 intent，设置 intent 的动作 action 为 Intent.ACTION_SEND、设置 intent 的类型为文本类型、设置邮件的主题、邮件的内容和邮件的接收邮箱，然后通过 startactivity 启动邮件发送界面，最后用户可以发送邮件。

4．实例扩展

在此实例中当用户单击 Button 的时候，会打开邮件发送界面，如果手机中有多个邮件发送的客户端，会弹出选择某个邮箱客户端发送的选项进行选择。

范例 099 启动多媒体播放

1．实例简介

在我们使用应用的过程中，经常会使用到用户单击某按钮的时候启动系统的声音播放软件播放音乐的效果。例如，在一些下载的客户端中，当用户下载完一首 mp3 之后希望直接单击播放。遇到这样的功能，我们一般是当用户进行某个操作的时候，例如，单击某个按钮的时候，通过 Intent 打开系统中的音乐播放软件进行音乐的播放。本例子就带领大家来实现一个单击打开系统播放软件播放音乐的实例。

2．运行效果

该实例运行效果如图 5.8 所示。

3．实例程序讲解

在本实例中，提供提示用户在 sdcard 下放置相应的 mp3 文件的标签框，一个播放音乐的按钮，然后当用户单击播放音乐的按钮时，程序打开系统播放音乐的程序进行音乐播放。

想要实现本实例效果，首先修改 res/layout/activity_main.xml 文件，代码如下：

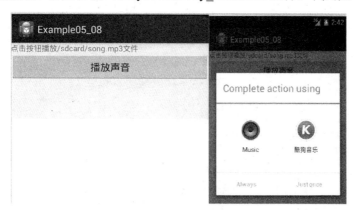

图 5.8 启动多媒体播放

```
01  <?xml version="1.0" encoding="utf-8"?>
02  <!-- 定义当前布局的基本 LinearLayout -->
03  <LinearLayout xmlns:android="http://schemas.android.com/apk/res/android"
04      android:layout_width="fill_parent"
05      android:layout_height="fill_parent"
06      android:orientation="vertical" >
07  
08      <!-- 定义提示用户播放 mp3 的显示控件 -->
09      <TextView
10          android:id="@+id/Tv"
11          android:layout_width="match_parent"
12          android:layout_height="wrap_content"
13          android:hint="单击按钮播放/sdcard/song.mp3 文件"
14          />
15  
16      <!-- 定义用户单击播放声音的按钮控件 -->
17      <Button
18          android:id="@+id/Btn"
19          android:layout_width="match_parent"
20          android:layout_height="wrap_content"
21          android:text="播放声音"
22          />
23  </LinearLayout>
```

这是我们的 Activity 的布局文件。在其中第 8～14 行定义了一个 TextView 控件，用来提示用户存储 sdcard 下的 mp3 文件。在第 17～22 行定义一个 Button 控件，当用户单击的时候打开系统的音乐播放程序进行音乐播放。

然后修改 src/com.wyl.example/MainActivity.java 文件，代码如下：

```
01  //定义了本实例的主要 Activity
02  public class MainActivity extends Activity {
03      //定义布局中的播放声音的 Button 控件
04      private Button btn;
05      //定义显示标签的控件
06      private TextView Tv;
07  
08      @Override
09      protected void onCreate(Bundle savedInstanceState) {
10          super.onCreate(savedInstanceState);
```

```
11          //设置当前 Activity 的布局文件为 activity_main
12          setContentView(R.layout.activity_main);
13          //得到浏览器中的控件对象
14          findView();
15          //设置对象的监听器
16          setListener();
17      }
18
19      private void setListener() {
20          //设置 btn 的单击监听器
21          btn.setOnClickListener(new OnClickListener() {
22              @Override
23              public void onClick(View v) {
24                  //定义 intent 对象,设置 action 属性为 Intent.ACTION_VIEW
25                  Intent it = new Intent(Intent.ACTION_VIEW);
26                  //定义 sdcard 下的 song.mp3 文件的 uri
27                  Uri uri = Uri.parse("file:///sdcard/song.mp3");
28                  //设置 intent 的数据类型为 audio/mp3,这样就可以启动系统程序打开 mp3
                    文件了
29                  it.setDataAndType(uri, "audio/mp3");
30                  //通过 intent 打开 activity
31                  startActivity(it);
32              }
33          });
34      }
35
36      private void findView() {
37          //得到布局中的开始加载的 Button 对象
38          btn = (Button) findViewById(R.id.Btn);
39          //得到布局中的开始加载的 EditText 对象
40          Tv = (TextView) findViewById(R.id.Tv);
41      }
42  }
```

此文件是 Activity 的代码文件,在其中第 4 行定义了一个 Button 对象,在第 6 行定义了提醒用户 mp3 文件路径的 TextView 对象。在第 14 行的 findView 方法中得到了布局中的所有控件,具体实现在第 36～41 行。在第 16 行通过 setListener 方法设置控件的监听器,具体方法在 19～34 行实现。其中当用户单击 Btn 按钮的时候定义 intent 对象,设置 action 属性为 Intent.ACTION_VIEW,然后定义 sdcard 下的 song.mp3 的 uri,然后设置 intent 的数据类型为 audio/mp3,然后通过 startactivity 启动系统的音乐播放软件进行音乐播放。

4. 实例扩展

在此实例中当用户单击 Button 的时候,会打开系统的音乐播放软件,如果手机中有多个音乐播放的客户端,会弹出选择某个音乐播放软件进行播放。注意需要打开的软件在本地对应目录下一定要有相应的 mp3 文件。

范例 100 安装指定的应用程序

1. 实例简介

在我们使用应用的过程中,经常会使用到用户单击某按钮的时候安装某个应用程序的效果。例如，在一些软件市场的客户端中，当用户下载完一个 Android 应用程序后之后，希望直接单击安装此程序。遇到这样的功能，一般是当用户进行某个操作的时候，例如：单击某个按钮的时候，通过 Intent 打开系统中安装应用程序的界面。本例子就带领大家来实现一个单击打开系统安装应用程序界面的实例。

2．运行效果

该实例运行效果如图 5.9 所示。

图 5.9　安装指定的应用程序

3．实例程序讲解

在本实例中，提供提示用户在 sdcard 下放置相应的 apk 文件标签框和一个安装软件的按钮，然后当用户单击"安装软件"的按钮时，程序打开系统的软件安装的界面进行软件安装。想要实现本实例效果，首先修改 res/layout/activity_main.xml 文件，代码如下：

```xml
01  <?xml version="1.0" encoding="utf-8"?>
02  <!-- 定义当前布局的基本 LinearLayout -->
03  <LinearLayout xmlns:android="http://schemas.android.com/apk/res/android"
04      android:layout_width="fill_parent"
05      android:layout_height="fill_parent"
06      android:orientation="vertical" >
07
08      <!-- 定义提示用户安装文件路径的显示控件 -->
09      <TextView
10          android:id="@+id/Tv"
11          android:layout_width="match_parent"
12          android:layout_height="wrap_content"
13          android:hint="单击安装 file:///sdcard/XMNotes.apk 文件"
14          />
15
16      <!-- 定义用户单击安装软件的按钮控件 -->
17      <Button
18          android:id="@+id/Btn"
19          android:layout_width="match_parent"
20          android:layout_height="wrap_content"
21          android:text="安装软件"
22          />
23  </LinearLayout>
```

这是我们的 Activity 的布局文件。在其中第 8～14 行定义了一个 TextView 控件，用来提示用户存储 sdcard 下的 apk 文件。在第 17～22 行定义一个 Button 控件，当用户单击的时候打开系统的安装应用程序的界面进行应用程序安装。

然后修改 src/com.wyl.example/MainActivity.java 文件，代码如下：

```java
01  //定义了本实例的主要 Activity
02  public class MainActivity extends Activity {
03    //定义布局中的安装软件的 Button 控件
04    private Button btn;
05    //定义布局中的提示用户安装软件目录的控件
06    private TextView Tv;
07
08    @Override
09    protected void onCreate(Bundle savedInstanceState) {
10        super.onCreate(savedInstanceState);
11        //设置当前 Activity 的布局文件为 activity_main
12        setContentView(R.layout.activity_main);
13        //得到浏览器中的控件对象
14        findView();
15        //设置对象的监听器
16        setListener();
17    }
18
19    private void setListener() {
20        //设置 btn 的单击监听器
21        btn.setOnClickListener(new OnClickListener() {
22            @Override
23            public void onClick(View v) {
24                //定义 intent 对象，设置 action 为 Intent.ACTION_VIEW
25                Intent intent = new Intent(Intent.ACTION_VIEW);
26                //获取 sdcard 下的 XMNotes.apk 的 uri 对象
27                //设置 intent 的类型为 application/vnd.android.package-archive
28                //代表使操作系统进行软件安装
29                intent.setDataAndType(Uri.parse("file://"
30                        + "/sdcard/XMNotes.apk"),
31                        "application/vnd.android.package-archive");
32                //通过 intent 启动 activity
33                startActivity(intent);
34            }
35        });
36    }
37
38    private void findView() {
39        //得到布局中的开始加载的 Button 的对象
40        btn = (Button) findViewById(R.id.Btn);
41        //得到布局中的开始加载的 EditText 的对象
42        Tv = (TextView) findViewById(R.id.Tv);
43    }
44  }
```

此文件是 Activity 的代码文件，在其中第 4 行定义了一个安装软件的 Button 对象，在第 6 行定义了提醒用户 apk 文件路径的 TextView 对象。在第 14 行的 findView 方法中得到了布局中的所有控件，具体实现在第 38～43 行。在第 16 行通过 setListener 方法设置控件的监听器，具体方法在 19～36 行实现。其中当用户单击 Btn 按钮的时候定义 intent 对象，设置 action 属性为 Intent.ACTION_VIEW，然后定义 sdcard 下的 XMNotes.apk 文件的 uri，然后设置 intent 的数据类型 application/vnd.android.package-archive，最后通过 startactivity 启动系统的音乐播放软件进行音乐播放。

4．实例扩展

在此实例中当用户单击 Button 的时候，会打开系统的安装软件的界面。注意需要打开的 apk 文件在本地对应目录下一定要有相应存储。

范例 101　卸载指定的应用程序

1．实例简介

在我们使用应用的过程中，经常会使用到用户单击某按钮的时候卸载某个应用程序的效果。例如，在一些软件市场的客户端中，当用户下载完一个 Android 应用程序安装之后，用户希望直接单击卸载此程序。遇到这样的功能，一般是当用户进行某个操作的时候，例如：单击某个按钮的时候，通过 Intent 打开系统中卸载应用程序的界面。本例子就带领大家来实现一个单击打开系统卸载应用程序界面的实例。

2．运行效果

该实例运行效果如图 5.10 所示。

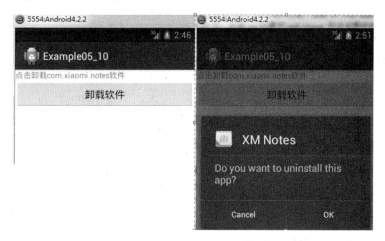

图 5.10　卸载指定的应用程序

3．实例程序讲解

在本实例中，提供提示用户要卸载的软件的报名标签框和一个单击卸载的按钮，然后当用户单击"卸载软件"的按钮时，程序打开系统的软件卸载界面去卸载软件了。想要实现本实例效果，首先修改 res/layout/activity_main.xml 文件，代码如下：

```
01  <?xml version="1.0" encoding="utf-8"?>
02  <!-- 定义当前布局的基本 LinearLayout -->
03  <LinearLayout xmlns:android="http://schemas.android.com/apk/res/android"
04      android:layout_width="fill_parent"
05      android:layout_height="fill_parent"
```

```
06         android:orientation="vertical" >
07
08         <!-- 定义提示用户卸载文件的包名显示控件 -->
09         <TextView
10             android:id="@+id/Tv"
11             android:layout_width="match_parent"
12             android:layout_height="wrap_content"
13             android:hint="单击卸载 com.xiaomi.notes 软件"
14             />
15
16         <!-- 定义用户单击卸载软件的按钮控件 -->
17         <Button
18             android:id="@+id/Btn"
19             android:layout_width="match_parent"
20             android:layout_height="wrap_content"
21             android:text="卸载软件"
22             />
23 </LinearLayout>
```

这是我们的 Activity 的布局文件。在其中第 8～14 行定义了一个 TextView 控件，用来提示用户卸载软件的包名。在第 17～22 行定义一个 Button 控件，当用户单击的时候打开系统的卸载应用程序的界面进行应用程序卸载。

然后修改 src/com.wyl.example/MainActivity.java 文件，代码如下：

```
01 //定义了本实例的主要 Activity
02 public class MainActivity extends Activity {
03 //定义布局中的卸载软件的 Button 控件
04 private Button btn;
05 //定义布局中给用户的提示内容的控件
06 private TextView Tv;
07
08 @Override
09 protected void onCreate(Bundle savedInstanceState) {
10     super.onCreate(savedInstanceState);
11     //设置当前 Activity 的布局文件为 activity_main
12     setContentView(R.layout.activity_main);
13     //得到浏览器中的控件对象
14     findView();
15     //设置对象的监听器
16     setListener();
17 }
18
19 private void setListener() {
20     //设置 btn 的单击监听器
21     btn.setOnClickListener(new OnClickListener() {
22         @Override
23         public void onClick(View v) {
24             //定义需要卸载的软件的包名 uri
25             Uri packageURI = Uri.parse("package:com.xiaomi.notes");
26             //设置 intent 的 action 为 Intent.ACTION_DELETE
27             //并且设置了 packageURI 为携带数据
28             Intent uninstallIntent = new Intent(Intent.ACTION_DELETE,
29                 packageURI);
30             //通过之前的 intent，启动 activity
31             startActivity(uninstallIntent);
32         }
33     });
```

```
34    }
35
36    private void findView() {
37        //得到布局中的开始加载 Button 的对象
38        btn = (Button) findViewById(R.id.Btn);
39        //得到布局中的开始加载 EditText 的对象
40        Tv = (TextView) findViewById(R.id.Tv);
41    }
42 }
```

此文件是 Activity 的代码文件，在其中第 4 行定义了一个卸载软件的 Button 对象，在第 6 行定义了提醒用户卸载软件的包名 TextView 对象。在第 14 行的 findView 方法中得到了布局中的所有控件，具体实现在第 36~41 行。在第 16 行通过 setListener 方法设置的控件的监听器，具体方法在 19~34 行实现。其中当用户单击 Btn 按钮的时候定义 intent 对象，得到系统中 package:com.xiaomi.notes 包名的 uri，然后通过 startactivity 启动系统的软件卸载界面进行软件卸载。

4．实例扩展

在此实例中当用户单击 Button 的时候，会打开系统的卸载软件的界面。注意需要卸载的软件的包名一定要存在，否则程序会提示异常。

范例 102 打开照相机获取图片

1．实例简介

在我们使用应用的过程中，经常会使用到用户单击某按钮的时候打开了手机的照相机软件进行拍照，然后将照片返回应用程序界面的效果。例如，在一些社交软件的客户端中，当用户希望设置自己的应用头像，而且直接进行相机拍照得到自己的头像照片。遇到这样的功能，一般是当用户进行某个操作的时候，例如，单击某个按钮的时候，通过 Intent 打开系统中照相机程序进行拍照，然后将得到的照片返回应用中显示。本例子就带领大家来实现一个单击按钮打开系统照相机拍照的功能，并且显示照片的实例效果。

2．运行效果

该实例运行效果如图 5.11 所示。

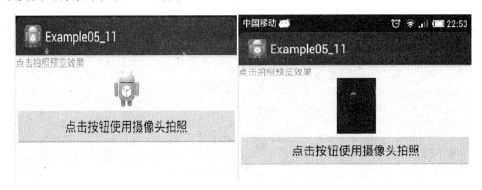

图 5.11 打开照相机获取照片的应用程序

3. 实例程序讲解

在本实例中，提供提示用户要单击按钮进行拍照的标签框、一个照相后显示照片的图片框和一个单击打开照相机拍照的按钮，然后当用户单击打开照相机的按钮时，程序打开系统的照相机软件进行拍照,并且把得到的照片返回之前的 ImageView 控件进行展示效果。想要实现本实例效果，首先修改 res/layout/activity_main.xml 文件，代码如下：

```xml
01  <?xml version="1.0" encoding="utf-8"?>
02  <!-- 定义当前布局的基本 LinearLayout -->
03  <LinearLayout xmlns:android="http://schemas.android.com/apk/res/android"
04      android:layout_width="fill_parent"
05      android:layout_height="fill_parent"
06      android:orientation="vertical" >
07
08      <!-- 定义提示用户单击拍照的标签控件 -->
09      <TextView
10          android:id="@+id/Tv"
11          android:layout_width="match_parent"
12          android:layout_height="wrap_content"
13          android:hint="单击拍照预览效果"
14          />
15
16      <!-- 定义显示照相结果的图片控件 -->
17      <ImageView
18          android:id="@+id/Iv"
19          android:layout_width="match_parent"
20          android:layout_height="wrap_content"
21          android:src="@drawable/ic_launcher"
22          />
23
24      <!-- 定义用户单击拍照的按钮控件 -->
25      <Button
26          android:id="@+id/Btn"
27          android:layout_width="match_parent"
28          android:layout_height="wrap_content"
29          android:text="单击按钮使用摄像头拍照"
30          />
31
32  </LinearLayout>
```

这是我们的 Activity 的布局文件。在其中第 8~14 行定义了一个 TextView 控件，用来提示用户单击按钮进行拍照。在第 17~22 行定义一个 ImageView 控件，用来显示拍照后的照片。在第 24~30 行定义一个 Button 控件，当用户单击的时候打开系统的照相机软件进行拍照并将得到的照片返回。

然后修改 src/com.wyl.example/MainActivity.java 文件，代码如下：

```java
01  //定义了本实例的主要 Activity
02  public class MainActivity extends Activity {
03      //定义布局中单击拍照的 Button 控件
04      private Button btn;
05      //定义布局中给用户提示内容的控件
06      private TextView Tv;
07      //定义布局中显示的图片控件
08      private ImageView Iv;
09
10      @Override
```

```java
11  protected void onCreate(Bundle savedInstanceState) {
12      super.onCreate(savedInstanceState);
13      //设置当前Activity的布局文件为activity_main
14      setContentView(R.layout.activity_main);
15      //得到浏览器中的控件对象
16      findView();
17      //设置对象的监听器
18      setListener();
19  }
20
21  private void setListener() {
22      //设置btn的单击监听器
23      btn.setOnClickListener(new OnClickListener() {
24          @Override
25          public void onClick(View v) {
26              //设置Intent的参数为通过摄像头获取的ACTION_IMAGE_CAPTURE
27              Intent intent = new Intent(MediaStore.ACTION_IMAGE_CAPTURE);
28              //启动activity返回照片结果，设置返回的requestCode为1
29              startActivityForResult(intent, 1);
30          }
31      });
32  }
33
34  private void findView() {
35      //得到布局中开始加载的Button的对象
36      btn = (Button) findViewById(R.id.Btn);
37      //得到布局中开始加载的EditText的对象
38      Tv = (TextView) findViewById(R.id.Tv);
39      //得到布局中开始加载的ImageView的对象
40      Iv = (ImageView) findViewById(R.id.Iv);
41  }
42
43  /*
44   * 系统的intent结果返回回调函数
45   */
46  @Override
47  protected void onActivityResult(int requestCode, int resultCode, Intent data) {
48      //接受用户通过其他activity返回的数据
49      super.onActivityResult(requestCode, resultCode, data);
50      //如果请求的requestCode为1的话，进行处理
51      if (requestCode == 1) {
52          //得到返回的处理状态，如果是成功得到了照片返回RESULT_OK值
53          if (resultCode == RESULT_OK) {
54              //成功得到照片后，得到data对象中的data值，并且转换为Bitmap对象
55              Bitmap bmPhoto = (Bitmap) data.getExtras().get("data");
56              //设置Iv的显示对象为此Bitmap
57              Iv.setImageBitmap(bmPhoto);
58          }
59      }
60  }
61  }
```

此文件是Activity的代码文件，在其中第4行定义了一个打开系统照相机进行拍照的Button对象。在第6行定义了提醒用户单击拍照按钮的TextView对象。在第8行定义了显示通过照相机得到的照片的ImageView对象。在第16行的findView方法中得到了布局中的所有控件，具体实现在第34～41行。在第18行通过setListener方法设置控件的监听器，

具体方法在 21~32 行实现。其中当用户单击 Btn 按钮的时候定义 intent 对象，设置 intent 的 action 为 MediaStore.ACTION_IMAGE_CAPTURE，然后通过 startactivityForResult 启动系统的照相机界面进行拍照，当照相完毕后，系统会自动回调 onActivityResult 方法，其中 requestCode 参数为请求码，resultCode 为返回请求的结果码，RESULT_OK 代表成功获得数据，data 代表返回给当前 activity 的数据携带对象，然后从 data 中获取拍照照片的 Bitmap 对象，最后设置给 ImageView 对象就可以显示拍照的照片效果了。

4．实例扩展

在此实例中当用户单击 Button 的时候，会打开系统的拍照的界面。注意需要此程序只可以在真机上进行效果展示，因为在 AVD 上是无法模拟手机的照相机效果的。

范例 103　打开系统图库获取图片

1．实例简介

在我们使用应用的过程中，经常会使用到用户单击某按钮的时候打开了手机的图片浏览软件选择照片，然后将选择的照片返回应用程序界面的效果。例如，在一些社交软件的客户端中，当用户希望设置自己的应用头像，而且本地图库中已经有了你想要的头像照片。遇到这样的功能，一般是当用户进行某个操作的时候，例如，单击某个按钮的时候，通过 Intent 打开系统中的图库浏览器进行图片选择，然后将选中的照片返回应用中显示。本例子就带领大家来实现一个单击按钮打开系统的图库软件选择照片，并且将选中的照片返回程序显示照片的实例效果。

2．运行效果

该实例运行效果如图 5.12 所示。

图 5.12　打开系统图库获取图片的应用程序

3．实例程序讲解

在本实例中，提供提示用户要单击按钮进行图片选择的标签框、一个选中图片后显示图片的图片框和一个单击打开系统图库选择照片的按钮，然后当用户单击此按钮时，程序打开系统的图片浏览软件选择照片,并且把选择的照片返回之前的 ImageView 控件进行展示效果。想要实现本实例效果,首先修改 res/layout/activity_main.xml 文件,代码如下:

```xml
01 <?xml version="1.0" encoding="utf-8"?>
02 <!-- 定义当前布局的基本 LinearLayout -->
03 <LinearLayout xmlns:android="http://schemas.android.com/apk/res/android"
04     android:layout_width="fill_parent"
05     android:layout_height="fill_parent"
06     android:orientation="vertical" >
07
08     <!-- 定义提示用户单击浏览照片的显示控件 -->
09     <TextView
10         android:id="@+id/Tv"
11         android:layout_width="match_parent"
12         android:layout_height="wrap_content"
13         android:hint="单击浏览系统照片效果"
14         />
15
16     <!-- 定义显示图片结果的图片控件 -->
17     <ImageView
18         android:id="@+id/Iv"
19         android:layout_width="match_parent"
20         android:layout_height="wrap_content"
21         android:src="@drawable/ic_launcher"
22         />
23
24     <!-- 定义用户单击选择照片的按钮控件 -->
25     <Button
26         android:id="@+id/Btn"
27         android:layout_width="match_parent"
28         android:layout_height="wrap_content"
29         android:text="单击按钮选择系统照片"
30         />
31
32 </LinearLayout>
```

这是我们的 Activity 的布局文件。在其中第 8~14 行定义了一个 TextView 控件,用来提示用户单击按钮进行图片选择,在第 17~22 行定义一个 ImageView 控件,用来显示通过浏览选择的图片,在第 24~30 行定义一个 Button 控件,当用户单击的时候打开系统的图片浏览软件选择图片并将选择的照片返回。

然后修改 src/com.wyl.example/MainActivity.java 文件,代码如下:

```java
01 //定义了本实例的主要 Activity
02 public class MainActivity extends Activity {
03     //定义布局中浏览图片的 Button 控件
04     private Button btn;
05     //定义布局中给用户提示内容的控件
06     private TextView Tv;
07     //定义布局中显示的图片控件
08     private ImageView Iv;
09
10     @Override
11     protected void onCreate(Bundle savedInstanceState) {
12         super.onCreate(savedInstanceState);
13         //设置当前 Activity 的布局文件为 activity_main
14         setContentView(R.layout.activity_main);
15         //得到浏览器中的控件对象
```

```java
16          findView();
17          //设置对象的监听器
18          setListener();
19      }
20
21      private void setListener() {
22          //设置btn的单击监听器
23          btn.setOnClickListener(new OnClickListener() {
24              @Override
25              public void onClick(View v) {
26                  Intent intent = new Intent();
27                  //定义intent的打开类型为图片类型
28                  intent.setType("image/*");
29                  //设置action为ACTION_GET_CONTENT，代表获得图片内容
30                  intent.setAction(Intent.ACTION_GET_CONTENT);
31                  //得到照片后返回当前页面
32                  startActivityForResult(intent, 1);
33              }
34          });
35      }
36
37      private void findView() {
38          //得到布局中开始加载的Button的对象
39          btn = (Button) findViewById(R.id.Btn);
40          //得到布局中开始加载的EditText的对象
41          Tv = (TextView) findViewById(R.id.Tv);
42          //得到布局中开始加载的ImageView的对象
43          Iv = (ImageView) findViewById(R.id.Iv);
44      }
45
46      /*
47       * 从系统中选择相应照片后的回调函数
48       */
49      @Override
50      protected void onActivityResult(int requestCode,
51              int resultCode, Intent data) {
52          //接受用户通过其他activity返回的数据
53          super.onActivityResult(requestCode, resultCode, data);
54          //如果请求的requestCode为1的话，进行处理
55          if (requestCode == 1) {
56              //得到返回的处理状态，如果是成功得到了照片返回RESULT_OK值
57              if (resultCode == RESULT_OK) {
58                  //得到返回的data数据
59                  Uri uri = data.getData();
60                  //通过当前的activity的content，得到ContentResolver
61                  ContentResolver cr = this.getContentResolver();
62                  try {
63                      //通过ContentResolver得到对应的图片Bitmap
64                      Bitmap bitmap =
65                              BitmapFactory.decodeStream(cr.openInputStream(uri));
66                      //将Bitmap设定到ImageView上
67                      Iv.setImageBitmap(bitmap);
68                  } catch (FileNotFoundException e) {
69                      Log.e("Exception", e.getMessage(),e);
70                  }
71              }
72          }
```

```
73    }
74  }
```

此文件是 Activity 的代码文件，在其中第 4 行定义了一个打开系统图片浏览软件进行图片选择的 Button 对象。在第 6 行定义了提醒用户单击浏览图片按钮的 TextView 对象。在第 8 行定义了显示通过图库选择得到的照片的 ImageView 对象。在第 16 行的 findView 方法中得到了布局中的所有控件，具体实现在第 37~44 行。在第 18 行通过 setListener 方法设置的控件的监听器，具体方法在 21~35 行实现。其中当用户单击 Btn 按钮的时候定义 intent 对象，设置 intent 的 action 为 Intent.ACTION_GET_CONTENT，设置 intent 的类型为 image/*，获取图片浏览类型，然后通过 startactivityForResult 启动系统的照片浏览程序进行图片选择，当选择完毕后，系统会自动回调 onActivityResult 方法，其中 requestCode 参数为请求码，resultCode 为返回请求的结果码，RESULT_OK 代表成功获得数据，data 代表返回给当前 activity 的数据携带对象，然后从 data 中获取选择的图片的 ContentResolver 对象，然后通过 openInputStream 方法得到 Bitmap 对象，最后设置给 ImageView 对象就可以显示之前选择的图片了。

4．实例扩展

注意在图库浏览软件选择图片后，可以自动回调 onActivityResult 方法，但是如果用户打开图库浏览图片后没有选择图片，这时候就可能会有问题，所以大家需要在 onActivityResult 方法中通过 resultCode 进行判断，如果是 RESULT_OK，那么说明用户选择了图片，否则说明用户没有选择图片，而是通过返回键返回当前页面。

范例 104 打开录音程序录音

1．实例简介

在我们使用应用的过程中，经常会使用到用户单击某按钮的时候打开了手机的录音软件进行录音的效果。例如，在一些特殊的记事儿的客户端中，当用户希望记录语音留言。遇到这样的功能，我们一般是当用户进行某个操作的时候，例如，单击某个按钮的时候，通过 Intent 打开系统中的录音软件进行录音。本例子就带领大家来实现一个单击按钮打开系统的录音软件进行录音的实例效果。

2．运行效果

该实例运行效果如图 5.13 所示。

图 5.13 打开录音程序录音

3. 实例程序讲解

在本实例中，提供提示用户要单击按钮进行录音的标签框和一个单击打开系统录音软件的按钮，然后当用户单击此按钮时，程序打开系统的录音机软件开始录音，并且把录音的文件返回到当前的 Activity 的效果。想要实现本实例效果，首先修改 res/layout/activity_main.xml 文件，代码如下：

```xml
01 <?xml version="1.0" encoding="utf-8"?>
02 <!-- 定义当前布局的基本 LinearLayout -->
03 <LinearLayout xmlns:android="http://schemas.android.com/apk/res/android"
04     android:layout_width="fill_parent"
05     android:layout_height="fill_parent"
06     android:orientation="vertical" >
07
08     <!-- 定义提示用户单击打开录音机的显示控件 -->
09     <TextView
10         android:id="@+id/Tv"
11         android:layout_width="match_parent"
12         android:layout_height="wrap_content"
13         android:hint="单击下面的打开录音机的按钮"
14         />
15
16     <!-- 定义用户单击打开录音机的按钮控件 -->
17     <Button
18         android:id="@+id/Btn"
19         android:layout_width="match_parent"
20         android:layout_height="wrap_content"
21         android:text="单击打开录音机"
22         />
23
24 </LinearLayout>
```

这是我们的 Activity 的布局文件。在其中第 8~14 行定义了一个 TextView 控件，用来提示用户单击按钮进行录音。在第 16~22 行定义一个 Button 控件，当用户单击的时候打开系统的录音软件进行录音，并且将录音的信息返回。

然后修改 src/com.wyl.example/MainActivity.java 文件，代码如下：

```java
01 //定义了本实例的主要 Activity
02 public class MainActivity extends Activity {
03     //定义布局中的打开录音机的 Button 控件
04     private Button btn;
05     //定义布局中给用户的提示内容的控件
06     private TextView Tv;
07
08     @Override
09     protected void onCreate(Bundle savedInstanceState) {
10         super.onCreate(savedInstanceState);
11         //设置当前 Activity 的布局文件为 activity_main
12         setContentView(R.layout.activity_main);
13         //得到浏览器中的控件对象
14         findView();
15         //设置对象的监听器
```

```java
16          setListener();
17    }
18
19    private void setListener() {
20        //设置 btn 的单击监听器
21        btn.setOnClickListener(new OnClickListener() {
22            @Override
23            public void onClick(View v) {
24                //设置 intent 的属性为录音设置, Media.RECORD_SOUND_ACTION
25                Intent intent = new Intent(Media.RECORD_SOUND_ACTION);
26                //通过 intent 启动 activity
27                startActivityForResult(intent, 1);
28            }
29        });
30    }
31
32    private void findView() {
33        //得到布局中开始加载的 Button 的对象
34        btn = (Button) findViewById(R.id.Btn);
35        //得到布局中开始加载的 TextView 的对象
36        Tv = (TextView) findViewById(R.id.Tv);
37    }
38
39    /*
40     * 提交 startactivityforresult 的回调函数
41     */
42    @Override
43    protected void onActivityResult(int requestCode, int resultCode, Intent data) {
44        //接受用户通过其他 activity 返回的数据
45        super.onActivityResult(requestCode, resultCode, data);
46        //如果请求的 requestCode 为 1 的话，进行处理
47        if (requestCode == 1) {
48            //得到返回的处理状态，如果是成功得到了音频，返回 RESULT_OK 值
49            if (resultCode == RESULT_OK) {
50                //得到录音的音频文件及路径
51                String dataFile=data.getDataString() ;
52                String dataUri=getIntent().getDataString();
53                Log.e("dataFile", dataFile);
54                Log.e("dataUri", dataUri);
55            }
56        }
57    }
58 }
```

此文件是 Activity 的代码文件，在其中第 4 行定义了一个打开系统录音机软件进行录音的 Button 对象，在第 6 行定义了提醒用户单击录音按钮的 TextView 对象。在第 14 行的 findView 方法中得到了布局中的所有控件，具体实现在第 32~37 行。在第 16 行通过 setListener 方法设置控件的监听器，具体方法在 19~30 行实现。其中当用户单击 Btn 按钮的时候定义 intent 对象，并且设置 intent 的 action 为 Media.RECORD_SOUND_ACTION，代表需要系统的录音软件，然后通过 startactivityForResult 启动系统的录音软件进行录音，当录音完毕后，系统会自动回调 onActivityResult 方法，其中 requestCode 参数为请求码，resultCode 为返回请求的结果码，RESULT_OK 代表成功获得数据，data 代表返回给当前 activity 的数据携带对象，然后从 data 中获取录音文件的存储位置，最后就可以对录音文件

进行处理了。

4．实例扩展

注意在录音完成后在 Activity 中仍然可以进行处理。例如，播放录音的音频文件，或者通过网络发送音频文件等，这里为了举例子，只是得到了录音文件的保存路径。

范例 105　打开已安装的应用程序信息

1．实例简介

在我们使用应用的过程中，经常会使用到用户单击某按钮的时候打开了手机的某个软件的应用程序信息的效果。例如，在一些软件管理的客户端中，用户希望查看已经安装的应用的信息。遇到这样的功能，我们一般是当用户进行某个操作的时候，例如，单击某个按钮的时候，通过 Intent 打开系统中的查看应用信息的界面。本例子就带领大家来实现一个单击按钮打开系统来查看软件信息界面的实例效果。

2．运行效果

该实例运行效果如图 5.14 所示。

图 5.14　打开已安装的应用程序信息

3．实例程序讲解

在本实例中，提供提示用户要单击按钮进行应用信息查看的标签框、一个输入需要查看的应用程序包名的输入框和一个单击打开系统软件信息的按钮，然后当用户单击此按钮时，程序打开系统的查看应用信息界面。想要实现本实例效果，首先修改 res/layout/activity_main.xml 文件，代码如下：

```
01  <?xml version="1.0" encoding="utf-8"?>
02  <!-- 定义当前布局的基本 LinearLayout -->
03  <LinearLayout xmlns:android="http://schemas.android.com/apk/res/android"
```

```
04      android:layout_width="fill_parent"
05      android:layout_height="fill_parent"
06      android:orientation="vertical" >
07
08      <!-- 定义提示用户单击打开软件信息的显示控件 -->
09      <TextView
10          android:id="@+id/Tv"
11          android:layout_width="match_parent"
12          android:layout_height="wrap_content"
13          android:hint="单击下面的打开软件信息的按钮"
14          />
15
16      <!-- 定义输入要查看的包名的控件 -->
17      <EditText
18          android:id="@+id/Et"
19          android:layout_width="match_parent"
20          android:layout_height="wrap_content"
21          android:text="com.wyl.example"
22          />
23
24      <!-- 定义用户单击打开应用信息的按钮控件 -->
25      <Button
26          android:id="@+id/Btn"
27          android:layout_width="match_parent"
28          android:layout_height="wrap_content"
29          android:text="单击打开应用信息"
30          />
31
32  </LinearLayout>
```

这是我们的 Activity 的布局文件。在其中第 8～14 行定义了一个 TextView 控件，用来提示用户单击按钮进行应用信息查看。在第 16～22 行定义一个 EditText 控件，用来得到用户需要查看的应用程序包名。在第 24～30 行定义一个 Button 控件，当用户单击的时候打开系统的查看应用程序信息界面。

然后修改 src/com.wyl.example/MainActivity.java 文件，代码如下：

```
01  //定义了本实例的主要 Activity
02  public class MainActivity extends Activity {
03      //定义布局中的打开软件详细信息的 Button 控件
04      private Button btn;
05      //定义布局中给用户提示内容的控件
06      private EditText Et;
07
08      @Override
09      protected void onCreate(Bundle savedInstanceState) {
10          super.onCreate(savedInstanceState);
11          //设置当前 Activity 的布局文件为 activity_main
12          setContentView(R.layout.activity_main);
13          //得到浏览器中的控件对象
14          findView();
15          //设置对象的监听器
16          setListener();
17      }
18
19      private void setListener() {
20          //设置 btn 的单击监听器
21          btn.setOnClickListener(new OnClickListener() {
```

```
22          @Override
23          public void onClick(View v) {
24              //定义 intent,设置 intent 的 action 为
25              //Settings.ACTION_APPLICATION_DETAILS_SETTINGS
26              //代表系统的设置应用程序信息
27              Intent intent = new Intent(Settings.ACTION_APPLICATION_
                DETAILS_SETTINGS);
28              //得到用户需要查看的包名,封装成 uri 对象
29              Uri uri = Uri.fromParts("package",Et.getText().toString(),
                null);
30              //设置 intent 对象的数据为上面的 uri
31              intent.setData(uri);
32              //通过 startActivity 启动界面
33              startActivity(intent);
34          }
35      });
36  }
37
38  private void findView() {
39      //得到布局中的开始加载的 Button 的对象
40      btn = (Button) findViewById(R.id.Btn);
41      //得到布局中的开始加载的 TextView 的对象
42      Et = (EditText) findViewById(R.id.Et);
43  }
44  }
```

此文件是 Activity 的代码文件,在其中第 4 行定义了一个打开系统查看软件信息的 Button 对象,在第 6 行定义了用户输入希望查看的软件的包名输入框。在第 14 行的 findView 方法中得到了布局中的所有控件,具体实现在第 38~43 行。在第 16 行通过 setListener 方法设置控件的监听器,具体方法在 19~36 行实现。其中当用户单击 Btn 按钮的时候定义 intent 对象,并且设置 intent 的 action 为 Settings.ACTION_APPLICATION_DETAILS_SETTINGS,代表需要打开系统的查看应用具体信息的界面,然后得到用户输入的查看的应用的包名,并且得到 uri 对象,设置给 intent。通过 startActivity 启动系统查看应用程序信息界面。

4. 实例扩展

一般查看应用信息的功能会在一些手机的管理软件中使用。例如,360 手机软件管家和豌豆荚等。这些都是通过此方法来查看系统应用信息的。

范例 106 打开软件市场搜索应用

1. 实例简介

在我们使用应用的过程中,经常会使用到用户单击某按钮的时候打开了手机上的软件市场去查找对应应用并进行下载的效果。例如,在一些社交软件的客户端中,可能会进行一些软件的推荐,QQ 软件可能希望推荐用户使用 QQ 空间单独的客户端。遇到这样的功能,我们一般是当用户进行某个操作的时候,例如,单击某个按钮的时候,通过 Intent 打

开手机中的软件市场（这里以安卓市场为例）搜索应用。本例子就带领大家来实现一个单击按钮打开系统的软件市场搜索应用的实例效果。

2. 运行效果

该实例运行效果如图 5.15 所示。

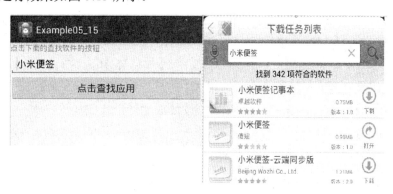

图 5.15　打开软件市场搜索应用

3. 实例程序讲解

在本实例中，提供提示用户要单击按钮查找应用的标签框、一个输入需要在市场查找的应用程序名称的输入框和一个单击打开软件市场查找软件的按钮，然后当用户单击此按钮时，程序打开系统的软件市场查找软件。想要实现本实例效果，首先修改 res/layout/activity_main.xml 文件，代码如下：

```
01  <?xml version="1.0" encoding="utf-8"?>
02  <!-- 定义当前布局的基本 LinearLayout -->
03  <LinearLayout xmlns:android="http://schemas.android.com/apk/res/android"
04      android:layout_width="fill_parent"
05      android:layout_height="fill_parent"
06      android:orientation="vertical" >
07
08      <!-- 定义提示用户单击查找软件信息的显示控件 -->
09      <TextView
10          android:id="@+id/Tv"
11          android:layout_width="match_parent"
12          android:layout_height="wrap_content"
13          android:hint="单击下面的查找软件的按钮"
14          />
15
16      <!-- 定义输入要查找的软件的输入的控件 -->
17      <EditText
18          android:id="@+id/Et"
19          android:layout_width="match_parent"
20          android:layout_height="wrap_content"
21          android:text="小米便签"
22          />
23
24      <!-- 定义用户单击查找应用的按钮控件 -->
25      <Button
```

```
26          android:id="@+id/Btn"
27          android:layout_width="match_parent"
28          android:layout_height="wrap_content"
29          android:text="单击查找应用"
30          />
31
32  </LinearLayout>
```

这是我们的 Activity 的布局文件。在其中第 8～14 行定义了一个 TextView 控件，用来提示用户单击按钮进行软件搜索。在第 16～22 行定义了一个 EditText 控件，用来得到用户需要搜索的应用程序的名称。在第 24～30 行定义了一个 Button 控件，当用户单击的时候打开系统的软件市场进行软件搜索。

然后修改 src/com.wyl.example/MainActivity.java 文件，代码如下：

```java
01  //定义了本实例的主要 Activity
02  public class MainActivity extends Activity {
03      //定义布局中的在软件市场查找应用的 Button 控件
04      private Button btn;
05      //定义布局中给用户提示内容的控件
06      private EditText Et;
07
08      @Override
09      protected void onCreate(Bundle savedInstanceState) {
10          super.onCreate(savedInstanceState);
11          //设置当前 Activity 的布局文件为 activity_main
12          setContentView(R.layout.activity_main);
13          //得到浏览器中的控件对象
14          findView();
15          //设置对象的监听器
16          setListener();
17      }
18
19      private void setListener() {
20          //设置 btn 的单击监听器
21          btn.setOnClickListener(new OnClickListener() {
22              @Override
23              public void onClick(View v) {
24                  //设置 Uri, 参数为 market://search?q=pname:, 后面加上要查找的应用的名字
25                  Uri uri = Uri.parse("market://search?q=pname:"
26                          +Et.getText().toString());
27                  //通过 uri 产生 intent
28                  Intent it = new Intent(Intent.ACTION_VIEW, uri);
29                  //通过 intent, 启动 activity
30                  startActivity(it);
31              }
32          });
33      }
34
35      private void findView() {
36          //得到布局中的开始加载的 Button 的对象
37          btn = (Button) findViewById(R.id.Btn);
38          //得到布局中的开始加载的 EditText 的对象
39          Et = (EditText) findViewById(R.id.Et);
```

```
40    }
41 }
```

此文件是 Activity 的代码文件,在其中第 4 行定义了一个打开系统中的软件市场进行搜索的 Button 对象,在第 6 行定义了用户输入希望查找的软件名称的输入框。在第 14 行的 findView 方法中得到了布局中的所有控件,具体实现在第 35~40 行。在第 16 行通过 setListener 方法设置控件的监听器,具体方法在 19~33 行实现。其中当用户单击 Btn 按钮的时候,通过用户输入的搜索应用程序的名称构造 uri 对象,定义 intent 对象,设置 intent 的 action 为 Intent.ACTION_VIEW,并且传入 uri 对象,通过 startActivity 启动系统中的软件市场查找相关软件。

4.实例扩展

想要实现此效果一般需要在手机上安装 Android 的软件市场。例如,Google Play、豌豆荚和安卓市场等软件。这些软件需要的 uri 参数可能有所不同,本实例在安卓市场下搜索测试完成,其他市场可能需要修改 uri 的参数。

范例 107 选择联系人功能

1.实例简介

在我们使用应用的过程中,经常会使用到用户单击某按钮的时候打开了手机上联系人菜单进行联系人选择的效果。例如,在一些社交软件的客户端中,需要通过手机中的联系人号码找到好友。遇到这样的功能,我们一般是当用户进行某个操作的时候,例如,单击某个按钮的时候,通过 Intent 打开手机中的联系人程序。本例子就带领大家来实现一个单击按钮打开系统的联系人程序。

2.运行效果

该实例运行效果如图 5.16 所示。

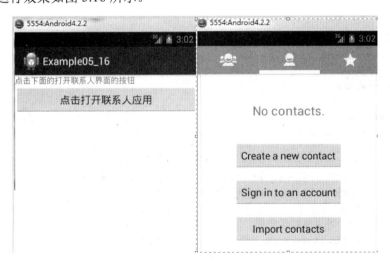

图 5.16 选择联系人功能

3. 实例程序讲解

在本实例中，提供提示用户要单击按钮查找联系人的标签框和一个单击打开联系人软件的按钮，然后当用户单击此按钮时，程序打开系统的联系人软件进行联系人的查看。想要实现本实例效果，首先修改 res/layout/activity_main.xml 文件，代码如下：

```xml
01 <?xml version="1.0" encoding="utf-8"?>
02 <!-- 定义当前布局的基本 LinearLayout -->
03 <LinearLayout xmlns:android="http://schemas.android.com/apk/res/android"
04     android:layout_width="fill_parent"
05     android:layout_height="fill_parent"
06     android:orientation="vertical" >
07
08     <!-- 定义提示用户单击打开联系人界面的显示控件 -->
09     <TextView
10         android:id="@+id/Tv"
11         android:layout_width="match_parent"
12         android:layout_height="wrap_content"
13         android:hint="单击下面的打开联系人界面的按钮"
14         />
15
16     <!-- 定义用户单击打开联系人应用的按钮控件 -->
17     <Button
18         android:id="@+id/Btn"
19         android:layout_width="match_parent"
20         android:layout_height="wrap_content"
21         android:text="单击打开联系人应用"
22         />
23
24 </LinearLayout>
```

这是我们的 Activity 的布局文件。在其中第 8~14 行定义了一个 TextView 控件，用来提示用户单击按钮进行联系人查看。在第 16~22 行定义一个 Button 控件，当用户单击的时候打开系统的联系人程序。

然后修改 src/com.wyl.example/MainActivity.java 文件，代码如下：

```java
01 //定义了本实例的主要 Activity
02 public class MainActivity extends Activity {
03 //定义布局中的打开联系人应用的 Button 控件
04 private Button btn;
05
06 @Override
07 protected void onCreate(Bundle savedInstanceState) {
08     super.onCreate(savedInstanceState);
09     //设置当前 Activity 的布局文件为 activity_main
10     setContentView(R.layout.activity_main);
11     //得到浏览器中的控件对象
12     findView();
13     //设置对象的监听器
14     setListener();
15 }
16
```

```
17  private void setListener() {
18      //设置 btn 的单击监听器
19      btn.setOnClickListener(new OnClickListener() {
20          @Override
21          public void onClick(View v) {
22              //定义 intent 对象
23              Intent intent = new Intent();
24              //设置 intnet 的 action 为 Intent.ACTION_VIEW
25              intent.setAction(Intent.ACTION_VIEW);
26              //设置 intent 的数据为 Contacts.People.CONTENT_URI
27              //代表查看联系人信息
28              intent.setData(Contacts.People.CONTENT_URI);
29              //启动 activity
30              startActivity(intent);
31          }
32      });
33  }
34
35  private void findView() {
36      //得到布局中的开始加载的 Button 的对象
37      btn = (Button) findViewById(R.id.Btn);
38  }
39  }
```

此文件是 Activity 的代码文件，在其中第 4 行定义了一个打开系统中的联系人程序的 Button 对象。在第 12 行的 findView 方法中得到了布局中的所有控件，具体实现在第 35～38 行。在第 14 行通过 setListener 方法设置的控件的监听器，具体方法在 17～33 行实现。其中当用户单击 Btn 按钮的时候，定义 intent 对象，设置 intent 的 action 为 Intent.ACTION_VIEW，设置 intent 的数据为 People.CONTENT_URI，通过 startActivity 启动系统中的联系人软件进行查看。

4．实例扩展

本实例只实现了查看联系人列表的效果，还可以在此实例的基础上实现单击某联系人后返回到 activity 中进行显示或者处理，请读者自行完成。

范例 108 添加联系人功能

1．实例简介

在我们使用应用的过程中，经常会使用到用户单击某按钮的时候打开了手机上添加联系人的界面。例如，在一些社交软件的客户端中，经常见到的是通过二维码得到用户的联系人信息，然后快速的添加到手机的联系人中。遇到这样的功能，我们一般是当用户进行某个操作的时候，例如，单击某个按钮的时候，通过 Intent 打开手机中的添加联系人界面。本例子就带领大家来实现一个单击按钮打开系统的添加联系人界面。

2．运行效果

该实例运行效果如图 5.17 所示。

Android 开发范例实战宝典

图 5.17　添加系统联系人

3．实例程序讲解

在本实例中，提供提示用户要单击按钮添加联系人的标签框，定义了四个 EditText，分别接收用户输入的联系人的姓名、联系人的电话、联系人的邮箱及联系人的公司信息。一个单击添加联系人的按钮，然后当用户单击此按钮时，打开系统的添加联系人界面，进行联系人的添加。想要实现本实例效果，首先修改 res/layout/activity_main.xml 文件，代码如下：

```
01  <?xml version="1.0" encoding="utf-8"?>
02  <!-- 定义当前布局的基本 LinearLayout -->
03  <LinearLayout xmlns:android="http://schemas.android.com/apk/res/android"
04      android:layout_width="fill_parent"
05      android:layout_height="fill_parent"
06      android:orientation="vertical" >
07
08      <!-- 定义提示用户单击添加联系人界面的显示控件 -->
09      <TextView
10          android:id="@+id/Tv"
11          android:layout_width="match_parent"
12          android:layout_height="wrap_content"
13          android:hint="单击下面的添加联系人界面的按钮"
14          />
15      <!-- 定义用户添加联系人姓名的输入控件 -->
16      <EditText
17          android:id="@+id/EtName"
18          android:layout_width="match_parent"
19          android:layout_height="wrap_content"
20          android:hint="联系人姓名"
21          />
22
23      <!-- 定义用户添加联系人电话的输入控件 -->
24      <EditText
25          android:id="@+id/EtPhone"
```

```
26          android:layout_width="match_parent"
27          android:layout_height="wrap_content"
28          android:hint="联系人电话"
29          />
30
31      <!-- 定义用户添加联系人邮箱的输入控件 -->
32      <EditText
33          android:id="@+id/EtEmail"
34          android:layout_width="match_parent"
35          android:layout_height="wrap_content"
36          android:hint="联系人邮箱"
37          />
38
39      <!-- 定义用户添加联系人公司的输入控件 -->
40      <EditText
41          android:id="@+id/EtCompany"
42          android:layout_width="match_parent"
43          android:layout_height="wrap_content"
44          android:hint="联系人公司"
45          />
46
47      <!-- 定义用户单击添加联系人应用的按钮控件 -->
48      <Button
49          android:id="@+id/Btn"
50          android:layout_width="match_parent"
51          android:layout_height="wrap_content"
52          android:text="单击添加联系人应用"
53          />
54
55  </LinearLayout>
```

这是我们的 Activity 的布局文件。在其中第 16、24、32 和 40 行分别定义了四个 EditText 控件，用来接收用户输入的联系人姓名、联系人电话、联系人公司和联系人邮箱信息。在第 48～53 行定义一个 Button 控件，当用户单击的时候打开系统的添加联系人界面。

然后修改 src/com.wyl.example/MainActivity.java 文件，代码如下：

```
01  //定义了本实例的主要 Activity
02  public class MainActivity extends Activity {
03  //定义布局中的添加联系人应用的 Button 控件
04  private Button btn;
05  //定义布局中的联系人名称的输入控件
06  private EditText EtName;
07  //定义布局中的联系人电话的输入控件
08  private EditText EtPhone;
09  //定义布局中的联系人邮件的输入控件
10  private EditText EtEmail;
11  //定义布局中的联系人公司的输入控件
12  private EditText EtCompany;
13
14  @Override
15  protected void onCreate(Bundle savedInstanceState) {
16      super.onCreate(savedInstanceState);
17      //设置当前 Activity 的布局文件为 activity_main
18      setContentView(R.layout.activity_main);
19      //得到浏览器中的控件对象
20      findView();
```

```
21            //设置对象的监听器
22            setListener();
23     }
24
25     private void setListener() {
26            //设置btn的单击监听器
27            btn.setOnClickListener(new OnClickListener() {
28                @Override
29                public void onClick(View v) {
30                    //设置intent的action为
31                    //Contacts.Intents.Insert.ACTION,插入联系人的动作
32                    Intent it = new Intent(Contacts.Intents.Insert.ACTION);
33                    it.setType(Contacts.People.CONTENT_TYPE);
34                    //添加联系人的名字
35                    it.putExtra(Contacts.Intents.Insert.NAME, EtName.getText().
                       toString());
36                    //添加联系人的电话
37                    it.putExtra(Contacts.Intents.Insert.PHONE,EtPhone.getText().
                       toString());
38                    //添加联系人的邮箱
39                    it.putExtra(Contacts.Intents.Insert.EMAIL, EtEmail.getText().
                       toString());
40                    //添加联系人的公司
41                    it.putExtra(Contacts.Intents.Insert.COMPANY,EtCompany.getText().
                       toString());
42
43                    //启动activity
44                    startActivity(it);
45                }
46            });
47     }
48
49     private void findView() {
50            //得到布局中的开始加载的Button的对象
51            btn = (Button) findViewById(R.id.Btn);
52            //得到布局中的姓名输入框
53            EtName = (EditText) findViewById(R.id.EtName);
54            //得到布局中的电话输入框
55            EtPhone = (EditText) findViewById(R.id.EtPhone);
56            //得到布局中的邮箱输入框
57            EtEmail = (EditText) findViewById(R.id.EtEmail);
58            //得到布局中的公司输入框
59            EtCompany = (EditText) findViewById(R.id.EtCompany);
60     }
61 }
```

此文件是 Activity 的代码文件，在其中第 4 行定义了一个打开系统中的添加联系人界面的 Button 对象。在第 5～12 行分别定义了接收联系人姓名、联系人号码、联系人邮箱和联系人公司的 EditText 对象。在第 20 行的 findView 方法中得到了布局中的所有控件，具体实现在第 49～60 行。在第 22 行通过 setListener 方法设置控件的监听器，具体方法在 25～47 行实现。其中当用户单击 Btn 按钮的时候，定义 intent 对象，设置 intent 的 action 为 Contacts.Intents.Insert.ACTION，并且在 intent 中分别加入了联系人姓名、邮箱、公司和号码等信息，通过 startActivity 启动系统中的添加联系人功能进行添加。

4．实例扩展

本实例涉及到联系人的添加操作，所以最好在真机上进行测试，因为有些 AVD 对于此部分的支持不太完善。

范例 109　程序内部启动外部程序

1．实例简介

在我们使用应用的过程中，经常会使用到用户单击某按钮的时候打开了手机上其他应用程序的情况。例如，一些在 QQ 软件的客户端中，如果用户希望访问 QQ 空间，单击相应按钮，系统会打开手机上的 QQ 空间客户端。遇到这样的功能，我们一般是当用户进行某个操作的时候，例如，单击某个按钮的时候，通过 Intent 打开手机中的其他应用程序。本例子就带领大家来实现一个单击按钮打开系统的外部程序的实例。

2．运行效果

该实例运行效果如图 5.18 所示。

图 5.18　程序内部启动外部程序

3．实例程序讲解

在本实例中，提供提示用户要单击打开小米便签的标签框和一个单击打开小米便签的按钮，然后当用户单击此按钮时，打开系统安装的小米便签应用程序，进行操作。想要实现本实例效果，首先修改 res/layout/activity_main.xml 文件，代码如下：

```
01  <?xml version="1.0" encoding="utf-8"?>
02  <!-- 定义当前布局的基本 LinearLayout -->
03  <LinearLayout xmlns:android="http://schemas.android.com/apk/res/android"
04      android:layout_width="fill_parent"
05      android:layout_height="fill_parent"
06      android:orientation="vertical" >
07  
08      <!-- 定义提示用户单击启动小米便签的标签控件 -->
09      <TextView
10          android:id="@+id/Tv"
11          android:layout_width="match_parent"
```

```
12          android:layout_height="wrap_content"
13          android:hint="单击下面的启动小米便签的按钮"
14          />
15
16      <!-- 定义用户单击添加联系人应用的按钮控件 -->
17      <Button
18          android:id="@+id/Btn"
19          android:layout_width="match_parent"
20          android:layout_height="wrap_content"
21          android:text="单击启动小米便签"
22          />
23
24  </LinearLayout>
```

这是我们的 Activity 的布局文件。在其中第 9~14 行分别定义了一个 TextView 控件，用来提醒用户单击启动小米便签。在第 17~22 行定义了一个 Button 控件，当用户单击的时候打开系统安装的小米便签应用，进行操作。

然后修改 src/com.wyl.example/MainActivity.java 文件，代码如下：

```
01  //定义了本实例的主要 Activity
02  public class MainActivity extends Activity {
03      //定义布局中的启动小米便签的 Button 控件
04      private Button btn;
05
06      @Override
07      protected void onCreate(Bundle savedInstanceState) {
08          super.onCreate(savedInstanceState);
09          //设置当前 Activity 的布局文件为 activity_main
10          setContentView(R.layout.activity_main);
11          //得到浏览器中的控件对象
12          findView();
13          //设置对象的监听器
14          setListener();
15      }
16
17      private void setListener() {
18          //设置 btn 的单击监听器
19          btn.setOnClickListener(new OnClickListener() {
20              @Override
21              public void onClick(View v) {
22                  //定义 intent 对象
23                  Intent intent = new Intent();
24                  //设置 intent 为包管理器的 com.fjsoft.xhx.miui.notes 包的 intent
25                  intent = getPackageManager().
26                          getLaunchIntentForPackage("com.fjsoft.xhx.miui.
                            notes");
27                  //通过 intent 启动 activity
28                  startActivity(intent);
29              }
30          });
31      }
32
33      private void findView() {
34          //得到布局中的开始加载的 Button 的对象
35          btn = (Button) findViewById(R.id.Btn);
36      }
37  }
```

此文件是 Activity 的代码文件，在其中第 4 行定义了一个打开系统中的小米便签的 Button 对象。在第 12 行的 findView 方法中得到了布局中的所有控件，具体实现在第 33～36 行。在第 14 行通过 setListener 方法设置控件的监听器，具体方法在 17～31 行实现。其中当用户单击 Btn 按钮的时候，定义 intent 对象，设置 intent 为 getPackagemanager 服务得到 com.fjsoft.xhx.miui.notes 包的 intent，通过 startActivity 启动系统中对应包的应用程序，也就是小米便签启动。

4．实例扩展

本实例涉及打开系统中的应用程序，想实现本例效果前提是系统中安装了 com.fjsoft.xhx.miui.notes 包对应的程序才可以打开，这个包对应了小米便签应用，如果希望打开其他应用程序则填入其他应用程序的包名即可。

注意一点，一般情况下在一个应用程序中调用其他应用程序这样的功能使用的不多，因为首先要保证对应的包的应用程序存在，而且还要对应包名去找应用程序，这样比较复杂。这个功能一般情况下用在同一个公司的不同项目中使用，例如，QQ 客户端和 QQ 控件客户端等，这样的话在程序内部已经了解相互的包名，而且如果用户没有安装希望的客户端，还可以通过包名提示用户去下载，一举两得。

范例 110　启动 Google 地图显示某个位置

1．实例简介

在我们使用应用的过程中，经常会使用到用户单击某按钮的时候打开了手机上的 Google 地图，并显示了某个固定的位置。例如，在一些在团购客户端中，用户购买相应的餐券，希望去此店消费，但是又不知道具体的位置，这时候单击地图按钮，在地图中显示了此商店的具体位置。遇到这样的功能，我们一般是当用户进行某个操作的时候，例如：单击某个按钮的时候，通过 Intent 打开手机中的 Google 地图，并且以相应的位置为中心。本例子就带领大家来实现一个单击按钮打开系统的 Google 地图程序的实例。

2．运行效果

该实例运行效果如图 5.19 所示。

图 5.19　启动 Google 地图显示某个固定位置

3. 实例程序讲解

在本实例中，提供提示用户要单击启动 Google 地图的标签框和一个单击打开 Google 地图的按钮，然后当用户单击此按钮时，打开系统安装的 Google 地图程序。想要实现本实例效果，首先修改 res/layout/activity_main.xml 文件，代码如下：

```xml
01  <?xml version="1.0" encoding="utf-8"?>
02  <!-- 定义当前布局的基本 LinearLayout -->
03  <LinearLayout xmlns:android="http://schemas.android.com/apk/res/android"
04      android:layout_width="fill_parent"
05      android:layout_height="fill_parent"
06      android:orientation="vertical" >
07
08      <!-- 定义提示用户单击启动谷歌地图的标签控件 -->
09      <TextView
10          android:id="@+id/Tv"
11          android:layout_width="match_parent"
12          android:layout_height="wrap_content"
13          android:hint="单击下面的启动谷歌地图"
14          />
15
16      <!-- 定义用户单击启动谷歌地图的按钮控件 -->
17      <Button
18          android:id="@+id/Btn"
19          android:layout_width="match_parent"
20          android:layout_height="wrap_content"
21          android:text="单击启动谷歌地图"
22          />
23
24  </LinearLayout>
```

这是我们的 Activity 的布局文件。在其中第 9～14 行定义了一个 TextView 控件，用来提醒用户单击启动 Google 地图。在第 17～22 行定义了一个 Button 控件，当用户单击的时候打开系统安装的 Google 地图应用进行位置显示。

然后修改 src/com.wyl.example/MainActivity.java 文件，代码如下：

```java
01  //定义了本实例的主要 Activity
02  public class MainActivity extends Activity {
03  //定义布局中的启动 Google 地图的 Button 控件
04  private Button btn;
05
06  @Override
07  protected void onCreate(Bundle savedInstanceState) {
08      super.onCreate(savedInstanceState);
09      //设置当前 Activity 的布局文件为 activity_main
10      setContentView(R.layout.activity_main);
11      //得到浏览器中的控件对象
12      findView();
13      //设置对象的监听器
14      setListener();
15  }
16
17  private void setListener() {
18      //设置 btn 的单击监听器
19      btn.setOnClickListener(new OnClickListener() {
20          @Override
```

```
21        public void onClick(View v) {
22            //定义uri, uri的字符串为geo: 代表经纬度
23            //39.888402,116.409561是北京天坛公园的经纬度坐标
24            Uri uri = Uri.parse("geo:39.888402,116.409561");
25            //定义intent为读取当前的uri中的经纬度显示
26            Intent it = new Intent(Intent.ACTION_VIEW,uri);
27            //启动activity
28            startActivity(it);
29        }
30    });
31 }
32
33 private void findView() {
34     //得到布局中的开始加载的Button的对象
35     btn = (Button) findViewById(R.id.Btn);
36 }
37 }
```

此文件是 Activity 的代码文件，在其中第 4 行定义了一个打开系统中 Google 地图的 Button 对象。在第 12 行的 findView 方法中得到了布局中的所有控件，具体实现在第 33~36 行。在第 14 行通过 setListener 方法设置控件的监听器，具体方法在 17~31 行实现。其中当用户单击 Btn 按钮的时候，通过北京天坛公园的经纬度定义了一个 Uri 对象，然后定义 intent，设置 action 为 Intent.ACTION_VIEW，然后数据设置了之前的 uri 对象，通过 startActivity 启动系统中的 Google 地图，并且以天坛公园的经纬度坐标为地图的中心的坐标。

4．实例扩展

本实例可以根据经纬度打开 Google 地图进行显示，需要注意的地方有如下两点：

- 希望在 Google 地图上显示地图，所以要保证你的手机上安装了 Google 地图应用。
- 经纬度如何获得，这里提示一下，如果希望在 Google 地图上显示，那么地点的经纬度最好在网页的 Google 地图上通过右击地图的某一个点，显示经纬度的方式获得。这里说明一下，对于经纬度来说 Google 地图和百度地图的计算是有一定的差别的，也就是相同的一个位置点在 Google 地图上和在百度地图上经纬度可能有些偏差，这点希望注意。

范例 111　启动 Google 地图进行路径规划

1．实例简介

在我们使用应用的过程中，经常会使用到用户单击某按钮的时候打开了手机上的 Google 地图，并进行了路径规划。例如，在一些在团购客户端中，用户购买相应的餐券，希望去此店消费，但是又不知道如何去，这时候单击导航按钮，在地图中显示了从指定位置到此商家的规划路线图。遇到这样的功能，我们一般是当用户进行某个操作的时候，例如，单击某个按钮的时候，通过 Intent 打开手机中的 Google 地图，并且根据起始点和终止点的坐标进行路径规划。本例子就带领大家来实现一个单击按钮打开系统的 Google 地图程序进行路径规划。

2. 运行效果

该实例运行效果如图 5.20 所示。

图 5.20　启动 Google 地图进行路径规划

3. 实例程序讲解

在本实例中,提供提示用户要单击启动 Google 地图进行路径规划的标签框和一个单击打开 Google 地图进行路径规划的按钮,然后当用户单击此按钮时,打开系统安装的 Google 地图进行路径规划。想要实现本实例效果,首先修改 res/layout/activity_main.xml 文件,代码如下:

```xml
01 <?xml version="1.0" encoding="utf-8"?>
02 <!-- 定义当前布局的基本 LinearLayout -->
03 <LinearLayout xmlns:android="http://schemas.android.com/apk/res/android"
04     android:layout_width="fill_parent"
05     android:layout_height="fill_parent"
06     android:orientation="vertical" >
07
08     <!-- 定义提示用户单击启动谷歌地图进行路径规划的标签控件 -->
09     <TextView
10         android:id="@+id/Tv"
11         android:layout_width="match_parent"
12         android:layout_height="wrap_content"
13         android:hint="单击启动谷歌地图进行路径规划"
14         />
15
16     <!-- 定义用户启动谷歌地图进行路径规划的按钮控件 -->
17     <Button
18         android:id="@+id/Btn"
19         android:layout_width="match_parent"
20         android:layout_height="wrap_content"
21         android:text="启动谷歌地图进行路径规划"
22         />
23
24 </LinearLayout>
```

这是我们的 Activity 的布局文件。在其中第 9~14 行定义了一个 TextView 控件,用来提醒用户单击启动 Google 地图进行路径规划。在第 17~22 行定义一个 Button 控件,当用户单击的时候打开系统安装的 Google 地图进行路径规划。

然后修改 src/com.wyl.example/MainActivity.java 文件，代码如下：

```java
01  //定义了本实例的主要 Activity
02  public class MainActivity extends Activity {
03      //定义布局中的启动谷歌地图路径规划的 Button 控件
04      private Button btn;
05
06      @Override
07      protected void onCreate(Bundle savedInstanceState) {
08          super.onCreate(savedInstanceState);
09          //设置当前 Activity 的布局文件为 activity_main
10          setContentView(R.layout.activity_main);
11          //得到浏览器中的控件对象
12          findView();
13          //设置对象的监听器
14          setListener();
15      }
16
17      private void setListener() {
18          //设置 btn 的单击监听器
19          btn.setOnClickListener(new OnClickListener() {
20              @Override
21              public void onClick(View v) {
22                  //构造路径规划的 uri,saddr 为起始点的经纬度,daddr 为终止点的经纬度,
                        hl 为语言
23                  Uri uri =
                        Uri.parse("https://maps.google.com/maps?f=d&saddr=
                        39.88445+116.257217&daddr=39.985538+116.544921&hl=cn");
24                  //定义 intent 对象设置 action 为 Intent.ACTION_VIEW
25                  //并且数据设置为定义的 uri
26                  Intent it = new Intent(Intent.ACTION_VIEW,uri);
27                  //启动 activity
28                  startActivity(it);
29              }
30          });
31      }
32
33      private void findView() {
34          //得到布局中的开始加载的 Button 的对象
35          btn = (Button) findViewById(R.id.Btn);
36      }
37  }
```

此文件是 Activity 的代码文件，在其中第 4 行定义了一个打开系统中 Google 地图进行路径规划的 Button 对象。在第 12 行的 findView 方法中得到了布局中的所有控件，具体实现在第 33~36 行。在第 14 行通过 setListener 方法设置控件的监听器，具体方法在 17~31 行实现。其中当用户单击 Btn 按钮的时候，通过 https://maps.google.com/maps 加上起始坐标和终止坐标构成了 Google 导航的 Uri，然后定义 intent，设置 action 为 Intent.ACTION_VIEW，然后数据设置了之前的 uri 对象，通过 startActivity 启动系统中的 Google 地图，并且进行了给出的起始点和终止点的路径规划。

4．实例扩展

本实例可以根据经纬度打开 Google 地图进行路径规划，需要注意的地方有如下两点：
❑ 希望在 Google 地图上进行路径规划，所以要保证你的手机上安装了 Google 地图

应用。
- 经纬度如何获得，如果希望在 Google 地图上显示，那么地点的经纬度最好在网页的 Google 地图上通过右击地图的某一个点，显示经纬度的方式获得。这里说明一下，对于经纬度来说 Google 地图和百度地图的计算是有一定的差别的，也就是相同的一个位置点在 Google 地图上和在百度地图上经纬度可能有些偏差，这点希望注意。

5.2 Android 中自定义 Intent 使用

范例 112 登录页面功能

1．实例简介

在我们使用应用的过程中，经常会使用到用户登录的功能，当用户输入正确的用户名密码后，程序跳转到应用首页显示欢迎信息，如果用户输入错误的用户名密码，则程序不进行跳转，并且提示用户名密码错误。这样的功能，我们一般是当用户单击登录按钮的时候得到用户名密码并进行判断是否为合法，然后跳转相应的 Activity 进行页面显示。本例子就带领大家来实现一个具有用户名密码判断的登录功能的实例。

2．运行效果

该实例运行效果如图 5.21 所示。

图 5.21 登录页面功能的实现

3．实例程序讲解

在本实例中，首先显示登录页面，提示用户输入用户名密码，然后单击登录按钮进行用户名密码合法性判断，如果正确则跳转到欢迎页面。想要实现本实例效果，首先修改 res/layout/activity_main.xml 文件，代码如下：

```
01  <?xml version="1.0" encoding="utf-8"?>
02  <!-- 定义当前布局的基本 LinearLayout -->
03  <LinearLayout xmlns:android="http://schemas.android.com/apk/res/android"
04      android:layout_width="fill_parent"
05      android:layout_height="fill_parent"
```

```xml
06        android:orientation="vertical" >
07
08    <LinearLayout
09        android:layout_width="match_parent"
10        android:layout_height="wrap_content"
11        android:orientation="horizontal" >
12
13        <!-- 定义用户名标签控件 -->
14        <TextView
15            android:layout_width="match_parent"
16            android:layout_height="wrap_content"
17            android:layout_weight="2"
18            android:text="用户名:" />
19
20        <!-- 定义用户名输入控件 -->
21        <EditText
22            android:id="@+id/EtName"
23            android:layout_width="match_parent"
24            android:layout_height="wrap_content"
25            android:layout_weight="1"
26            android:hint="用户名" />
27    </LinearLayout>
28
29    <LinearLayout
30        android:layout_width="match_parent"
31        android:layout_height="wrap_content"
32        android:orientation="horizontal" >
33
34        <!-- 定义密码标签控件 -->
35        <TextView
36            android:layout_width="match_parent"
37            android:layout_height="wrap_content"
38            android:layout_weight="2"
39            android:text="密码:" />
40        <!-- 定义密码输入控件 -->
41        <EditText
42            android:id="@+id/EtPwd"
43            android:layout_width="match_parent"
44            android:layout_height="wrap_content"
45            android:layout_weight="1"
46            android:numeric="integer"
47            android:password="true"
48            android:hint="密码" />
49    </LinearLayout>
50
51    <!-- 定义用户登录的按钮控件 -->
52    <Button
53        android:id="@+id/Btn"
54        android:layout_width="match_parent"
55        android:layout_height="wrap_content"
56        android:text="登录" />
57
58 </LinearLayout>
```

这是我们的 Activity 的布局文件。在其中第 8~27 行定义了用户名的输入框。在第 29~49 行定义了密码的输入框。在第 52~56 行定义了登录的 Button 控件，当用户单击的时候打开获取用户输入的用户名和密码进行登录。

然后修改 src/com.wyl.example/MainActivity.java 文件，代码如下：

```java
01  //定义了本实例的主要Activity
02  public class MainActivity extends Activity {
03      //定义布局中的登录Button控件
04      private Button btn;
05      //定义布局中的用户名输入框控件
06      private EditText EtName;
07      //定义布局中的密码输入框控件
08      private EditText EtPwd;
09
10      @Override
11      protected void onCreate(Bundle savedInstanceState) {
12          super.onCreate(savedInstanceState);
13          //设置当前Activity的布局文件为activity_main
14          setContentView(R.layout.activity_main);
15          //得到浏览器中的控件对象
16          findView();
17          //设置对象的监听器
18          setListener();
19      }
20
21      private void setListener() {
22          //设置btn的单击监听器
23          btn.setOnClickListener(new OnClickListener() {
24              @Override
25              public void onClick(View v) {
26                  //得到用户输入的用户名
27                  String name = EtName.getText().toString();
28                  //得到用户输入的密码
29                  String pwd = EtPwd.getText().toString();
30
31                  //判断用户名密码是否合法
32                  if (isUser(name,pwd)) {
33                      //如果合法,程序跳转到欢迎界面
34                      Intent i = new Intent(MainActivity.this,SecActivity.class);
35                      //设置传递的参数为用户名
36                      i.putExtra("USERNAME", name);
37                      //启动activity
38                      startActivity(i);
39                  }
40                  else{
41                      //如果用户名密码不合法,显示toast提示用户
42                      Toast.makeText(MainActivity.this, "用户名密码错误",
43                              Toast.LENGTH_SHORT).show();
44                  }
45              }
46          });
47      }
48
49      //判断用户名密码是否合法
50      protected boolean isUser(String name, String pwd) {
51          //这里简单的模拟了用户名密码的判断,实际应该从数据库中判断
52          if ("admin".equals(name) && "123".equals(pwd)) {
53              return true;
54          }
55          else{
56              return false;
57          }
58  }
```

```
59
60   private void findView() {
61       //得到布局中的开始加载的 Button 的对象
62       btn = (Button) findViewById(R.id.Btn);
63       //得到布局中的用户名 EditText 的对象
64       EtName = (EditText) findViewById(R.id.EtName);
65       //得到布局中的密码 EditText 的对象
66       EtPwd = (EditText) findViewById(R.id.EtPwd);
67   }
68  }
```

此文件是 Activity 的代码文件,在其中第 4 行定义了登录的 Button 对象。在第 6、8 行分别定义了用户名和密码的输入 EditText 对象。在第 16 行通过 findView 方法得到了布局中的所有控件,具体实现在第 60~67 行。在第 18 行通过 setListener 方法设置控件的监听器,具体方法在 21~47 行实现。其中当用户单击 Btn 按钮的时候,首先获得用户输入的用户名和密码,然后进行检查是否为合法用户,这里通过 isUser 方法进行检查在第 50~58 行。当用户名和密码合法时跳转到 secActivity 页面显示,并且携带用户名参数过去,否则通过 Toast 提醒用户名密码错误。

然后新建 res/layout/activity_sec.xml 文件,代码如下:

```
01  <?xml version="1.0" encoding="utf-8"?>
02  <!-- 定义当前布局的基本 LinearLayout -->
03  <LinearLayout xmlns:android="http://schemas.android.com/apk/res/android"
04      android:layout_width="fill_parent"
05      android:layout_height="fill_parent"
06      android:orientation="vertical" >
07
08      <!-- 定义欢迎标签控件 -->
09
10      <TextView
11          android:id="@+id/Tv"
12          android:layout_width="match_parent"
13          android:layout_height="wrap_content"
14          android:text="恭喜您已登录" />
15
16  </LinearLayout>
```

此文件为欢迎界面的布局,这里为了简单只设置了一个 TextView 控件。

然后新建 src/com.wyl.example/SecActivity 文件,代码如下:

```
01  //定义了本实例的主要 Activity
02  public class SecActivity extends Activity {
03
04  @Override
05  protected void onCreate(Bundle savedInstanceState) {
06      super.onCreate(savedInstanceState);
07      //设置当前 Activity 的布局文件为 activity_sec
08      setContentView(R.layout.activity_sec);
09
10      //得到传递过来的 Intent
11      Intent i = getIntent();
12      //取得传递过来的 Intent 中携带的 USERNAME 数据
13      String name = i.getStringExtra("USERNAME");
14      //得到布局中的 TextView 控件
15      TextView tv = (TextView)findViewById(R.id.Tv);
```

```
16          //设置 TextView 控件的显示文字
17          tv.setText(name+",欢迎您登录本系统");
18    }
19 }
```

这是我们的欢迎界面,在其中主要的功能是通过 getIntent 得到传过来的用户名参数,然后给当前页面的 TextView 控件设置相应的内容显示。

4. 实例扩展

本实例涉及到了两个 Activity,通过 Intent 来传递输入,当然每当应用程序中添加了一个 Activity 的时候都要在 Manifest 文件中添加如下代码,声明新添加的 Activity:

```
<activity android:name=".SecActivity"></activity>
```

其中 name 参数后面是加入的 Activity 的类名。

范例 113 注册页面功能

1. 实例简介

在我们第一次使用一个应用程序的时候,经常会使用到用户注册的功能,当用户输入希望注册的用户名密码后,程序跳转到应用首页显示注册成功。这样的功能,我们一般是当用户单击注册按钮的时候得到用户名密码并构造合法的用户对象,然后跳转相应的 Activity,并且携带用户数据对象。本例子就带领大家来实现一个注册页面功能的实例。

2. 运行效果

该实例运行效果如图 5.22 所示。

图 5.22 注册页面功能的实现

3. 实例程序讲解

在本实例中,首先显示注册信息页面,提示用户输入用户名密码邮箱,然后单击注册按钮进行用户注册,然后携带 user 对象到信息显示页面。想要实现本实例效果,首先修改 res/layout/activity_main.xml 文件,代码如下:

```xml
01 <?xml version="1.0" encoding="utf-8"?>
02 <!-- 定义当前布局的基本 LinearLayout -->
03 <LinearLayout xmlns:android="http://schemas.android.com/apk/res/android"
04     android:layout_width="fill_parent"
05     android:layout_height="fill_parent"
06     android:orientation="vertical" >
07
08     <LinearLayout
09         android:layout_width="match_parent"
10         android:layout_height="wrap_content"
11         android:orientation="horizontal" >
12
13         <!-- 定义用户名标签控件 -->
14         <TextView
15             android:layout_width="match_parent"
16             android:layout_height="wrap_content"
17             android:layout_weight="2"
18             android:text="用户名: " />
19
20         <!-- 定义用户名输入控件 -->
21         <EditText
22             android:id="@+id/EtName"
23             android:layout_width="match_parent"
24             android:layout_height="wrap_content"
25             android:layout_weight="1"
26             android:hint="用户名" />
27     </LinearLayout>
28
29     <LinearLayout
30         android:layout_width="match_parent"
31         android:layout_height="wrap_content"
32         android:orientation="horizontal" >
33
34         <!-- 定义密码标签控件 -->
35         <TextView
36             android:layout_width="match_parent"
37             android:layout_height="wrap_content"
38             android:layout_weight="2"
39             android:text="密码: " />
40         <!-- 定义密码输入控件 -->
41         <EditText
42             android:id="@+id/EtPwd"
43             android:layout_width="match_parent"
44             android:layout_height="wrap_content"
45             android:layout_weight="1"
46             android:hint="密码" />
47     </LinearLayout>
48
49     <LinearLayout
50         android:layout_width="match_parent"
51         android:layout_height="wrap_content"
52         android:orientation="horizontal" >
53
54         <!-- 定义邮箱标签控件 -->
55         <TextView
56             android:layout_width="match_parent"
57             android:layout_height="wrap_content"
58             android:layout_weight="2"
59             android:text="邮箱: " />
```

```
60        <!-- 定义邮箱输入控件 -->
61        <EditText
62            android:id="@+id/EtEmail"
63            android:layout_width="match_parent"
64            android:layout_height="wrap_content"
65            android:layout_weight="1"
66            android:hint="邮箱" />
67    </LinearLayout>
68
69
70    <!-- 定义用户注册的按钮控件 -->
71    <Button
72        android:id="@+id/Btn"
73        android:layout_width="match_parent"
74        android:layout_height="wrap_content"
75        android:text="注册" />
76
77 </LinearLayout>
```

这是我们的 Activity 的布局文件。在其中第 8～27 行定义了用户名的输入框。在第 29～47 行定义了密码的输入框。在第 49～67 行定义了邮箱的输入框。在第 71～75 行定义了注册的 Button 控件，当用户单击的时候打开用户信息显示页面。

然后修改 src/com.wyl.example/MainActivity.java 文件，代码如下：

```
01 //定义了本实例的主要 Activity
02 public class MainActivity extends Activity {
03     //定义布局中的注册 Button 控件
04     private Button btn;
05     //定义布局中的用户名输入框控件
06     private EditText EtName;
07     //定义布局中的密码输入框控件
08     private EditText EtPwd;
09     //定义布局中的邮箱输入框控件
10     private EditText EtEmail;
11
12     @Override
13     protected void onCreate(Bundle savedInstanceState) {
14         super.onCreate(savedInstanceState);
15         //设置当前 Activity 的布局文件为 activity_main
16         setContentView(R.layout.activity_main);
17         //得到浏览器中的控件对象
18         findView();
19         //设置对象的监听器
20         setListener();
21     }
22
23     private void setListener() {
24         //设置 btn 的单击监听器
25         btn.setOnClickListener(new OnClickListener() {
26             @Override
27             public void onClick(View v) {
28                 //得到用户输入的用户名
29                 String name = EtName.getText().toString();
30                 //得到用户输入的密码
31                 String pwd = EtPwd.getText().toString();
32                 //得到用户输入的邮箱
33                 String email = EtEmail.getText().toString();
```

```
34
35                    //判断用户名密码邮箱是否为空
36                    if (!"".equals(name)
37                            && !"".equals(pwd)
38                            && !"".equals(email)) {
39                        //定义 User 对象
40                        User u = new User();
41                        //设置 Email
42                        u.setEamil(email);
43                        //设置用户名
44                        u.setName(name);
45                        //设置密码
46                        u.setPwd(pwd);
47
48                        //如果合法,程序跳转到欢迎界面
49                        Intent i = new Intent(MainActivity.this,SecActivity.class);
50                        //设置传递的参数为用户名
51                        i.putExtra("USER", u);
52                        //启动 activity
53                        startActivity(i);
54                    }
55                    else{
56                        //如果用户名密码不合法,显示 toast 提示用户
57                        Toast.makeText(MainActivity.this,
58                                "输入信息为空",Toast.LENGTH_SHORT).show();
59                    }
60                }
61           });
62      }
63
64      private void findView() {
65           //得到布局中的开始加载的 Button 的对象
66           btn = (Button) findViewById(R.id.Btn);
67           //得到布局中的用户名 EditText 的对象
68           EtName = (EditText) findViewById(R.id.EtName);
69           //得到布局中的密码 EditText 的对象
70           EtPwd = (EditText) findViewById(R.id.EtPwd);
71           //得到布局中的邮箱 EditText 的对象
72           EtEmail = (EditText) findViewById(R.id.EtEmail);
73      }
74 }
```

此文件是 Activity 的代码文件,在其中第 4 行定义了注册的 Button 对象。在第 6、8、10 行分别定义了用户名、密码和邮箱的输入 EditText 对象。在第 18 行通过 findView 方法中得到了布局中的所有控件,具体实现在第 64~73 行。在第 20 行通过 setListener 方法设置控件的监听器,具体方法在 23~62 行实现。其中当用户单击 Btn 按钮的时候,首先获得用户输入的用户名、密码和邮箱,然后进行检查是否为空,如果不为空的话,构成 User 对象,并且设置相应的属性值,并且通过 Intent 跳转到 SecActivity 页面。如果为空的话,提示用户输入信息。

然后新建 res/layout/activity_sec.xml 文件,代码如下:

```
01 <?xml version="1.0" encoding="utf-8"?>
02 <!-- 定义当前布局的基本 LinearLayout -->
03 <LinearLayout xmlns:android="http://schemas.android.com/apk/res/android"
04     android:layout_width="fill_parent"
05     android:layout_height="fill_parent"
```

```
06        android:orientation="vertical" >
07
08        <!-- 定义欢迎标签控件 -->
09
10        <TextView
11            android:id="@+id/Tv"
12            android:layout_width="match_parent"
13            android:layout_height="wrap_content"
14
15    </LinearLayout>
```

此文件为用户信息显示界面的布局，这里为了简单只设置了一个 TextView 控件。

然后新建 src/com.wyl.example/SecActivity 文件，代码如下：

```
01  //定义了本实例的主要 Activity
02  public class SecActivity extends Activity {
03
04  @Override
05  protected void onCreate(Bundle savedInstanceState) {
06      super.onCreate(savedInstanceState);
07      setContentView(R.layout.activity_sec);
08
09      //得到传递过来的 Intent
10      Intent i = getIntent();
11      //取得传递过来的 Intent 中携带的 USERNAME 数据
12      User u = (User) i.getSerializableExtra("USER");
13      //得到布局中的 TextView 控件
14      TextView tv = (TextView)findViewById(R.id.Tv);
15      //设置 TextView 控件的显示文字
16      tv.setText("用户的注册信息为："+u.toString());
17  }
18  }
```

这是我们的欢迎界面，在其中主要的功能是通过 getIntent 得到传过来的用户参数，然后给当前页面的 TextView 控件设置相应的内容显示。

然后新建 src/com.wyl.example/User.java 文件，代码如下：

```
01  //定义实体类 User 用户类
02  public class User implements Serializable{
03  //定义序列化的 id
04  private static final long serialVersionUID = 1L;
05  //用户名属性
06  public String name;
07  //用户密码属性
08  public String pwd;
09  //用户名邮箱属性
10  public String eamil;
11  /**
12   * 获取用户名
13   */
14  public String getName() {
15      return name;
16  }
17  /**
18   * 设置用户名
19   */
20  public void setName(String name) {
21      this.name = name;
```

```
22   }
23   /**
24    *获取用户密码
25    */
26   public String getPwd() {
27       return pwd;
28   }
29   /**
30    * 设置用户密码
31    */
32   public void setPwd(String pwd) {
33       this.pwd = pwd;
34   }
35   /**
36    * 获取 Email
37    */
38   public String getEamil() {
39       return eamil;
40   }
41   /**
42    * 设置用户邮箱
43    */
44   public void setEamil(String eamil) {
45       this.eamil = eamil;
46   }
47   /*
48    * 格式化显示 User 用户的所有属性
49    */
50   @Override
51   public String toString() {
52       return "User [name=" + name + ", pwd=" + pwd + ", eamil=" + eamil + "]";
53   }
54 }
```

这里定义了用户的实体类 User，在此类中定义了用户名、密码和邮箱三个属性，并且分别实现了 set 和 get 方法，并且重写了 tostring 方法。

4．实例扩展

本实例涉及到了两个 Activity，通过 Intent 来传递 User 类的对象，这时候在 Intent 中传递的话无法直接传递类的对象，所以在本例中 User 实现了 Serializable 接口，这样才可以在 Intent 中进行传递。

范例 114　获取随机验证码功能

1．实例简介

我们在使用某些应用程序的时候，经常会使用到从另外的页面获取验证码的功能。这样的功能，我们一般是当用户单击获取验证码按钮的时候，跳转到验证码生成界面，然后生成验证码后返回当前页面显示。本例子就带领大家来实现一个获取验证码功能的实例。

2．运行效果

该实例运行效果如图 5.23 所示。

图 5.23　获取随机验证码的功能

3．实例程序讲解

在本实例中，首先显示一个 TextView 的提示框，然后定义了一个获取验证码的 Button 控件，当用户单击按钮的时候获取验证码并显示。想要实现本实例效果，首先修改 res/layout/activity_main.xml 文件，代码如下：

```xml
01  <?xml version="1.0" encoding="utf-8"?>
02  <!-- 定义当前布局的基本 LinearLayout -->
03  <LinearLayout xmlns:android="http://schemas.android.com/apk/res/android"
04      android:layout_width="fill_parent"
05      android:layout_height="fill_parent"
06      android:orientation="vertical" >
07  
08      <!-- 定义显示随机数的标签控件 -->
09      <TextView
10          android:id="@+id/Tv"
11          android:layout_width="match_parent"
12          android:layout_height="wrap_content"
13          android:text="单击获取随机数验证码" />
14  
15      <!-- 定义获取随机数的按钮控件 -->
16      <Button
17          android:id="@+id/Btn"
18          android:layout_width="match_parent"
19          android:layout_height="wrap_content"
20          android:text="获取随机数验证码" />
21  
22  </LinearLayout>
```

这是我们的 Activity 的布局文件。在其中第 9～13 行定义了提示标签控件。在第 16～20 行定义了获取验证码的 Button 控件，当用户单击的时候获取随机验证码。

然后修改 src/com.wyl.example/MainActivity.java 文件，代码如下：

```
01  //定义了本实例的主要 Activity
02  public class MainActivity extends Activity {
03      //定义布局中的登录 Button 控件
04      private Button btn;
05      //定义布局中的标签 TextView 控件
```

```java
06    private TextView Tv;
07    //定义传递的请求码
08    protected static final int MyrequestCode = 100;
09
10    @Override
11    protected void onCreate(Bundle savedInstanceState) {
12        super.onCreate(savedInstanceState);
13        //设置当前Activity的布局文件为activity_main
14        setContentView(R.layout.activity_main);
15        //得到浏览器中的控件对象
16        findView();
17        //设置对象的监听器
18        setListener();
19    }
20
21    private void setListener() {
22        //设置btn的单击监听器
23        btn.setOnClickListener(new OnClickListener() {
24            @Override
25            public void onClick(View v) {
26                //程序跳转到获取随机验证码界面
27                Intent i = new Intent(MainActivity.this, SecActivity.class);
28                //启动Activity
29                startActivityForResult(i, MyrequestCode);
30            }
31        });
32    }
33
34    private void findView() {
35        //得到布局中的开始加载的Button的对象
36        btn = (Button) findViewById(R.id.Btn);
37        //得到布局中的TextView的对象
38        Tv = (TextView) findViewById(R.id.Tv);
39    }
40
41    /*
42     * 获取验证码后,返回的回调函数
43     */
44    @Override
45    protected void onActivityResult(int requestCode, int resultCode, Intent data) {
46        //TODO Auto-generated method stub
47        super.onActivityResult(requestCode, resultCode, data);
48        //如果请求码与当前页面的请求码相同
49        if (requestCode == MyrequestCode) {
50            //返回的RANDOM数据值
51            int res = data.getIntExtra("RANDOM", 0);
52            //在Tv中设置随机数值
53            Tv.setText("您获取的随机数是: "+res);
54        }
55    }
56 }
```

此文件是 Activity 的代码文件,在其中第 4 行定义了注册的 Button 对象。在第 6 行定义了提示 TextView 对象。在第 16 行通过 findView 方法中得到了布局中的所有控件,具体实现在第 34~39 行。在第 18 行通过 setListener 方法设置控件的监听器,具体方法在 21~

32 行实现。其中当用户单击 Btn 按钮的时候，定义 Intent 对象，并且设置跳转页面为 SecActivity 页面，然后使用 startActivityForResult 方法启动页面，获取随机数后程序回调 onActivityResult，在第 45～55 行实现，拿到随机数并在 TextView 中显示。

然后新建 res/layout/activity_sec.xml 文件，代码如下：

```xml
01 <?xml version="1.0" encoding="utf-8"?>
02 <!-- 定义当前布局的基本 LinearLayout -->
03 <LinearLayout xmlns:android="http://schemas.android.com/apk/res/android"
04     android:layout_width="fill_parent"
05     android:layout_height="fill_parent"
06     android:orientation="vertical" >
07
08     <!-- 定义标签控件 -->
09
10     <TextView
11         android:id="@+id/Tv"
12         android:layout_width="match_parent"
13         android:layout_height="wrap_content"/>
14
15 </LinearLayout>
```

此文件为用户信息显示界面的布局，这里为了简单只设置了一个 TextView 控件。

然后新建 src/com.wyl.example/SecActivity 文件，代码如下：

```java
01 //定义了本实例的主要 Activity
02 public class SecActivity extends Activity {
03
04 @Override
05 protected void onCreate(Bundle savedInstanceState) {
06     super.onCreate(savedInstanceState);
07     setContentView(R.layout.activity_sec);
08
09     //得到传递过来的 Intent
10     Intent i = getIntent();
11     //产生 0-1000 的整数随机数
12     int r=(int)(Math.random()*1000);
13     //设置得到的随机数为 Intent 的返回参数
14     i.putExtra("RANDOM", r);
15     //设置返回的 Intent
16     setResult(MainActivity.MyrequestCode, i);
17     //结束当前 Activity
18     finish();
19 }
20 }
```

这是我们的欢迎界面，在其中主要的功能是通过 getIntent 得到 Intent 对象，其次通过 random 方法获取随机数，然后设置到 Intent 中，然后调用 setResult 方法返回设置的 Intent，最后通过 finish 方法结束当前的 Activity。

4．实例扩展

本实例用到了 startActivityForResult 方法，此方法就是为了在另一个页面中获取信息，然后再返回当前页面的时候使用的，再返回后系统会自动调用 onActivityResult 方法来进行返回数据的处理。

范例 115 模拟站内搜索

1. 实例简介

在我们使用某些应用程序的时候，经常会使用到一些自定义 Action 启动程序内部的页面的效果。这样的功能，我们一般是当用户单击按钮的时候，通过一串固定的文字串，进行页面的跳转。本例子就带领大家来使用固定的 Action 串来实现页面跳转的实例。

2. 运行效果

该实例运行效果如图 5.24 所示。

图 5.24 模拟站内搜索功能

3. 实例程序讲解

在本实例中，首先显示一个 TextView 的提示框，然后定义了一个搜索关键字的输入框，然后定义了跳转搜索界面的 Button 控件，当用户单击按钮的时候跳转到搜索页面，并且携带搜索关键字。想要实现本实例效果，首先修改 res/layout/activity_main.xml 文件，代码如下：

```
01  <?xml version="1.0" encoding="utf-8"?>
02  <!-- 定义当前布局的基本 LinearLayout -->
03  <LinearLayout xmlns:android="http://schemas.android.com/apk/res/android"
04      android:layout_width="fill_parent"
05      android:layout_height="fill_parent"
06      android:orientation="vertical" >
07
08      <!-- 定义提示用户输入搜索关键字的标签控件 -->
09      <TextView
10          android:id="@+id/Tv"
11          android:layout_width="match_parent"
12          android:layout_height="wrap_content"
13          android:text="单击跳转站内搜索效果" />
14
15      <!-- 定义提示用户输入搜索关键字的标签控件 -->
16      <EditText
17          android:id="@+id/Et"
18          android:layout_width="match_parent"
```

```
19        android:layout_height="wrap_content"
20        android:hint="请输入搜索关键字" />
21
22    <!-- 定义单击跳转站内搜索页面的按钮控件 -->
23    <Button
24        android:id="@+id/Btn"
25        android:layout_width="match_parent"
26        android:layout_height="wrap_content"
27        android:text="单击跳转站内搜索页面" />
28
29 </LinearLayout>
```

这是我们的 Activity 的布局文件。在其中第 9~13 行定义了提示标签控件。在第 16~20 行定义了用户输入搜索关键字的输入框。第 23~27 行，定义了跳转搜索页面的 Button 控件，当用户单击的时候跳转到搜索页面。

然后修改 src/com.wyl.example/MainActivity.java 文件，代码如下：

```
01 //定义了本实例的主要 Activity
02 public class MainActivity extends Activity {
03    //定义布局中的登录 Button 控件
04    private Button btn;
05    //定义布局中的搜索关键字的 EditText 控件
06    private EditText Et;
07
08    @Override
09    protected void onCreate(Bundle savedInstanceState) {
10        super.onCreate(savedInstanceState);
11        //设置当前 Activity 的布局文件为 activity_main
12        setContentView(R.layout.activity_main);
13        //得到浏览器中的控件对象
14        findView();
15        //设置对象的监听器
16        setListener();
17    }
18
19    private void setListener() {
20        //设置 btn 的单击监听器
21        btn.setOnClickListener(new OnClickListener() {
22            @Override
23            public void onClick(View v) {
24                String key = Et.getText().toString();
25
26                if (!"".equals(key)) {
27                    //搜索关键字不为空，程序跳转到搜索界面
28                    Intent i = new Intent();
29                    i.setAction("wyl.com.test");
30                    i.putExtra("KEY",key);
31                    //启动 activity
32                    startActivity(i);
33                }
34                else{
35                    //如果用户名密码不合法，显示 Toast 提示用户
36                    Toast.makeText(MainActivity.this,
37                        "搜索关键字为空",Toast.LENGTH_SHORT).show();
38                }
39
40            }
```

```
41          });
42      }
43
44      private void findView() {
45          //得到布局中的开始加载的 Button 的对象
46          btn = (Button) findViewById(R.id.Btn);
47          //得到布局中的 EditText 的对象
48          Et = (EditText) findViewById(R.id.Et);
49      }
50  }
```

此文件是 Activity 的代码文件,在其中第 4 行定义了跳转搜索页面的 Button 对象。在第 6 行定义了用户输入搜索关键字的 EditText 框。在第 14 行通过 findView 方法中得到了布局中的所有控件,具体实现在第 44~49 行。在第 16 行通过 setListener 方法设置控件的监听器,具体方法在 19~42 行实现。其中当用户单击 Btn 按钮的时候,获取用户输入的关键字,如果为空则提示用户再次输入,如果不为空则定义 Intent 对象,设置 Intent 的 Action 为 wyl.com.test 字符串。

然后新建 res/layout/activity_sec.xml 文件,代码如下:

```xml
01  <?xml version="1.0" encoding="utf-8"?>
02  <!-- 定义当前布局的基本 LinearLayout -->
03  <LinearLayout xmlns:android="http://schemas.android.com/apk/res/android"
04      android:layout_width="fill_parent"
05      android:layout_height="fill_parent"
06      android:orientation="vertical" >
07
08      <!-- 定义标签控件 -->
09
10      <TextView
11          android:id="@+id/Tv"
12          android:layout_width="match_parent"
13          android:layout_height="wrap_content"/>
14
15  </LinearLayout>
```

此文件为用户信息显示界面的布局,这里为了简单只设置了一个 TextView 控件。

然后新建 src/com.wyl.example/SecActivity 文件,代码如下:

```
01  //定义了本实例的主要 Activity
02  public class SecActivity extends Activity {
03
04  @Override
05  protected void onCreate(Bundle savedInstanceState) {
06      super.onCreate(savedInstanceState);
07      setContentView(R.layout.activity_sec);
08
09      //得到传递过来的 intent
10      Intent i = getIntent();
11      //设置得到的传过来的 intent 的数据
12      String key = i.getStringExtra("KEY");
13      //得到页面中的 TextView 控件
14      TextView tv = (TextView)findViewById(R.id.Tv);
15      //设置 TextView 的值为传过来的搜索关键字
16      tv.setText("您要搜索的关键字是: "+key);
17  }
18  }
```

这是我们的搜索界面，在其中主要的功能是通过 getIntent 得到 Intent 对象，然后通过 getStringExtra 方法获取传递过来的搜索关键字的值，然后在当前页面的 TextView 控件中显示。

4．实例扩展

本实例最重要的一点就是，如何对应 wyl.com.test 和 SecActivity 的关系呢？之前给大家说过，只要在工程中添加一个 Activity，那么就要在 Manifest 文件中注册此 Activity，那么在注册 Activity 的是采用如下方法注册，就可以绑定一个 action 的字符串了。

```xml
<activity android:name=".SecActivity">
    <intent-filter>
        <action android:name="wyl.com.test" />
        <category android:name="android.intent.category.DEFAULT" />
    </intent-filter>
</activity>
```

其中 intent-filter 代表 activity 的过滤器，其中 action 节点就是绑定的 activity 的启动字符串，category 代表此 activity 的类型为默认类型。只有在 Manifest 中设置了 action 字符串的 Activity，才可以通过当前方法来启动。

注意，这种方法启动 Activity，主要是为了给外部应用程序提供访问界面的方法，在程序的内部的话通过 action 和类名.class 启动 activity 的方法没有什么区别。

5.3 小　　结

在本章节中主要介绍了 Android 中 Intent 的常见使用方法，其中常见的应用有两种，一种是通过 Intent 调用系统的对应功能界面。另一种是通过调用自定义 Intent 来实现自身程序内部的组件之间的信息传递。本章的内容在实际项目开发使用的地方很多，他可以使用户通过你的应用程序快速的操作系统的功能模块，例如，单击某个按钮直接拨打电话，直接发送邮件等，这也是 Android 作为开放式系统相对于 iOS 系统的一个优势所在。下一章我们会讲述在 Android 中如何长期保存数据。

第 6 章　Android 的数据存储

上一章了解了 Android 中 Intent 的常见使用方法,其应用方法主要有两种:一种是通过 Android 系统提供的 Intent 调用系统的对应功能界面;另一种是通过调用自定义 Intent 来实现自身程序内部的组件之间的信息传递,并且可以携带数据。所以通过之前几章的学习我们已经能够独立完成一些完整的应用了,但是这对于一个完整的应用来说还是完全不够的,因为应用程序大多数时间不会独立存在,它需要通过外部的资源来完善应用程序功能。例如,我们的应用程序数据不可能只保存在内存中,这样的话,我们在使用程序的时候就无法保存已有的数据;我们的应用程序有时候也需要获得系统的数据,如得到系统联系人和得到系统短信等。这也就说明应用程序不可能单独存在,它们需要和外部交互,包括文件或者系统的数据资源。那么本章就给大家介绍 Android 中的有关数据存储的内容。

Android 系统中的数据存储主要分为三大类:

第一类是保存程序的独立数据,包括保存配置信息、保存数据文件和保存数据库。这些用法使用起来比较简单,而且可以方便的保存用户数据下次使用。

第二类是调用系统的资源,这种方法使用的情况也比较多,包括得到系统联系人和得到系统短信等。这样就可以使应用和系统资源紧密结合起来,这样可以提升软件的可用性。

第三类是调用系统的资源文件,一般对于开发者来说,我们尽量希望程序的逻辑和程序的资源分开,这样在修改资源的时候不会影响逻辑的变化,反之亦然。所以我们在程序中可以包含资源文件,这样可以方便我们的使用。

本章主要通过各种实例来介绍 Android 中常用的数据资源的使用方法。希望读者阅读完本章内容后,能够在自己的应用中保存用户已有的数据,这是应用程序增加用户粘性的主要手段。

6.1　Android 中的文件操作

范例 116　可记住用户名密码的登录界面

1. 实例简介

在很多应用中当用户希望进行一些操作前首先需要用户登录,因为对于某些功能来说是指针对个别用户开放的。但是一般用户的用户名或者密码为了安全起见都是比较长或者比较复杂的,这对于用户在手机上输入用户名和密码造成了很大的障碍。例如,我在某网站的登录账号是 395928533@qq.com,而我的登录密码为 passw_ord111!,这样的话我要输入这样的用户名密码,就得一直在各种输入法之间进行切换。那我们就通过本实例带领大家一起来看制作一个可以记住用户输入的用户名密码的登录界面。

2. 运行效果

该实例运行效果如图 6.1 所示。

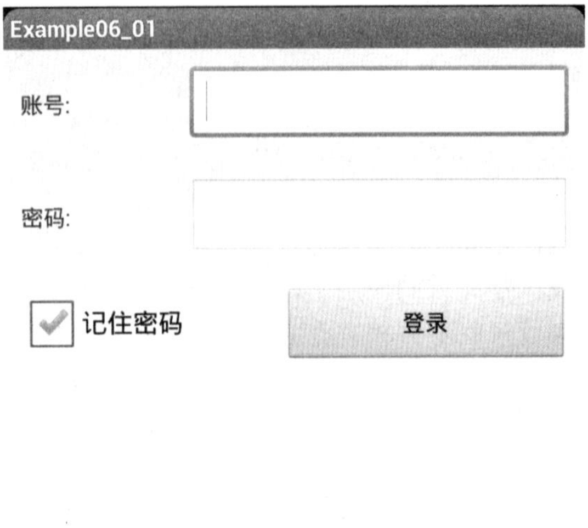

图 6.1 可记住用户名密码的登录界面

3. 实例程序讲解

在上例中用户输入框中输入希望用户名密码，然后单击登录即可登录，在登录之前如果用户勾选了记住密码的选择，则下次打开此界面的时候，程序会自动填写上次您输入的用户名密码。想要实现如上效果，首先修改我们建立的工程下的 res/layout/activity_main.xml 文件，代码如下：

```
01  <LinearLayout xmlns:android="http://schemas.android.com/apk/res/android"
02      xmlns:tools="http://schemas.android.com/tools"
03      android:id="@+id/LinearLayout1"
04      android:layout_width="match_parent"
05      android:layout_height="match_parent"
06      android:orientation="vertical"
07      tools:context=".MainActivity" >
08  <!-- 显示用户名输入行 -->
09      <LinearLayout
10          android:layout_width="match_parent"
11          android:layout_height="wrap_content"
12          android:layout_margin="10dp"
13          android:orientation="horizontal" >
14
15          <TextView
16              android:layout_width="wrap_content"
17              android:layout_height="wrap_content"
18              android:layout_weight="1"
19              android:layout_marginRight="5dp"
20              android:text="@string/tv_username" />
21
22          <EditText
23              android:id="@+id/et_username"
```

```xml
24          android:layout_weight="3"
25          android:layout_width="wrap_content"
26          android:layout_height="wrap_content" />
27      </LinearLayout>
28      <!-- 显示密码输入行 -->
29      <LinearLayout
30          android:layout_width="match_parent"
31          android:layout_height="wrap_content"
32          android:layout_margin="10dp"
33          android:orientation="horizontal" >
34
35          <TextView
36              android:layout_width="wrap_content"
37              android:layout_height="wrap_content"
38              android:layout_weight="1"
39              android:layout_marginRight="5dp"
40              android:text="@string/tv_password" />
41
42          <EditText
43              android:id="@+id/et_password"
44              android:layout_weight="3"
45              android:layout_width="wrap_content"
46              android:layout_height="wrap_content"
47              android:password="true" />
48      </LinearLayout>
49      <!-- 显示登录按钮行 -->
50      <LinearLayout
51          android:layout_width="match_parent"
52          android:layout_height="wrap_content"
53          android:layout_marginLeft="10dp"
54          android:layout_marginRight="10dp"
55          android:layout_marginTop="5dp"
56          android:orientation="horizontal" >
57
58          <CheckBox
59              android:id="@+id/cb_keeppsd"
60              android:layout_width="wrap_content"
61              android:layout_height="wrap_content"
62              android:layout_weight="1"
63              android:text="@string/cb_keeppsd"
64              android:checked="true"/>
65
66          <Button
67              android:id="@+id/btn_login"
68              android:layout_width="wrap_content"
69              android:layout_height="wrap_content"
70              android:layout_weight="3"
71              android:layout_marginLeft="20dp"
72              android:text="@string/btn_login" />
73      </LinearLayout>
74
75  </LinearLayout>
```

这是我们的 Activity 的布局文件，其中第 9～27 行定义了用户名输入的提示框控件和输入框控件，用来接收用户输入的用户名。第 29～48 行定义了密码输入的提示框控件和输入框控件，用来接收用户输入的密码。在第 58～64 行，定义了一个 CheckBox 控件，当用户单击登录按钮时如果用户勾选了此对话框，则可以记下用户输入的用户名和密码。

然后修改 src/com.wyl.example/MainActivity.java 文件，代码如下：

```java
001 //定义了本实例的主要Activity
002 public class MainActivity extends Activity {
003
004     private EditText mEtUserName;              //账号
005     private EditText mEtPassWord;              //密码
006     private CheckBox mCbKeepPsd;               //是否保存密码复选框
007     private Button mBtnLogin;                  //登录按钮
008
009     //声明一个SharedPreferences用于保存数据
010     private SharedPreferences mSpSettings = null;
011
012     private static final String PREFS_NAME = "NamePwd";
013
014     @Override
015     protected void onCreate(Bundle savedInstanceState) {
016         super.onCreate(savedInstanceState);
017         setContentView(R.layout.activity_main);
018         //得到布局中的控件
019         findView();
020         //绑定控件事件
021         setListener();
022         //获取数据
023         getData();
024     }
025
026     /**
027      * 绑定控件
028      */
029     private void findView() {
030         mEtUserName = (EditText) findViewById(R.id.et_username);
031         mEtPassWord = (EditText) findViewById(R.id.et_password);
032         mCbKeepPsd = (CheckBox) findViewById(R.id.cb_keeppsd);
033         mBtnLogin = (Button) findViewById(R.id.btn_login);
034     }
035
036     /**
037      * 为控件添加事件
038      */
039     private void setListener() {
040         //为登录按钮绑定事件
041         mBtnLogin.setOnClickListener(new OnClickListener() {
042
043             @Override
044             public void onClick(View arg0) {
045                 //判断用户名和密码
046                 if ("wyl".equals(mEtUserName.getText().toString())
047                         && "123".equals(mEtPassWord.getText()
048                                 .toString())) {
049                     //判断复选框是否选中
049                     if (mCbKeepPsd.isChecked()) {
050                         mSpSettings = getSharedPreferences(PREFS_NAME,
051                                 MODE_PRIVATE);
052                         //得到Editor对象
053                         Editor edit = mSpSettings.edit();
054                         //记录保存标记
055                         edit.putBoolean("isKeep", true);
056                         //记录用户名
057                         edit.putString("username", mEtUserName
```

```
058                        .getText()
059                        .toString());
060               //记录密码
061               edit.putString("password", mEtPassWord.getText()
062                        .toString());
063               edit.commit();            //一定记得提交
064          } else {
065               mSpSettings = getSharedPreferences(PREFS_NAME,
066                        MODE_PRIVATE);
067               //得到Editor对象
068               Editor edit = mSpSettings.edit();
069               //记录保存标记
070               edit.putBoolean("isKeep", false);
071               //记录用户名
072               edit.putString("username", "");
073               //记录密码
074               edit.putString("password", "");
075               edit.commit();            //一定记得提交
076          }
077          //跳转到首页
078          Intent intent = new Intent(MainActivity.this,
079                   SuccessActivity.class);
080          startActivity(intent);
081       } else {
082           //显示错误提示
083           Toast.makeText(getApplicationContext(),"用户名或密
                码错误",
084                   Toast.LENGTH_SHORT).show();
085       }
086     }
087   });
088 }
089
090 @Override
091 protected void onResume() {
092     //在界面显示数据之前得到之前存储的数据
093     super.onResume();
094     getData();
095 }
096
097 /**
098  * 获取存储是数据
099  */
100 private void getData() {
101     //得到sharedpreferences对象
102     mSpSettings = getSharedPreferences(PREFS_NAME, MODE_PRIVATE);
103     //判断是否之前存储过用户名密码
104     if (mSpSettings.getBoolean("isKeep", false)) {
105         //如果之前存储过,则显示在相应文本框内
106         mEtUserName.setText(mSpSettings.getString("username", ""));
107         mEtPassWord.setText(mSpSettings.getString("password", ""));
108     } else {
109         //否则显示空
110         mEtUserName.setText("");
111         mEtPassWord.setText("");
112     }
113 }
```

如以上代码的第 4～7 行定义了控件对象，在第 10 行定义了 SharedPreferences 对象，用它来存储用户输入的用户名和密码。在第 29 行的 findView 函数中得到布局中的对应控件对象。在第 39 行设置得到控件的监听器对象，当用户单击登录按钮时，首先判断用户名密码是否为合法用户，这里就进行了固定值的判断，如果是合法用户并且用户勾选了记住密码，则在第 50 行得到系统的 SheredPreferences 对象，在第 53 行得到对应的 Editor 对象，然后写入用户名密码信息。在第 62 行确认保存这些信息。如用户密码输入错误，则清空之前记录的用户名密码。然后跳转到 SuccessActivity 页面。

下面定义 SuccessActivity 页面的布局文件为：

```xml
01  <RelativeLayout xmlns:android="http://schemas.android.com/apk/
    res/android"
02      xmlns:tools="http://schemas.android.com/tools"
03      android:id="@+id/RelativeLayout1"
04      android:layout_width="match_parent"
05      android:layout_height="match_parent"
06      android:orientation="vertical"
07      tools:context=".MainActivity" >
08
09      <!-- 显示提示标签 -->
10      <TextView
11          android:layout_width="wrap_content"
12          android:layout_height="wrap_content"
13          android:layout_centerInParent="true"
14          android:text="@string/tv_success"
15          android:textSize="20sp" />
16
17  </RelativeLayout>
```

SuccessActivity 的代码为：

```java
1  //登录成功后的 Activity 注意要在 AndroidManifest 中注册
2  public class SuccessActivity extends Activity {
3
4    @Override
5    protected void onCreate(Bundle savedInstanceState) {
6        super.onCreate(savedInstanceState);
7        setContentView(R.layout.success_main);
8    }
9  }
```

定义完 SuccessActivity 后，记得在 manifest 文件中注册此 Activity，否则是无法打开的。在 manifest 文件中添加如下代码即可：

```xml
<activity android:name=".SuccessActivity" ></activity>
```

4．实例扩展

需要注意的就是 SharedPreferences 中的数据其实也是以文件的形式存储的，只不过 Android 系统封装了这些文件存储的过程和方式。大家可以在手机上的 /data/data/PACKAGE_NAME/shared_prefs 目录下查找，其中对应的文件内容是使用 xml 来进行存储的，大致如下：

```xml
<?xml version='1.0' encoding='utf-8' standalone='yes' ?>
<map>
    <int name=" isKeep " value="true" />
```

```
    <string name=" username ">zhangsan</string>
    <string name=" password ">1234</string>
</map>
```

范例 117 系统的设置界面

1. 实例简介

在我们使用应用的过程中，经常会发现程序不但是对用户文件保存有存储配置的功能，而且对于某些设置我们也会进行保存。例如，我们手机中的无线网络的开关状态、GPS 的开关状态和无线的开关状态等。一般遇到这样的功能，我们并不是每次都去获取系统对应硬件的状态，而是在当前页的控件中保存它的状态，这样会比较方便，本例子就带领大家来实现一个常见的 Android 中的设置页面的实例。

2. 运行效果

该实例运行效果如图 6.2 所示。

图 6.2 系统的设置页面

3. 实例程序讲解

在本实例中，首先显示一个设置按钮，当用户单击此设置按钮时打开系统设置页面，并且记录相应的设置信息。想要实现本实例效果，首先修改 res/layout/activity_main.xml 文件，代码如下：

```
01  <LinearLayout xmlns:android="http://schemas.android.com/apk/res/
    android"
02      xmlns:tools="http://schemas.android.com/tools"
03      android:layout_width="match_parent"
```

```
04        android:layout_height="match_parent"
05        android:paddingBottom="@dimen/activity_vertical_margin"
06        android:paddingLeft="@dimen/activity_horizontal_margin"
07        android:paddingRight="@dimen/activity_horizontal_margin"
08        android:paddingTop="@dimen/activity_vertical_margin"
09        tools:context=".MainActivity" >
10     <!-- 定义设置按钮 -->
11        <Button
12            android:id="@+id/btn_setting"
13            android:layout_width="match_parent"
14            android:layout_height="wrap_content"
15            android:layout_centerInParent="true"
16            android:text="@string/setting" />
17
18  </LinearLayout>
```

这是首页的 Activity 的布局文件。其中包含一个设置按钮控件。

然后修改 src/com.wyl.example/MainActivity.java 文件，代码如下：

```
01  //定义了本实例的主要Activity
02  public class MainActivity extends Activity {
03
04  @Override
05  protected void onCreate(Bundle savedInstanceState) {
06      super.onCreate(savedInstanceState);
07      setContentView(R.layout.activity_main);
08      //绑定Button控件
09      Button btnSetting = (Button)findViewById(R.id.btn_setting);
10      //设置控件事件
11      btnSetting.setOnClickListener(new OnClickListener() {
12
13          @Override
14          public void onClick(View arg0) {
15              //TODO Auto-generated method stub
16              //跳转到设置页面
17              Intent intent = new Intent(MainActivity.this,
                    SettingActivity.class);
18              startActivity(intent);
19          }
20      });
21  }
22  }
```

此文件中主要内容就是当用户单击按钮的时候跳转到 SettingActivity 页面显示。

SettingActivity 页面的代码为 src/com.wyl.example/SettingActivity.java 文件，代码如下：

```
01  //定义设置页面的Activity
02  public class SettingActivity extends Activity{
03
04  @SuppressLint("NewApi")
05  @Override
06  protected void onCreate(Bundle savedInstanceState) {
07      super.onCreate(savedInstanceState);
08      //得到当前页面的FragmentManager
09      FragmentManager fragmentManager = getFragmentManager();
10      //得到FragmentTransaction
11      FragmentTransaction fragmentTransaction = fragmentManager.
        beginTransaction();
12      //定义PreferenceSetting对象
```

```
13      PreferenceSetting setting = new PreferenceSetting(this);
14      //设置当前页面的设置属性
15      fragmentTransaction.replace(android.R.id.content, setting);
16      fragmentTransaction.addToBackStack(null);
17      //确认保存属性
18      fragmentTransaction.commit();
19    }
20  }
```

在此文件的第 9 行得到了页面的 FragmentManager 对象, 在第 11 行得到转换 Fragment 对象, 在第 13 行定义 PreferenceSetting 对象, 并在第 15 行设置给 Fragment 转换对象, 第 18 行确认保存。

PreferenceSetting 页面的代码为 src/com.wyl.example/ PreferenceSetting.java 文件, 代码如下:

```
001  //Android 3.0 之后用 PreferenceFragment 3.0 之前可参考 PreferenceActivity
002  @SuppressLint({ "NewApi", "ValidFragment" })
003  public class PreferenceSetting extends PreferenceFragment implements
004       OnPreferenceClickListener, OnPreferenceChangeListener {
005
006      private CheckBoxPreference mapply_wifiPreference;      //打开 wifi
007      private CheckBoxPreference mapply_internetPreference;
                                                                //Internet 共享
008      private ListPreference depart_valuePreference;         //部门设置
009      private EditTextPreference number_editPreference;      //输入电话号码
010      private Preference mwifi_settingPreference;            //wifi 设置
011      private Context mContext;
012
013      private static final String TAG = "PreferenceSetting";
014
015      //构造函数 context 用于实例化 intent
016      public PreferenceSetting(Context context) {
017          mContext = context;
018      }
019
020      @Override
021      public void onCreate(Bundle savedInstanceState) {
022          //TODO Auto-generated method stub
023          super.onCreate(savedInstanceState);
024          addPreferencesFromResource(R.xml.mypreference);
025          //得到布局中的控件
026          findView();
027          //绑定控件事件
028          setListener();
029
030      }
031
032      private void findView() {
033          //TODO Auto-generated method stub
034          //根据 key 值找到控件
035          mapply_wifiPreference = (CheckBoxPreference) findPreference
             ("apply_wifi");
036          mapply_internetPreference = (CheckBoxPreference) findPreference
             ("apply_internet");
037          depart_valuePreference = (ListPreference) findPreference
             ("depart_value");
038          number_editPreference = (EditTextPreference) findPreference
```

```
                ("number_edit");
039         mwifi_settingPreference = (Preference) findPreference("wifi_
                setting");
040     }
041
042     private void setListener() {
043         //TODO Auto-generated method stub
044         //设置监听器
045         mapply_internetPreference.setOnPreferenceClickListener
                (this);
046         mapply_internetPreference.setOnPreferenceChangeListener
                (this);
047         depart_valuePreference.setOnPreferenceClickListener(this);
048         depart_valuePreference.setOnPreferenceChangeListener(this);
049         number_editPreference.setOnPreferenceClickListener(this);
050         number_editPreference.setOnPreferenceChangeListener(this);
051         mwifi_settingPreference.setOnPreferenceClickListener(this);
052     }
053
054     /**
055      * 触发规则 1 先调用 onPreferenceClick()方法,
056      * 如果该方法返回 true,则不再调用 onPreferenceTreeClick 方法
057      * 如果 onPreferenceClick 方法返回 false,则继续调用
                onPreferenceTreeClick 方法
058      * 2onPreferenceChange 的方法独立与其他两种方法的运行。也就是说,它总是会运行
059      * 当 Preference 控件被单击时,触发该方法
060      */
061     @Override
062     public boolean onPreferenceClick(Preference preference) {
063         //TODO Auto-generated method stub
064         Log.e(TAG,
065                 "onPreferenceClick----->" + String.valueOf(preference.
                    getKey()));
066         return false;
067     }
068
069     /**
070      * 说明: 当 Preference 的元素值发生改变时,触发该事件
071      * 返回值: true 代表将新值写入 sharedPreference 文件中
072      * false 则不将新值写入 sharedPreference 文件
073      */
074     @Override
075     public boolean onPreferenceChange(Preference preference, Object
            objValue) {
076         //TODO Auto-generated method stub
077         Log.i(TAG,
078                 "onPreferenceChange----->"
079                         + String.valueOf(preference.getKey()));
080         if (preference == mapply_wifiPreference) {
081             Log.i(TAG, "Wifi CB, and isCheckd = " + String.
                    valueOf(objValue));
082         } else if (preference.getKey().equals("apply_internet")) {
083             Log.i(TAG,
084                     "internet CB, and isCheckd = " + String.valueOf
                        (objValue));
085             //return false;                        //不保存该新值
086         } else if (preference == depart_valuePreference) {
087             Log.i(TAG, " Old Value" + depart_valuePreference.
                    getValue()
```

第6章 Android 的数据存储

```
088                     + " NewDeptName" + objValue);
089            } else if (preference.getKey().equals("wifi_setting")) {
090                Log.i(TAG, "change" + String.valueOf(objValue));
091                mwifi_settingPreference.setTitle("its turn me.");
                                        //重新设置 title
092            } else if (preference == number_editPreference) {
093                Log.i(TAG, "Old Value = " + String.valueOf(objValue));
094            }
095            return true;              //保存更新后的值
096        }
097
098        /**
099         * 当 Preference 控件被单击时,触发该方法
100         */
101        @Override
102        public boolean onPreferenceTreeClick(PreferenceScreen preferenceScreen,
103                Preference preference) {
104            //TODO Auto-generated method stub
105            Log.i(TAG, "onPreferenceTreeClick----->" + preference.getKey());
106            //对控件进行操作
107            if (preference == mapply_wifiPreference) {
108                //单击了 "打开 wifi"
109                Log.e(TAG,
110                        " Wifi , and isCheckd ="
111                                + mapply_wifiPreference.isChecked());
112            } else if (preference.getKey().equals("apply_internet")) {
113                //单击了 "Internet 共享"
114                Log.e(TAG, " internet , and isCheckd = "
115                        + mapply_internetPreference.isChecked());
116            } else if (preference == depart_valuePreference) {
117                //单击了 "部门设置"
118                Log.e(TAG, " department CB,and selectValue = "
119                        + depart_valuePreference.getValue() + ", Text="
120                        + depart_valuePreference.getEntry());
121            } else if (preference.getKey().equals("wifi_setting")) {
122                //单击了 "wifi 设置"
123                Log.e(TAG,
124                        " wifi , and isCheckd = "
125                                + mapply_wifiPreference.isChecked());
126            } else if (preference == number_editPreference) {
127                //单击了 "输入电话号码"
128                Log.e(TAG, "Old Value=" + number_editPreference.getText()
129                        + ", New Value="
130                        + number_editPreference.getEditText().toString());
131            }
132
133            if (preference.getKey().equals("wifi_setting")) {
134                //创建一个新的 Intent,
135                //函数如果返回 true, 则跳转至该新的 Intent ;
136                //函数如果返回 false,则跳转至 xml 文件中配置的 Intent ;
137                Intent i = new Intent(mContext, WifiSettingActivity.class);
                    //MainActivity 只是一个简单的 Activity
138                startActivity(i);
139                return true;
140            }
141            return super.onPreferenceTreeClick(preferenceScreen, preference);
```

```
142    }
143 }
```

此文件中主要实现了一个 Setting 页面需要保存的状态信息,在代码第 6~10 行定义了页面的控件对象及 preference 对象。在第 32 行的 findView 方法中得到对应的控件状态,在第 42 行得到相应的设置属性单击的监听器,在第 62 行定义了当属性被单击时的回调函数,在第 75 行定义了当属性改变时的回调函数,在第 102 行定义了设置树被单击时的回调函数。在这里我只处理了设置 wifi 状态的函数,主要是跳转到 WifiSettingActivity 中。在此 Activity 中主要设置 wifi 的连接状态。布局如下:

```
01 <RelativeLayout xmlns:android="http://schemas.android.com/apk/res/android"
02     xmlns:tools="http://schemas.android.com/tools"
03     android:id="@+id/RelativeLayout1"
04     android:layout_width="match_parent"
05     android:layout_height="match_parent"
06     android:orientation="vertical"
07     tools:context=".MainActivity" >
08
09     <!-- 显示提示标签 -->
10     <TextView
11         android:layout_width="wrap_content"
12         android:layout_height="wrap_content"
13         android:layout_centerInParent="true"
14         android:text="WIFI 设置页面"
15         android:textSize="20sp" />
16
17 </RelativeLayout>
```

WifiSettingActivity 的代码如下:

```
1 //定义 wifi 设置的具体页面
2 public class WifiSettingActivity extends Activity {
3
4     @Override
5     protected void onCreate(Bundle savedInstanceState) {
6         super.onCreate(savedInstanceState);
7         setContentView(R.layout.wifisetting_main);
8     }
9 }
```

4. 实例扩展

在此实例中主要实现了系统的设置界面中保存控件状态的效果,在我们实际的应用中大家可以根据自己的需要来实现此部分功能。例如,应用程序中是否定期清理内存的设置,程序是否默认加载大图的设置等。

范例 118　系统图片剪裁

1. 实例简介

在我们使用应用的过程中,经常会使用到用户选择图片截取的功能。例如,用户希望拍照上传头像,但是用户拍照的照片尺寸和需要的尺寸大小不一致,所以用户拍照完毕后,可以通过 Android 内置的应用进行图片的剪裁。本例子就带领大家来实现一个通过 Android

系统进行照片剪裁的实例。

2. 运行效果

该实例运行效果如图 6.3 所示。

图 6.3 系统图片剪裁

3. 实例程序讲解

在本实例中,提供一个按钮,当用户单击此按钮时提示用户拍照或者通过本地浏览得到图片,然后进行图片剪裁。想要实现本实例效果,首先修改 res/layout/activity_main.xml 文件,代码如下:

```xml
01  <LinearLayout xmlns:android="http://schemas.android.com/apk/res/android"
02      xmlns:tools="http://schemas.android.com/tools"
03      android:id="@+id/LinearLayout1"
04      android:layout_width="match_parent"
05      android:layout_height="match_parent"
06      android:orientation="vertical"
07      tools:context=".MainActivity" >
08      <!-- 定义图片显示控件 -->
09      <ImageView
10          android:id="@+id/imageView"
11          android:layout_width="wrap_content"
12          android:layout_height="wrap_content" />
13      <!-- 定义选择图片按钮 -->
14      <Button
15          android:id="@+id/selectImageBtn"
16          android:layout_width="match_parent"
17          android:layout_height="wrap_content"
18          android:text="选择图片" />
19
20  </LinearLayout>
```

这是我们的 Activity 的布局文件。在其中第 14 行定义了一个 Button 控件，当用户单击此按钮的时候选择图片进行剪裁，然后把剪裁结果显示在第 9 行定义的 ImageView 控件中。

然后修改 src/com.wyl.example/MainActivity.java 文件，代码如下：

```
001  //定义了本实例的主要Activity
002  public class MainActivity extends Activity {
003      private static final int TAKE_PICTURE = 0;      //从相机获取
004      private static final int CHOOSE_PICTURE = 1;    //从图库获取
005      //状态码
006      private static final int CROP = 2;              //裁剪
007      private static final int CROP_PICTURE = 3;
008      private Button mBtnSlect;                       //选择按钮
009      private ImageView mIvShow;                      //显示框
010
011
012      @Override
013      public void onCreate(Bundle savedInstanceState) {
014          super.onCreate(savedInstanceState);
015          setContentView(R.layout.activity_main);
016          //得到布局中的控件
017          findView();
018          //绑定控件事件
019          setListener();
020      }
021
022      private void findView() {
023          //绑定控件
024          mBtnSlect = (Button) findViewById(R.id.selectImageBtn);
025          mIvShow = (ImageView) findViewById(R.id.imageView);
026      }
027
028      private void setListener() {
029          //添加事件
030          mBtnSlect.setOnClickListener(new OnClickListener() {
031              @Override
032              public void onClick(View arg0) {
033                  //单击按钮时显示选择图片的对话框
034                  showPicturePicker(MainActivity.this);
035              }
036          });
037      }
038
039      //主要逻辑选择图片裁剪并显示
040      public void showPicturePicker(Context context) {
041          //定义AlertDialog.Builder
042          AlertDialog.Builder builder = new AlertDialog.Builder(context);
043          //设置标题
044          builder.setTitle("图片来源");
045          builder.setNegativeButton("取消", null);
046          builder.setItems(new String[] { "拍照", "相册" },
047                  new DialogInterface.OnClickListener() {
048                      @Override
049                      public void onClick(DialogInterface dialog, int which) {
050                          switch (which) {
051                              //从相机拍照
```

```java
                            case TAKE_PICTURE:
                                Uri imageUri = null;
                                String fileName = null;
                                //打开照相机的intent
                                Intent openCameraIntent = new Intent(
                                        MediaStore.ACTION_IMAGE_CAPTURE);
                                //删除上一次截图的临时文件
                                SharedPreferences sharedPreferences =
                                getSharedPreferences(
                                        "temp", Context.MODE_WORLD
                                        _WRITEABLE);
                                ImageTools.deletePhotoAtPathAndName
                                (Environment
                                        .getExternalStorageDirectory()
                                        .getAbsolutePath(),
                                        sharedPreferences
                                        .getString("tempName", ""));

                                //保存本次截图临时文件名字
                                fileName = String.valueOf(System
                                        .currentTimeMillis()) + ".jpg";
                                Editor editor = sharedPreferences.edit();
                                editor.putString("tempName", fileName);
                                editor.commit();
                                imageUri = Uri.fromFile(new File
                                (Environment
                                        .getExternalStorageDirectory(),
                                        fileName));
                                //指定照片保存路径（SD卡），image.jpg为一个临时文件，
                                //每次拍照后这个图片都会被替换
                                openCameraIntent.putExtra(MediaStore
                                .EXTRA_OUTPUT,
                                        imageUri);
                                startActivityForResult(openCameraIntent,
                                        CROP);
                                break;
                            //从图库选择
                            case CHOOSE_PICTURE:
                                Intent openAlbumIntent = new Intent(
                                        Intent.ACTION_GET_CONTENT);
                                openAlbumIntent
                                        .setDataAndType(

                                MediaStore.Images.Media.EXTERNAL_CONTENT_URI,
                                "image/*");
                                startActivityForResult(openAlbumIntent,
                                        CROP);
                                break;

                            default:
                                break;
                        }
                    }
                });
        builder.create().show();
    }

    @Override
    protected void onActivityResult(int requestCode, int resultCode,
    Intent data) {
```

```java
102        super.onActivityResult(requestCode, resultCode, data);
103        if (resultCode == RESULT_OK) {
104            switch (requestCode) {
105            case CROP:
106                //选定截图范围
107                Uri uri = null;
108                if (data != null) {
109                    //得到uri的数据
110                    uri = data.getData();
111                } else {
112                    //得到临时保存的文件
113                    String fileName = getSharedPreferences("temp",
114                            Context.MODE_WORLD_WRITEABLE).getString
                                ("tempName",
115                            "");
116                    //得到保存文件的uri
117                    uri = Uri.fromFile(new File(Environment
118                            .getExternalStorageDirectory(), fileName));
119                }
120                //开始截图
121                cropImage(uri, 500, 500, CROP_PICTURE);
122                break;
123
124            case CROP_PICTURE:
125                //开始截图
126                Bitmap photo = null;
127                //得到uri中的数据
128                Uri photoUri = data.getData();
129                if (photoUri != null) {
130                    //得到bitmap
131                    photo = BitmapFactory.decodeFile(photoUri.getPath());
132                }
133                if (photo == null) {
134                    Bundle extra = data.getExtras();
135                    if (extra != null) {
136                        photo = (Bitmap) extra.get("data");
137                        //进行图片的压缩截取
138                        ByteArrayOutputStream stream = new
                                ByteArrayOutputStream();
139                        photo.compress(Bitmap.CompressFormat.JPEG,
                                100, stream);
140                    }
141                }
142                //显示截图结果
143                mIvShow.setImageBitmap(photo);
144                break;
145            default:
146                break;
147            }
148        }
149    }
150
151    //截取图片
152    public void cropImage(Uri uri, int outputX, int outputY, int requestCode) {
153        //发送截取图片的intent
154        Intent intent = new Intent("com.android.camera.action.CROP");
```

```
155            //设置截取的图片信息
156            intent.setDataAndType(uri, "image/*");
157            intent.putExtra("crop", "true");
158            intent.putExtra("aspectX", 1);              //裁剪框比例
159            intent.putExtra("aspectY", 1);
160            intent.putExtra("outputX", outputX);         //裁剪大小
161            intent.putExtra("outputY", outputY);
162            intent.putExtra("outputFormat", "JPEG");     //图片类型
163            intent.putExtra("noFaceDetection", true);
164            intent.putExtra("return-data", true);
165            //请求系统截图功能
166            startActivityForResult(intent, requestCode);
167        }
168    }
```

此文件是 Activity 的代码文件，在其中第 6~9 行定义了程序中需要用到的状态码，在第 22 行得到了系统用到的控件对象。在第 28 行的 setListener 方法中设置了按钮的单击事件，具体实现在第 40~98 行，在此函数中设置一个 AlertDialog 的创建对象，然后添加两个选择，拍照和从本地图片选择，单击后分别打开相机拍照得到图片或图库选择图片。在第 101 行中 onActivityResult 回调函数中根据之前的标示符判断是拍照还是本地选择得到的图片，然后进行图片截取。截取图片的方法在第 152 行，设置相应的 Intent 属相来得到图片。

4．实例扩展

在此实例中当用户单击 Button 的时候，打开系统选择图片，然后就可以进行截取了，这里需要注意一点，图片处理在 Android 中最常见的问题就是 OOM 问题了，因为手机的内存有限，所以大家在使用的过程中应该尽量避免处理大图片。

范例 119　SDCard 信息查询

1．实例简介

在我们使用应用的过程中，经常会使用到查看手机 SDCard 的信息的功能。例如，当用户希望保存图片时，或者用户希望对 SDCard 进行格式化时等。遇到这样的功能，我们首先要在程序中拿到手机的 SDCard 的信息，然后根据 SDCard 的信息进行判断，最后再进行相应的处理工作。本例子就带领大家来实现一个获取手机 SDCard 状态信息的实例。

2．运行效果

该实例运行效果如图 6.4 所示。

3．实例程序讲解

在本实例中，页面显示一个文本框、显示 SDCard 的 Block 的数量、每个 block 的容量、SDCard 的总容量和 SDCard 的剩余容量等信息。想要实现本实例效果，首先修改 res/layout/activity_main.xml 文件，代码如下：

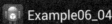

```
SDCard上BLOCK总数: 3493972
SDCard上每个bloc的SIZE:4096
可供程序使用的Block的数量 : 1755366
剩下的所有Block的数量: 1755366
SDCard 总容量大小MB: 13648MB
SDCard 剩余大小MB: 6856MB
```

图 6.4　SDCard 信息显示

```xml
01 <LinearLayout xmlns:android="http://schemas.android.com/apk/res/android"
02     xmlns:tools="http://schemas.android.com/tools"
03     android:id="@+id/LinearLayout1"
04     android:layout_width="match_parent"
05     android:layout_height="match_parent"
06     android:orientation="vertical"
07     tools:context=".MainActivity" >
08     <!-- 显示 block 的数量的标签控件 -->
09     <TextView
10         android:id="@+id/tv_TotalBlocks"
11         android:layout_width="wrap_content"
12         android:layout_height="wrap_content" />
13     <!-- 显示 block 的大小的标签控件 -->
14     <TextView
15         android:id="@+id/tv_BlocSize"
16         android:layout_width="wrap_content"
17         android:layout_height="wrap_content" />
18     <!-- 显示可用 block 的标签控件 -->
19     <TextView
20         android:id="@+id/tv_AvailaBlock"
21         android:layout_width="wrap_content"
22         android:layout_height="wrap_content" />
23     <!-- 显示空 block 的标签控件 -->
24     <TextView
25         android:id="@+id/tv_FreeBlock"
26         android:layout_width="wrap_content"
27         android:layout_height="wrap_content" />
28     <!-- 显示 SDCard 的总大小的标签控件 -->
29     <TextView
30         android:id="@+id/tv_SDTotalSize"
31         android:layout_width="wrap_content"
32         android:layout_height="wrap_content" />
33     <!-- 显示 SDCard 的剩余大小的标签控件 -->
34     <TextView
35         android:id="@+id/tv_SDFreeSize"
36         android:layout_width="wrap_content"
37         android:layout_height="wrap_content" />
38
```

这是我们的 Activity 的布局文件。在其中分别定义了显示 SD 卡信息的标签控件。

然后修改 src/com.wyl.example/MainActivity.java 文件，代码如下：

```java
01  //定义了本实例的主要 Activity
02  public class MainActivity extends Activity {
03
04      private TextView mTvTotalBlocks;        //SDCard 上 Block 总数
05      private TextView mTvBlocSize;           //SDCard 上每个 block 的 Size
06      private TextView mTvAvailaBlock;        //可供程序使用的 Block 的数量
07      private TextView mTvFreeBlock;          //剩下的所有 Block 的数量(包括预留
                                                //的一般程序无法使用的块)
08      private TextView mTvSDTotalSize;        //SDCard 总容量大小 MB
09      private TextView mTvSDFreeSize;         //SDCard 剩余大小 MB
10
11      @Override
12      public void onCreate(Bundle savedInstanceState) {
13          super.onCreate(savedInstanceState);
14          setContentView(R.layout.activity_main);
15          //得到布局中的控件
16          findView();
17          //绑定控件事件
18          SDCardSizeTest();
19      }
20
21      private void findView() {
22          //绑定控件
23          mTvTotalBlocks = (TextView)findViewById(R.id.tv_TotalBlocks);
24          mTvBlocSize = (TextView)findViewById(R.id.tv_BlocSize);
25          mTvAvailaBlock = (TextView)findViewById(R.id.tv_AvailaBlock);
26          mTvFreeBlock = (TextView)findViewById(R.id.tv_FreeBlock);
27          mTvSDTotalSize = (TextView)findViewById(R.id.tv_SDTotalSize);
28          mTvSDFreeSize = (TextView)findViewById(R.id.tv_SDFreeSize);
29      }
30
31
32      public void SDCardSizeTest() {
33          //取得 SDCard 当前的状态
34          String sDcString = android.os.Environment
                  .getExternalStorageState();
35
36          //如果当前系统有 SDcard 存在
37          if (sDcString.equals(android.os.Environment.MEDIA_MOUNTED)) {
38              //取得 SDcard 文件路径
39              File pathFile = android.os.Environment
40                      .getExternalStorageDirectory();
41              //得到 SDcard 的状态
42              android.os.StatFs statfs = new android.os.StatFs(pathFile
                      .getPath());
43
44              //获取 SDCard 上 BLOCK 总数
45              long nTotalBlocks = statfs.getBlockCount();
46              mTvTotalBlocks.setText("SDCard 上 BLOCK 总数: "+nTotalBlocks);
47
48              //获取 SDCard 上每个 block 的 SIZE
49              long nBlocSize = statfs.getBlockSize();
50              mTvBlocSize.setText("SDCard 上每个 bloc 的 SIZE:"+nBlocSize);
51
52              //获取可供程序使用的 Block 的数量
53              long nAvailaBlock = statfs.getAvailableBlocks();
```

```
54          mTvAvailaBlock.setText("可供程序使用的 Block 的数量 : " +
            nAvailaBlock);
55
56          //获取剩下的所有 Block 的数量(包括预留的一般程序无法使用的块)
57          long nFreeBlock = statfs.getFreeBlocks();
58          mTvFreeBlock.setText("剩下的所有 Block 的数量: " + nFreeBlock);
59
60          //计算 SDCard 总容量大小 MB
61          long nSDTotalSize = nTotalBlocks * nBlocSize / 1024 / 1024;
62          mTvSDTotalSize.setText("SDCard 总容量大小 MB: " + nSDTotalSize
            +"MB");
63
64          //计算 SDCard 剩余大小 MB
65          long nSDFreeSize = nAvailaBlock * nBlocSize / 1024 / 1024;
66          mTvSDFreeSize.setText(" SDCard 剩余大小 MB: " + nSDFreeSize
            +"MB");
67      }
68  }
69  }
```

此文件是 Activity 的代码文件，在其中第 4~9 行定义了一系列的 TextView 控件，来显示对应的 SDCard 信息。在第 21 行的 findView 方法中得到了布局中的所有控件。在第 32 行的 SDCardSizeTest 方法中得到了 SDCard 的相应信息并且填入相应的 TextView 控件中，在此方法中首先判断 SDCard 是否挂载，然后得到 SDCard 的根目录，通过 StatFs 方法得到根目录的状态信息，然后依次得到相应的参数指标显示即可。

4．实例扩展

在此实例中仅仅显示了 SDCard 的基本信息，当然还有一些信息大家可以获取。例如，获取 SDCard 的读写状况和是否已经挂载等。详细内容大家可以参看 Google 的开发者文档。

范例 120 图片旋转保存

1．实例简介

在我们使用应用的过程中，经常会使用到图片旋转的效果。例如，用户拍照的时候，手机的方向有可能是多个角度的，这样我们在使用的时候如果不做处理的话，只能倾斜来看。遇到这样的情况，我们希望用户在使用照片前对照片做处理，然后再使用照片。本例子就带领大家来实现一个图片旋转的实例。

2．运行效果

该实例运行效果如图 6.5 所示。

3．实例程序讲解

在本实例中，页面显示了原始的背景图，以及旋转 30°后的效果图。想要实现本实例效果，首先新建 src/com.wyl.example/BitmapView.java 文件，代码如下：

图 6.5 图片旋转保存

```
01  //自定义的展示View类
02  public class BitmapView extends View {
03      //图片的旋转矩阵
04      private Matrix matrix = null;
05
06      public BitmapView(Context context) {
07          super(context);
08      }
09
10      public void onDraw(Canvas canvas) {
11          //获取资源文件的引用res
12          Resources res = getResources();
13          //获取图形资源文件
14          Bitmap bmp = BitmapFactory.decodeResource(res, R.drawable.h);
15          //设置canvas画布背景为白色
16          canvas.drawColor(Color.BLACK);
17          canvas.drawBitmap(bmp, 0, 0, null);
18          //定义矩阵对象
19          matrix = new Matrix();
20          //旋转30度
21          matrix.postRotate(30);
22          Bitmap bitmap = Bitmap.createBitmap(bmp, 0, 50, bmp.getWidth(),
23                  bmp.getHeight() / 2, matrix, true);
24          canvas.drawBitmap(bitmap, 0, 250, null);
25          SaveBitmap(bitmap);
26      }
27
28      //保存到本地
29      public void SaveBitmap(Bitmap bmp) {
30          Bitmap bitmap = Bitmap.createBitmap(800, 600, Config.ARGB_8888);
31          Canvas canvas = new Canvas(bitmap);
32          //加载背景图片
```

```
33      Bitmap bmps = BitmapFactory.decodeResource(getResources(),
34              R.drawable.playerbackground);
35      canvas.drawBitmap(bmps, 0, 0, null);
36      //加载要保存的画面
37      canvas.drawBitmap(bmp, 10, 100, null);
38      //保存全部图层
39      canvas.save(Canvas.ALL_SAVE_FLAG);
40      canvas.restore();
41      //存储路径
42      File file = new File("/sdcard/wyl/");
43      if (!file.exists())
44          file.mkdirs();
45      try {
46          //注意添加读写权限
47          FileOutputStream fileOutputStream = new FileOutputStream(
48                  file.getPath() + "/wyl.jpg");
49          bitmap.compress(Bitmap.CompressFormat.JPEG, 100,
                fileOutputStream);
50          fileOutputStream.close();
51          System.out.println("saveBmp is here");
52      } catch (Exception e) {
53          Log.e("error", "noSave");
54          e.printStackTrace();
55      }
56  }
57 }
```

这是我们自定义的 View 展示类。在其中第 4 行定义了一个矩阵对象，在下面的代码中用来旋转图片。在第 10~26 行定义了 onDraw 方法，在此 View 进行绘制的时候自动调用，其中首先得到工程的资源图片，然后通过 Matrix 类对图片进行旋转，旋转后通过 Bitmap 类的 createBitmap 方法得到旋转后的 Bitmap 对象，然后保存图片。在第 29 行定义了 SaveBitmap 方法保存 Bitmap 对象到本地。

然后修改 src/com.wyl.example/MainActivity.java 文件，代码如下：

```
01 //定义了本实例的主要 Activity
02 public class MainActivity extends Activity {
03     //定义自定义类的对象
04     private BitmapView bitmapView = null;
05
06     @Override
07     public void onCreate(Bundle savedInstanceState) {
08         super.onCreate(savedInstanceState);
09         //初始化 bitmapView 对象
10         bitmapView = new BitmapView(this);
11         //设置当前 Activity 的显示视图为 bitmapView
12         setContentView(bitmapView);
13     }
14 }
```

此文件是 Activity 的代码文件，在此文件中设置当前 Activity 的布局视图为之前定义的 BitmapView 即可。

4．实例扩展

在此实例中仅仅是固定的将图片旋转 30°，大家可以通过程序进行选择角度的设置，动态产生旋转后的效果。

范例 121 学生成绩管理系统

1．实例简介

到目前为止我们就可以做一些稍微简单的单机应用了，其中对于数据的增删改查我们都可以加入，但是之前我们的所有的实例操作都是文件相应的操作，文件操作对于数据的查找、删除和添加来说不是一个好的选择，如果你的程序由大量的规范化数据需要进行存储，那么最好是使用数据库。本例就带领大家来完成一个简单的学生管理系统的实现。

2．运行效果

该实例运行效果如图 6.6 所示。

3．实例程序讲解

在本实例中，展示了一个简单的学生管理系统的所有功能，如添加学生信息、修改学生信息、删除学生信息和查询学生信息等。本实例涉及三个文件，如下所示。

- MainActivity：是程序的主界面，包括了所有的增删改查的界面所在。
- MySqliteHelper：是程序操纵数据库的辅助类。
- State：是程序中辅助的状态类。

本实例主要的 Activity 页面是 MainActivity 文件，首先修改 src/com.wyl.example/MainActivity.java 文件，代码如下：

图 6.6 学生成绩管理系统

```
001  //定义了本实例的主要Activity
002  public class MainActivity extends TabActivity {
003      //各种变量的声明
004      private static final String QUERY_TAG = "查询TAG";
005      private static final String INSERT_TAG = "添加TAG";
006      private static final String UPDATE_TAG = "修改TAG";
007      private static final String DELETE_TAG = "删除TAG";
008      private TabHost mTabHost;
009      private View mViews;
010      private EditText mEditText_Name;          //姓名
011      private EditText mEditText_Number;        //学号
012      //省略其他控件对象的定义
013
014      private Cursor mCursor_query;
015      private String name, number, score;
016
017      @Override
018      public void onCreate(Bundle savedInstanceState) {
019          super.onCreate(savedInstanceState);
020          //设置标题颜色和文字
```

```
021         this.setTitle(getResources().getString(R.string.title));
022         this.setTitleColor(Color.MAGENTA);
023
024         //获得 TabHost 对象
025         mTabHost = this.getTabHost();
026         //通过布局选择器获得所有的组件
027         mViews = LayoutInflater.from(this).inflate(R.layout
                .activity_main,
028                 mTabHost.getTabContentView(), true);
029         this.findViewsAndSetListener();        //获得组件 Widget
030         this.addTabs();                        //添加 TAB
031
032         //创建 SQLiteOpenHelper 对象的引用
033         database = new MySqliteHelper(this);
034         showUiAdapter();                       //更新界面（设置适配器）
035     }
036
037     //绑定 Tabs
038     public void addTabs() {
039         //添加查询 Tab
040         mTabHost.addTab(mTabHost
041                 .newTabSpec(QUERY_TAG)
042                 .setContent(R.id.queryLayout_id)
043                 .setIndicator(getResources().getString(R.string
                    .query_str),
044                         getResources().getDrawable(R.drawable.query)));
045         //添加学生 Tab
046         mTabHost.addTab(mTabHost
047                 .newTabSpec(INSERT_TAG)
048                 .setContent(R.id.insertLayout_id)
049                 .setIndicator(getResources().getString(R.string
                    .insert_str),
050                         getResources().getDrawable(R.drawable.add)));
051         //省略删除 Tab 代码，省略刷新 Tab 代码
052
053         mTabHost.setCurrentTab(1);              //设置当前显示第 2 个 Tab
054         //表单切换时的事件处理
055         mTabHost.setOnTabChangedListener(new OnTabChangeListener() {
056
057             public void onTabChanged(String tabId) {
058                 //TODO Auto-generated method stub
059                 if (tabId.equalsIgnoreCase(MainActivity.INSERT_TAG)) {
060                     //设置输入框的状态
061                     mEditText_Name.setEnabled(true);
062                     mEditText_Number.setEnabled(true);
063                     mEditText_Score.setEnabled(true);
064                     mButton_insert.setText(getResources().getString(
065                             R.string.insert_str));
066
067                     State.setSearch(false);
068                     State.setInsert(true);
069                     State.setDelete(false);
070                     setHintText();
071                     insert_ListView
072                             .setOnItemClickListener
                                (myOnItemClickListener);
073                 } else if (tabId.equalsIgnoreCase(MainActivity
                    .UPDATE_TAG)) {
074                     //省略更新、删除和查找的状态修改信息
```

```
075                  }
076
077                  Toast.makeText(getApplicationContext(),
078                          "现在是>>" + tabId + "<<选项", Toast.LENGTH_SHORT)
                             .show();
079              }
080         });
081     }
082
083     //找到组件
084     public void findViewsAndSetListener() {
085         //省略得到控件并且设置监听器的代码
086     }
087
088     //按键监听事件
089     OnKeyListener keyListener = new OnKeyListener() {
090
091         public boolean onKey(View v, int keyCode, KeyEvent event) {
092             //TODO Auto-generated method stub
093             if (!mEditText_query.getText().toString()
                     .equalsIgnoreCase("")) {
094                 mButton_query.setEnabled(true);
095             } else {
096                 mButton_query.setEnabled(false);
097             }
098
099             return false;
100         }
101
102     };
103
104
105     //新增数据的方法
106     public void insertMethod() {
107         //修改状态为插入状态
108         State.setInsert(true);
109         //查看是否可用
110         if (isValid()) {
111             //判断是否有此用户
112             Cursor c = database.isHaveThisStu(Integer
                     .valueOf(number));
113             if (c.getCount() > 0 && c != null) {
114                 Toast.makeText(this,
115                         getResources().getString(R.string
                             .haveNumber), 1000)
116                         .show();
117             } else {
118                 database.insertData(name, Integer.valueOf(number),
119                         Float.valueOf(score));
120                 update();                      //告知适配器更新
121             }
122
123         }
124         //update();                             //告知适配器更新
125     }
126
127     //告知适配器更新
128     public void update() {
129         _id = 0;
```

```java
130         mEditText_Name.setText("");
131         mEditText_Number.setText("");
132         mEditText_Score.setText("");
133
134         //判断是否为修改或删除表单,用于是否设置提示语
135         if (State.isUpdataOrDelete()) {
136             setHintEmptyText();
137         } else {
138             setHintText();
139         }
140
141         mCursor.requery();
142         adapter.notifyDataSetChanged();
143         //设置文本编辑区不可以编辑
144         Log.i("//////", "" + State.isInsert());
145         if (State.isInsert()) {
146             mEditText_Name.setEnabled(true);
147             mEditText_Number.setEnabled(true);
148             mEditText_Score.setEnabled(true);
149         } else {
150             mEditText_Name.setEnabled(false);
151             mEditText_Number.setEnabled(false);
152             mEditText_Score.setEnabled(false);
153         }
154
155     }
156
157     //查询指定学号的学生成绩信息
158     public void searchSpecificStuMethod() {
159         //设置为查询状态
160         State.setSearch(true);
161         String number = mEditText_query.getText().toString().trim();
162         if (number.equalsIgnoreCase("")) {
163             return;
164         }
165         //得到查询结果
166         mCursor_query = database.searchSpecific(Integer.valueOf
            (number));
167         //更新数据展示界面
168         if (mCursor_query.getCount() > 0) {
169             showUiAdapter();
170         } else {
171             Toast.makeText(this, getResources().getString(R.string
                .noinfo),
172                     Toast.LENGTH_SHORT).show();
173         }
174         mEditText_query.setHint(getResources().getString(
175                 R.string.query_edit_hint));
176     }
177
178     //判断是否输入有合法
179     public boolean isValid() {
180         name = null;
181         number = null;
182         score = null;
183         name = mEditText_Name.getText().toString().trim();
184         number = mEditText_Number.getText().toString().trim();
185         score = mEditText_Score.getText().toString().trim();
186         if (name.equalsIgnoreCase("") || number.equalsIgnoreCase("")
187                 || score.equalsIgnoreCase("")) {
```

```
188                 Toast.makeText(MainActivity.this, "输入有误,请核对！", Toast
                        .LENGTH_SHORT)
189                         .show();
190             return false;
191         } else {
192             return true;
193         }
194     }
195
196     //修改数据
197     public void upDateMethod() {
198         State.setUpdataOrDelete(true);
199         if (isValid()) {
200             database.upDateinfo(_id, name, Integer.valueOf(number),
201                     Float.valueOf(score));
202         }
203         update();                              //告知适配器更新
204
205     }
206
207     //删除数据的 function
208     public void deleteDataMethod() {
209         State.setUpdataOrDelete(true);
210         if (isValid()) {
211             database.deleteStuInfo(_id);
212         }
213         update();                              //告知适配器更新
214         //删除按钮始终为不可编辑
215         mEditText_Name.setEnabled(false);
216         mEditText_Number.setEnabled(false);
217         mEditText_Score.setEnabled(false);
218     }
219 }
220
```

这是我们的本例子的主要代码，因为其中代码量很大，所以这里截取了部分代码进行展示和讲解。在上面代码第 4~8 行定义了一个 TabHost 标记文字，在第 10 行定义了所有需要的控件对象，在第 25 行得到 TabHost 对象，在第 33 行初始化 MySqliteHelper 对象，在第 34 行更新 UI 界面。在第 38 行用来添加 Tab，分别加入添加学生、删除学生信息、修改学生信息和更新学生信息的 Tab，并且设置相应的监听器。在第 106 行定义了插入学生信息的方法，首先修改状态为插入状态，判断用户是否可用，然后插入信息后更新 UI。在第 128 行定义了更新学生信息的 update 方法，在第 158 行定义了 search 方法，查找学生信息。在第 197 行定义了更新学生信息的方法，思路都是类似的。

然后添加 src/com.wyl.example/ MySqliteHelper.java 文件，代码如下：

```
001 //定义了数据库访问对象
002 public class MySqliteHelper extends SQLiteOpenHelper {
003     private static final String DATABASE_NAME = "stu_db";
                                                            //数据库的名字
004     private static final int DATABASEVERSION = 1;  //版本号
005     private static final String TABLE_NAME = "stu_table"; //表名
006
007     private SQLiteDatabase db;                           //数据库
008     private static final String TAG = "MyDataBase";
009
```

```
010        //4个字段
011        public static final String ID = "_id";
012        public static final String stuName = "stu_name";
013        public static final String stuNumber = "stu_number";
014        public static final String stuScore = "stu_score";
015
016        public MySqliteHelper(Context context) {
017            //TODO Auto-generated constructor stub
018            super(context, DATABASE_NAME, null, DATABASEVERSION);
019            //打开或新建数据库(第一次时创建)获得SQLiteDatabase对象,为了读取和写
                入数据
020            db = this.getWritableDatabase();
021        }
022
023        @Override
024        public void onCreate(SQLiteDatabase db) {
025            //TODO Auto-generated method stub
026            Log.i(TAG, "onCreate()");
027            //创建表的SQL语句
028            String sql = "CREATE TABLE " + TABLE_NAME + " (" + ID
029                    + " INTEGER PRIMARY KEY AUTOINCREMENT," + stuName + " TEXT,"
030                    + stuNumber + " INTEGER," + stuScore + " FLOAT)";
031            db.execSQL(sql);
032
033        }
034        //更新数据库
035        @Override
036        public void onUpgrade(SQLiteDatabase db, int oldVersion, int
           newVersion) {
037            //TODO Auto-generated method stub
038            Log.i(TAG, " onUpgrade() ");
039            //删除表的SQL
040            String sql = "DROP TABLE IF EXITS " + TABLE_NAME;
041            db.execSQL(sql);
042            onCreate(db);
043        }
044        //关闭数据库
045        @Override
046        public synchronized void close() {
047            //TODO Auto-generated method stub
048            Log.i(TAG, "close()");
049            //关闭数据库
050            db.close();
051            super.close();
052        }
053
054        //查询所有的数据,返回Cursor对象(按照id的升序排列)
055        public Cursor searchAllData()
056        {
057            Log.i(TAG, " searchAllData()");
058            //数据查询
059            //asc是升序 desc为降序(默认为asc)
060            return db.query(TABLE_NAME, null, null, null, null, null,
                MySqliteHelper.ID+" ASC" );
061        }
062        //插入数据
063        public void  insertData(String name,int number,float score )
064        {
065            //在数据库中插入记录
```

```
066        ContentValues values=new ContentValues();
067        values.put(MySqliteHelper.stuName, name);
068        values.put(MySqliteHelper.stuNumber, number);
069        values.put(MySqliteHelper.stuScore, score);
070        long row=db.insert(TABLE_NAME, null, values);
071        Log.i(TAG, "insertData row="+row);
072    }
073    //查询指定的信息
074    public Cursor searchSpecific(int number)
075    {
076        //按条件查询
077        String[] columns={
078                MySqliteHelper.ID,
079                MySqliteHelper.stuName,
080                MySqliteHelper.stuNumber,
081                MySqliteHelper.stuScore
082        };
083
084        Cursor cur=db.query(TABLE_NAME, columns,
            MySqliteHelper.stuNumber+"="+number, null, null, null, null);
085        Log.i("searchSpecific()", " cur.getCount()="+cur.getCount());
086        return cur;
087    }
088    //修改数据
089    public void upDateinfo(int id ,String name,int number,float score )
090    {
091        //修改数据库中的某条记录
092        ContentValues values=new ContentValues();
093        values.put(MySqliteHelper.stuName, name);
094        values.put(MySqliteHelper.stuNumber, number);
095        values.put(MySqliteHelper.stuScore, score);
096        String whereClause=MySqliteHelper.ID+" = ? ";
097        String whereArgs[]={Integer.toString(id)};
098
099        int rowaffected =db.update(TABLE_NAME, values, whereClause,
            whereArgs);
100        Log.i(TAG, "upDateinfo()  rowaffected="+rowaffected);
101    }
102
103    //删除数据
104    public void deleteStuInfo(int id)
105    {
106        int rowaffected =db.delete(TABLE_NAME, MySqliteHelper.ID+"="
            +id, null);
107        Log.i(TAG, "deleteStuInfo()  rowaffected="+rowaffected);
108    }
109    //判断是否存在该学生的信息
110    public Cursor isHaveThisStu(int number)
111    {
112        String[] columns={
113                MySqliteHelper.ID,
114                MySqliteHelper.stuName,
115                MySqliteHelper.stuNumber,
116                MySqliteHelper.stuScore
117        };
118
119        Cursor cur=db.query(TABLE_NAME, columns,
            MySqliteHelper.stuNumber+"="+number, null, null, null, null);
120        Log.i("isHaveThisStu()", " cur.getCount()="+cur.getCount());
121        return cur;
```

```
122     }
123 }
```

此文件类定义了我们方便访问数据库的操作类，在此类中的第 3~14 行定义了数据库的基本信息，包括数据库名字、版本号和字段名等。在第 18~20 行初始化此类的对象时，会初始化 db 对象。然后在第 24 行的 onCreate 方法中，在 db 中执行新建表的操作，在第 36 行定义了通过 SQL 语句更新表的操作。在第 46 行定义了关闭数据库的操作，在第 55 行定义了查询表的操作，在第 63 行定义了插入表的操作，在第 89 行定义了更新表的操作，在第 104 行定义了删除表记录的操作。

在本程序中还有一个辅助类，新建 src/com.wyl.example/ State.java 文件，代码如下：

```
01 //定义数据库的状态类
02 public class State {
03 public static boolean isSearch = false; //判断是否处于查询的表单的标志(flag)
04 public static boolean isInsert = false; //判断是否处于添加的表单的标志(flag)
05 public static boolean isUpdataOrDelete = false;
                              //判断是否处于删除或更新的表单的标志(flag)
06 public static boolean isDelete = false;//判断是否处于删除的表单的标志(flag)
07 //判断查找状态
08 public static boolean isSearch() {
09     return isSearch;
10 }
11 //设置查找状态
12 public static void setSearch(boolean isSearch) {
13     State.isSearch = isSearch;
14 }
15 //判断插入状态
16 public static boolean isInsert() {
17     return isInsert;
18 }
19 //设置插入状态
20 public static void setInsert(boolean isInsert) {
21     State.isInsert = isInsert;
22 }
23 //判断更新状态
24 public static boolean isUpdataOrDelete() {
25     return isUpdataOrDelete;
26 }
27 //设置更新状态
28 public static void setUpdataOrDelete(boolean isUpdataOrDelete) {
29     State.isUpdataOrDelete = isUpdataOrDelete;
30 }
31 //判断删除状态
32 public static boolean isDelete() {
33     return isDelete;
34 }
35  //设置删除状态
36 public static void setDelete(boolean isDelete) {
37     State.isDelete = isDelete;
38 }
39 }
```

在此类中主要用来标记程序对于数据库的操作状态主要是添加、删除、更新和查找等状态的切换。

第 6 章　Android 的数据存储

4．实例扩展

此实例相对是一个比较完整的实例，包括了一个学生管理系统的简单的添加、删除、修改和查找。并且可以简单的体现界面和数据库访问的结合性。能够完成此实例，可以说大家学习 Android 的一个小小的里程碑。

6.2　Android 中的 ContentProvider

范例 122　音乐播放器

1．实例简介

在使用应用的过程中，经常会发现我们可以得到系统中的所有音频的效果。例如，音乐播放软件得到系统的所有音频等，或者要对音频进行筛选等等。遇到这样的功能，我们一般是通过系统提供的 ContentProvider 来实现的，系统对于常用的数据给开发者提供了方便的获取方式，例如，手机中的音频文件列表、手机中的短信获取和手机的电话本获取等。本例子就带领大家来实现一个得到系统筛选的音乐并且进行播放的实例。

2．运行效果

该实例运行效果如图 6.7 所示。

图 6.7　音乐播放器

3．实例程序讲解

在本实例中，提供四个按钮，分别是播放、上一曲、下一曲和暂停按钮。想要实现本

实例效果,首先修改 res/layout/activity_main.xml 文件,代码如下:

```xml
01 <?xml version="1.0" encoding="utf-8"?>
02 <LinearLayout xmlns:android="http://schemas.android.com/apk/res/android"
03     android:layout_width="fill_parent"
04     android:layout_height="fill_parent"
05     android:orientation="vertical" >
06
07     <TextView
08         android:layout_width="fill_parent"
09         android:layout_height="wrap_content"
10         android:text="wyl music player" />
11     <!-- 显示四个按钮 -->
12     <LinearLayout
13         android:layout_width="fill_parent"
14         android:layout_height="wrap_content"
15         android:orientation="horizontal" >
16
17         <!-- 上一首按钮 -->
18         <Button
19             android:id="@+id/previous"
20             android:layout_width="wrap_content"
21             android:layout_height="fill_parent"
22             android:layout_weight="1"
23             android:text="上一首" />
24         <!-- 播放按钮 -->
25         <Button
26             android:id="@+id/play"
27             android:layout_width="wrap_content"
28             android:layout_height="fill_parent"
29             android:layout_weight="1"
30             android:text="播放" />
31         <!-- 下一首按钮 -->
32         <Button
33             android:id="@+id/next"
34             android:layout_width="wrap_content"
35             android:layout_height="fill_parent"
36             android:layout_weight="1"
37             android:text="下一首" />
38         <!-- 暂停按钮 -->
39         <Button
40             android:id="@+id/pause"
41             android:layout_width="wrap_content"
42             android:layout_height="fill_parent"
43             android:layout_weight="1"
44             android:text="暂停" />
45     </LinearLayout>
46
47 </LinearLayout>
```

这是我们的 Activity 的布局文件。定义了四个按钮分别为,上一首、播放、下一首和暂停。当用户单击某个按钮的时候调用对应的事件响应。

然后修改 src/com.wyl.example/MainActivity.java 文件,代码如下:

```java
01 //定义了本实例的主要 Activity
02 public class MainActivity extends Activity implements OnClickListener {
```

```
03    private Button mBtnPrevious;                    //上一首
04    private Button mBtnPlay;                        //播放
05    private Button mBtnNext;                        //下一首
06    private Button mBtnPause;                       //暂停
07    private ComponentName component;                //用于启动服务
08
09    public void onCreate(Bundle savedInstanceState) {
10        super.onCreate(savedInstanceState);
11        setContentView(R.layout.activity_main);
12        //得到布局中的控件
13        findView();
14        //绑定控件事件
15        setListener();
16    }
17
18    //得到布局中的控件
19    private void findView() {
20
21        component = new ComponentName(this, MusicService.class);
22        mBtnPrevious = (Button) findViewById(R.id.previous);
23        mBtnPlay = (Button) findViewById(R.id.play);
24        mBtnNext = (Button) findViewById(R.id.next);
25        mBtnPause = (Button) findViewById(R.id.pause);
26    }
27
28    //绑定控件事件
29    private void setListener() {
30        mBtnPrevious.setOnClickListener(this);
31        mBtnPlay.setOnClickListener(this);
32        mBtnNext.setOnClickListener(this);
33        mBtnPause.setOnClickListener(this);
34    }
35
36    //按钮单击事件响应
37    public void onClick(View v) {
38        //如果单击前一首歌,就在intent中传递前一首歌参数
39        if (v == mBtnPrevious) {
40            Intent mIntent = new Intent(MusicService.PREVIOUS_ACTION);
41            mIntent.setComponent(component);
42            startService(mIntent);
43        //如果单击前播放歌曲,就在intent中传递播放当前歌参数
44        } else if (v == mBtnPlay) {
45            Intent mIntent = new Intent(MusicService.PLAY_ACTION);
46            mIntent.setComponent(component);
47            startService(mIntent);
48        //如果单击前一首歌,就在intent中传递下一首歌参数
49        } else if (v == mBtnNext) {
50            Intent mIntent = new Intent(MusicService.NEXT_ACTION);
51            mIntent.setComponent(component);
52            startService(mIntent);
53        //如果单击前一首歌,就在intent中传递暂停首歌参数
54        } else {
55            Intent mIntent = new Intent(MusicService.PAUSE_ACTION);
56            mIntent.setComponent(component);
57            startService(mIntent);
58        }
59    }
60 }
```

此文件是Activity的代码文件，在第3～6行定义了四个按钮。在第19行定义的findView方法中得到响应的控件对象，在第29行给相应的控件绑定了相应的单击监听器对象。在第37行实现了通过不同的按钮单击传递不同的参数给Service。

还有自定义的 Service 类，新建 src/com.wyl.example/ MusicService.java 文件，代码如下：

```
001    //定义音乐服务类
002    public class MusicService extends Service {
003        //定义需要显示的音乐的字段
004        String[] mCursorCols = new String[] {
005            "audio._id AS _id", //index must match IDCOLIDX below
006            MediaStore.Audio.Media.ARTIST, MediaStore.Audio
                .Media.ALBUM,
007            MediaStore.Audio.Media.TITLE, MediaStore.Audio.Media
                .DATA,
008            MediaStore.Audio.Media.MIME_TYPE, MediaStore.Audio.Media
                .ALBUM_ID,
009            MediaStore.Audio.Media.ARTIST_ID, MediaStore.Audio.Media
                .DURATION };
010        private MediaPlayer mMediaPlayer;              //声明播放器
011        private Cursor mCursor;                        //声明游标
012        private int mPlayPosition = 0;                 //当前播放的歌曲
013
014        //注册意图
015        public static final String PLAY_ACTION = "com.wyl.music.PLAY
               _ACTION";
016        public static final String PAUSE_ACTION = "com.wyl.music.PAUSE
               _ACTION";
017        public static final String NEXT_ACTION = "com.wyl.music.NEXT
               _ACTION";
018        public static final String PREVIOUS_ACTION = "com.wyl.music
               .PREVIOUS_ACTION";
019
020        @Override
021        public IBinder onBind(Intent arg0) {
022            //TODO Auto-generated method stub
023            return null;
024        }
025
026        @Override
027        public void onCreate() {
028            super.onCreate();
029            mMediaPlayer = new MediaPlayer();
030            //通过一个URI可以获取所有音频文件
031            Uri MUSIC_URL = MediaStore.Audio.Media.EXTERNAL_CONTENT_URI;
032            //这里我过滤了一下，因为我机里有些音频文件是游戏音频，很短
033            //我这里作了处理，默认大于10秒的可以看作是系统音乐
034            mCursor = getContentResolver().query(MUSIC_URL, mCursorCols,
035                "duration > 10000", null, null);
036        }
037
038        @Override
039        public void onStart(Intent intent, int startId) {
040            super.onStart(intent, startId);
041            //根据不同的action，做不同的响应
042            String action = intent.getAction();
043            //播放
```

```
044            if (action.equals(PLAY_ACTION)) {
045                play();
046            //暂停
047            } else if (action.equals(PAUSE_ACTION)) {
048                pause();
049            //下一首
050            } else if (action.equals(NEXT_ACTION)) {
051                next();
052            //前一首
053            } else if (action.equals(PREVIOUS_ACTION)) {
054                previous();
055            }
056        }
057
058        //播放音乐
059        public void play() {
060            //初始化音乐播放器
061            inite();
062        }
063
064        //暂停时,结束服务
065        public void pause() {
066            //暂停音乐播放
067            stopSelf();
068        }
069
070        //上一首
071        public void previous() {
072            //得到前一首的歌曲
073            if (mPlayPosition == 0) {
074                mPlayPosition = mCursor.getCount() - 1;
075            } else {
076                mPlayPosition--;
077            }
078            //开始播放
079            inite();
080        }
081
082        //下一首
083        public void next() {
084            //得到后一首歌曲
085            if (mPlayPosition == mCursor.getCount() - 1) {
086                mPlayPosition = 0;
087            } else {
088                mPlayPosition++;
089            }
090            //开始播放
091            inite();
092        }
093
094        //初始化播放器
095        public void inite() {
096            //充值 MediaPlayer
097            mMediaPlayer.reset();
098            //获取歌曲位置
099            String dataSource = getDateByPosition(mCursor, mPlayPosition);
100            //歌曲信息
101            String info = getInfoByPosition(mCursor, mPlayPosition);
```

```
102        //用 Toast 显示歌曲信息
103        Toast.makeText(getApplicationContext(), info, Toast.LENGTH_SHORT)
104                .show();
105        try {
106            //播放器绑定资源
107            mMediaPlayer.setDataSource(dataSource);
108            //播放器准备
109            mMediaPlayer.prepare();
110            //播放
111            mMediaPlayer.start();
112        } catch (IllegalArgumentException e1) {
113            e1.printStackTrace();
114        } catch (IllegalStateException e1) {
115            e1.printStackTrace();
116        } catch (IOException e1) {
117            e1.printStackTrace();
118        }
119    }
120
121    //根据位置来获取歌曲位置
122    public String getDateByPosition(Cursor c, int position) {
123        c.moveToPosition(position);
124        int dataColumn = c.getColumnIndex(MediaStore.Audio.Media.DATA);
125        String data = c.getString(dataColumn);
126        return data;
127    }
128
129    //获取当前播放歌曲演唱者及歌名
130    public String getInfoByPosition(Cursor c, int position) {
131        c.moveToPosition(position);
132        int titleColumn = c.getColumnIndex(MediaStore.Audio.Media.TITLE);
133        int artistColumn = c.getColumnIndex(MediaStore.Audio
            .Media.ARTIST);
134        String info = c.getString(artistColumn) + " "
135                + c.getString(titleColumn);
136        return info;
137
138    }
139
140    //服务结束时要释放 MediaPlayer
141    public void onDestroy() {
142        super.onDestroy();
143        mMediaPlayer.release();
144    }
145 }
```

此类是自定义的服务类，在代码的第 4~9 行定义了获取音频数据的字段名称，在第 15~18 行定义了启动服务所能做的一些服务操作。在第 27 行 onCreate 方法中初始化 MediaPlayer 对象，并且通过第 34 行的 getContentResolver 得到了系统中所有的音乐，这里做了一个限制就是播放时间在 10 秒以上的可以得到。在第 39 行 onStart 方法中判断得到的 intent 中的参数，调用相应的方法，在第 59、65、71 和 83 行分别定义了音乐的播放、暂停、前一首和下一首的实现内容，基本思路都是一样，得到需要播放的音频的位置，然后调用 init 方法播放根据位置得到的音频文件。这样一个音乐播放器就完成了。

4. 实例扩展

在此实例中仅仅通过 ContentProvider 得到了系统中的音频并且进行播放，其实我们可以得到更多的系统音频的信息，如音乐的专辑和音乐的歌唱者等信息，所以基本上现在市面上的音乐播放器都是基于 ContentProvider 来实现的。

范例 123 系统图片选择预览

1. 实例简介

在我们使用应用的过程中，经常会遇到一些功能使用户选择某张图片，然后预览效果然后提交。例如，用户发送新浪微博时会选择系统内部的照片，插入微博后预览效果，没问题的时候提交服务器。遇到这样的功能，我们一般是当用户进行某个操作的时候，例如：单击某个按钮的时候，打开系统的图片浏览工具进行图片选择，然后在页面中可以预览选择的图片。本例子就带领大家来实现一个单击打开系统图片相册选择图片并且预览的实例。

2. 运行效果

该实例运行效果如图 6.8 所示。

图 6.8 系统图片选择浏览

3. 实例程序讲解

在本实例中，首先一个页面中包含一个 ImageView 控件和一个 Button 控件，当单击按钮控件的时候打开系统的浏览图片的工具选择图片，然后在 ImageView 控件中预览选择的图片效果。想要实现本实例效果，首先修改 res/layout/activity_main.xml 文件，代码如下：

```xml
01  <?xml version="1.0" encoding="utf-8"?>
02  <LinearLayout xmlns:android="http://schemas.android.com/apk/res/android"
03      android:layout_width="fill_parent"
04      android:layout_height="fill_parent"
05      android:orientation="vertical" >
06      <!-- 定义显示图片的控件 -->
07      <ImageView
08          android:id="@+id/imgView"
09          android:layout_width="fill_parent"
10          android:layout_height="wrap_content"
11          android:layout_weight="1" >
12      </ImageView>
13
14      <!--定义一个选择图片的按钮 -->
15      <Button
16          android:id="@+id/buttonLoadPicture"
17          android:layout_width="match_parent"
18          android:layout_height="wrap_content"
19          android:layout_gravity="center"
20          android:layout_weight="0"
21          android:text="选择图片" >
22      </Button>
23
24  </LinearLayout>
```

这是我们的 Activity 的布局文件。在其中第 7～12 行定义了一个 ImageView 控件，用来预览选择的图片。在第 15～22 行定义一个 Button 控件，当用户单击的时候打开系统的图片浏览工具浏览图片。

然后修改 src/com.wyl.example/MainActivity.java 文件，代码如下：

```java
01  //定义了本实例的主要Activity
02  public class MainActivity extends Activity {
03
04      //设置类型
05      private static int RESULT_LOAD_IMAGE = 1;
06      private Button BtnLoadImage;
07      private ImageView IvLoadImage;
08
09      @Override
10      public void onCreate(Bundle savedInstanceState) {
11          super.onCreate(savedInstanceState);
12          setContentView(R.layout.activity_main);
13
14          //得到布局中的控件
15          findView();
16          //绑定控件事件
17          setListener();
18      }
19
20      private void findView() {
21          //绑定控件
22          BtnLoadImage = (Button) findViewById(R.id.buttonLoadPicture);
23          IvLoadImage = (ImageView) findViewById(R.id.imgView);
24      }
25
26      private void setListener() {
27          //设置事件
28          BtnLoadImage.setOnClickListener(new View.OnClickListener() {
```

```
29
30          @Override
31          public void onClick(View arg0) {
32              //初始化 intent 控件,设置属性为 ACTION_PICK,可以从系统中选择图片
33              Intent i = new Intent(
34                  Intent.ACTION_PICK,
35                  android.provider.MediaStore.Images.Media
                    .EXTERNAL_CONTENT_URI);
36              //启动 intent
37              startActivityForResult(i, RESULT_LOAD_IMAGE);
38          }
39      });
40
41 }
42
43 //获取返回结果
44 @Override
45 protected void onActivityResult(int requestCode, int resultCode, Intent data) {
46      super.onActivityResult(requestCode, resultCode, data);
47
48      if (requestCode == RESULT_LOAD_IMAGE && resultCode == RESULT_OK
49          && null != data) {
50          //获取数据
51          Uri selectedImage = data.getData();
52          String[] filePathColumn = { MediaStore.Images.Media.DATA };
53          //查询游标
54          Cursor cursor = getContentResolver().query(selectedImage,
55              filePathColumn, null, null, null);
56          cursor.moveToFirst();
57          //获取数据
58          int columnIndex = cursor.getColumnIndex(filePathColumn[0]);
59          String picturePath = cursor.getString(columnIndex);
60          cursor.close();
61
62          //设置图片
63          IvLoadImage.setImageBitmap(BitmapFactory.decodeFile
                (picturePath));
64      }
65 }
66 }
```

此文件是 Activity 的代码文件,在其中第 6~7 行定义了一个 ImageView 对象和一个 Button 对象,在第 20 行得到了布局中的相应对象。在第 26 行设置了相应的按钮监听器,当单击选择图片按钮的时候,初始化一个 Intent,设置方法为 ACTION_PICK,然后启动此 intent,打开系统的浏览图片的工具选择图片。在第 45 行定义了 onActivityResult 的回调方法,得到回调的 intent 参数中包含着用户选择的图片信息,这里我们就可以通过 ContentProvider 得到此回调信息中的图片了,然后设置给 ImageView 控件,就可以看到预览效果了。

4. 实例扩展

用户通过内置的 ContentProvider 不但可以得到图片的内容,而且还可以得到图片的详细信息。例如,图片尺寸、图片创建时间和图片大小等信息,我们常见的图片浏览器基本都是依据 ContentProvider 来实现的。

范例 124　系统的联系人

1．实例简介

在我们使用应用的过程中，经常会使用得到系统联系人的功能，例如，你在话费充值的时候希望直接选择系统的联系人，而不是输入电话号码；在大家通过手机客户端订购团购商品时也要留下联系人电话等，这样方便大家可以直接从系统的联系人中选择得到电话信息。本例子就带领大家来实现一个获得系统联系人列表的实例。

2．运行效果

该实例运行效果如图 6.9 所示。

3．实例程序讲解

在本实例中，首先显示一个页面列出了所有手机的联系人列表。想要实现本实例效果，首先修改 res/layout/activity_main.xml 文件，代码如下：

图 6.9　系统联系人列表

```
01  <RelativeLayout xmlns:android="http://schemas.android.com/apk/res/android"
02      xmlns:tools="http://schemas.android.com/tools"
03      android:id="@+id/RelativeLayout1"
04      android:layout_width="match_parent"
05      android:layout_height="match_parent"
06      android:orientation="vertical"
07      tools:context=".MainActivity" >
08      <!-- 定义联系人的显示列表控件 -->
09      <ListView
10          android:id="@+id/Lv_show"
11          android:layout_width="match_parent"
12          android:layout_height="wrap_content"
13          android:layout_alignParentLeft="true"
14          android:layout_alignParentTop="true"
15          android:visibility="gone" >
16      </ListView>
17      <!-- 定义加载进度条的控件 -->
18      <LinearLayout
19          android:id="@+id/Llay_progressBar"
20          android:layout_width="wrap_content"
21          android:layout_height="wrap_content"
22          android:layout_alignParentBottom="true"
23          android:layout_centerHorizontal="true" >
24
25          <ProgressBar
26              android:id="@+id/progressBar1"
27              android:layout_width="wrap_content"
28              android:layout_height="wrap_content" />
29
```

```
30          <TextView
31              android:layout_width="wrap_content"
32              android:layout_height="wrap_content"
33              android:paddingTop="20dp"
34              android:text="加载中..." />
35      </LinearLayout>
36
37  </RelativeLayout>
```

这是我们的 Activity 的布局文件。在其中第 9~16 行定义了一个 ListView 控件，用来显示用户的联系人信息，在第 18~35 行定义了在联系人加载过程中的 ProgressBar 控件。

然后修改 src/com.wyl.example/MainActivity.java 文件，代码如下：

```
001  //定义了本实例的主要 Activity
002  public class MainActivity extends Activity {
003
004      private static final String TAG = "MainActivity";
005      private List<String> data = new ArrayList<String>();
                                                      //存储联系人数据
006      private ListView mLvShow;                    //用于显示的列表
007      private LinearLayout mLlay_progressBar;      //进度框
008
009      Handler mHandler = new Handler() {
010          public void handleMessage(Message msg) {
011              //隐藏进度条
012              mLlay_progressBar.setVisibility(View.GONE);
013              //并显示绑定数据
014              mLvShow.setAdapter(new ArrayAdapter<String>(MainActivity
                     .this,
015                  android.R.layout.simple_list_item_1, data));
016              mLvShow.setVisibility(View.VISIBLE);
017          }
018      };
019
020      @Override
021      public void onCreate(Bundle savedInstanceState) {
022          super.onCreate(savedInstanceState);
023          setContentView(R.layout.activity_main);
024
025          //绑定布局
026          findView();
027          //获取联系人信息
028          new Thread(new myRunnable()).start();
029      }
030
031      private void findView() {
032          //得到布局中的控件对象
033          mLvShow = (ListView) findViewById(R.id.Lv_show);
034          mLlay_progressBar = (LinearLayout) findViewById(R.id.Llay
                 _progressBar);
035          mLlay_progressBar.setVisibility(View.VISIBLE);
036      }
037
038      //获取号码的代码
039      private void getContacts() {
040          //得到 ContentResolver 对象
041          ContentResolver cr = this.getContentResolver();
042          //取得电话本中开始一项的游标,主要就是查询"contacts"表
```

```
043        Cursor cursor = cr.query(ContactsContract.Contacts.CONTENT_
           URI,null,
044                null, null, null);
045        while (cursor.moveToNext()) {
046            StringBuilder sbLog = new StringBuilder();
047
048            //取得联系人名字 (显示出来的名字),实际内容在 ContactsContract
                 .Contacts 中
049            int nameIndex = cursor
050                    .getColumnIndex(ContactsContract.Contacts
                        .DISPLAY_NAME);
051            String name = cursor.getString(nameIndex);
052            sbLog.append(name + ":");
053
054            //取得联系人 ID
055            String contactId = cursor.getString(cursor
056                    .getColumnIndex(ContactsContract.Contacts._ID));
057
058            //根据联系人 ID 查询对应的电话号码
059            Cursor phoneNumbers = cr.query(
060                    ContactsContract.CommonDataKinds.Phone.CONTENT
                        _URI, null,
061                    ContactsContract.CommonDataKinds.Phone
                        .CONTACT_ID + " = "
062                        + contactId, null, null);
063            //取得电话号码(可能存在多个号码)
064            while (phoneNumbers.moveToNext()) {
065                String strPhoneNumber = phoneNumbers
066                        .getString(phoneNumbers
067                        .getColumnIndex(ContactsContract.CommonDataKinds
                            .Phone.NUMBER));
068                sbLog.append(strPhoneNumber + ";");
069            }
070            phoneNumbers.close();
071
072            //根据联系人 ID 查询对应的 email
073            Cursor emails = cr.query(
074                    ContactsContract.CommonDataKinds.Email.CONTENT
                        _URI, null,
075                    ContactsContract.CommonDataKinds.Email.
                        CONTACT_ID + " = "
076                        + contactId, null, null);
077            //取得 email(可能存在多个 email)
078            while (emails.moveToNext()) {
079                String strEmail = emails
080                        .getString(emails
081                            .getColumnIndex(ContactsContract
                            .CommonDataKinds.Email.DATA));
082                sbLog.append("Email=" + strEmail + ";");
083            }
084            emails.close();
085            //打印联系人
086            Log.e(TAG, sbLog.toString());
087            data.add(sbLog.toString());
088        }
089        cursor.close();
090
091    }
092
093    class myRunnable implements Runnable {
```

```
094            public void run() {
095                Message message = new Message();
096                //得到系统中的联系人数据
097                getContacts();
098                //发送 handler 消息
099                mHandler.sendMessage(message);
100            }
101        }
102 }
```

此文件是 Activity 的代码文件，在其中第 5 行定义了一个存储联系人的 List 对象，在第 9～18 行定义了接收用户信息的 handler 对象，当联系人信息加载完毕后会发送信息给 handler，并且显示 list 对象。在第 26～28 行就是通过 findView 方法得到了布局中的控件对象，然后启动一个线程去加载联系人信息。在第 93～101 行定义了一个 Runnable 对象用来加载联系人信息，其中主要调用了 getContacts 方法。在第 39 行定义了 getContacts 方法，在其中首先获得 ContentResolver 对象，然后通过 query 方法得到联系人的结果 cursor，通过循环遍历得到每个联系人的信息，并且存储在 data 链表中。

4．实例扩展

在此实例中仅仅获得了联系人的名字，联系人的电话信息，其实可以获得的信息还很多，例如：联系人的邮箱、工作单位和分组等信息。详细内容请大家查看 Google 开发者文档。

范例 125　得到系统的音频文件

1．实例简介

在我们使用应用的过程中，经常会去查找系统的一些资源，如，查找系统的音频资源和查找系统的图片资源。遇到这样的功能，我们一般是当用户进行某个操作的时候，例如，单击某个按钮的时候，通过 Android 提供的相应的 ContentProvider 得到系统的音频或者视频资源。本例子就带领大家来实现一个得到系统中音频文件的实例。

2．运行效果

该实例运行效果如图 6.10 所示。

3．实例程序讲解

在本实例中，通过 ContentProvider 得到系统中的音频文件并且放入 ListView 进行显示。想要实现本实例效果，首先修改 res/layout/activity_main.xml 文件。代码如下：

图 6.10　得到系统的音频文件

```
01 <RelativeLayout xmlns:android="http://schemas.android.com/apk/res/
   android"
```

```xml
02      xmlns:tools="http://schemas.android.com/tools"
03      android:id="@+id/RelativeLayout1"
04      android:layout_width="match_parent"
05      android:layout_height="match_parent"
06      android:orientation="vertical"
07      tools:context=".MainActivity" >
08      <!-- 显示音频文件列表 -->
09      <ListView
10          android:id="@+id/Lv_show"
11          android:layout_width="match_parent"
12          android:layout_height="wrap_content"
13          android:layout_alignParentLeft="true"
14          android:layout_alignParentTop="true"
15          android:visibility="gone" >
16      </ListView>
17      <!-- 定义的加载进度条 -->
18      <LinearLayout
19          android:id="@+id/Llay_progressBar"
20          android:layout_width="wrap_content"
21          android:layout_height="wrap_content"
22          android:layout_centerInParent="true" >
23
24          <ProgressBar
25              android:id="@+id/progressBar"
26              style="?android:attr/progressBarStyleLarge"
27              android:layout_width="wrap_content"
28              android:layout_height="wrap_content" />
29      </LinearLayout>
30
31  </RelativeLayout>
```

这是我们的 Activity 的布局文件。在其中第 9~16 行定义了一个 ListView 控件，用来显示得到的音频文件列表。在第 18~29 行定义一个 ProgressBar 控件用来显示在加载系统音频过程中的进度条。

然后修改 src/com.wyl.example/MainActivity.java 文件，代码如下：

```java
01  //定义了本实例的主要 Activity
02  public class MainActivity extends Activity {
03      private ListView mLvShow;                           //用于显示的列表
04      private LinearLayout mLlay_progressBar;             //进度框
05      private List<String> data = new ArrayList<String>();//存储多媒体数据
06
07      Handler mHandler = new Handler() {
08          public void handleMessage(Message msg) {
09              //隐藏进度条
10              mLlay_progressBar.setVisibility(View.GONE);
11              //并显示绑定数据
12              mLvShow.setAdapter(new ArrayAdapter<String>(MainActivity
                  .this,
13                  android.R.layout.simple_list_item_1, data));
14              mLvShow.setVisibility(View.VISIBLE);
15          }
16      };
17
18      @Override
19      public void onCreate(Bundle savedInstanceState) {
20          super.onCreate(savedInstanceState);
21          setContentView(R.layout.activity_main);
```

```
22        //得到控件对象
23        findView();
24        //获取联信息
25        new Thread(new myRunnable()).start();
26    }
27
28    private void findView() {
29        mLvShow = (ListView) findViewById(R.id.Lv_show);
30        mLlay_progressBar = (LinearLayout) findViewById(R.id.Llay
          _progressBar);
31        mLlay_progressBar.setVisibility(View.VISIBLE);
32    }
33
34    private void getData() {
35        //TODO Auto-generated method stub
36        ContentResolver cr = getContentResolver();
37        //ContentProvider 只能由 ContentResolver 发送请求
38        Uri AUDIO_URI = MediaStore.Audio.Media.EXTERNAL_CONTENT_URI;
39        //获取音频文件的 URI,
40        //视频 MediaStore.Video.Media.EXTERNAL_CONTENT_URI
41        //图片 MediaStore.Images.Media.EXTERNAL_CONTENT_URI
42        String[] columns = new String[] { MediaStore.Audio.Media.TITLE,
43               MediaStore.Audio.Media.DATA };
44        //要读的列名,这些常量可以查 GOOGLE 官方开发文档,TITLE 是标题 DATA 是路径
45        Cursor cursor = cr.query(AUDIO_URI, columns,
46               MediaStore.Audio.Media.DURATION + ">?",
47               new String[] { "1000" }, null);
48        Log.e("cursor", cursor.getCount()+"");
49        //跟查询 SQL 一样了,除了第一个参数不同外。后面根据时长过滤小于 1 秒的文件
50        while (cursor.moveToNext()) {
51            //循环读取第一列,即文件路径,0 列是标题
52            Log.e("ContentProvider 的使用 读取手机内音频 视频 图片文件", cursor
              .getString(1));
53            data.add(cursor.getString(1));
54        }
55        cursor.close();
56    }
57    class myRunnable implements Runnable {
58        public void run() {
59            Message message = new Message();
60            //通过 contentprovider 获取数据
61            getData();
62            mHandler.sendMessage(message);
63        }
64    }
65 }
```

此文件是 Activity 的代码文件,在其中第 3~4 行定义 ListView 对象用来显示音频文件的列表,在第 7~16 行定义了当前页面的 handler 对象,在接下来线程处理完毕的时候发送消息给此 handler,然后显示 ListView 控件。在第 19 行的 onCreate 方法中调用 findview 方法得到布局中的控件对象,然后启动线程加载系统的音频文件数据。在第 57~64 行定义了一个 Runnable 类,在此 run 回调方法中调用 getData 方法获取系统的音频文件。在第 34~56 行实现了 getData 方法,在其中定义了得到音频文件的 URI,然后通过 ContentResolver 类的 query 方法得到所有的音频文件,并且加入到 data 列表中。当线程执行完毕后发送 handler 消息给 handler 对象进行处理,显示音频文件列表。

4．实例扩展

在此实例中我们仅仅得到了系统音频文件的名字，其实我们还有很多音频文件的信息可以得到，如音频文件的播放长度、大小、歌唱者和专辑等。详细参数请参考 Google 开发者文档。

6.3 Android 中的资源文件

范例 126　全屏界面

1．实例简介

在我们使用应用的过程中，经常会需要应用全屏的效果，也就是隐藏上面的状态栏和标题栏。例如，在一些浏览器中支持全屏效果，在大部分游戏中都是全屏效果。遇到这样的功能，我们一般是通过当前窗口属性来进行设置的。本例子就带领大家来实现一个全屏的应用实例效果。

2．运行效果

该实例运行效果如图 6.11 所示。

图 6.11　全屏界面

3．实例程序讲解

在本实例中，当用户打开当前的程序后会看到一个隐藏了标题栏和状态栏的界面。想要实现本实例效果，首先修改 res/layout/activity_main.xml 文件，代码如下：

```
01  <LinearLayout xmlns:android="http://schemas.android.com/apk/res/android"
02      xmlns:tools="http://schemas.android.com/tools"
03      android:id="@+id/LinearLayout1"
04      android:layout_width="match_parent"
05      android:layout_height="match_parent"
06      android:orientation="vertical"
07      tools:context=".MainActivity" >
08      <!-- 定义显示的标签控件 -->
09      <TextView
10          android:text="我是全屏效果"
11          android:gravity="center"
12          android:layout_width="match_parent"
13          android:layout_height="match_parent"
14          android:textSize="30sp"
15          />
16  </LinearLayout>
```

这是我们的 Activity 的布局文件。在其中定义了一个 TextView 控件主要是为了测试全屏效果。

然后修改 src/com.wyl.example/MainActivity.java 文件，代码如下：

```
01  //定义了本实例的主要 Activity
02  public class MainActivity extends Activity {
03
04  @Override
05  public void onCreate(Bundle savedInstanceState) {
06      super.onCreate(savedInstanceState);
07      //代码方式全屏
08      requestWindowFeature(Window.FEATURE_NO_TITLE);
09      getWindow().setFlags(WindowManager.LayoutParams.FLAG_FULLSCREEN,
10      WindowManager.LayoutParams.FLAG_FULLSCREEN);
11
12      setContentView(R.layout.activity_main);
13  }
14  }
```

此文件是 Activity 的代码文件，这个文件决定了全屏显示的本质，在第 8～9 行通过 getWindow 得到了当前界面，通过 setFlags 设置当前窗口的显示形式，如果把宽高都设置为 FLAG_FULLSCREEN，则可以全屏显示此 Activity。

4．实例扩展

在此实例中我们显示了一个全屏的 Activity，但是在实际应用中可能要求显示的需求不太一样，如可以只显示标题栏隐藏状态栏，或者自定义状态栏等。这些功能都可以通过此方式来实现，具体实现参数请参考 Google 开发者文档。

范例 127　小图堆积背景

1．实例简介

在我们使用应用的过程中，经常使用到的控件背景是通过相同的小图片连续堆积而成的。例如，我们在玩儿一些 RPG 游戏的时候的游戏背景，或者我们在使用一些应用的时候的特殊背景等。遇到这样的功能，我们一般是通过定义相应的背景布局，然后设置给相应的 View 来实现。本例子就带领大家来实现一个小图堆积背景的实例效果。

2. 运行效果

该实例运行效果如图 6.12 所示。

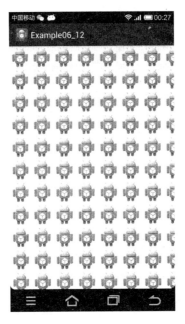

图 6.12　小图堆积背景

3. 实例程序讲解

在本实例中，当用户打开本应用程序时就可以看到由小图堆积后的效果作为页面的背景。想要实现本实例效果，首先修改 res/layout/activity_main.xml 文件，代码如下：

```xml
01  <!-- 定义activity的布局，设置背景为自定义文件 -->
02  <LinearLayout xmlns:android="http://schemas.android.com/apk/res/android"
03      android:id="@+id/MainLayout"
04      android:layout_width="fill_parent"
05      android:layout_height="fill_parent"
06      android:orientation="vertical"
07      android:background="@drawable/backrepeat">
08
09  </LinearLayout>
```

这是我们的 Activity 的布局文件。在其中第 7 行设置了当前 LinearLayout 的背景为 backrepeat，这不是一张固定的资源图片，而是我们定义好的一种 drawable 文件。

在我们工程目录下的 res/drawable/backrepeat.xml 文件定义了背景的显示样式，代码如下：

```xml
01  <!-- 定义背景重复显示的效果 -->
02  <bitmap xmlns:android="http://schemas.android.com/apk/res/android"
03      android:dither="true"
04      android:src="@drawable/ic_launcher"
05      android:tileMode="repeat" />
```

在此文件中相当于定义了一个 bitmap 对象，其中主要应用资源文件 ic_laucher，显示的模式为 repeat，也就是重复累积显示。所以本例子的关键就在这里。

然后修改 src/com.wyl.example/MainActivity.java 文件，代码如下：

```
01  //定义了本实例的主要Activity
02  public class MainActivity extends Activity {
03    @Override
04    public void onCreate(Bundle savedInstanceState) {
05      super.onCreate(savedInstanceState);
06      //在布局文件中设置了小图重复堆积的效果
07      setContentView(R.layout.activity_main);
08    }
09  }
```

此文件是 Activity 的代码文件，给当前的 Activity 设置刚才定义好的布局文件即可显示重复背景的页面效果了。

4．实例扩展

注意在本实例中实现了一种定义重复背景的操作，但是通过类似的思路可以实现更多的特殊效果。例如，不同状态下背景的切换，或者背景的各种操作对应的展示图片等。

范例 128　自定义 EditText 样式

1．实例简介

在我们使用应用的过程中，经常会发现应用中的控件样式不是系统默认的控件样式。例如，在某个应用中所有的按钮都是蓝色背景，字体是斜体的；在某个应用中所有的 Spinner 弹出框都有着与众不同的样式。遇到这样的功能，我们一般是通过自定义按钮的样式来实现的。本例子就带领大家来实现一个自定义样式的 EditText 的实例效果。

2．运行效果

该实例运行效果如图 6.13 所示。

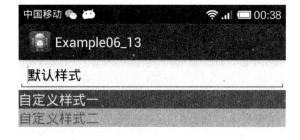

图 6.13　自定义样式的 EditText 控件

3. 实例程序讲解

在本实例中,当用户打开此应用程序时就可以看到三个 EditText,其中第一个是默认样式的,第二个和第三个都是使用自定义样式。想要实现本实例效果,首先修改 res/layout/activity_main.xml 文件,代码如下:

```xml
01 <LinearLayout xmlns:android="http://schemas.android.com/apk/res/android"
02     xmlns:tools="http://schemas.android.com/tools"
03     android:id="@+id/LinearLayout1"
04     android:layout_width="match_parent"
05     android:layout_height="match_parent"
06     android:orientation="vertical"
07     tools:context=".MainActivity" >
08     <!-- 定义了一个默认 EditText -->
09     <EditText
10         android:id="@+id/et1"
11         android:layout_width="fill_parent"
12         android:layout_height="wrap_content"
13         android:text="默认样式" >
14     </EditText>
15     <!-- 定义了一个应用样式的 EditText -->
16     <EditText
17         android:id="@+id/et2"
18         style="@style/et1"
19         android:layout_width="fill_parent"
20         android:layout_height="wrap_content"
21         android:text="自定义样式一" >
22     </EditText>
23     <!-- 定义了一个应用样式的 EditText -->
24     <EditText
25         android:id="@+id/et3"
26         style="@style/et2"
27         android:layout_width="fill_parent"
28         android:layout_height="wrap_content"
29         android:text="自定义样式二" >
30     </EditText>
31
32 </LinearLayout>
```

这是我们的 Activity 的布局文件。在其中第 8~14 行定义了一个 TextView 控件,没有设置样式。第 15~22 行定义了一个使用样式一的 EditText 控件。第 24~30 行定义了一个使用样式二的 EditText 控件。其中样式一和样式二分别定义了相应的样式文件。

新建 res/values/activity_main.xml 文件,其中定义了 EditText 的两种样式,代码如下:

```xml
01 <?xml version="1.0" encoding="utf-8"?>
02 <resources>
03     <!-- 自定义 EditText 的样式 -->
04     <style name="et1" parent="@android:style/Widget.EditText">
05         <item name="android:background">#1A4EA4</item>
06         <item name="android:textColor">#FFF111</item>
07     </style>
08
09     <style name="et2" parent="@android:style/Widget.EditText">
10         <item name="android:background">#A6C60F</item>
11         <item name="android:textColor">#EC02C3</item>
12     </style>
```

```
13
14  </resources>
```

此文件是用户定义的 style 文件,其中以 style 为节点,每个节点定义一个 view 的样式,在 style 节点中包括这个 style 的各种属性的值。上例中设置了 background 和 textColor 两个属性的值。

然后修改 src/com.wyl.example/MainActivity.java 文件,代码如下:

```
01  //定义了本实例的主要 Activity
02  public class MainActivity extends Activity {
03  //自定义样式
04  @Override
05  public void onCreate(Bundle savedInstanceState) {
06      super.onCreate(savedInstanceState);
07      setContentView(R.layout.activity_main);
08  }
09  }
```

此文件是 Activity 的代码文件,在此文件中设置相应的布局文件,这样就可以显示自定义的 EditText 样式了。

4.实例扩展

注意在本实例中仅仅定义了 EditText 的两个样式即背景和字体颜色。大家根据应用的需求定义多种样式,实现自己的控件自己做主的效果。

范例 129 透明背景的 Activity

1.实例简介

在我们使用应用的过程中,可能会遇到一些要求背景为透明 Activity。例如,在使用程序的时候希望能够看到之前 Activity 的内容。遇到这样的功能,我们一般是在程序的 manifest 文件中做配置。本例子就带领大家来实现一个背景完全透明的 Activity 的实例效果。

2.运行效果

该实例运行效果如图 6.14 所示。

图 6.14 透明背景的 Activity

3. 实例程序讲解

在本实例中，当用户打开应用程序会发现程序的整个背景为全透明。想要实现本实例效果，首先修改 res/layout/activity_main.xml 文件，代码如下：

```
01  <LinearLayout xmlns:android="http://schemas.android.com/apk/res/android"
02      xmlns:tools="http://schemas.android.com/tools"
03      android:id="@+id/LinearLayout1"
04      android:layout_width="match_parent"
05      android:layout_height="match_parent"
06      android:orientation="vertical"
07      tools:context=".MainActivity" >
08
09      <!-- 定义显示的标签控件 -->
10      <TextView
11          android:layout_width="wrap_content"
12          android:layout_height="wrap_content"
13          android:text="在 Manifest 中设置 Activity theme 为 Theme
                .Translucent 使 Activity 背景透明"
14          android:textSize="20sp"
15          android:textColor="@android:color/widget_edittext_dark"
16          />
17
18  </LinearLayout>
```

这是我们的 Activity 的布局文件。在其中第 10 行定义了一个 TextView 控件，显示提示信息。

然后修改 src/com.wyl.example/MainActivity.java 文件，代码如下：

```
01  //定义了本实例的主要 Activity
02  public class MainActivity extends Activity {
03      //Activity 透明 在 Manifest 中设置 Activity theme 为 Theme.
04      //Translucent 可使 Activity 背景透明
05      @Override
06      public void onCreate(Bundle savedInstanceState) {
07          super.onCreate(savedInstanceState);
08          setContentView(R.layout.activity_main);
09      }
10  }
```

此文件是 Activity 的代码文件，在其中设置了当前 Activity 的布局为 activity_main。

到现在为止我们还没有发现与之前代码的不同之处，接下来要在 manifest 文件中添加如下代码：

```
01  <!-- 设置 Activity 背景色为透明 -->
02      <application
03          android:allowBackup="true"
04          android:icon="@drawable/ic_launcher"
05          android:label="@string/app_name"
06          android:theme="@android:style/Theme.Translucent">
07          ...
08      </application>
```

在如上代码的第 6 行定义了当前应用程序的主题为 Theme.Translucent，也就是背景为透明样式。

4.实例扩展

一般的应用有很多种设置主题和样式方法,大部分都是在 manifest 文件中进行设置,大家可以根据自己的需求自定义主题和样式然后进行设置即可。

范例 130　圆角控件的制作

1.实例简介

在我们使用应用的过程中,经常会看到圆角的控件效果。例如,在一些社交软件的客户端中,常见的控件有圆角矩形、圆角 ImageView 控件和圆角 ListView 等。遇到这样的功能,我们一般是通过自定义 View 的背景来实现的。本例子就带领大家来实现一个圆角 ListView 的实例效果。

2.运行效果

该实例运行效果如图 6.15 所示。

3.实例程序讲解

在本实例中,当用户打开此应用程序时就会显示一个圆角的 ListView。想要实现本实例效果,首先修改 res/layout/activity_main.xml 文件,代码如下:

图 6.15　仿 IOS 的圆角 ListView

```
01  <LinearLayout xmlns:android="http://schemas.android.com/apk/res/android"
02      xmlns:tools="http://schemas.android.com/tools"
03      android:id="@+id/LinearLayout1"
04      android:layout_width="match_parent"
05      android:layout_height="match_parent"
06      android:orientation="vertical"
07      tools:context=".MainActivity" >
08
09      <!-- 自定义背景是圆角的原因 -->
10      <ListView
11          android:layout_margin="10dp"
12          android:id="@+id/lv_show"
13          android:layout_width="match_parent"
14          android:layout_height="wrap_content"
15          android:background="@drawable/list_bg"
16      ></ListView>
17
18  </LinearLayout>
```

这是我们的 Activity 的布局文件。在其中第 10～16 行定义了一个 ListView 控件,用来显示圆角的背景效果,在其中设置了 background 为 list_bg 文件,此文件为自定义的背景文件。

新建 res/drawable/list_bg.xml 文件,代码如下:

```
01  <?xml version="1.0" encoding="utf-8"?>
02  <shape xmlns:android="http://schemas.android.com/apk/res/android" >
```

```
03        <!-- 定义间隔 -->
04        <stroke
05            android:width="1dp"
06            android:color="@color/gray" />
07        <!-- 定义形状的背景颜色 -->
08        <solid android:color="@color/white" />
09        <!-- 定义圆角的角度 -->
10        <corners android:radius="8dp" />
11    </shape>
```

在这个文件中定义了一个 shape，其中背景为浅灰色，边线为白色，角度为 8dp 的角度。此文件是本实例的关键所在，也就是定义 View 的背景文件。

然后修改 src/com.wyl.example/MainActivity.java 文件，代码如下：

```
01  //定义了本实例的主要Activity
02  public class MainActivity extends Activity {
03
04  //listview 圆角背景
05  private ListView mLvShow;                        //展示 listvie
06
07  @Override
08  public void onCreate(Bundle savedInstanceState) {
09      super.onCreate(savedInstanceState);
10      setContentView(R.layout.activity_main);
11      //得到布局中的控件
12      findView();
13  }
14
15  private void findView() {
16      //绑定控件
17      mLvShow = (ListView) findViewById(R.id.lv_show);
18      mLvShow.setAdapter(new ArrayAdapter<String>(this,
19          android.R.layout.simple_expandable_list_item_1, getData()));
20  }
21
22  private List<String> getData() {
23      //加入模拟数据
24      List<String> data = new ArrayList<String>();
25      data.add("圆角背景测试数据 1");
26      data.add("圆角背景测试数据 2");
27      data.add("圆角背景测试数据 3");
28      data.add("圆角背景测试数据 4");
29
30      return data;
31  }
32  }
```

此文件是 Activity 的代码文件，在其中第 10 行设置了当前 Activity 的布局文件。在第 15 行定义了 findView 方法得到布局中的控件对象，在第 22 行定义了 getData 方法，设置了相应的模拟数据。

4．实例扩展

本实例的效果主要反映了在 Android 中我们可以自定义任何一个 View 的背景，这样在你的应用中就可以加入你的个性化设计了。

范例 131 程序的国际化

1．实例简介

在我们使用应用的过程中，一些程序可能有其他国家的开发者完成的，所以需要做国家化和本地化。例如，当用户设置手机为简体中文时我们的应用就要以简体中文的形式显示；如果用户设置为繁体中文，我们的应用就要以繁体中文显示。这种功能一般是由Android程序的固定目录方法来解决的。本例子就带领大家来实现一个国际化的界面效果。

2．运行效果

该实例运行效果如图 6.16 所示。

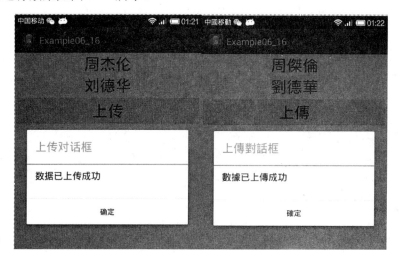

图 6.16 程序的国际化

3．实例程序讲解

当用户打开本实例时，本实例会根据用户当前的系统语言来显示对应的文字。想要实现本实例效果，首先修改 res/layout/activity_main.xml 文件，代码如下：

```xml
01 <?xml version="1.0" encoding="utf-8"?>
02 <LinearLayout xmlns:android="http://schemas.android.com/apk/res/android"
03     android:layout_width="fill_parent"
04     android:layout_height="fill_parent"
05     android:orientation="vertical" >
06 <!-- 定义显示的标签控件 -->
07     <TextView
08         android:layout_width="fill_parent"
09         android:layout_height="wrap_content"
10         android:gravity="center_horizontal"
11         android:textSize="30sp"
12         android:text="@string/text_a" />
13 <!-- 定义显示的标签控件 -->
14     <TextView
15         android:layout_width="fill_parent"
```

```xml
16        android:layout_height="wrap_content"
17        android:gravity="center_horizontal"
18        android:textSize="30sp"
19        android:text="@string/text_b" />
20  <!-- 定义显示的按钮控件 -->
21    <Button
22        android:id="@+id/flag_button"
23        android:layout_width="fill_parent"
24        android:layout_height="wrap_content"
25        android:textSize="30sp"
26        android:layout_gravity="center"
27        android:text="@string/btn_text"
28        />
29
30  </LinearLayout>
```

这是我们的 Activity 的布局文件。在其中定义了两个 TextView 控件和一个 Button 控件，这里需要注意一点，我们的布局文件中所有的文字属性，全部都是用了@string/的形式，这是保证我们程序国际化的重要前提。

然后修改 src/com.wyl.example/MainActivity.java 文件，代码如下：

```java
01  //定义了本实例的主要Activity
02  public class MainActivity extends Activity {
03  //国际化
04  @Override
05  public void onCreate(Bundle savedInstanceState) {
06      super.onCreate(savedInstanceState);
07      setContentView(R.layout.activity_main);
08
09      //绑定控件
10      Button b;
11      b = (Button) findViewById(R.id.flag_button);
12
13      //定义弹出框
14      AlertDialog.Builder builder = new AlertDialog.Builder(this);
15      builder.setMessage(R.string.dialog_text)
16              .setCancelable(false)
17              .setTitle(R.string.dialog_title)
18              .setPositiveButton(R.string.dialog_ok,
19                  new DialogInterface.OnClickListener() {
20                      public void onClick(DialogInterface dialog, int id) {
21                          dialog.dismiss();
22                      }
23                  });
24
25      final AlertDialog alert = builder.create();
26      //绑定事件
27      b.setOnClickListener(new View.OnClickListener() {
28          public void onClick(View v) {
29              alert.show();
30          }
31      });
32  }
33  }
```

此文件是 Activity 的代码文件，在其中第 7 行设置了当前 Activity 的布局文件为 activity_main，然后定义了一个 AlertDialog.Builder 对象并且设置了单击监听事件，其中 AlertDialog 的标题内容使用了 R.string.的形式来访问 String 中的内容。

这时候我们就可以在工程下进行国际化了，在我们的工程的 res 目录下建立 values-zh-rCN 和 values-zh-rTW 两个目录，然后分别建立相应的 strings.xml。

Values-zh-rCN 目录下的 strings.xml 内容为：

```xml
<?xml version="1.0" encoding="utf-8"?>
<resources>
    <string name="app_name">Example06_16</string>
    <string name="action_settings">Settings</string>
    <string name="hello_world">Hello world!</string>
    <string name="text_a">周杰伦</string>
    <string name="text_b">刘德华</string>
    <string name="dialog_ok">确定</string>
    <string name="dialog_title">上传对话框</string>
    <string name="dialog_text">数据已上传成功</string>
    <string name="btn_text">上传</string>
</resources>
```

Values-zh-rTW 目录下的 strings.xml 内容为：

```xml
<?xml version="1.0" encoding="utf-8"?>
<resources>
    <string name="app_name">Example06_16</string>
    <string name="action_settings">Settings</string>
    <string name="hello_world">Hello world!</string>
    <string name="text_a">周傑倫</string>
    <string name="text_b">劉德華</string>
    <string name="dialog_ok">確定</string>
    <string name="dialog_title">上傳對話框</string>
    <string name="dialog_text">數據已上傳成功</string>
    <string name="btn_text">上傳</string>
</resources>
```

这样当用户启动本程序时，应用程序就会自动根据用户当前的系统语言来选择对应的 string 文件显示了。

4．实例扩展

本实例只实现了简体中文和繁体中文的国际化，还有其他很多语言的国际化都可以实现，有具体需求的读者可以查看 Google 开发者文档。

6.4 小　　结

在本章节中主要介绍了 Android 中数据存储的使用方法，其中常见的应用有三种，一种是通过文件存储数据。第二种是通过 ContentProvider 来访问系统的数据资源。第三种是工程资源文件的使用。本章的内容在实际项目开发中使用很多，它可以使你的应用程序保存用户之前的数据，增加用户粘性。例如，记录用户之前输入的密码和得到系统短信等，这也让你的应用能够和系统融为一体，增加用户粘性。下一章我们会讲述 Android 中服务和广播的使用。

第 7 章　Android 中的服务和广播

上一章了解了 Android 中数据存储的三种方式，第一种是通过系统文件存储应用数据。第二种是通过 ContentProvider 得到系统的资源，第三种是程序内部的资源文件的使用。有了之前章节的内容讲解，大家基本上可以完成一些常见的 Android 应用了。但是对于 Android 开发来说还有两个基本的组件——服务和广播，它们在用户使用的时候是无法看到的，但这两个组件适用于一些特殊场合，如发送短信的通知和后台播放音乐等。那么本章就给大家介绍 Android 中的服务和广播的使用，通过本章的学习大家就可以完成一些特殊的功能要求了。

本章主要通过各种实例来介绍 Android 中的服务和广播的使用方法。希望读者阅读完本章内容后，能够在自己的应用程序中灵活运用广播和服务，为自己的应用实现一些特殊的亮点功能。

7.1　Android 中的服务的使用

范例 132　查看手机运行的进程列表

1．实例简介

在 Android 的手机中我们经常希望能够看到程序中有哪些进程在运行。例如，现在的手机管理软件都可以查看运行中的进程。我们通过本实例带领大家一起来制作一个查看手机正在运行的进程的应用。

2．运行效果

该实例运行效果如图 7.1 所示。

图 7.1　查看手机运行的进程

3. 实例程序讲解

在上例中程序运行后可以看到两个按钮,第一个按钮是查看可用的内存,第二个按钮的功能是查看现在手机中运行的进程。想要实现如上效果,首先修改我们建立的工程下的 res/layout/activity_main.xml 文件,代码如下:

```xml
01 <LinearLayout xmlns:android="http://schemas.android.com/apk/res/android"
02     xmlns:tools="http://schemas.android.com/tools"
03     android:layout_width="match_parent"
04     android:layout_height="match_parent"
05     android:orientation="vertical" >
06     <!-- 定义获取可用内存的按钮 -->
07     <Button
08         android:id="@+id/btn_main_ablememory"
09         android:layout_width="fill_parent"
10         android:layout_height="wrap_content"
11         android:text="可用 的内存"/>
12     <!-- 定义查看正在运行的进程的按钮 -->
13     <Button
14         android:id="@+id/btn_main_lookruningmemory"
15         android:layout_width="fill_parent"
16         android:layout_height="wrap_content"
17         android:text="查看正在运行的进程"/>
18     <!-- 定义显示当前进程数的标签控件 -->
19     <TextView
20     android:id="@+id/tv_main_currentprocessnum"
21     android:layout_width="fill_parent"
22     android:layout_height="wrap_content" />
23     <!-- 定义进程列表控件 -->
24     <ListView
25     android:id="@+id/lv_main_list"
26     android:layout_width="fill_parent"
27     android:layout_height="wrap_content"/>
28     <!-- 定义获取内容显示的标签控件 -->
29     <TextView
30         android:id="@+id/tv_main_text"
31         android:layout_width="wrap_content"
32         android:layout_height="wrap_content" />
33
34 </LinearLayout>
```

这是我们的 Activity 的布局文件,其中第 7、13 行定义了两个 Button 控件,分别进行手机可用内存的查询和运行的进程查询。在第 24 行,定义了一个 ListView 控件,用来显示手机中运行的进程的列表,还有显示可用内存大小的 TextView 控件。

然后修改 src/com.wyl.example/MainActivity.java 文件,代码如下:

```java
01//定义了本实例的主要 Activity
02public class MainActivity extends Activity {
03    //省略定义的控件对象
04
05    //定义 WifiManager 对象
06    private WifiManager mWifiManager;
07    //扫描出的网络连接列表
08    private List<ScanResult> mWifiList=new ArrayList<ScanResult>();
```

```
09  //网络连接的 string 列表
10  private List<String> wifiList=new ArrayList<String>();
11
12  @Override
13  public void onCreate(Bundle savedInstanceState) {
14      super.onCreate(savedInstanceState);
15      setContentView(R.layout.activity_main);
16      //取得 WifiManager 对象
17      mWifiManager = (WifiManager)getSystemService(Context.WIFI_SERVICE);
18      //取得 WifiInfo 对象
19      mWifiInfo = mWifiManager.getConnectionInfo();
20
21      //得到布局中的所有对象
22      findView();
23      //设置对象的监听器
24      setListener();
25  }
26
27  //省略 findview 的实现
28  //省略 setListener 的实现
29  //省略 setAdapter 的实现
30  //定义当前页面的单击事件监听器对象
31  OnClickListener listener=new OnClickListener() {
32
33      @Override
34      public void onClick(View v) {
35          //TODO Auto-generated method stub
36          switch (v.getId()) {
37
38          //打开 WIFI
39          case R.id.btn_openwifi:
40              //检查 wifi 是否开启, isWifiEnabled()等于 true 说明开启
41              if (!mWifiManager.isWifiEnabled())
42              {
                      //设置开启 wifi
43                  mWifiManager.setWifiEnabled(true);
                      //通过 Toast 提示 wifi 已开启
44                  Toast.makeText(MainActivity.this,
45                      "WIFI 已经开启", Toast.LENGTH_SHORT).show();
46              }
47              break;
48
49          //关闭 WIFI
50          case R.id.btn_closewifi:
51              //检查 wifi 是否开启, isWifiEnabled()等于 true 说明开启]
52              if (mWifiManager.isWifiEnabled())
53              {
                      //设置关闭 wifi
54                  mWifiManager.setWifiEnabled(false);
                      //通过 Toast 提示 wifi 关闭
55                  Toast.makeText(MainActivity.this,
56                      "WIFI 已经关闭", Toast.LENGTH_SHORT).show();
57              }
58              break;
59
60          //得到 WIFI 列表
61          case R.id.btn_getwifilist:
62              //检查 wifi 是否开启, isWifiEnabled()等于 true 说明开启
```

```
63              if(mWifiManager.isWifiEnabled())
64              {
65                  //扫描 WIFI
66                  mWifiManager.startScan();
67                  //得到扫描结果
68                  mWifiList = mWifiManager.getScanResults();
69                  //把 mWifiList 里面的内容放在 wifiList 里面
70                  for(int i=0;i<mWifiList.size();i++)
71                  {
72                      wifiList.add(mWifiList.get(i).SSID);
73                  }
                    //设置 adapter 对象
74                  setAdapter();
75              }
76              else{
77                  Toast.makeText(MainActivity.this,
78                      "WIFI 没有开启,无法扫描", Toast.LENGTH_SHORT).show();
79              }
80              break;
81          case R.id.btn_getwifiinfo:
82              //得到接入点的 BSSID
83              String bssid=mWifiInfo.getBSSID();
84              //得到 MAC 地址
85              String macAddress=mWifiInfo.getMacAddress();
86              //得到 IP 地址
87              int ipAddress=mWifiInfo.getIpAddress();
88              //得到连接的 ID
89              int netWorkId=mWifiInfo.getNetworkId();
90                  tvWifiInfo.setText("接入点的 BSSID :"+bssid+"\nMAC 地址 :"
91                      +macAddress+"\nIP 地址 :"+ipAddress+"\n 连接的
                        ID :"+netWorkId);
92
93          default:
94              break;
95          }
96      }
97  };
98}
```

此文件是当前 Activity 的关键代码,在第 6 行定义了 WifiManager 管理对象,在第 17 行初始化 WiFi 服务管理对象,在第 19 行得到了 WiFi 信息。关键代码在第 34 行,当用户单击开启 WIFI 按钮时首先判断 WiFi 是否可用如果可用则设置 WifiEnable 状态。如果用户单击关闭 WiFi,则设置 WiFi 状态为 false。当用户单击获取 WiFi 列表时则调用 wifimanager 的 getScanResult 方法,得到所有可连接的 WiFi,当用户单击查看当前的 WiFi 连接状态,则通过 wifiinfo 类来获取信息并且进行显示。

4. 实例扩展

本实例使用到了系统的 WIFI 服务,所以需要在 manifest 文件中加入对应的权限代码如下:

```
<uses-permission android:name="android.permission.ACCESS_WIFI_STATE">
</uses-permission>
<uses-permission android:name="android.permission.CHANGE_WIFI_STATE">
</uses-permission>
```

范例 133　得到系统的唤醒服务

1. 实例简介

在我们使用应用的过程中，经常会遇到当我们正在查看某个页面，或者大段文字的时候，我们在一段时间内没有触屏屏幕，然后屏幕就黑屏锁屏了，这样对于某些应用来说是不太合理的。例如，某些图书阅读客户端和某些视频播放软件客户端等。一般遇到这样的功能，我们都是在程序开始得到系统的唤醒服务，在程序退出时关闭程序的唤醒服务。本例子就带领大家来实现一个得到系统的唤醒锁和释放系统的唤醒锁的实例。

2. 运行效果

该实例运行效果如图 7.2 所示。

图 7.2　得到系统的唤醒服务

3. 实例程序讲解

在本实例中，显示两个按钮，单击第一个按钮得到系统的唤醒锁，这时候系统不会进行休眠和锁屏，单击第二个按钮，释放系统的唤醒锁，系统经过默认的时间屏幕就会进入黑屏锁屏状态。想要实现本实例效果，首先修改 res/layout/activity_main.xml 文件，代码如下：

```
01  <LinearLayout xmlns:android="http://schemas.android.com/apk/res/android"
02      xmlns:tools="http://schemas.android.com/tools"
03      android:layout_width="match_parent"
04      android:layout_height="match_parent"
05      android:orientation="vertical" >
06      <!-- 定义获取唤醒锁的按钮 -->
07      <Button
08          android:id="@+id/BtnGet"
09          android:layout_width="fill_parent"
10          android:layout_height="wrap_content"
11          android:text="点击获取唤醒锁"
12          />
13
```

```
14          <!-- 定义释放唤醒锁的按钮 -->
15          <Button
16              android:id="@+id/BtnRelease"
17              android:layout_width="fill_parent"
18              android:layout_height="wrap_content"
19              android:text="点击释放唤醒锁"
20              />
21  </LinearLayout>
```

这是我们的 Activity 的布局文件。在其中第 7~12 行定义了一个 Button 控件，当用户单击此按钮的时候得到系统的唤醒锁服务。在第 14~20 行定义了一个 Button 对象，当用户单击此按钮的时候就释放唤醒锁服务，系统可以进入黑屏的锁屏状态。

然后修改 src/com.wyl.example/MainActivity.java 文件，代码如下：

```
01  //定义了本实例的主要 Activity
02  public class MainActivity extends Activity {
03      //得到唤醒锁对象
04      private WakeLock wakeLock;
05      //定义得到唤醒锁对象的按钮
06      private Button BtnGet;
07      //定义释放唤醒锁对象的按钮
08      private Button BtnRelease;
09
10      @Override
11      public void onCreate(Bundle savedInstanceState) {
12          super.onCreate(savedInstanceState);
13          //定义当前页面的布局文件
14          setContentView(R.layout.activity_main);
15          //得到布局中的所有对象
16          findView();
17          //设置对象的监听器
18          setListener();
19      }
20
21      private void setListener() {
22          //设置对象的监听器
23          BtnGet.setOnClickListener(mylistener);
24          BtnRelease.setOnClickListener(mylistener);
25      }
26
27      private void findView() {
28          //得到视图中的控件对象
29          BtnGet = (Button)findViewById(R.id.BtnGet);
30          BtnRelease = (Button)findViewById(R.id.BtnRelease);
31      }
32
33      OnClickListener mylistener = new OnClickListener() {
34
35          @Override
36          public void onClick(View v) {
37              //TODO Auto-generated method stub
38              switch (v.getId()) {
39              case R.id.BtnGet:
40                  //开始获得唤醒锁
41                  acquireWakeLock();
42                  break;
43              case R.id.BtnRelease:
44                  //释放锁
```

```
45                    releaseWakeLock();
46                    break;
47
48                default:
49                    break;
50            }
51        }
52    };
53
54    //开始获得唤醒锁
55    private void acquireWakeLock() {
56        if (wakeLock == null) {
57            Toast.makeText(this, "得到唤醒锁", Toast.LENGTH_SHORT)
58                    .show();
59            //得到电源管理服务
60            PowerManager pm = (PowerManager) getSystemService(Context
                    .POWER_SERVICE);
61            //加锁方便控制电源的状态
62            wakeLock = pm.newWakeLock(PowerManager.FULL_WAKE_LOCK, this
63                    .getClass().getCanonicalName());
64            //获取相应的锁,屏幕将停留在设定的状态,一般为亮、暗状态
65            wakeLock.acquire();
66        }
67    }
68
69    private void releaseWakeLock() {
70        if (wakeLock != null && wakeLock.isHeld()) {
71            //释放掉正在运行的cpu或关闭屏幕
72            wakeLock.release();
73            wakeLock = null;
74        }
75        Toast.makeText(MainActivity.this, "释放唤醒锁",Toast.LENGTH_SHORT)
             .show();
76    }
77 }
```

此文件是Activity的代码文件,在其中第4行定义了wakeLock对象,此对象代表系统的唤醒服务,在第6~8行定义了两个Button对象。在第16行的findView方法中得到了布局中的所有控件,具体实现在第27~31行。在第18行通过setListener方法设置控件的监听器,具体方法在21~25行实现。在第33~52行实现了自定义的OnClickListener类的对象,当用户单击的是获得唤醒锁按钮时调用acquireWakeLock方法,当用户单击释放唤醒锁按钮时调用releaseWakeLock方法。acquireWakeLock方法在第54~67行实现,再启动通过getSystemService得到系统的PowerService,然后设置屏幕的加锁状态。releaseWakeLock方法在第69~76行实现,通过调用之前的wakeLock对象的release方法来释放屏幕锁的状态。

4. 实例扩展

在此实例中是通过按钮来进行屏幕锁的获得和释放的,一般情况下,在一个应用Activity的onresume方法中获得屏幕锁,在onstop方法中释放屏幕锁。

还需要注意一点,本实例得到了系统的电源服务,所以需要在manifest文件中添加获得系统的唤醒锁的权限,所以要在manifest文件中添加如下代码:

```
<uses-permission android:name="android.permission.WAKE_LOCK"/>
```

保证程序可以获得唤醒锁权限。

范例 134　定时任务启动

1．实例简介

在我们使用应用的过程中，经常会使用到定时启动某个任务。例如，基本上每个人的手机中常用的功能就是闹钟，也就是设定一个时间点，当系统到达此时间点后启动闹铃，或者定时生日提醒，或者时间倒计时在当前时间后的 30 分钟后启动闹铃等，这些都是当到达某个预订的时间后程序就会自动完成某个任务。遇到这样的功能，我们一般是当用户进行某个操作的时候，例如：单击某个按钮的时候，建立一个定时任务，当时间满足设置的条件时就启动相应的程序。本例子就带领大家来实现一个通过定时任务启动程序的实例。

2．运行效果

该实例运行效果如图 7.3 所示。

图 7.3　定时任务启动

3．实例程序讲解

本实例中，在程序启动后能够看到四个按钮，分别是定时发送广播、定时启动服务、停止定时服务和停止定时广播，当用户单击定时发送广播时，系统会设置一个时间任务，当到达此时间后自动发送广播，在这期间如果你单击停止定时广播则此定时任务取消，定时服务是相似的功能。想要实现本实例效果，首先修改 res/layout/activity_main.xml 文件，代码如下：

```
01  <LinearLayout xmlns:android="http://schemas.android.com/apk/res/android"
02      xmlns:tools="http://schemas.android.com/tools"
03      android:layout_width="match_parent"
04      android:layout_height="match_parent"
05      android:orientation="vertical" >
06      <!-- 发送定时广播按钮 -->
07      <Button
08          android:id="@+id/brodcast"
```

```
09          android:layout_width="match_parent"
10          android:layout_height="wrap_content"
11          android:text="定时发送广播" />
12      <!-- 发送定时服务按钮 -->
13      <Button
14          android:id="@+id/service"
15          android:layout_width="match_parent"
16          android:layout_height="wrap_content"
17          android:text="定时启动服务" />
18      <!-- 停止定时广播按钮 -->
19      <Button
20          android:id="@+id/brodcaststop"
21          android:layout_width="match_parent"
22          android:layout_height="wrap_content"
23          android:text="停止定时广播" />
24      <!-- 停止定时服务按钮 -->
25      <Button
26          android:id="@+id/servicestop"
27          android:layout_width="match_parent"
28          android:layout_height="wrap_content"
29          android:text="停止定时服务" />
30
31  </LinearLayout>
```

这是我们的 Activity 的布局文件。在其中第 7、13、19 和 25 行分别定义了四个 Button 控件，分别用来启动定时广播、启动定时服务、取消定时服务和取消定时广播。当用户单击某个按钮的时候发出某个事件。

然后修改 src/com.wyl.example/MainActivity.java 文件，代码如下：

```
01  //定义了本实例的主要Activity
02  public class MainActivity extends Activity {
03      //定义发送广播的按钮
04      private Button btnBrodcast;
05      //定义启动服务的按钮
06      private Button btnService;
07      //定义闹钟管理对象
08      private AlarmManager am;
09      //定义广播闹钟停止的按钮
10      private Button btnBrodcastStop;
11      //定义服务闹钟停止的按钮
12      private Button btnServiceStop;
13      //定义延迟发送请求
14      private PendingIntent pi1;
15      private PendingIntent pi2;
16
17      @Override
18      public void onCreate(Bundle savedInstanceState) {
19          super.onCreate(savedInstanceState);
20          setContentView(R.layout.activity_main);
21          //获得AlermManager服务
22          am = (AlarmManager) getSystemService(ALARM_SERVICE);
23          //得到布局中的所有对象
24          findView();
25          //设置对象的监听器
26          setListener();
27      }
28
```

```
29  private void findView() {
30      //得到布局中的所有对象
31      btnBrodcast = (Button) findViewById(R.id.brodcast);
32      btnService = (Button) findViewById(R.id.service);
33      btnBrodcastStop = (Button) findViewById(R.id.brodcaststop);
34      btnServiceStop = (Button) findViewById(R.id.servicestop);
35  }
36
37  private void setListener() {
38      //设置对象的监听器
39      btnBrodcast.setOnClickListener(listener);
40      btnBrodcastStop.setOnClickListener(listener);
41      btnService.setOnClickListener(listener);
42      btnServiceStop.setOnClickListener(listener);
43  }
44
45  //定义了监听器对象
46  OnClickListener listener = new OnClickListener() {
47
48      @Override
49      public void onClick(View v) {
50          //TODO Auto-generated method stub
51          switch (v.getId()) {
52
53          //开启闹钟广播
54          case R.id.brodcast:
55              pi1 = PendingIntent.getBroadcast(MainActivity.this, 0,
56                      new Intent(MainActivity.this, ActionBroadCast
                          .class),
57                      Intent.FLAG_ACTIVITY_NEW_TASK);
58              //得到当前时间
59              long now = System.currentTimeMillis();
60              //设置重复的闹钟
61              am.setInexactRepeating(AlarmManager.RTC_WAKEUP, now, 3000,
                      pi1);
62              break;
63
64          //停止闹钟广播
65          case R.id.brodcaststop:
66              //关闭闹钟
67              am.cancel(pi1);
68              break;
69
70          //开启闹钟服务
71          case R.id.service:
72              pi2 = PendingIntent.getService(MainActivity.this, 0,
73                      new Intent(MainActivity.this, ActionService.class),
74                      Intent.FLAG_ACTIVITY_NEW_TASK);
75              //得到当前时间
76              long now1 = System.currentTimeMillis();
77              //设置重复的闹钟
78              am.setInexactRepeating(AlarmManager.RTC_WAKEUP, now1,
                      3000, pi2);
79              break;
80
81          //关闭闹钟服务
82          case R.id.servicestop:
83              //关闭闹钟
84              am.cancel(pi2);
```

```
85              break;
86
87          default:
88              break;
89          }
90      }
91  };
92  }
```

此文件是 Activity 的代码文件，在其中第 3~12 行分别定义了四个按钮对象。在第 22 行通过系统提供的 getSystemService 得到系统的闹铃服务，然后第 24 行得到所有的布局中的控件对象，具体实现在 29~35 行。在第 26 行设置所有控件的监听器，具体实现在 37~43 行，所有的控件都绑定了自定义的监听器。在第 45~91 行，其中如果用户单击开启定时广播的按钮，那就执行第 55 行定义的 PendingIntent，并且设置 am 的闹钟时间为当前时间后 3 秒开始发送。如果单击停止广播，则执行 67 行停止 am 的延迟任务。单击开启定时服务和取消定时服务的原理相同。

在本例子中为了能够模拟启动服务和发送广播，本例子定义了一个服务类和一个广播类，在工程目录新建 src/com.wyl.example/ActionService.java 文件，代码如下：

```
01  //自定义 Service
02  public class ActionService extends Service{
03      //定义计数变量
04      private static int num = 0;
05      @Override
06      public IBinder onBind(Intent intent) {
07          //TODO Auto-generated method stub
08          return null;
09      }
10      @Override
11      public void onCreate() {
12          //TODO Auto-generated method stub
13          super.onCreate();
14          Log.e("ActionService", "New Message !" + num++);
15      }
16  }
```

此文件定义了一个服务类，用来接收定时任务后的 intent。

在工程目录新建 src/com.wyl.example/ActionBroadCast.java 文件，代码如下：

```
01  //自定义广播接收器
02  public class ActionBroadCast extends BroadcastReceiver {
03
04      private static int num = 0;
05      /* (non-Javadoc)
06       * @see android.content.BroadcastReceiver#onReceive(android.content
         .Context, android.content.Intent)
07       */
08      //如果收到广播信息，回调 onReceive
09      @Override
10      public void onReceive(Context context, Intent intent) {
11          //当接收到广播的时候，自动回到此方法
12          Log.e("ActionBroadCast", "New Message !" + num++);
13      }
14
15  }
```

此文件定义了一个广播接收器类，用来接收定时发送的广播。

4．实例扩展

在此实例中我们设置了定时任务可以定时启动程序内的广播或者服务，在我们本例中时间设定为当前时间的后 3 秒，在实际的应用中既可以设置延后时间，也可以设置固定的时间，如 12:23 分发送广播消息等等。

范例 135　发送状态栏信息

1．实例简介

在我们使用应用的过程中，经常会看到系统的状态栏的通知信息。例如，当用户接收到短信的时候、当用户某个应用下载成功的时候以及当某个无线网络连接成功的时候。遇到这样的功能，我们一般是当用户进行某个操作的时候，例如：单击某个按钮的时候，发送系统的状态栏通知信息。本例子就带领大家来实现一个发送状态栏通知的实例。

2．运行效果

该实例运行效果如图 7.4 所示。

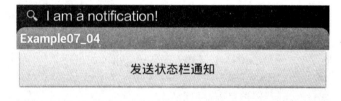

图 7.4　发送状态栏信息

3．实例程序讲解

在本实例中，用户启动程序可以看到一个按钮，当单击此按钮的时候发送状态栏通知。想要实现本实例效果，首先修改 res/layout/activity_main.xml 文件，代码如下：

```
01  <RelativeLayout xmlns:android="http://schemas.android.com/apk/res
    /android"
02      xmlns:tools="http://schemas.android.com/tools"
03      android:layout_width="match_parent"
04      android:layout_height="match_parent" >
05      <!-- 定义发送通知的按钮 -->
06      <Button
07          android:id="@+id/show"
08          android:layout_width="match_parent"
09          android:layout_height="wrap_content"
10          android:text="发送状态栏通知" />
11
```

```
12      </RelativeLayout>
13
```

这是我们的 Activity 的布局文件。在其中第 6~10 行定义了一个 Button 控件,当用户单击此按钮的时候发送状态栏通知。

然后修改 src/com.wyl.example/MainActivity.java 文件,代码如下:

```
01  //定义了本实例的主要 Activity
02  public class MainActivity extends Activity {
03      //定义了通知状态栏按钮
04      private Button btnNotification;
05      //定义了通知对象
06      private Notification baseNF;
07      //定义了通知管理对象
08      private NotificationManager nm;
09
10      @Override
11      public void onCreate(Bundle savedInstanceState) {
12          super.onCreate(savedInstanceState);
13          setContentView(R.layout.activity_main);
14          //得到 NotificationManager 服务
15          nm = (NotificationManager) getSystemService(NOTIFICATION_SERVICE);
16          //得到布局中的所有对象
17          findView();
18          //设置对象的监听器
19          setListener();
20
21      }
22
23      private void findView() {
24          //TODO Auto-generated method stub
25          //得到布局中所有对象
26          btnNotification = (Button) findViewById(R.id.show);
27      }
28
29      private void setListener() {
30          //TODO Auto-generated method stub
31          //设置对象的监听器
32          btnNotification.setOnClickListener(new OnClickListener() {
33              @Override
34              public void onClick(View v) {
35                  //TODO Auto-generated method stub
36                  //设置发送状态栏通知
37                  setNotification();
38              }
39          });
40      }
41
42      //设置状态栏
43      @SuppressWarnings("deprecation")
44      protected void setNotification() {
45          //TODO Auto-generated method stub
46          //新建状态栏通知
47          baseNF = new Notification();
48
49          //设置通知在状态栏显示的图标
50          baseNF.icon = R.drawable.ic_action_search;
51
```

```
52        //通知时在状态栏显示的内容
53        baseNF.tickerText = "I am a notification!";
54
55        //通知的默认参数 DEFAULT_SOUND,DEFAULT_VIBRATE,DEFAULT_LIGHTS
56        //如果要全部采用默认值,用 DEFAULT_ALL
57        //此处采用默认声音
58        baseNF.defaults = Notification.DEFAULT_SOUND;
59
60        //当用户拉下notify显示列表,并且单击对应的项的时候,才会触发系统跳转到该activity
61        PendingIntent pi = PendingIntent.getActivity(MainActivity.this, 0,
62              new Intent(this, MainActivity.class), 0);
63
64        //第二个参数:下拉状态栏时显示的消息标题 expandedmessagetitle
65        //第三个参数:下拉状态栏时显示的消息内容 expandedmessagetext
66        //第四个参数:单击该通知时执行页面跳转
67        //新的API17 不建议使用 setLatestEventInfo
68        baseNF.setLatestEventInfo(MainActivity.this, "Title01", "Content01", pi);
69
70        //发出状态栏通知
71        //第一个参数是发出请求的 ID,第二个参数是 Notification
72        nm.notify(0, baseNF);
73    }
74 }
```

此文件是 Activity 的代码文件,在其中第 4 行定义了一个 Button 对象,在第 6 行定义了一个通知对象,在第 8 行定义了一个通知管理对象。在第 15 行通过 getSystemService 得到了系统的发送通知的服务,第 17 行得到所有的控件对象,第 19 行设置相应的监听器。主要是在第 37 行,当用户单击某个按钮的时候发送 Notification 信息,这里的 setNotification 方法在第 44~73 行,首先在第 47 行初始化了通知对象,然后设置通知对象的标题、图片和文字,然后在第 61 行定义延迟的 intent,然后在第 68 行设置单击通知信息后的时间,最后在第 72 行发送通知,这时候我们就可以看到状态栏的通知信息了。

4. 实例扩展

在此实例中设置的 Notification 是 Android 系统默认的格式,包括标题、图标和内容等,当然用户也可以自定义通知的布局,这样你就可以有个性的通知栏信息样式了。

范例 136 得到屏幕状态

1. 实例简介

在我们使用应用的过程中,有时候会需要判断屏幕的状态是正常、是锁屏、还是休眠。例如,但我们在玩儿游戏的过程中,不能让我们的程序进入休眠;当我们程序锁屏的时候,关闭某些服务等。遇到这样的功能,我们一般是当用户进行某个操作的时候,例如,单击某个按钮的时候,得到系统的屏幕状态服务进行修改和判断。本例子就带领大家来实现一个通过系统服务得到屏幕状态的实例。

2. 运行效果

该实例运行效果如图 7.5 所示。

图 7.5　屏幕的状态

3．实例程序讲解

在本实例中，提供一个得到系统的屏幕状态的按钮，当用户单击此按钮的时候显示屏幕的锁屏状态。想要实现本实例效果，首先修改 res/layout/activity_main.xml 文件，代码如下：

```xml
01 <LinearLayout xmlns:android="http://schemas.android.com/apk/res/android"
02     xmlns:tools="http://schemas.android.com/tools"
03     android:layout_width="match_parent"
04     android:layout_height="match_parent"
05     android:orientation="vertical" >
06     <!-- 定义获得屏幕状态的按钮 -->
07     <Button
08         android:id="@+id/reenableKeyguard"
09         android:layout_width="match_parent"
10         android:layout_height="wrap_content"
11         android:text="屏幕的状态" />
12
13 </LinearLayout>
```

这是我们的 Activity 的布局文件。在其中第 7~11 行定义了一个 Button 控件，当用户单击的时候获取系统屏幕的状态。

然后修改 src/com.wyl.example/MainActivity.java 文件，代码如下：

```java
01 //定义了本实例的主要 Activity
02 public class MainActivity extends Activity {
03     //定义锁屏的按钮
04     private Button btnKeyguard;
05     //声明 KeyguardManager 对象
06     private KeyguardManager keyguardManager;
07
08
09     @Override
10     public void onCreate(Bundle savedInstanceState) {
11         super.onCreate(savedInstanceState);
12         setContentView(R.layout.activity_main);
13         //获得 KeyguardManager 服务
14         keyguardManager=(KeyguardManager)getSystemService(Context
    .KEYGUARD_SERVICE);
15         //得到布局中的所有对象
16         findView();
```

```
17        //设置对象的监听器
18        setListener();
19    }
20
21    private void findView() {
22        //得到布局中的所有对象
23        btnKeyguard = (Button) findViewById(R.id.reenableKeyguard);
24    }
25
26    private void setListener() {
27        //设置对象的监听器
28        btnKeyguard.setOnClickListener(new OnClickListener() {
29
30            @Override
31            public void onClick(View arg0) {
32                //TODO Auto-generated method stub
33                //判断当前屏幕的状态
34                if(keyguardManager.isKeyguardLocked())
35                {
36                    Toast.makeText(MainActivity.this, "锁屏", Toast.LENGTH
                        _SHORT).show();
37                }
38                else
39                {
40                    Toast.makeText(MainActivity.this, "没有锁屏", Toast
                        .LENGTH_SHORT).show();
41                }
42            }
43        });
44    }
45 }
```

此文件是 Activity 的代码文件，在其中第 4 行定义了一个 Button 对象，当用户单击的时候检查屏幕的状态。在第 6 行定义了一个键盘管理对象，在第 14 行通过 getSystemService 得到系统的锁屏键的服务。在第 16 行得到布局中的所有的控件，在第 18 行设置所有控件的监听器对象。其中关键代码在第 34 行，通过 keyguardManager 的 iskeyguardLocked 方法得到屏幕是否为锁屏状态，然后屏幕显示对应提示。

4．实例扩展

在此实例中仅仅通过得到的锁屏服务进行了屏幕状态的查看，当然通过此服务可以实现更多的功能，如单击某个按钮进行锁屏等。需要注意一点，此程序需要得到系统的锁屏权限，所以需要在 manifest 文件中加入得到权限代码如下：

```
<uses-permission android:name="android.permission.DISABLE_KEYGUARD"/>
```

范例 137　程序中得到经纬度

1．实例简介

在我们使用应用的过程中，经常会使用到获取当前手机经纬度的功能，这是一些基于位置的功能的基础。例如，得到当前位置附近的好友，得到当前 1000 米内的公交站牌等。遇到这样的功能，我们一般是当用户进行某个操作的时候，例如，单击某个按钮的时候，得到系统中的位置服务,然后得到当前的经纬度。本例子就带领大家来实现一个得到当前经纬度的实例。

2. 运行效果

该实例运行效果如图 7.6 所示。

图 7.6 程序中得到经纬度

3. 实例程序讲解

在本实例中,提供一个按钮,当用户单击此按钮的时候尝试获取最后一次得到的经纬度信息,然后当经纬度信息改变的时候自动更新下面的值。想要实现本实例效果,首先修改 res/layout/activity_main.xml 文件,代码如下:

```xml
01  <LinearLayout xmlns:android="http://schemas.android.com/apk/res/android"
02      xmlns:tools="http://schemas.android.com/tools"
03      android:layout_width="match_parent"
04      android:layout_height="match_parent"
05      android:orientation="vertical" >
06      <!-- 得到当前位置的按钮 -->
07      <Button
08          android:id="@+id/position"
09          android:layout_width="match_parent"
10          android:layout_height="wrap_content"
11          android:text="当前的位置" />
12      <!-- 显示位置信息的标签控件 -->
13      <TextView
14          android:id="@+id/tv"
15          android:layout_width="match_parent"
16          android:layout_height="wrap_content"
17          android:text="" />
18  
19  </LinearLayout>
```

这是我们的 Activity 的布局文件。在其中第 7~11 行定义了一个 Button 控件,当用户单击时获取经纬度。在第 13~17 行定义了一个 TextView 控件,显示当前的经纬度信息。

然后修改 src/com.wyl.example/MainActivity.java 文件,代码如下:

```java
01  //定义了本实例的主要 Activity
02  public class MainActivity extends Activity {
03      //定义位置的按钮
04      private Button btnPosition;
05      //声明 LocationManager 对象
06      private LocationManager locationManager;
07      private TextView tv;
08
09
10      @Override
11      public void onCreate(Bundle savedInstanceState) {
12          super.onCreate(savedInstanceState);
13          setContentView(R.layout.activity_main);
14          //获得 locationManager 服务
15          locationManager=(LocationManager)getSystemService(Context
            .LOCATION_SERVICE);
16          //得到布局中的所有对象
17          findView();
18          //设置对象的监听器
19          setListener();
20
21      }
22
23      private void findView() {
24          //得到布局中的所有对象
25          btnPosition = (Button) findViewById(R.id.position);
26          tv = (TextView) findViewById(R.id.tv);
27      }
28
29      private void setListener() {
30          //监听位置变化,2 秒一次,距离 10 米以上
31          locationManager.requestLocationUpdates(LocationManager.GPS
            _PROVIDER, 2000, 10,
32                  locationListener);
33
34          //设置对象的监听器
35          btnPosition.setOnClickListener(new OnClickListener() {
36
37              @Override
38              public void onClick(View arg0) {
39                  //TODO Auto-generated method stub
40                  //getLatitude()获得当前的经度, getLongitude(); 获得当前的纬度
41                  tv.setText("当前的经度是:39.926325,"+"\n 当前的纬度是:
                    116.38945");
42              }
43          });
44      }
45
46      //位置监听器
47      private final LocationListener locationListener = new LocationListener() {
48          @Override
```

```
49      public void onStatusChanged(String provider, int status,
50              Bundle extras) {
51      }
52
53      @Override
54      public void onProviderEnabled(String provider) {
55      }
56
57      @Override
58      public void onProviderDisabled(String provider) {
59      }
60
61      //当位置变化时触发
62      @Override
63      public void onLocationChanged(Location location) {
64          //getLatitude()获得当前的经度,getLongitude();获得当前的纬度
65          tv.setText("当前的经度是"+location.getLatitude()+"当前的纬度是
            "+location.getLongitude());
66      }
67  };
68  }
```

此文件是 Activity 的代码文件,在其中第 4 行定义了一个 Button 对象,在第 6 行定义了系统的服务定位管理器,在第 15 行通过 getSystemService 方法得到系统的服务管理器对象,在第 17 行得到布局中的所有控件对象,在第 19 行设置控件对象的所有监听器。在第 31 行,在设置监听器的时候给布局管理对象也设置了更新监听器,这里设置了更新时间为 2 秒,最小位置变化距离为 10 米时更新,此监听器在第 47 行定义,其中涉及到如下回调方法。

❑ onStatusChanged 方法:当使用的位置提供者状态改变时回调。
❑ onProviderEnabled 方法:当某个服务提供者可以使用时回调。
❑ onProviderDisabled 方法:当某个服务提供者不可用时回调。
❑ onLocationChanged 方法:当位置改变时回调。

本实例只写了当位置改变时修改 TextView 的显示内容。

4. 实例扩展

在此实例中我们需要获得系统的位置服务信息,所以需要在 manifest 文件中添加如下代码获取权限:

```
<uses-permission android:name="android.permission.ACCESS_FINE_LOCATION" />
```

范例 138 振动器应用

1. 实例简介

在我们使用应用的过程中,经常会见到当用户单击某个按钮的时候手机会有的效果。例如,当用户打游戏过关的时候,或者当用户单击某个屏幕的时候。遇到这样的功能,我们一般是当用户进行某个操作的时候,例如,单击某个按钮的时候,获取系统的振动服务,

然后进行振动的设置。本例子就带领大家来实现一个振动器的实例。

2. 运行效果

该实例运行效果如图 7.7 所示。

图 7.7 振动器应用

3. 实例程序讲解

在本实例中，用户打开应用，显示两个按钮控件，当用户单击第一个按钮控件时，手机开始振动；单击第二个按钮时手机停止振动。想要实现本实例效果，首先修改 res/layout/activity_main.xml 文件。代码如下：

```
01  <LinearLayout xmlns:android="http://schemas.android.com/apk/res/android"
02      xmlns:tools="http://schemas.android.com/tools"
03      android:layout_width="match_parent"
04      android:layout_height="match_parent"
05      android:orientation="vertical" >
06
07      <!-- 定义开始振动的按钮 -->
08      <Button
09          android:id="@+id/start"
10          android:layout_width="match_parent"
11          android:layout_height="wrap_content"
12          android:text="开始振动" />
13
14      <!-- 定义停止振动的按钮 -->
15      <Button
16          android:id="@+id/stop"
17          android:layout_width="match_parent"
18          android:layout_height="wrap_content"
19          android:text="停止振动" />
20
21
22  </LinearLayout>
```

这是我们的 Activity 的布局文件。在其中第 8～12 行定义了开始振动的按钮控件，在第 15～19 行定义了停止振动的按钮。

然后修改 src/com.wyl.example/MainActivity.java 文件，代码如下：

```
01  //定义了本实例的主要 Activity
```

```java
02  public class MainActivity extends Activity {
03      //定义振动开始的按钮
04      private Button btnStart;
05      //声明 Vibrator 对象
06      private Vibrator vibrator;
07      //声明振动停止按钮
08      private Button btnStop;
09
10
11      @Override
12      public void onCreate(Bundle savedInstanceState) {
13          super.onCreate(savedInstanceState);
14          setContentView(R.layout.activity_main);
15          //获得 locationManager 服务
16          vibrator=(Vibrator)getSystemService(Context.VIBRATOR_SERVICE);
17          //得到布局中的所有对象
18          findView();
19          //设置对象的监听器
20          setListener();
21
22      }
23
24      private void findView() {
25          //得到布局中的所有对象
26          btnStart = (Button) findViewById(R.id.start);
27          btnStop=(Button)findViewById(R.id.stop);
28      }
29
30      private void setListener() {
31          //设置对象的监听器
32          btnStart.setOnClickListener(new OnClickListener() {
33
34              @Override
35              public void onClick(View arg0) {
36                  //TODO Auto-generated method stub
37                  //long 的第一个参数是单击按钮后经过多长时间振动
38                  //第二个参数是每次振动的时间
39                  //第三个参数是每次振动之间相隔的时间
40                  //vibrate 的第二个参数 0 是重复振动,-1 不重复振动
41                  vibrator.vibrate(new long[] {1000,1000,1000}, 0);
42              }
43          });
44          btnStop.setOnClickListener(new OnClickListener() {
45
46              @Override
47              public void onClick(View v) {
48                  //TODO Auto-generated method stub
49                  //停止振动
50                  vibrator.cancel();
51              }
52          });
53
54      }
55  }
```

此文件是 Activity 的代码文件,在其中第 4 行定义了一个 Button 对象,在第 6 行定义了 Vibrator 对象用来获得系统的振动器对象,在第 16 行对 Vibrator 对象进行了初始化,得到了系统的振动服务。第 18 行得到布局中的所有控件对象,在第 20 行设置所有控件的监

听器,其中当用户单击了开始振动的按钮则调用第 41 行中 vibrator 的 vibrator 方法,设置振动的时间、频率和间隔,然后程序即开始振动。当用户单击停止振动的按钮时调用 vibrator 的 cancel 方法来取消振动。

4. 实例扩展

在此实例中需要用到系统的振动服务,所以在 manifest 文件中要添加对应的权限,代码如下:

```
<uses-permission android:name="android.permission.VIBRATE"/>
```

范例 139 获得当前网络状态

1. 实例简介

在我们使用应用的过程中,经常会使用到判断用户网络状态的功能。例如,在用户开始下载东西前,需要判断用户是否连接网络;当用户打开应用时,可以设置,当使用 WiFi 连接网络时下载大图,3G 连接时显示小图来实现减少流量消耗的功能。遇到这样的功能,我们一般是当用户进行某个操作的时候,例如:单击某个按钮的时候,得到系统的网络连接对象,然后查看当前网络连接对象的信息。本例子就带领大家来实现一个获得当前网络状态的实例。

2. 运行效果

该实例运行效果如图 7.8 所示。

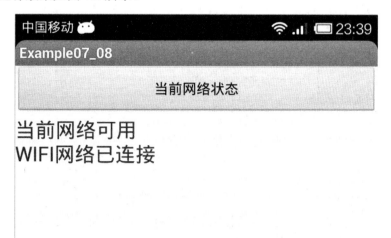

图 7.8 获得网络状态

3. 实例程序讲解

在本实例中,程序打开后显示一个获取当前网络状态的按钮,当用户单击此按钮时下面的标签框中显示了当前网络的连接状况。想要实现本实例效果,首先修改 res/layout/activity_main.xml 文件,代码如下:

```xml
01  <LinearLayout xmlns:android="http://schemas.android.com/apk/res/android"
02      xmlns:tools="http://schemas.android.com/tools"
03      android:layout_width="match_parent"
04      android:layout_height="match_parent"
05      android:orientation="vertical" >
06
07      <!-- 得到当前网络状态的按钮 -->
08      <Button
09          android:id="@+id/btn_network"
10          android:layout_width="match_parent"
11          android:layout_height="wrap_content"
12          android:text="当前网络状态" />
13      <!-- 显示网络状态的标签控件 -->
14      <TextView
15          android:id="@+id/tv_networkshow"
16          android:layout_width="match_parent"
17          android:layout_height="wrap_content"
18          android:textSize="20dp" />
19
20  </LinearLayout>
```

这是我们的 Activity 的布局文件。在其中第 8～12 行定义了一个 Button 控件，当用户单击时得到当前状态。在第 14～18 行定义了一个 TextView 控件显示当前网络状态的标签控件。

然后修改 src/com.wyl.example/MainActivity.java 文件，代码如下：

```java
01  //定义了本实例的主要 Activity
02  public class MainActivity extends Activity {
03      //定义当前网络状态的按钮
04      private Button btnNetWork;
05      //声明网络连接管理器
06      ConnectivityManager connManager ;
07      //声明代表连网状态的 NetworkInfo 对象
08      private NetworkInfo networkInfo;
09      //定义显示当前网络连接状态的文本框
10      private TextView tvNetWorkShow;
11      //定义当前网络连接的 String
12      private String strNetWork;
13
14
15      @Override
16      public void onCreate(Bundle savedInstanceState) {
17          super.onCreate(savedInstanceState);
18          setContentView(R.layout.activity_main);
19          //获得网络连接服务
20          connManager=(ConnectivityManager)getSystemService(Context
             .CONNECTIVITY_SERVICE);
21
22          //获取代表联网状态的 NetWorkInfo 对象
23          networkInfo = connManager.getActiveNetworkInfo();
24
25          //得到布局中的所有对象
26          findView();
27          //设置对象的监听器
28          setListener();
29
30      }
```

```java
31
32  private void findView() {
33      //得到布局中的所有对象
34      btnNetWork = (Button) findViewById(R.id.btn_network);
35      tvNetWorkShow=(TextView)findViewById(R.id.tv_networkshow);
36  }
37
38  private void setListener() {
39      //设置对象的监听器
40      btnNetWork.setOnClickListener(new OnClickListener() {
41
42          @Override
43          public void onClick(View arg0) {
44              //TODO Auto-generated method stub
45              //当前网络是否可用
46              if(networkInfo.isAvailable())
47              {
48                  strNetWork="当前网络可用\n";
49              }
50              else{
51                  strNetWork="当前网络不可用\n";
52              }
53
54              //获取GPRS网络模式连接的描述
55              State state =
                      connManager.getNetworkInfo(ConnectivityManager
                      .TYPE_MOBILE).getState();
56              //State.CONNECTED 表示当前GPRS已连接
57              if(state==State.CONNECTED)
58              {
59                  strNetWork+="GPRS 网络已连接\n";
60              }
61
62              //获取WIFI网络模式连接的描述
63              state = connManager.getNetworkInfo(ConnectivityManager
                      .TYPE_WIFI).getState();
64              //State.CONNECTED 表示当前WIFI已连接
65              if(state==State.CONNECTED)
66              {
67                  strNetWork+="WIFI 网络已连接\n";
68              }
69              //设置textview的text属性
70              tvNetWorkShow.setText(strNetWork);
71          }
72      });
73  }
74  }
```

此文件是 Activity 的代码文件,在其中第 4 行定义了一个 Button 对象,在第 6 行定义了网络连接管理器对象,在第 8 行定义了网络状态信息对象。在第 20 行得到了系统的网络连接服务管理器对象,在第 23 行得到了当前的网络连接信息,在第 26 行得到所有的布局兑现,在第 28 行设置所有控件的监听器。关键代码在第 46 行得到 networkInfo 对象是否可用,然后在第 55 行得到当前已连接的网路类型,然后在第 57 和 65 行分别判断当前网络的连接类型,并且构造字符串在 textview 中进行展示。

4．实例扩展

在此实例中我们需要得到系统的网络状态权限，所以需要在 manifest 文件中添加如下代码：

```
<uses-permission android:name="android.permission.ACCESS_NETWORK_STATE"/>
```

范例 140　获得手机 SIM 卡信息

1．实例简介

在我们使用应用的过程中，经常会使用到用户单击某按钮的时候安装某个应用程序的效果。例如，在一些软件市场的客户端中，当用户下载完一个 Android 应用程序之后，希望直接单击安装此程序。遇到这样的功能，我们一般是当用户进行某个操作的时候，例如，单击某个按钮的时候，通过 Intent 打开系统中安装应用程序的界面。本例子就带领大家来实现一个单击打开系统安装应用程序界面的实例。

2．运行效果

该实例运行效果如图 7.9 所示。

图 7.9　获得手机 SIM 卡信息

3．实例程序讲解

在本实例中，当用户单击获取当前 SIM 卡信息的按钮是在下面的 TextView 显示所有的 SIM 卡信息。想要实现本实例效果，首先修改 res/layout/activity_main.xml 文件，代码如下：

```
01  <LinearLayout xmlns:android="http://schemas.android.com/apk/res/android"
02      xmlns:tools="http://schemas.android.com/tools"
03      android:layout_width="match_parent"
04      android:layout_height="match_parent"
05      android:orientation="vertical" >
```

```
06      <!-- 获得手机卡信息的按钮 -->
07      <Button
08          android:id="@+id/btn_siminfo"
09          android:layout_width="match_parent"
10          android:layout_height="wrap_content"
11          android:text="当前SIM卡信息" />
12      <!-- 显示手机卡信息的标签 -->
13      <TextView
14          android:id="@+id/tv_siminfoshow"
15          android:layout_width="match_parent"
16          android:layout_height="wrap_content"
17          android:textSize="20dp"/>
18
19  </LinearLayout>
```

这是我们的 Activity 的布局文件。在其中定义了一个按钮控件和一个标签控件。

然后修改 src/com.wyl.example/MainActivity.java 文件，代码如下：

```
01  //定义了本实例的主要 Activity
02  public class MainActivity extends Activity {
03  //省略定义控件的对象代码
04  //声明 TelephonyManager 对象
05  private TelephonyManager tm;
06  //定义 SIM 卡的状态
07  private String [] simState={"状态未知","无SIM卡",
08  "别PIN加锁","被PUK加锁",
09  "被NetWork PIN加锁","已准备好"};
10  //定义手机的制式
11  private String [] phoneTypes={"未知","GSM","CDMA"};
12
13  @Override
14  public void onCreate(Bundle savedInstanceState) {
15      super.onCreate(savedInstanceState);
16      setContentView(R.layout.activity_main);
17      //获得 locationManager 服务
18      tm=(TelephonyManager)getSystemService(Context.TELEPHONY
          _SERVICE);
19
20      //得到布局中的所有对象
21      findView();
22      //设置对象的监听器
23      setListener();
24  }
25  //省略得到控件代码
26  private void setListener() {
27      //设置对象的监听器
28      btnNetWork.setOnClickListener(new OnClickListener() {
29
30          @Override
31          public void onClick(View arg0) {
32              //TODO Auto-generated method stub
33              //获得设备的编号
34              String deviceId=tm.getDeviceId();
35              //获得 SIM 的国别
36              String Country=tm.getSimCountryIso();
37              //获取 SIM 卡序列号
38              String SIMSerial=tm.getSimSerialNumber();
```

```
39              //获取SIM卡状态
40              String SIMState=simState[tm.getSimState()];
41              //获取网络运营商代号
42              String networkOperator= tm.getNetworkOperator();
43              //获取网络运营商名称
44              String networkOperatorName= tm.getNetworkOperatorName();
45              //获得手机的制式
46              String phoneType=phoneTypes[tm.getPhoneType()];
47
48              strSimInfo="设备编号: "+deviceId+
49                  "\nSIM卡的国别: "+Country+"\nSIM卡序列号: "+SIMSerial
50                  +"\nSIM卡状态: "+SIMState+"\n网络运营商代号 : 
                     "+networkOperator
51                  +"\n网络运营商名称: "+networkOperatorName
52                  +"\n手机的制式: "+phoneType;
53              //设置textview的text属性
54              tvSimInfoShow.setText(strSimInfo);
55          };
56      });
57  }
58 }
```

此文件是 Activity 的代码文件，在其中第 4 行定义了电话管理器对象，第 7 行定义了一些状态字符串，第 18 行初始化了电话服务，在第 34~54 行，都是调用 tm 对象的相关方法，得到电话服务类的相关信息参数，然后显示在 TextView 控件中。

4．实例扩展

在此实例中需要得到系统的电话权限，所以需要在 manifest 文件中添加此代码行：

```
<uses-permission android:name="android.permission.READ_PHONE_STATE" />
```

范例 141　WiFi 管理器

1．实例简介

在我们使用应用的过程中，经常会用到 WiFi 的相关功能。例如，有些游戏支持 WiFi 对战，游戏软件支持 WiFi 传输等。遇到这样的功能，我们一般是当用户进行某个操作的时候，例如，单击某个按钮的时候，得到系统的 WiFi 管理对象，然后进行一次设置。本例子就带领大家来实现一个 WiFi 管理器的实例。

2．运行效果

该实例运行效果如图 7.10 所示。

3．实例程序讲解

在本实例中，提供四个按钮分别是开启 WiFi、关闭 WiFi、获取可用 WiFi 列表和获得具体 WiFi 信息。想要实现本实例效果，首先修改 res/layout/activity_main.xml 文件，代码如下：

接入点的BSSID :00:1f:9f:e2:16:52
MAC地址 :a0:0b:ba:c6:97:b0
IP地址 :1090627776
连接的ID :7

图 7.10　WiFi 管理器

```
01  <LinearLayout xmlns:android="http://schemas.android.com/apk/res/android"
02      xmlns:tools="http://schemas.android.com/tools"
03      android:layout_width="match_parent"
04      android:layout_height="match_parent"
05      android:orientation="vertical" >
06      <!--定义开启 wifi 按钮 -->
07      <Button
08          android:id="@+id/btn_openwifi"
09          android:layout_width="match_parent"
10          android:layout_height="wrap_content"
11          android:text="开启 WIFI" />
12      <!--定义关闭 wifi 的按钮 -->
13      <Button
14          android:id="@+id/btn_closewifi"
15          android:layout_width="match_parent"
16          android:layout_height="wrap_content"
17          android:text="关闭 WIFI" />
18      <!--定义获取 wifi 列表的按钮 -->
19      <Button
20          android:id="@+id/btn_getwifilist"
21          android:layout_width="match_parent"
22          android:layout_height="wrap_content"
23          android:text="获取 WIFI 列表" />
24
25      <Button
26          android:id="@+id/btn_getwifiinfo"
27          android:layout_width="match_parent"
28          android:layout_height="wrap_content"
29          android:text="得到 WIFI 信息" />
30      <!--定义获取显示结果的文本标签 -->
31      <TextView
32          android:id="@+id/tv_getwifiinfo"
33          android:layout_width="match_parent"
34          android:layout_height="wrap_content" />
35      <!--定义 wifi 列表的 listview -->
36      <ListView
37          android:id="@+id/lv_wifilist"
38          android:layout_width="match_parent"
39          android:layout_height="wrap_content" />
```

```
40
41    </LinearLayout>
```

这是我们的 Activity 的布局文件。在其中定义了四个 Button 对象，一个 TextView 对象用来展示信息，定义了一个 ListView 对象，用来显示获取 WiFi 列表。

然后修改 src/com.wyl.example/MainActivity.java 文件，其中关键代码如下：

```
01  //定义了本实例的主要 Activity
02  public class MainActivity extends Activity {
03      //省略定义的控件对象
04
05      //定义 WifiManager 对象
06      private WifiManager mWifiManager;
07      //扫描出的网络连接列表
08      private List<ScanResult> mWifiList=new ArrayList<ScanResult>();
09      //网络连接的 string 列表
10      private List<String> wifiList=new ArrayList<String>();
11
12      @Override
13      public void onCreate(Bundle savedInstanceState) {
14          super.onCreate(savedInstanceState);
15          setContentView(R.layout.activity_main);
16          //取得 WifiManager 对象
17          mWifiManager = (WifiManager)getSystemService(Context.WIFI
              _SERVICE);
18          //取得 WifiInfo 对象
19          mWifiInfo = mWifiManager.getConnectionInfo();
20
21          //得到布局中的所有对象
22          findView();
23          //设置对象的监听器
24          setListener();
25      }
26
27      //省略 findview 的实现
28      //省略 setListener 的实现
29      //省略 setAdapter 的实现
30
31      OnClickListener listener=new OnClickListener() {
32
33          @Override
34          public void onClick(View v) {
35              //TODO Auto-generated method stub
36              switch (v.getId()) {
37
38              //打开 WIFI
39              case R.id.btn_openwifi:
40                  //检查 wifi 是否开启,isWifiEnabled()等于 true 说明开启,等于 false
                    说明关闭
41                  if (!mWifiManager.isWifiEnabled())
42                  {
43                      mWifiManager.setWifiEnabled(true);
44                      Toast.makeText(MainActivity.this,"WIFI 已经开启", Toast
                        .LENGTH_SHORT).show();
45                  }
46                  break;
47
48              //关闭 WIFI
```

```
49          case R.id.btn_closewifi:
50              //检查wifi是否开启,isWifiEnabled()等于true说明开启,等于false
                  说明关闭
51              if (mWifiManager.isWifiEnabled())
52              {
53                  mWifiManager.setWifiEnabled(false);
54                  Toast.makeText(MainActivity.this, "WIFI已经关闭",
                        Toast.LENGTH_SHORT).show();
55              }
56              break;
57
58          //得到WIFI列表
59          case R.id.btn_getwifilist:
60              //检查wifi是否开启,isWifiEnabled()等于true说明开启,等于false
                  说明关闭
61              if(mWifiManager.isWifiEnabled())
62              {
63                  //扫描WIFI
64                  mWifiManager.startScan();
65                  //得到扫描结果
66                  mWifiList = mWifiManager.getScanResults();
67                  //把mWifiList里面的内容放在wifiList里面
68                  for(int i=0;i<mWifiList.size();i++)
69                  {
70                      wifiList.add(mWifiList.get(i).SSID);
71                  }
72                  setAdapter();              //设置页面adapter
73              }
74              else{
75                  Toast.makeText(MainActivity.this,
76                      "WIFI没有开启,无法扫描", Toast.LENGTH_SHORT).show();
77              }
78              break;
79          case R.id.btn_getwifiinfo:
80              //得到接入点的BSSID
81              String bssid=mWifiInfo.getBSSID();
82              //得到MAC地址
83              String macAddress=mWifiInfo.getMacAddress();
84              //得到IP地址
85              int ipAddress=mWifiInfo.getIpAddress();
86              //得到连接的ID
87              int netWorkId=mWifiInfo.getNetworkId();
88              tvWifiInfo.setText("接入点的BSSID :"+bssid+"\nMAC地址 :"
89              +macAddress+"\nIP地址 :"+ipAddress+"\n连接的
                ID :"+netWorkId);
90
91          default:
92              break;
93          }
94      }
95  };
96 }
97
```

此文件是 Activity 的关键代码文件,其中省略了之前反复讲过的控件的获取,控件设置监听器的代码,在其中第 6 行定义了一个 WifiManager 对象,用来管理手机中 Wifi 的所有信息。在程序第 17 行得到系统的 WifiManager 对象,然后通过此对象的 getConnectionInfo

方法得到当前的 WiFi 信息对象。然后在第 31 行定义了一个单击监听器对象，其中分别对不同的按钮做出不同的响应，当用户单击开启 WiFi 的时候，在第 43 行调用的 setWifiEnabled 为 true，同理，当用户单击关闭 WiFi 的时候，在第 53 行设置为 false，当然当用户单击扫描附近的 WiFi 的时候，调用了第 64 行的 startScan 方法，得到了附近的 WiFi 列表。如果用户当前已经连接某个 WiFi，这时候调用第 83~85 行的代码，得到连接的 Mac 地址和 IP 地址。

4．实例扩展

在此实例中需要程序获得手机的 WiFi 管理权限，所以需要在文件的 manifest 文件中添加相应的权限，代码如下：

```
<uses-permission android:name="android.permission.ACCESS_WIFI_STATE" >
</uses-permission>
<uses-permission android:name="android.permission.CHANGE_WIFI_STATE" >
</uses-permission>
```

范例 142 系统软键盘显示

1．实例简介

在我们使用应用的过程中，经常会使用到用户单击某按钮的时候打开了手机的照相机软件进行拍照，然后将照片返回应用程序界面的效果。例如，在一些社交软件的客户端中，当用户希望设置自己的应用头像，而且直接进行相机拍照得到自己的头像照片。遇到这样的功能，我们一般是当用户进行某个操作的时候，例如：单击某个按钮的时候，通过 Intent 打开系统中照相机程序进行拍照，然后将得到的照片返回应用中显示。本例子就带领大家来实现一个单击按钮打开系统照相机拍照，并且显示照片的实例效果。

2．运行效果

该实例运行效果如图 7.11 所示。

3．实例程序讲解

在本实例中，页面中包含了一个显示系统软键盘的按钮，当用户单击此按钮的时候显示系统的软键盘。想要实现本效果，首先修改 res/layout/activity_main.xml 文件，代码如下：

图 7.11 系统软键盘显示

```
01  <LinearLayout xmlns:android="http://schemas.android.com/apk/
    res/android"
02      xmlns:tools="http://schemas.android.com/tools"
03      android:layout_width="match_parent"
04      android:layout_height="match_parent"
05      android:orientation="vertical" >
06      <!-- 定义显示软键盘按钮 -->
07      <Button
08          android:id="@+id/btn_soft"
09          android:layout_width="match_parent"
```

```
10            android:layout_height="wrap_content"
11            android:text="弹出/关闭软键盘" />
12
13 </LinearLayout>
```

这是我们的 Activity 的布局文件。在此文件中定义了一个显示软键盘的按钮。

然后修改 src/com.wyl.example/MainActivity.java 文件，代码如下：

```
01 //定义了本实例的主要 Activity
02 public class MainActivity extends Activity {
03    //省略控件对象的声明
04    //声明 InputMethodManager 对象
05    InputMethodManager imm ;
06
07    @Override
08    public void onCreate(Bundle savedInstanceState) {
09        super.onCreate(savedInstanceState);
10        setContentView(R.layout.activity_main);
11        //取得 InputMethodManager 对象
12        imm = (InputMethodManager)getSystemService(Context.INPUT_METHOD
          _SERVICE);
13        //得到布局中的所有对象
14        findView();
15        //设置对象的监听器
16        setListener();
17    }
18    //省略 findview 方法和 setlistener 方法的定义
19
20    OnClickListener listener=new OnClickListener() {
21
22        @Override
23        public void onClick(View v) {
24            //TODO Auto-generated method stub
25            //第一次调用显示，再次调用则隐藏，如此反复
26            //触发软键盘，InputMethodManager.HIDE_NOT_ALWAYS 能够正常的隐藏
27            imm.toggleSoftInput(0, InputMethodManager.HIDE_NOT_ALWAYS);
28        }
29    };
30 }
```

此文件是 Activity 的代码文件，在其中第 5 行定义了 InputMethodManager 的对象，在第 12 行通过系统提供的 getSystemService 方法得到系统的输入法管理对象，当用户单击显示软键盘的按钮时，调用第 27 行显示或者隐藏软键盘。

4. 实例扩展

在此实例中主要是针对于软键盘的操作，在我们实际应用中有一些操作是针对于特殊的需求的，如自定义软键盘等。

范例 143 打开系统行车模式

1. 实例简介

在我们使用应用的过程中，经常会在不同的情况下使用手机。例如，在我们开车的时

候使用手机,这样我就希望有电话拨入自动接听;当我在夜晚使用手机的时候我就希望手机屏幕亮度暗一些,以免刺眼等。遇到这样的功能,我们一般是当用户进行某个操作的时候,例如,单击某个按钮的时候,得到系统的模式切换服务,然后进行切换。本例子就带领大家来实现一个单击按钮切换系统模式的实例效果。

2. 运行效果

该实例运行效果如图 7.12 所示。

图 7.12　打开系统行车模式

3. 实例程序讲解

在本实例中的四个按钮分别是打开行车模式、关闭行车模式、开启夜间模式和关闭夜间模式。想要实现本实例效果,首先修改 res/layout/activity_main.xml 文件,在其中包含对应的四个按钮。代码省略。

然后修改 src/com.wyl.example/MainActivity.java 文件,代码如下:

```
01  //定义了本实例的主要 Activity
02  public class MainActivity extends Activity {
03      //省略控件对象的定义
04      //定义 UiModeManager 对象
05      private UiModeManager uiModeManager;
06
07      @Override
08      public void onCreate(Bundle savedInstanceState) {
09          super.onCreate(savedInstanceState);
10          setContentView(R.layout.activity_main);
11          //取得 uiModeManager 对象
12          uiModeManager = (UiModeManager)getSystemService(Context.UI_MODE_
            SERVICE);
13          //得到布局中的所有对象
14          findView();
15          //设置对象的监听器
16          setListener();
```

```
17  }
18
19  //省略findview方法和setlistener方法定义
20
21  OnClickListener listener=new OnClickListener() {
22
23      @Override
24      public void onClick(View v) {
25          //TODO Auto-generated method stub
26          switch (v.getId()) {
27
28          //打开行车模式
29          case R.id.btn_opendrivercar:
30              //启用行车模式
31              uiModeManager.enableCarMode(
                    UiModeManager.ENABLE_CAR_MODE_GO_CAR_HOME);
32              Toast.makeText(MainActivity.this,
33                  "已经开启行车模式", Toast.LENGTH_SHORT).show();
34              break;
35
36          //关闭行车模式
37          case R.id.btn_closedrivercar:
38              //调用uiModeManager的关闭行车模式方法
39              uiModeManager.disableCarMode(UiModeManager.DISABLE_CAR
                    _MODE_GO_HOME);
40              Toast.makeText(MainActivity.this,
41                  "已经关闭行车模式", Toast.LENGTH_SHORT).show();
42              break;
43
44          //开启夜间模式
45          case R.id.btn_opennight:
46              uiModeManager.setNightMode(UiModeManager.MODE_
                    NIGHT_YES);
47              Toast.makeText(MainActivity.this,
48                  "已经开启夜间模式", Toast.LENGTH_SHORT).show();
49              break;
50          //关闭夜间模式
51          case R.id.btn_closenight:
52              uiModeManager.setNightMode(UiModeManager.MODE_NIGHT_NO);
53              Toast.makeText(MainActivity.this,
54                  "已经关闭夜间模式", Toast.LENGTH_SHORT).show();
55              break;
56          default:
57              break;
58          }
59      }
60  };
```

此文件是 Activity 的代码文件,在其中第 5 行定义了 UiModeManager 对象,在第 12 行得到了系统的界面模式管理对象。然后在第 21 行定义的监听器对象中,当用户单击某个按钮的时候设置对应的模式,当用户单击启用行车模式的时候设置值为 UiModeManager.ENABLE_CAR_MODE_GO_CAR_HOME。

4. 实例扩展

注意一点,在本例中只进行了模式的切换,在实际应用当中,可能这些切换只有在特殊情况下使用的比较多。例如，如果我们开发一款车载软件，那么行车模式是你一定要掌握的。

范例 144 音量控制器

1．实例简介

在我们使用应用的过程中，经常会希望在程序内部进行音量的调节。例如，在一些游戏的设置界面可以去修改应用的音量，或者显示当前的音量值。遇到这样的功能，我们一般是当用户进行某个操作的时候，例如，单击某个按钮的时候，获取到系统的音量管理服务，然后得到相应的参数，或者修改相应的参数。本例子就带领大家来实现一个音量控制器的实例效果。

2．运行效果

该实例运行效果如图 7.13 所示。

图 7.13 音量控制器

3．实例程序讲解

在本实例中，提供了三个 Button 控件和一个显示音量信息的 TextView 控件。想要实现本实例效果，首先修改 res/layout/activity_main.xml 文件，在其中加入对应的控件。代码省略。

然后修改 src/com.wyl.example/MainActivity.java 文件，代码如下：

```
01  //定义了本实例的主要 Activity
02  public class MainActivity extends Activity {
03      //省略控件兑现的定义
04      //定义 AudioManager 对象
05      private AudioManager audioManager ;
06
07      @Override
08      public void onCreate(Bundle savedInstanceState) {
09          super.onCreate(savedInstanceState);
```

```java
10      setContentView(R.layout.activity_main);
11      //取得 audiomanage 对象
12      audioManager = (AudioManager)getSystemService(Context.AUDIO
        _SERVICE);
13      //得到布局中的所有对象
14      findView();
15      //设置对象的监听器
16      setListener();
17  }
18  //省略 findView 和 setListener 方法的定义
19
20  OnClickListener listener=new OnClickListener() {
21
22      @Override
23      public void onClick(View v) {
24          //TODO Auto-generated method stub
25          switch (v.getId()) {
26
27          //当前的音量
28          case R.id.btn_currentvolume:
29              //得到当前的通话音量
30              int currentCall = audioManager.
                      getStreamVolume(AudioManager.STREAM_VOICE_CALL );
31              //得到当前的系统音量
32              int currentSystem = audioManager.getStreamVolume
                  (AudioManager.STREAM_SYSTEM );
33              //得到当前的音乐音量
34              int currentMusic = audioManager.getStreamVolume
                  (AudioManager.STREAM_MUSIC );
35              //得到当前的提示声音音量
36              int currentTip = audioManager.getStreamVolume
                  (AudioManager.STREAM_ALARM );
37              tvVolume.setText("当前的通话音量: "+currentCall+
38                      "\n 当前的系统音量: "+currentSystem+
39                      "\n 当前的音乐音量: "+currentMusic+
40                      "\n 当前的提示声音音量: "+currentTip);
41              break;
42
43          //增加系统音量
44          case R.id.btn_increaselvolume:
45              //参数 1: 声音类型, 可取为 STREAM_VOICE_CALL (通话)
46              //                       STREAM_SYSTEM (系统声音)
47              //                       STREAM_RING (铃声)
48              //                       STREAM_MUSIC (音乐)
49              //                       STREAM_ALARM (闹铃声)
50              //参数 2: 调整音量的方向, 可取 ADJUST_LOWER (降低)
51              //                            ADJUST_RAISE (升高)
52              //                            ADJUST_SAME
53              //参数 3: 可选的标志位
54              audioManager.adjustStreamVolume(AudioManager
                  .STREAM_SYSTEM,
55                      AudioManager.ADJUST_RAISE,
56                      AudioManager.FX_FOCUS_NAVIGATION_UP);
57              break;
58
59          //减小系统音量
60          case R.id.btn_declinevolume:
61              audioManager.adjustStreamVolume(
```

```
                       AudioManager.STREAM_SYSTEM,AudioManager.ADJUST_LOWER,
62
63                     AudioManager.FX_FOCUS_NAVIGATION_UP);
64             break;
65         }
66     }
67 };
68 }
```

此文件是 Activity 的代码文件，在其中第 5 行定义了一个音频管理对象，在第 12 行得到系统的音频管理对象，在第 20 行定义了 OnClickListener，当用户单击某个按钮执行相应的代码，当用户单击得到当前的音量信息时，调用第 30～40 行代码，得到 audioManager 对象的各种声音参数。当用户单击增加系统音量时，调用第 54～56 行，通过 audioManager 增加系统音量，同理单击降低系统音量时设置降低音量的参数，降低系统音量。

4．实例扩展

在本实例中对于音量的操作仅仅是限于系统的多媒体音量，当然还可以设置其他的音量，如通话声音、闹铃声音和音乐声音。

范例 145　短信群发软件

1．实例简介

在我们使用应用的过程中，经常会是一个短信后台群发的效果。例如，一般软件的客户端都会提供群发功能。遇到这样的功能，我们一般是当用户进行某个操作的时候，例如，单击某个按钮的时候，得到系统的短信服务，然后进行短信的发送。本例子就带领大家来实现一个单击按钮进行短信群发的实例效果。

2．运行效果

该实例运行效果如图 7.14 所示。

图 7.14　短信群发软件

3. 实例程序讲解

在本实例中，提供提示用户输入需要发送短信的电话号码和短信内容，然后当用户单击此发送按钮时，程序开始依次发送短信。想要实现本实例效果，首先修改 res/layout/activity_main.xml 文件，定义如上效果。代码省略。

然后修改 src/com.wyl.example/MainActivity.java 文件，代码如下：

```
01  //定义了本实例的主要Activity
02  public class MainActivity extends Activity {
03      //省略控件定义
04
05      @Override
06      protected void onCreate(Bundle savedInstanceState) {
07          super.onCreate(savedInstanceState);
08          setContentView(R.layout.activity_main);
09          //得到布局中的所有对象
10          findView();
11          //设置对象的监听器
12          setListener();
13      }
14      //省略findview方法定义
15
16      private void setListener() {
17          //TODO Auto-generated method stub
18          button.setOnClickListener(new OnClickListener() {
19              @Override
20              public void onClick(View v) {
21                  //从文本框中得到要发送的号码
22                  String p = edit_no.getText().toString();
23                  //从文本框中得到要发送的内容
24                  body = edit_body.getText().toString();
25                  //不同电话号码用,隔开 如果是中文,请修改否则无法拆分
26                  address = p.split(",");
27                  //定义一个String类型的HASHSET
28                  Set<String> addr = new HashSet<String>();
29                  //把号码拆分然后放在hashset里面
30                  for (int i = 0; i < address.length; i++) {
31                      addr.add(address[i]);
32                  }
33                  //定义发送短信的函数
34                  sendSMS(addr, body);
35                  //清空输入号码和内容的文本框
36                  edit_no.setText("");
37                  edit_body.setText("");
38              }
39          });
40      }
41
42      //发送短信的函数
43      public void sendSMS(Set<String> phone, String body) {
44          //实例化SmsManager
45          SmsManager msg = SmsManager.getDefault();
46          //设置要跳转的intent
47          Intent send = new Intent(SENT_SMS_ACTION);
48          //短信发送广播
49          PendingIntent sendPI = PendingIntent.getBroadcast(this, 0, send, 0);
```

```java
50      for (String pno : phone) {
51          //sendTextMessage 各个参数
52          //第一个参数 发送短信的地址（也就是号码）
53          //第二个参数 短信服务中心，如果为null，就是用当前默认的短信服务中心
54          //第三个参数 短信内容
55          //第四个参数
56          //当短信发送成功或者失败时，产生下面这些错误之一
57          //RESULT_ERROR_GENERIC_FAILURE
58          //RESULT_ERROR_RADIO_OFF RESULT_ERROR_NULL_PDU
59          //第五个参数是否生产状态报告生成的pdu
60          msg.sendTextMessage(pno, null, body, sendPI, null);
61      }
62
63  }
64
65  protected void onResume() {
66      super.onResume();
67      //注册监听，接收广播的信息
68      registerReceiver(sendMessage, new IntentFilter(SENT_SMS_ACTION));
69  }
70
71  BroadcastReceiver sendMessage = new BroadcastReceiver() {
72      @Override
73      public void onReceive(Context c, Intent intent) {
74          //判断短信是否成功，Activity.RESULT_OK，短信发送成功
75          switch (getResultCode()) {
76          case Activity.RESULT_OK:
77              Toast.makeText(MainActivity.this, "发送成功！",
78                      Toast.LENGTH_SHORT)
79                      .show();
80              break;
81          default:
82              Toast.makeText(MainActivity.this, "发送失败！",
83                      Toast.LENGTH_SHORT)
84                      .show();
85              break;
86          }
87      }
88  };
89  }
```

此文件是 Activity 的代码文件，在其中第 10 行得到页面中的所有布局控件，在第 12 行设置所有控件的监听器对象。当用户单击发送短信按钮的时候，执行 12～37 行，首先得到用户输入的电话号码和短信内容，如果电话号码为多个，则进行处理后放入 addr 数组中，然后调用 sendSMS 方法进行短信发送，具体实现在第 43～61 行，首先得到 SmsManager 对象，然后建立一个延迟的 Intent 对象，再通过 msg 对象的 sendTextMessage 方法依次进行短信发送，为了能够得到短信是否送达，我们在第 68 行注册了广播监听器，监听器的实现在第 71～88 行。当短信发送完毕后，可以在此监听器中得到一个广播，通过 getResultCode 方法得到短信是否发送成功，然后进行提示。

4. 实例扩展

本实例使用到了系统发送短信的权限，所以需要在你的 manifest 文件中加入如下代码获取权限：

```xml
<uses-permission android:name="android.permission.SEND_SMS" />
```

范例 146 电池状态查看器

1. 实例简介

在我们使用应用的过程中，经常会需要获得手机电池的电量。例如，当手机电量不足时，某些功能无法开启等。遇到这样的功能，我们一般是当用户进行某个操作的时候，例如，单击某个按钮的时候，开启系统电量改变广播监听器。本例子就带领大家来实现一个单击按钮得到手机电池状态的实例效果。

2. 运行效果

该实例运行效果如图 7.15 所示。

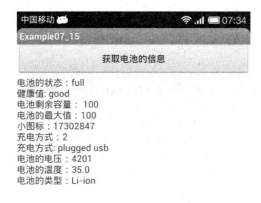

图 7.15 电池状态查看器

3. 实例程序讲解

在本实例中，提供提示一个查看电池状态的按钮，当用户单击按钮时显示当前手机电池状态。想要实现本实例效果，首先修改 res/layout/activity_main.xml 文件，添加相应的控件。代码省略。

然后修改 src/com.wyl.example/MainActivity.java 文件，代码如下：

```
001  //定义了本实例的主要Activity
002  public class MainActivity extends Activity {
003
004      @Override
005      public void onCreate(Bundle savedInstanceState) {
006          super.onCreate(savedInstanceState);
007          setContentView(R.layout.activity_main);
008          //得到布局中的所有对象
009          findView();
010          //设置对象的监听器
011          setListener();
012      }
013      //省略findview和setlistener方法的定义
014
015      OnClickListener listener = new OnClickListener() {
```

```java
016
017        @Override
018        public void onClick(View v) {
019            //TODO Auto-generated method stub
020            switch (v.getId()) {
021
022            //当前的音量
023            case R.id.btn_battery:
024                //注册广播接收器
025                IntentFilter filter = new IntentFilter();
026                filter.addAction(Intent.ACTION_BATTERY_CHANGED);
027                registerReceiver(mBroadcastReceiver, filter);
028                break;
029            }
030        }
031    };
032    //声明广播接收者对象
033    private BroadcastReceiver mBroadcastReceiver = new BroadcastReceiver() {
034
035        @Override
036        public void onReceive(Context context, Intent intent) {
037            //TODO Auto-generated method stub
038            String action = intent.getAction();
039            if (action.equals(Intent.ACTION_BATTERY_CHANGED)) {
040                //得到电池状态：
041                //BatteryManager.BATTERY_STATUS_CHARGING: 充电状态
042                //BatteryManager.BATTERY_STATUS_DISCHARGING: 放电状态
043                //BatteryManager.BATTERY_STATUS_NOT_CHARGING: 未充满
044                //BatteryManager.BATTERY_STATUS_FULL: 充满电
045                //BatteryManager.BATTERY_STATUS_UNKNOWN: 未知状态
046                int status = intent.getIntExtra("status", 0);
047                //得到健康状态：
048                //BatteryManager.BATTERY_HEALTH_GOOD: 状态良好
049                //BatteryManager.BATTERY_HEALTH_DEAD: 电池没有电
050                //BatteryManager.BATTERY_HEALTH_OVER_VOLTAGE: 电池电压过高
051                //BatteryManager.BATTERY_HEALTH_OVERHEAT: 电池过热
052                //BatteryManager.BATTERY_HEALTH_UNKNOWN: 未知状态
053                int health = intent.getIntExtra("health", 0);
054                //boolean 类型
055                boolean present = intent.getBooleanExtra("present", false);
056                //得到电池剩余容量
057                int level = intent.getIntExtra("level", 0);
058                //得到电池最大值。通常为100
059                int scale = intent.getIntExtra("scale", 0);
060                //得到图标ID
061                int icon_small = intent.getIntExtra("icon-small", 0);
062                //充电方式：
                //      BatteryManager.BATTERY_PLUGGED_AC: AC 充电
                //      BatteryManager.BATTERY_PLUGGED_USB: USB 充电
063                int plugged = intent.getIntExtra("plugged", 0);
064                //得到电池的电压
065                int voltage = intent.getIntExtra("voltage", 0);
066                //得到电池的温度,0.1度单位。例如 表示197 的时候，意思为19.7度
067                int temperature = intent.getIntExtra("temperature", 0);
068                //得到电池的类型
069                String technology = intent.getStringExtra("technology");
```

```
070                    //得到电池状态
071                    String statusString = "";
072                    //根据状态id,得到状态字符串,代码省略
073
074                    //得到电池的寿命状态,代码省略
075
076                    //得到充电模式
077                    String acString = "";
078                    //根据充电状态id,得到充电模式,代码省略
079
080                    //显示电池信息
081                    tvBattery.setText("电池的状态: " + statusString
082                            + "\n 健康值: "+ healthString
083                            + "\n 电池剩余容量: " + level
084                            + "\n 电池的最大值: " + scale
085                            + "\n 小图标: " + icon_small
086                            + "\n 充电方式: " + plugged
087                            + "\n 充电方式: " + acString
088                            + "\n 电池的电压: " + voltage
089                            + "\n 电池的温度: " + (float) temperature * 0.1
090                            + "\n 电池的类型: " + technology);
091                }
092            }
093        };
094
095        @Override
096        protected void onPause() {
097            super.onPause();
098            //解除注册监听
099            unregisterReceiver(mBroadcastReceiver);
100        }
101    }
```

此文件是 Activity 的代码文件,在其中第 9 行得到布局中的所有控件对象,第 11 行设置相应的监听器对象,按钮被单击的时候执行第 25～27 行,注册广播监听器。在第 33 行定义了广播监听器的具体实现,主要代码在第 38 行得到广播监听器接收到的广播类型,如果是电池状态改变的广播,那么就通过 Intent 得到相应的状态,包括电池的充电状态、电池容量最大值和电池的电压等信息,然后进行显示即可。

4. 实例扩展

本实例实现的原理就是基于广播的接收器,当设置此广播接收器后,即可监听系统的某些状态的改变了。

7.2 Android 中的广播的使用

范例 147 飞行模式的切换

1. 实例简介

在我们使用应用的过程中,经常会使用到需要切换系统飞行模式的功能。例如,当用

户希望进行省电模式时。遇到这样的功能，我们一般是当用户进行某个操作的时候，例如，单击某个按钮的时候，发送关闭飞行模式的广播。本例子就带领大家来实现一个单击按钮打开关闭飞行模式的程序。

2. 运行效果

该实例运行效果如图 7.16 所示。

图 7.16 飞行模式的切换

3. 实例程序讲解

在本实例中，定义了两个按钮对象，分别是打开飞行模式和关闭飞行模式。想要实现本实例效果，首先修改 res/layout/activity_main.xml 文件。代码省略。

然后修改 src/com.wyl.example/MainActivity.java 文件，代码如下：

```
01  //定义了本实例的主要Activity
02  public class MainActivity extends Activity {
03      //省略定义控件对象代码
04      private ContentResolver cr;
05
06      @Override
07      public void onCreate(Bundle savedInstanceState) {
08          super.onCreate(savedInstanceState);
09
10          setContentView(R.layout.activity_main);
11          //得到数据共享ContentResolver实例
12          cr = getContentResolver();
13          //得到布局中的所有对象
14          findView();
15          //设置对象的监听器
16          setListener();
17      }
18  //省略findview和setlistener方法的实现
19
20  OnClickListener listener = new OnClickListener() {
21
22      @SuppressWarnings("deprecation")
23      @Override
24      public void onClick(View v) {
25          //TODO Auto-generated method stub
26          switch (v.getId()) {
27
```

```
28            //开启飞行模式的按钮
29            case R.id.btn_startflying:
30                //开启飞行模式
31                if (Settings.System.getString(cr,
32                        Settings.System.AIRPLANE_MODE_ON).equals("0")) {
33                    Log.e("AIRPLANE_MODE_ON", "0");
34                    //获取当前飞行模式状态,返回的是String值0,或1.0为关闭飞行,1
                        为开启飞行
35                    //如果关闭飞行,则打开飞行
36                    Settings.System.putString(cr,
37                            Settings.System.AIRPLANE_MODE_ON, "1");
38                    //设置需要启动的ACTION
39                    Intent intent = new Intent(
40                            Intent.ACTION_AIRPLANE_MODE_CHANGED).
41                            putExtra("state", true);;
42                    //发送飞行广播
43                    sendBroadcast(intent);
44                    Toast.makeText(MainActivity.this,
45                        "已开启飞行模式",
46                        Toast.LENGTH_SHORT).show();
47                }
48
49                break;
50            //关闭飞行模式的按钮
51            case R.id.btn_stopflying:
52                if (Settings.System.getString(cr,
53                        Settings.System.AIRPLANE_MODE_ON).equals("1")) {
54                    Log.e("AIRPLANE_MODE_ON", "1");
55                    //获取当前飞行模式状态,返回的是String值0,或1.0为关闭飞行,1
                        为开启飞行
56                    //如果打开飞行,则关闭飞行
57                    Settings.System.putString(cr,
58                            Settings.System.AIRPLANE_MODE_ON, "0");
59                    //设置需要启动的ACTION
60                    Intent intent = new Intent(
61                            Intent.ACTION_AIRPLANE_MODE_CHANGED).
62                            putExtra("state", false);;
63                    //发送飞行广播
64                    sendBroadcast(intent);
65                    Toast.makeText(MainActivity.this, "已关闭飞行模式",
66                            Toast.LENGTH_SHORT).show();
67                }
68                break;
69            }
70        }
71    };
72 }
```

此文件是 Activity 的代码文件,在其中第 4 行定义了一个 ContentResolver 对象,在第 12 行对其进行初始化。当用户单击开启飞行模式和关闭飞行模式的按钮后,分别执行第 31~46 行,首先判断当前系统的设置值是否为开启了飞行模式,容纳后通过 Settings.System 设置飞行模式的状态,最后通过 Intent 发送广播给系统,然后就可以实现控制系统飞行模式的功能了。关闭飞行模式原理相同。

4. 实例扩展

本实例实现了飞行模式状态的改变，所以我们需要在 manifest 文件中添加获取飞行模式的代码如下：

```xml
<uses-permission android:name="android.permission.WRITE_SETTINGS" />
```

注意一点，此权限功能在 Android 4.0 版本后已经被修改，所以在 Android 4.0 后飞行模式的权限在一般应用中无法获取。此实例只能在 4.0 之前版本的手机运行成功。

范例 148　创建桌面快捷方式

1. 实例简介

在我们使用应用的过程中，经常会发现这样的一个功能，就是在程序安装完毕后会在手机桌面创建一个快捷方式，这样方便用户调用本应用程序，同时也可以删除桌面的快捷方式。遇到这样的功能，我们一般是当用户进行某个操作的时候，例如，单击某个按钮的时候，创建桌面快捷方式，同时也可以单击按钮删除桌面快捷方式。本例子就带领大家来实现一个桌面快捷方式创建的实例。

2. 运行效果

该实例运行效果如图 7.17 所示。

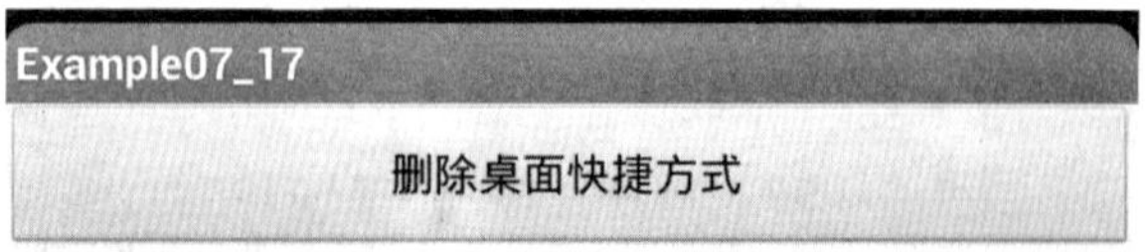

图 7.17　创建桌面快捷方式

3. 实例程序讲解

在本实例中，当程序启动的时候就会在桌面创建快捷方式，程序界面中有删除快捷方式的按钮，当用户单击此按钮时删除桌面快捷方式。想要实现本实例效果，首先修改 res/layout/activity_main.xml 文件，创建如图的布局。代码省略。

然后修改 src/com.wyl.example/MainActivity.java 文件，代码如下：

```java
01  //定义了本实例的主要Activity
02  public class MainActivity extends Activity {
03      //定义创建桌面快捷方式的action字符串
04      private static final String CREATE_SHORTCUT_ACTION =
                      "com.android.launcher.action.INSTALL_SHORTCUT";
05      //定义删除桌面快捷方式的action字符串
```

```
06    private static final String DROP_SHORTCUT_ACTION =
                   "com.android.launcher.action.UNINSTALL_SHORTCUT";
07
08    @Override
09    public void onCreate(Bundle savedInstanceState) {
10        super.onCreate(savedInstanceState);
11        addShortcut(MainActivity.this);
12
13        setContentView(R.layout.activity_main);
14        //得到布局中的所有对象
15        findView();
16        //设置对象的监听器
17        setListener();
18
19    }
20    //省略findview和setlistener方法定义
21
22    OnClickListener listener = new OnClickListener() {
23
24        @Override
25        public void onClick(View v) {
26            //TODO Auto-generated method stub
27            switch (v.getId()) {
28
29            //删除桌面快捷方式
30            case R.id.btn_del:
31                //调用删除快捷方式函数
32                delShortcut(MainActivity.this);
33                Toast.makeText(MainActivity.this,
34                    "删除成功", Toast.LENGTH_SHORT)
35                        .show();
36            }
37        }
38    };
39
40    //创建桌面快捷方式
41    public void addShortcut(Context cx) {
42        //定义intent
43        Intent shortcut = new Intent(CREATE_SHORTCUT_ACTION);
44        //得到当前的包名
45        Intent shortcutIntent = cx.getPackageManager()
46                .getLaunchIntentForPackage(cx.getPackageName());
47        //设置快捷方式的单击指向
48        shortcut.putExtra(Intent.EXTRA_SHORTCUT_INTENT, shortcutIntent);
49        //获取当前应用名称
50        String title = null;
51        try {
52            final PackageManager pm = cx.getPackageManager();
53            //得到当前应用程序的lable名称
54            title = pm.getApplicationLabel(
55                pm.getApplicationInfo(cx.getPackageName(),
56                    PackageManager.GET_META_DATA)).toString();
57        } catch (Exception e) {
58        }
59        //快捷方式名称
60        shortcut.putExtra(Intent.EXTRA_SHORTCUT_NAME, title);
```

```
61      //不允许重复创建（不一定有效）
62      shortcut.putExtra("duplicate", false);
63      //快捷方式的图标
64      Parcelable iconResource = Intent.ShortcutIconResource
                .fromContext(cx,
65                  R.drawable.ic_launcher);
66      //设置快捷方式的图标
67      shortcut.putExtra(Intent.EXTRA_SHORTCUT_ICON_RESOURCE, iconResource);
68
69      cx.sendBroadcast(shortcut);
70  }
71
72  public void delShortcut(Context cx) {
73      //定义删除快捷方式的 intent
74      Intent shortcut = new Intent(DROP_SHORTCUT_ACTION);
75
76      //获取当前应用名称
77      String title = null;
78      try {
79          final PackageManager pm = cx.getPackageManager();
80          //得到当前应用程序的 lable
81          title = pm.getApplicationLabel(
82                  pm.getApplicationInfo(cx.getPackageName(),
83                      PackageManager.GET_META_DATA)).toString();
84      } catch (Exception e) {
85      }
86      //快捷方式名称
87      shortcut.putExtra(Intent.EXTRA_SHORTCUT_NAME, title);
88      Intent shortcutIntent = cx.getPackageManager()
89              .getLaunchIntentForPackage(cx.getPackageName());
90      shortcut.putExtra(Intent.EXTRA_SHORTCUT_INTENT, shortcutIntent);
91      cx.sendBroadcast(shortcut);
92  }
93  }
```

此文件是 Activity 的代码文件，在第 4 和 6 行定义了要发送的广播 Action 字符串。在第 11 行调用了 addShortcut 方法来创建桌面快捷方式，第 15 行得到所有的控件对象，在第 17 行设置控件对象的监听器。在第 22 行定义的单击事件监听器对象中，当用户单击删除按钮的时候调用 delShortcut 来删除桌面控件。在第 41 行定义了 addShortcut 方法，在其中定义一个 Intent，然后得到应用程序的包名，再通过 PackageManager 对象的 sendBroadcast 方法发送创建快捷方式的广播。同样道理在第 72 行定义了 delShortcut 方法，其中首先创建 Intent 对象，然后得到 PackageManager 对象，通过发送广播来发送删除桌面快捷方式的广播。

4．实例扩展

本实例涉及桌面快捷方式的创建和删除，所以需要获取系统的相应权限，在 manifest 文件中添加：

```
<uses-permission android:name="com.android.launcher.permission.INSTALL_SHORTCUT" />
<uses-permission android:name="com.android.launcher.permission.UNINSTALL_SHORTCUT" />
```

范例 149　程序开机自动启动

1．实例简介

在我们使用应用的过程中，经常会使用到某些软件希望用户开机的时候自动启动。例如，网络流量监听器和短信黑名单监听等。遇到这样的功能，我们一般是监听系统启动的广播，然后启动我们的后台服务或者广播监听器。本例子就带领大家来实现一个监听手机启动的实例。

2．运行效果

该实例运行效果如图 7.18 所示。

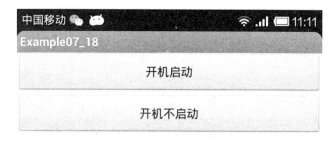

图 7.18　程序开机启动广播

3．实例程序讲解

在本实例中，提供两个按钮，当用户单击第一个按钮的时候监听手机重启的广播，当单击第二个按钮的时候取消广播监听。想要实现本实例效果，首先修改 res/layout/activity_main.xml 文件。代码省略。

然后修改 src/com.wyl.example/MainActivity.java 文件，代码如下：

```
01  //定义了本实例的主要 Activity
02  public class MainActivity extends Activity {
03      //定义布局控件对象
04      //声明 broadcastReceiver 对象
05      private BootCompletedReceiver receiver;
06
07      @Override
08      public void onCreate(Bundle savedInstanceState) {
09          super.onCreate(savedInstanceState);
10          setContentView(R.layout.activity_main);
11          //给 broadcastReceiver 对象赋值
12          receiver=new BootCompletedReceiver();
13          //得到布局中的所有对象
14          findView();
```

```
15        //设置对象的监听器
16        setListener();
17    }
18 //省略findview和setlistener方法的定义
19
20 OnClickListener listener=new OnClickListener() {
21
22     @Override
23     public void onClick(View v) {
24         //TODO Auto-generated method stub
25         switch (v.getId()) {
26
27         //注册广播
28         case R.id.btn_startafterboot:
29             //注册监听者,第一个参数是需要绑定的监听器,第二个是需要监听的广播
30             registerReceiver(receiver,
31                 new IntentFilter("android.intent.action.BOOT_
                  COMPLETED"));
32             Toast.makeText(MainActivity.this,
33                 "开机自启动已完成", Toast.LENGTH_SHORT).show();
34             break;
35         case R.id.btn_notstartafterboot:
36             if(receiver!=null)
37             {
38                 //取消监听者
39                 unregisterReceiver(receiver);
40                 Toast.makeText(MainActivity.this,
41                     "取消开机自启动已完成", Toast.LENGTH_SHORT).show();
42             }
43
44             break;
45         }
46     }
47 };
48 }
```

此文件是 Activity 的代码文件,在其中第 5 行定义了一个广播接收器对象,在第 12 行进行了初始化。当用户单击开启广播按钮的时候调用第 30~31 行注册广播接收器,当用户单击取消广播的时候调用第 39 行取消此广播接收器。

本实例中还需要一个广播接收器类。新建 src/com.wyl.example/ BootCompletedReceiver.java 文件,代码如下:

```
01 //自定义广播接收器
02 public class BootCompletedReceiver extends BroadcastReceiver {
03
04     @Override
05     public void onReceive(Context arg0, Intent arg1) {
06         //TODO Auto-generated method stub
07         //检测手机是否是重启广播
08         if (arg1.getAction().equals(Intent.ACTION_BOOT_COMPLETED)) {
09             //检测到手机启动后,启动 MainActivity
10             Intent newIntent = new Intent(arg0, MainActivity.class);
11             //让 activity 启动一个新任务
12             //注意,必须添加这个标记,否则启动会失败
13             newIntent.addFlags(Intent.FLAG_ACTIVITY_NEW_TASK);
14
15             //启动 MainActivity
```

```
16              arg0.startActivity(newIntent);
17          }
18      }
19  }
```

此类定义了一个广播接收器,用来接收系统重启的广播,并且在接收到此广播后可以通过 Intent 启动 Activity,启动 Service 等。

4. 实例扩展

本实例涉及到接收系统重启的广播,所以需要在 manifest 文件中添加如下代码得到相应权限:

```
<uses-permission android:name="android.permission.RECEIVE_BOOT_COMPLETED" />
```

范例 150　拍照物理键的功能定制

1. 实例简介

在我们使用应用的过程中,经常会使用到某个物理键按键的监听,其中有一个比较特殊的物理键。就是拍照键,例如,在一些拍照软件中为了方便打开摄像头,或者方便自拍时使用。遇到这样的功能,我们一般是定义广播接收器对象来接收物理拍照键按下的广播,然后做出处理。本例子就带领大家来实现一个监听物理拍照键的实例。

2. 运行效果

该实例运行效果如图 7.19 所示。

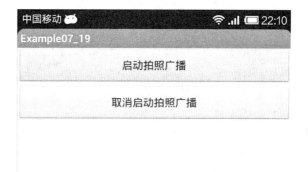

图 7.19　物理拍照键的监听

3. 实例程序讲解

在本实例中,当程序打开的时候显示两个按钮,第一个是开启物理拍照键的监听,第二个是取消监听物理拍照键的广播。想要实现本实例效果,首先修改 res/layout/activity_main.xml 文件。代码省略。

然后修改 src/com.wyl.example/MainActivity.java 文件,代码如下:

```
01  //定义了本实例的主要 Activity
02  public class MainActivity extends Activity {
```

```
03     //省略定义布局中的控件的代码
04     //声明 TakePhotoReceiver 对象
05     private TakePhotoReceiver receiver;
06
07     @Override
08     public void onCreate(Bundle savedInstanceState) {
09         super.onCreate(savedInstanceState);
10         setContentView(R.layout.activity_main);
11         //给 broadcastReceiver 对象赋值
12         receiver=new TakePhotoReceiver();
13         //得到布局中的所有对象
14         findView();
15         //设置对象的监听器
16         setListener();
17     }
18
19     //省略 findview 和 setlistener 方法
20     OnClickListener listener=new OnClickListener() {
21
22         @Override
23         public void onClick(View v) {
24             //TODO Auto-generated method stub
25             switch (v.getId()) {
26             //按下手机的物理拍照键
27             case R.id.btn_starttake:
28                 //注册监听者,第一个参数是需要绑定的监听器,第二个是需要监听的广播
29                 registerReceiver(receiver,
30                     new IntentFilter(Intent.ACTION_CAMERA_BUTTON));
31                 Toast.makeText(MainActivity.this,
32                     "拍照广播启动已完成", Toast.LENGTH_SHORT).show();
33                 break;
34             case R.id.btn_notstarttake:
35                 if(receiver!=null)
36                 {
37                     //取消监听者
38                     unregisterReceiver(receiver);
39                     Toast.makeText(MainActivity.this,
40                         "取消拍照广播已完成", Toast.LENGTH_SHORT).show();
41                 }
42                 break;
43             }
44         }
45     };
46 }
```

此文件是 Activity 的代码文件,在其中第 5 行定义了一个拍照按键的广播接收器对象,在第 12 行初始化广播接收器对象,第 14 行得到布局中的所有的控件对象,第 16 行设置所有控件对象的监听器。其中关键代码是在第 20 行定义单击的事件监听器,当用户按下开启广播的按钮的时候,在第 29～30 行注册广播监听器,当用户按下关闭广播的时候,在第 38 行取消广播接收器。

本实例中最主要的代码是新建 src/com.wyl.example/TakePhotoReceiver.java 文件,代码如下:

```
01 //定义广播接收器对象
02 public class TakePhotoReceiver extends BroadcastReceiver {
03
```

```
04  @Override
05  public void onReceive(Context arg0, Intent arg1) {
06      //TODO Auto-generated method stub
07      //检测是否按下拍照键
08      if (arg1.getAction().equals(Intent.ACTION_CAMERA_BUTTON)) {
09          //检测按下拍照键
10          Toast.makeText(arg0,
11              "检测到拍照广播", Toast.LENGTH_SHORT).show();
12      }
13  }
14  }
```

此文件定义了一个广播接收器，当用户按下手机上的物理拍照键时系统就会发送一个广播，在此广播接收器中可以进行广播判断然后进行相应的操作，这里只是做了 Toast 的提示。

4．实例扩展

需要注意本实例中需要在真机上进行测试，而且需要你的手机有物理的拍照按钮。

范例 151　锁屏广播接收器

1．实例简介

在我们使用应用的过程中，经常会遇到用户在使用应用程序的过程中锁屏的现象。例如，用户在阅读某本电子书的时候锁屏，这时候需要记录电子书的阅读位置，下次用户再次打开此软件应该显示上次阅读的位置。遇到这样的功能，我们一般是定义一个广播接收器用来接收系统的锁屏的广播。本例子就带领大家来实现一个接收系统锁屏广播的实例。

2．运行效果

该实例运行效果如图 7.20 所示。

图 7.20　系统锁屏广播接收器

3．实例程序讲解

在本实例中，显示了四个按钮，分别是启动锁屏广播、启动解锁广播、取消锁屏广播和取消解锁广播。想要实现本实例效果，首先修改 res/layout/activity_main.xml 文件，定义这四个按钮控件。代码省略。

然后修改 src/com.wyl.example/MainActivity.java 文件，代码如下：

```java
01  //定义了本实例的主要Activity
02  public class MainActivity extends Activity {
03      //省略控件对象的定义代码
04      //声明LockScreenReceiver对象
05      private LockScreenReceiver lockReceiver;
06      //声明LockScreenReceiver对象
07      private LockScreenReceiver openReceiver;
08
09      @Override
10      public void onCreate(Bundle savedInstanceState) {
11          super.onCreate(savedInstanceState);
12          setContentView(R.layout.activity_main);
13          //给broadcastReceiver对象赋值
14          lockReceiver=new LockScreenReceiver();
15          openReceiver=new LockScreenReceiver();
16          //得到布局中的所有对象
17          findView();
18          //设置对象的监听器
19          setListener();
20      }
21
22  //省略findview和setlistener的方法实现
23
24  OnClickListener listener=new OnClickListener() {
25
26      @Override
27      public void onClick(View v) {
28          //TODO Auto-generated method stub
29          switch (v.getId()) {
30          //锁屏广播启动按钮
31          case R.id.btn_startlockscreen:
32              //注册监听者
33              registerReceiver(lockReceiver,
34                  new IntentFilter(Intent.ACTION_SCREEN_OFF));
35              Toast.makeText(MainActivity.this,
36                  "锁屏广播启动已完成", Toast.LENGTH_SHORT).show();
37              break;
38          //取消锁屏广播按钮
39          case R.id.btn_cancellockscreen:
40              if(lockReceiver!=null)
41              {
42                  //取消锁屏广播监听者
43                  unregisterReceiver(lockReceiver);
44                  Toast.makeText(MainActivity.this,
45                      "取消锁屏广播已完成", Toast.LENGTH_SHORT).show();
46              }
47              break;
48          //解锁广播启动按钮
49          case R.id.btn_startopenscreen:
50              //注册解锁广播监听器
51              registerReceiver(openReceiver,
52                  new IntentFilter(Intent.ACTION_SCREEN_ON));
53              Toast.makeText(MainActivity.this,
54              "解锁广播启动已完成", Toast.LENGTH_SHORT).show();
55          break;
```

```
56              //取消解锁广播按钮
57              case R.id.btn_cancelopenscreen:
58                  if(openReceiver!=null)
59                  {
60                      //取消解锁监听者
61                      unregisterReceiver(openReceiver);
62                      Toast.makeText(MainActivity.this,
63                          "取消解锁广播已完成", Toast.LENGTH_SHORT).show();
64                  }
65                  break;
66              }
67          }
68      };
69  }
```

此文件是 Activity 的代码文件，在其中第 4~6 行定义了两个广播接收器对象，一个用来接受解锁广播，一个用来接受锁屏广播。在第 14~15 行对它们进行初始化，当用户单击接受锁屏广播的时候，调用第 33 行注册广播接收器，当用户单击取消锁屏广播的时候，调用第 43 行取消广播接收器，当用户单击接受解锁广播的时候，调用第 51~52 行，当用户单击取消解锁广播的时候，调用第 61 行取消解锁广播的监听。

在本例子中还要自定义一个广播接收器，新建代码 src/com.wyl.example/MainActivity.java 文件，代码如下：

```
01  //自定义广播接收器
02  public class LockScreenReceiver extends BroadcastReceiver {
03
04      @Override
05      public void onReceive(Context arg0, Intent arg1) {
06          //TODO Auto-generated method stub
07          //Intent.ACTION_SCREEN_ON 解锁的 ACTION
08          //Intent.ACTION_SCREEN_OFF 锁屏的 ACTION
09          if (arg1.getAction().equals(Intent.ACTION_SCREEN_ON)) {
10              //检测到解锁的广播
11              Toast.makeText(arg0,
12                  "检测到解锁广播", Toast.LENGTH_SHORT).show();
13          }else if (arg1.getAction().equals(Intent.ACTION_SCREEN_OFF)) {
14              //检测到锁屏广播
15              Toast.makeText(arg0,
16                  "检测到锁屏广播", Toast.LENGTH_SHORT).show();
17          }
18      }
19  }
```

此广播接收器接收系统有两种广播，一个是锁屏的广播，一个是解锁屏幕的广播，当接收到不同广播的时候做不同的处理。

4. 实例扩展

本实例可以接受屏幕锁屏和解锁的广播，这时候我们就可以自定义一些应用的例子，或者在用户锁屏的时候暂时保存当前应用程序的数据所在。

范例 152　系统设置信息改变的广播

1. 实例简介

在我们使用应用的过程中，经常会使用到用户登录的功能，当用户输入正确的用户名密码后，程序跳转到应用首页显示欢迎信息。如果用户输入错误的用户名密码，则程序不进行跳转，并且提示用户名密码错误。这样的功能，我们一般是当用户单击登录按钮的时候得到用户名密码并进行判断是否为合法，然后跳转相应的 Activity 进行页面显示。本例子就带领大家来实现一个具有用户名密码判断的登录功能的实例。

2. 运行效果

该实例运行效果如图 7.21 所示。

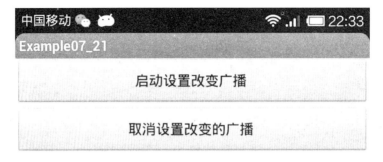

图 7.21　系统设置改变的广播接收器

3. 实例程序讲解

在本实例中，显示两个按钮，功能分别是启动设置改变广播和取消设置改变的广播。想要实现本实例效果，首先修改 res/layout/activity_main.xml 文件定义两个按钮。代码省略。

然后修改 src/com.wyl.example/MainActivity.java 文件，代码如下：

```
01  //定义了本实例的主要Activity
02  public class MainActivity extends Activity {
03      //省略控件对象定义的代码
04
05      //声明SetChangeReceiver对象
06      private SetChangeReceiver SetingReceiver;
07
08      @Override
09      public void onCreate(Bundle savedInstanceState) {
10          super.onCreate(savedInstanceState);
11          setContentView(R.layout.activity_main);
12          //给broadcastReceiver对象赋值
13          SetingReceiver=new SetChangeReceiver();
14          //得到布局中的所有对象
15          findView();
```

```
16          //设置对象的监听器
17          setListener();
18      }
19
20  //省略findview和setlistener方法实现
21  OnClickListener listener=new OnClickListener() {
22
23      @Override
24      public void onClick(View v) {
25          //TODO Auto-generated method stub
26          switch (v.getId()) {
27          //开启设置改变广播
28          case R.id.btn_startsetchange:
29              //注册监听者,第一个参数是需要绑定的监听器,第二个是需要监听的广播
30              registerReceiver(SetingReceiver,
31                  new IntentFilter(Intent.ACTION_CONFIGURATION
                      _CHANGED));
32              Toast.makeText(MainActivity.this,
33                  "设置改变广播启动已完成", Toast.LENGTH_SHORT).show();
34              break;
35
36          //取消监听者
37          case R.id.btn_cancelsetchange:
38              if(SetingReceiver!=null)
39              {
40                  //取消监听者
41                  unregisterReceiver(SetingReceiver);
42                  Toast.makeText(MainActivity.this,
43                      "取消设置改变广播已完成", Toast.LENGTH
                          _SHORT).show();
44              }
45              break;
46          }
47      }
48  };
49  }
```

此文件是 Activity 的代码文件,在其中第 6 行定义一个广播接收器对象,用来接收系统的设置信息改变的广播。在第 13 行进行了此对象的初始化,当用户单击开启设置广播时调用第 28~31 行注册广播接收器,当用户单击取消广播的时候调用第 41 行取消广播接收器的注册。

然后在工程目录下要建立广播接收者类,新建 src/com.wyl.example/SetChangeReceiver.java 文件,代码如下:

```
01  public class SetChangeReceiver extends BroadcastReceiver {
02
03      @Override
04      public void onReceive(Context arg0, Intent arg1) {
05          //TODO Auto-generated method stub
06          //Intent.ACTION_CONFIGURATION_CHANGED
07          //设备当前设置被改变时发出的广播
08          //(包括的改变:界面语言,设备方向等,请参考Configuration.java)
09
10          if (arg1.getAction().equals(Intent.ACTION_CONFIGURATION_CHANGED))
    {
11              //检测到设置改变的广播
12              Toast.makeText(arg0,
```

```
13                        "检测到设置改变广播", Toast.LENGTH_SHORT).show();
14                }
15        }
16 }
```

在此广播中判断接收到的广播是否为配置信息改变的广播，然后做出对应的处理，这里仅仅是通过 Toast 显示接收广播的提示信息。

4．实例扩展

本实例中仅仅得到了配置改变的广播，具体哪些配置进行了更改，更改了哪些信息都包含在 onReceive 方法中的第二个参数 arg1 中，大家可以取出相应的信息得到设置改变的详细信息。

范例 153　系统内存不足提醒

1．实例简介

在我们使用一个应用程序的时候，可能会遇到检测系统内存不足的情况。例如，当系统的可用内存不足的情况下，我们最好不要进行消耗内存的操作，或者自动释放某些内存。这样的功能，我们一般是需要接收系统中内存不足的广播，然后做出相应操作。本例子就带领大家来实现一个系统内存不足提示的实例。

2．运行效果

该实例运行效果如图 7.22 所示。

图 7.22　系统内存不足的广播监听

3．实例程序讲解

在本实例中，显示了两个按钮分别是开启系统内存不足的广播监听和取消系统内存不足的广播监听。想要实现本实例效果，首先修改 res/layout/activity_main.xml 文件，加入两个按钮对象。代码省略。

然后修改 src/com.wyl.example/MainActivity.java 文件，代码如下：

```
01  //定义了本实例的主要 Activity
```

```
02  public class MainActivity extends Activity {
03      //省略按钮对象的定义
04
05      //声明 MemorylessReceiver 对象
06      private MemoryLessReceiver memoryLessReceiver;
07
08      @Override
09      public void onCreate(Bundle savedInstanceState) {
10          super.onCreate(savedInstanceState);
11          setContentView(R.layout.activity_main);
12          //给 broadcastReceiver 对象赋值
13          memoryLessReceiver=new MemoryLessReceiver();
14          //得到布局中的所有对象
15          findView();
16          //设置对象的监听器
17          setListener();
18
19      }
20
21      //省略 findview 和 setlistener 方法的实现
22
23      OnClickListener listener=new OnClickListener() {
24
25          @Override
26          public void onClick(View v) {
27              //TODO Auto-generated method stub
28              switch (v.getId()) {
29
30              //开启内存不足广播
31              case R.id.btn_startmemoryless:
32                  //注册监听者,第一个参数是需要绑定的监听器,第二个是需要监听的广播
33                  registerReceiver(memoryLessReceiver,
34                      new IntentFilter(Intent.ACTION_DEVICE_STORAGE_LOW));
35                  Toast.makeText(MainActivity.this,
36                      "设置内存不足广播启动已完成", Toast.LENGTH_SHORT).show();
37                  break;
38
39              //取消监听者
40              case R.id.btn_cancelmemoryless:
41                  if(memoryLessReceiver!=null)
42                  {
43                      //取消监听者
44                      unregisterReceiver(memoryLessReceiver);
45                      Toast.makeText(MainActivity.this,
46                          "取消内存不足广播已完成", Toast.LENGTH_SHORT)
                                .show();
47                  }
48                  break;
49              }
50          }
51      };
52  }
53
```

此文件是 Activity 的代码文件,在其中第 6 行定义了广播接收器对象,在第 13 行进行了初始化,当用户单击开启广播监听的时候,调用第 33~34 行启动广播监听器,当用户单击取消广播监听器按钮的时候,调用第 44 行取消广播监听器。

然后定义广播接收器类，新建 src/com.wyl.example/ MemoryLessReceiver.java 文件，代码如下：

```
01  //自定义广播接收器类，接收系统内存不足的广播
02  public class MemoryLessReceiver extends BroadcastReceiver {
03
04      @Override
05      public void onReceive(Context arg0, Intent arg1) {
06          //TODO Auto-generated method stub
07          //Intent.ACTION_DEVICE_STORAGE_LOW
08          //设备内存不足时发出的广播，此广播只能由系统使用
09
10          if (arg1.getAction().equals(Intent.ACTION_DEVICE_STORAGE_LOW)) {
11              //检测到内存不足的广播
12              Toast.makeText(arg0,
13                  "检测到内存不足广播", Toast.LENGTH_SHORT).show();
14          }
15      }
16  }
```

在此类中接收系统的内存不足的广播，并且做出提示。

4．实例扩展

本实例可以接收系统中内存不足的广播，一般情况下这种功能是在一个完善的应用程序中所必备的，在系统内存不足的情况下，要进行内存的清理，或者不再做需要大量内存的工作。

范例 154　接收耳机插入广播

1．实例简介

在我们使用某些应用程序的时候，经常会需要接受耳机插入的事件。例如，在音乐播放的时候当耳机插入手机是需要关闭播放效果，开启耳机模式，或者某些收音机软件，当耳机插入的时候才可以接收信号，这些都需要接收耳机插入的广播。一般我们都是通过设置一个广播接收器来进行此事件的监听。本例子就带领大家来实现一个接收耳机插入事件的广播实例。

2．运行效果

该实例运行效果如图 7.23 所示。

图 7.23　接收耳机插入广播

3. 实例程序讲解

在本实例中，页面显示两个按钮控件，当用户单击第一个按钮的时候开启耳机插入广播接收器，单击第二个按钮的时候取消耳机插入广播接收器。想要实现本实例效果，首先修改 res/layout/activity_main.xml 文件，定义两个按钮。代码省略。

然后修改 src/com.wyl.example/MainActivity.java 文件，代码如下：

```
01  //定义了本实例的主要 Activity
02  public class MainActivity extends Activity {
03      //省略控件定义代码
04
05      //声明 EerphoneInsertReceiver 对象
06      private EarphoneInsertReceiver earphoneInsertReceiver;
07
08      @Override
09      public void onCreate(Bundle savedInstanceState) {
10          super.onCreate(savedInstanceState);
11          setContentView(R.layout.activity_main);
12          //给 earphoneInsertReceiver 对象赋值
13          earphoneInsertReceiver = new EarphoneInsertReceiver();
14          //得到布局中的所有对象
15          findView();
16          //设置对象的监听器
17          setListener();
18
19      }
20
21  //省略 findview 和 setlistener 方法的实现代码
22
23  OnClickListener listener = new OnClickListener() {
24
25      @Override
26      public void onClick(View v) {
27          //TODO Auto-generated method stub
28          switch (v.getId()) {
29
30          //开启耳机插入广播
31          case R.id.btn_startearphoneinsert:
32              //注册监听者，第一个参数是需要绑定的监听器，第二个是需要监听的广播
33              registerReceiver(earphoneInsertReceiver, new IntentFilter(
34                  Intent.ACTION_HEADSET_PLUG));
35              Toast.makeText(MainActivity.this, "设置耳机插入广播启动已完成",
36                  Toast.LENGTH_SHORT).show();
37              break;
38              //取消监听者
39          case R.id.btn_cancelearphoneinsert:
40              if (earphoneInsertReceiver != null) {
41                  //取消监听者
42                  unregisterReceiver(earphoneInsertReceiver);
43                  Toast.makeText(MainActivity.this, "取消耳机插入广播已完成",
44                      Toast.LENGTH_SHORT).show();
45              }
46              break;
47          }
48      }
```

```
49    };
50  }
```

此文件是 Activity 的代码文件,在其中第 6 行定义了广播接收器对象,在第 13 行进行了初始化,当用户单击启动耳机插入广播的时候调用第 33~34 行,当用户单击取消耳机插入广播时调用第 42 行,取消广播接收器。

然后定义广播接收器类,新建 src/com.wyl.example/ EarphoneInsertReceiver.java 文件,代码如下:

```
01  //定义广播接收耳机插入的广播
02  public class EarphoneInsertReceiver extends BroadcastReceiver {
03
04      @Override
05      public void onReceive(Context arg0, Intent arg1) {
06          //TODO Auto-generated method stub
07          //Intent.ACTION_HEADSET_PLUG; 在耳机口上插入耳机时发出的广播
08          if (arg1.getAction().equals(Intent.ACTION_HEADSET_PLUG)) {
09              //检测到耳机插入的广播
10              Toast.makeText(arg0,
11                  "检测到耳机插入广播", Toast.LENGTH_SHORT).show();
12          }
13  }
```

此文件为广播接收器类,在此类中接受耳机插入的广播并且可以做出相应操作。

4. 实例扩展

本实例一般情况使用在特定的领域中,例如,音乐播放器中或收音机软件中等。

范例 155　手机区域设置更改监听器

1. 实例简介

在我们使用某些应用程序的时候,经常需要知道手机的设置区域是否更改,如果系统的手机设置区域更改了,那么程序中的对应参数可能需要进行调整。这样的功能,我们一般是通过广播接收器,接收系统区域更改设置的广播。本例子就带领大家来使用广播接收器来接受系统区域设置更改的实例。

2. 运行效果

该实例运行效果如图 7.24 所示。

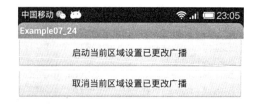

图 7.24　手机区域设置更改监听器

3. 实例程序讲解

在本实例中，首先显示两个按钮控件，当用户单击启动区域设置广播时开启广播监听，当用户单击取消区域设置广播时，取消对应的广播接收器。想要实现本实例效果，首先修改 res/layout/activity_main.xml 文件。代码省略。

然后修改 src/com.wyl.example/MainActivity.java 文件，代码如下：

```java
01  //定义了本实例的主要 Activity
02  public class MainActivity extends Activity {
03      //省略控件对象的定义
04  
05      //声明 AreaChangeReceiver 对象
06      private AreaChangeReceiver areaChangeReceiver;
07  
08      @Override
09      public void onCreate(Bundle savedInstanceState) {
10          super.onCreate(savedInstanceState);
11          setContentView(R.layout.activity_main);
12          //给 areaChangeReceiver 对象赋值
13          areaChangeReceiver = new AreaChangeReceiver();
14          //得到布局中的所有对象
15          findView();
16          //设置对象的监听器
17          setListener();
18      }
19  
20      //省略 findview 和 setlistener 方法的定义
21  
22      OnClickListener listener = new OnClickListener() {
23  
24          @Override
25          public void onClick(View v) {
26              //TODO Auto-generated method stub
27              switch (v.getId()) {
28  
29              //开启当前区域设置已更改广播
30              case R.id.btn_startedistancechange:
31                  //注册监听者
32                  registerReceiver(areaChangeReceiver, new IntentFilter(
33                          Intent.ACTION_LOCALE_CHANGED));
34                  Toast.makeText(MainActivity.this,
35                      "设置当前区域设置已更改广播启动已完成",
36                          Toast.LENGTH_SHORT).show();
37                  break;
38  
39              //取消监听者
40              case R.id.btn_canceldistancechange:
41                  if (areaChangeReceiver != null) {
42                      //取消监听者
43                      unregisterReceiver(areaChangeReceiver);
44                      Toast.makeText(MainActivity.this,
45                          "取消当前区域设置已更改广播已完成",
46                              Toast.LENGTH_SHORT).show();
47                  }
48                  break;
49              }
```

```
50      }
51   };
52 }
```

此文件是 Activity 的代码文件，在其中第 6 行定义了区域改变的广播接收器对象，在第 13 行进行了初始化，当用户单击启动广播按钮的时候调用第 32～33 行注册广播，当用户单击取消广播的时候，调用第 43 行取消广播接收器。

然后定义广播接收器类，新建 src/com.wyl.example/AreaChangeReceiver.java 文件，代码如下：

```
01  //定义接收区域改变的广播接收器
02  public class AreaChangeReceiver extends BroadcastReceiver {
03
04      @Override
05      public void onReceive(Context arg0, Intent arg1) {
06          //TODO Auto-generated method stub
07
08          //Intent.ACTION_LOCALE_CHANGED; 设备当前区域设置已更改时发出的广播
09          if (arg1.getAction().equals(Intent.ACTION_LOCALE_CHANGED)) {
10              //检测当前区域设置已更改的广播
11              Toast.makeText(arg0,
12                  "检测到当前区域设置已更改广播", Toast.LENGTH_SHORT).show();
13          }
14      }
15 }
```

此文件为广播接收器类，用来接收当前区域设置改变的广播。

4．实例扩展

本实例主要用在时区改变的时候对应的与时区相关的值的改变，如获得当前时间。

范例 156　SDCard 插入的广播

1．实例简介

在我们使用某些应用程序的时候，经常会需要检测手机 SDcard 是否插入，当 SDCard 插入时，我们系统的一些操作需要在 SDCard 上进行，否则就在手机内存上进行。这样的功能，我们一般是通过广播接收器，接收 SDCard 插入的广播。本例子就带领大家使用广播接收器来接收 SDCard 卡插入的广播的实例。

2．运行效果

该实例运行效果如图 7.25 所示。

3．实例程序讲解

在本实例中，首先显示两个按钮控件，当用户单击启动 SDCard 插入广播时开启广播监听，当用户单击取消 SDCard 插入手机的广播时，取消对应的广播接收器。想要实现本实例效果，首先修改 res/layout/activity_main.xml 文件。代码省略。

图 7.25　SDCard 插入的广播

然后修改 src/com.wyl.example/MainActivity.java 文件，代码如下：

```java
01  //定义了本实例的主要 Activity
02  public class MainActivity extends Activity {
03      //省略控件对象的定义代码
04
05      //声明 SdcardInsertReceiver 对象
06      private SdcardInsertReceiver sdcardInsertReceiver;
07
08      @Override
09      public void onCreate(Bundle savedInstanceState) {
10          super.onCreate(savedInstanceState);
11          setContentView(R.layout.activity_main);
12          //给 SdcardInsertReceiver 对象赋值
13          sdcardInsertReceiver=new SdcardInsertReceiver();
14          //得到布局中的所有对象
15          findView();
16          //设置对象的监听器
17          setListener();
18
19      }
20
21      //省略 findview 和 setlistener 方法的定义代码
22
23      OnClickListener listener=new OnClickListener() {
24
25          @Override
26          public void onClick(View v) {
27              //TODO Auto-generated method stub
28              switch (v.getId()) {
29
30              //开启 SD 卡插入广播
31              case R.id.btn_startedsdcardinsert:
32                  //注册监听者
33                  registerReceiver(sdcardInsertReceiver,
34                      new IntentFilter(Intent.ACTION_MEDIA_MOUNTED));
35                  Toast.makeText(MainActivity.this,
36                      "设置 SD 卡插入广播启动已完成", Toast.LENGTH_
                        SHORT).show();
37                  break;
38
39              //取消监听者
40              case R.id.btn_cancelsdcardinsert:
41                  if(sdcardInsertReceiver!=null)
42                  {
43                      //取消监听者
44                      unregisterReceiver(sdcardInsertReceiver);
45                      Toast.makeText(MainActivity.this,
46                          "取消 SD 卡插入广播已完成", Toast.LENGTH_SHORT).
                            show();
47                  }
48                  break;
49          }
50      }
51  };
52  }
```

此文件是 Activity 的代码文件，在其中第 6 行定义了 SDCard 插入的广播接收器对象，在第 13 行进行了初始化，当用户单击启动广播按钮的时候调用第 33~34 行注册广播，当用户单击取消广播的时候，调用第 44 行取消广播接收器。

然后定义广播接收器类，新建 src/com.wyl.example/SdcardInsertReceiver.java 文件，代码如下：

```
01  //定义接收SDcard正确插入的广播
02  public class SdcardInsertReceiver extends BroadcastReceiver {
03
04      @Override
05      public void onReceive(Context arg0, Intent arg1) {
06          //Intent.ACTION_MEDIA_MOUNTED; 插入SD卡并且已正确安装（识别）时发出的广播
07          if (arg1.getAction().equals(Intent.ACTION_MEDIA_MOUNTED)) {
08              Toast.makeText(arg0,
09                  "检测到SD卡插入广播", Toast.LENGTH_SHORT).show();
10          }
11      }
12  }
```

此文件为广播接收器类，用来接收 SDCard 插入的广播。

4．实例扩展

本实例主要用在 SDCard 插入与否会影响手机操作的应用中。

范例 157　SDCard 移除的广播

1．实例简介

在我们使用某些应用程序的时候，经常需要检测手机 SDcard 是否移除，当 SDCard 移除时，我们系统的一些操作就无法在 SDCard 上进行，只能在手机内存上进行。这样的功能，我们一般是通过广播接收器，接收 SDCard 移除的广播。本例子就带领大家来使用广播接收器来接收 SDCard 卡移除的广播的实例。

2．运行效果

该实例运行效果如图 7.26 所示。

图 7.26　SDCard 移除的广播

3. 实例程序讲解

在本实例中，首先显示两个按钮控件，当用户单击启动 SDCard 移除广播时开启广播监听，当用户单击取消 SDCard 移除手机的广播时，取消对应的广播接收器。想要实现本实例效果，首先修改 res/layout/activity_main.xml 文件。代码省略。

然后修改 src/com.wyl.example/MainActivity.java 文件，代码如下：

```
01  //定义了本实例的主要 Activity
02  public class MainActivity extends Activity {
03      //省略控件定义代码
04
05      //声明 SdcardDrawReceiver 对象
06      private SdcardDrawReceiver sdcardDrawReceiver;
07
08      @Override
09      public void onCreate(Bundle savedInstanceState) {
10          super.onCreate(savedInstanceState);
11          setContentView(R.layout.activity_main);
12          //给 SdcardDrawReceiver 对象赋值
13          sdcardDrawReceiver = new SdcardDrawReceiver();
14          //得到布局中的所有对象
15          findView();
16          //设置对象的监听器
17          setListener();
18
19      }
20
21  //省略 findview 和 setlistener 方法定义
22
23  OnClickListener listener = new OnClickListener() {
24      @Override
25      public void onClick(View v) {
26          //TODO Auto-generated method stub
27          switch (v.getId()) {
28
29          //开启 SD 卡拔出广播
30          case R.id.btn_startedsdcarddraw:
31              //注册监听者
32              registerReceiver(sdcardDrawReceiver, new IntentFilter(
33                  Intent.ACTION_MEDIA_EJECT));
34              Toast.makeText(MainActivity.this,
35                  "设置 SD 卡拔出广播启动已完成",
36                  Toast.LENGTH_SHORT).show();
37              break;
38
39          //取消监听者
40          case R.id.btn_cancelsdcarddraw:
41              if (sdcardDrawReceiver != null) {
42                  //取消监听者
43                  unregisterReceiver(sdcardDrawReceiver);
44                  Toast.makeText(MainActivity.this,
45                      "取消 SD 卡拔出广播已完成",
46                      Toast.LENGTH_SHORT).show();
47              }
48              break;
49          }
```

```
50          }
51      };
52  }
```

此文件是 Activity 的代码文件,在其中第 6 行定义了 SDCard 移除的广播接收器对象,在第 13 行进行了初始化,当用户单击启动广播按钮的时候调用第 32~33 行注册广播,当用户单击取消广播的时候,调用第 43 行取消广播接收器。

然后定义广播接收器类,新建 src/com.wyl.example/SdcardDrawReceiver.java 文件,代码如下:

```
01  //定义广播接收器来接收 SDcard 移除的广播
02  public class SdcardDrawReceiver extends BroadcastReceiver {
03  
04      @Override
05      public void onReceive(Context arg0, Intent arg1) {
06          //Intent.ACTION_MEDIA_EJECT;
07          //已拔掉外部大容量储存设备发出的广播(比如SD卡,或移动硬盘)
08          //不管有没有正确卸载都会发出此广播
09          if (arg1.getAction().equals(Intent.ACTION_MEDIA_EJECT)) {
10              Toast.makeText(arg0,
11                  "检测到SD卡拔出广播", Toast.LENGTH_SHORT).show();
12          }
13      }
14  }
15
```

此文件为广播接收器类,用来接收 SDCard 移除的广播。

4. 实例扩展

本实例主要应用在当 SDCard 移除会影响手机操作的应用中。

范例 158 APK 安装完成的广播

1. 实例简介

在我们使用某些应用程序的时候,经常会需要检测手机中是否有新的 APK 安装完成,如果有,则提示用户是否立刻进行打开。这样的功能,我们一般是通过广播接收器,接收 APK 的安装完成的广播。本例子就带领大家使用广播接收器来接收 APK 安装完成的广播的实例。

2. 运行效果

该实例运行效果如图 7.27 所示。

3. 实例程序讲解

在本实例中,首先显示两个按钮控件,当用户单击启动 APK 安装完成广播时开启广播监听,当用户单击取消 APK 安装完成广播时,取消对应的广播接收器。想要实现本实例效果,首先修改 res/layout/activity_main.xml 文件。代码省略。

图 7.27 APK 安装完成的广播

然后修改 src/com.wyl.example/MainActivity.java 文件，代码如下：

```
01  //定义了本实例的主要 Activity
02  public class MainActivity extends Activity {
03      //省略控件代码的定义
04
05      //声明 InstallApkReceiver 对象
06      private InstallApkReceiver installApkReceiver;
07
08      @Override
09      public void onCreate(Bundle savedInstanceState) {
10          super.onCreate(savedInstanceState);
11          setContentView(R.layout.activity_main);
12          //给 InstallApkReceiver 对象赋值
13          installApkReceiver=new InstallApkReceiver();
14          //得到布局中的所有对象
15          findView();
16          //设置对象的监听器
17          setListener();
18
19      }
20
21      //省略 findview 和 setlistener 方法的定义
22
23      OnClickListener listener=new OnClickListener() {
24
25          @Override
26          public void onClick(View v) {
27              //TODO Auto-generated method stub
28              switch (v.getId()) {
29                  //开启安装完成 APK 广播
30                  case R.id.btn_startedinstallapk:
31                      //注册监听者
32                      registerReceiver(installApkReceiver,
33                          new IntentFilter(Intent.ACTION_PACKAGE_ADDED));
34                      Toast.makeText(MainActivity.this,
35                          "设置安装完成 APK 广播启动已完成", Toast.LENGTH_
                            SHORT).show();
36                      break;
37
38                  //取消监听者
39                  case R.id.btn_cancelinstallapk:
40                      if(installApkReceiver!=null)
41                      {
42                          //取消监听者
43                          unregisterReceiver(installApkReceiver);
44                          Toast.makeText(MainActivity.this,
45                              "取消安装完成 APK 广播已完成", Toast.LENGTH_SHORT).
                                show();
46                      }
47                      break;
48              }
49          }
50      };
51  }
52
```

此文件是 Activity 的代码文件，在其中第 6 行定义了 APK 安装完成广播的广播接收器对象，在第 13 行进行了初始化，当用户单击启动广播按钮的时候调用第 32~33 行注册广播，当用户单击取消广播的时候，调用第 43 行取消广播接收器。

然后定义广播接收器类，新建 src/com.wyl.example/ InstallApkReceiver.java 文件，代码如下：

```java
01  //定义广播接收器接收 APK 安装成功的广播
02  public class InstallApkReceiver extends BroadcastReceiver {
03  
04      @Override
05      public void onReceive(Context arg0, Intent arg1) {
06          //Intent.ACTION_PACKAGE_ADDED;
07          //成功的安装 APK 之后
08          //广播：设备上新安装了一个应用程序包。
09          if (arg1.getAction().equals(Intent.ACTION_PACKAGE_ADDED)) {
10              Toast.makeText(arg0,
11                  "检测到安装完成 APK 广播", Toast.LENGTH_SHORT).show();
12          }
13      }
14  }
```

此文件为广播接收器类，用来接收 APK 安装完成广播。

4．实例扩展

本实例主要应用在软件市场中，当用户下载完某个 APK 后直接进行安装打开功能。

范例 159 APK 卸载完成的广播

1．实例简介

在我们使用某些应用程序的时候，经常需要检测手机中是否有新的 APK 卸载完成，如果有，则提示用户卸载完毕。这样的功能，我们一般是通过广播接收器，接收 APK 的卸载完成的广播。本例子就带领大家使用广播接收器来接收 APK 卸载完成的广播的实例。

2．运行效果

该实例运行效果如图 7.28 所示。

图 7.28 APK 卸载完成的广播

3. 实例程序讲解

在本实例中,首先显示两个按钮控件,当用户单击启动 APK 卸载完成广播时开启广播监听,当用户单击取消 APK 卸载完成广播时,取消对应的广播接收器。想要实现本实例效果,首先修改 res/layout/activity_main.xml 文件。代码省略。

然后修改 src/com.wyl.example/MainActivity.java 文件,代码如下:

```java
01  //定义了本实例的主要 Activity
02  public class MainActivity extends Activity {
03      //省略控件对象定义的代码
04
05      //声明 UnstallApkReceiver 对象
06      private UnstallApkReceiver unstallApkReceiver;
07
08      @Override
09      public void onCreate(Bundle savedInstanceState) {
10          super.onCreate(savedInstanceState);
11          setContentView(R.layout.activity_main);
12          //给 unstallApkReceiver 对象赋值
13          unstallApkReceiver=new UnstallApkReceiver();
14          //得到布局中的所有对象
15          findView();
16          //设置对象的监听器
17          setListener();
18      }
19
20      //省略 findview 和 setlistener 方法定义
21
22      OnClickListener listener=new OnClickListener() {
23
24          @Override
25          public void onClick(View v) {
26              //TODO Auto-generated method stub
27              switch (v.getId()) {
28              //开启卸载 APK 广播
29              case R.id.btn_startedunstallapk:
30                  //注册监听者
31                  registerReceiver(unstallApkReceiver,
32                      new IntentFilter(Intent.ACTION_PACKAGE_REMOVED));
33                  Toast.makeText(MainActivity.this,
34                      "设置卸载 APK 广播启动已完成", Toast.LENGTH_SHORT).show();
35                  break;
36              //取消监听者
37              case R.id.btn_cancelunstallapk:
38                  if(unstallApkReceiver!=null)
39                  {
40                      //取消监听者
41                      unregisterReceiver(unstallApkReceiver);
42                      Toast.makeText(MainActivity.this,
43                          "取消卸载 APK 广播已完成", Toast.LENGTH_SHORT).
                            show();
44                  }
45                  break;
46              }
47          }
48      };
```

```
49  }
50
```

此文件是 Activity 的代码文件,在其中第 6 行定义了 APK 卸载完成广播的广播接收器对象,在第 13 行进行了初始化,当用户单击启动广播按钮的时候调用第 31~32 行注册广播,当用户单击取消广播的时候,调用第 41 行取消广播接收器。

然后定义广播接收器类,新建 src/com.wyl.example/ UnstallApkReceiver.java 文件,代码如下:

```
01  //定义广播接收器,接收下载 APK 的广播
02  public class UnstallApkReceiver extends BroadcastReceiver {
03
04  @Override
05  public void onReceive(Context arg0, Intent arg1) {
06      //Intent.ACTION_PACKAGE_REMOVED;
07      //成功的删除某个 APK 之后发出的广播
08      //一个已存在的应用程序包已经从设备上移除
09      if (arg1.getAction().equals(Intent.ACTION_PACKAGE_REMOVED)) {
10          Toast.makeText(arg0,
11              "检测到卸载 APK 广播", Toast.LENGTH_SHORT).show();
12      }
13  }
14  }
```

此文件为广播接收器类,用来接收 APK 卸载完成广播。

4. 实例扩展

本实例主要应用在软件市场中,当用户卸载完某个 APK 后进行用户提示功能。

范例 160　外部电源接入的广播

1. 实例简介

在我们使用某些应用程序的时候,经常需要检测手机中是否有外部电源接入。这样的功能,我们一般是通过广播接收器,接收外部电源接入的广播。本例子就带领大家使用广播接收器来接收外部电源接入的广播的实例。

2. 运行效果

该实例运行效果如图 7.29 所示。

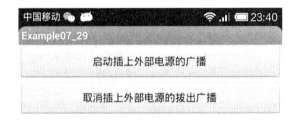

图 7.29　外部电源接入的广播

3. 实例程序讲解

在本实例中，首先显示两个按钮控件，当用户单击启动外部电源接入广播时开启广播监听，当用户单击取消外部电源接入广播时，取消对应的广播接收器。想要实现本实例效果，首先修改 res/layout/activity_main.xml 文件。代码省略。

然后修改 src/com.wyl.example/MainActivity.java 文件，代码如下：

```
01  //定义了本实例的主要 Activity
02  public class MainActivity extends Activity {
03      //省略控件对象的定义代码
04
05      //声明 PlugInReceiver 对象
06      private PlugInReceiver plugInReceiver;
07
08      @Override
09      public void onCreate(Bundle savedInstanceState) {
10          super.onCreate(savedInstanceState);
11          setContentView(R.layout.activity_main);
12          //给 PlugInReceiver 对象赋值
13          plugInReceiver = new PlugInReceiver();
14          //得到布局中的所有对象
15          findView();
16          //设置对象的监听器
17          setListener();
18      }
19
20      //省略 findview 和 setlistener 方法的定义
21
22      OnClickListener listener = new OnClickListener() {
23
24          @Override
25          public void onClick(View v) {
26              //TODO Auto-generated method stub
27              switch (v.getId()) {
28
29              //开启插上外部电源广播
30              case R.id.btn_startedplugin:
31                  //注册监听者，第一个参数是需要绑定的监听器，第二个是需要监听的广播
32                  registerReceiver(plugInReceiver, new IntentFilter(
33                      Intent.ACTION_POWER_CONNECTED));
34                  Toast.makeText(MainActivity.this, "设置插上外部电源广播启动已完成",
35                      Toast.LENGTH_SHORT).show();
36                  break;
37
38              //取消监听者
39              case R.id.btn_cancelplugin:
40                  if (plugInReceiver != null) {
41                      //取消监听者
42                      unregisterReceiver(plugInReceiver);
43                      Toast.makeText(MainActivity.this, "取消插上外部电源广播已完成",
44                          Toast.LENGTH_SHORT).show();
45                  }
46                  break;
47              }
```

```
48        }
49    };
50 }
```

此文件是 Activity 的代码文件,在其中第 6 行定义了外部电源接入广播的广播接收器对象,在第 13 行进行了初始化,当用户单击启动广播按钮的时候调用第 32~33 行注册广播,当用户单击取消广播的时候,调用第 42 行取消广播接收器。

然后定义广播接收器类,新建 src/com.wyl.example/ PlugInReceiver.java 文件,代码如下:

```
01 //定义广播接收器,接收外部电源接入的广播
02 public class PlugInReceiver extends BroadcastReceiver {
03
04     @Override
05     public void onReceive(Context arg0, Intent arg1) {
06         //Intent.ACTION_POWER_CONNECTED;
07         //插上外部电源时发出的广播
08         if (arg1.getAction().equals(Intent.ACTION_POWER_CONNECTED)) {
09             Toast.makeText(arg0,
10                 "检测到插上外部电源广播", Toast.LENGTH_SHORT).show();
11         }
12     }
13 }
```

此文件为广播接收器类,用来接收外部电源接入广播。

4. 实例扩展

本实例主要应用显示充电速度的软件中,例如:手机硬件检测软件等。

范例 161 重启系统的广播

1. 实例简介

在我们使用某些应用程序的时候,经常需要检测手机中是否在进行重启操作的广播。这样的功能,我们一般是通过广播接收器,接收手机重启的广播。本例子就带领大家使用广播接收器来接收手机重启的广播的实例。

2. 运行效果

该实例运行效果如图 7.30 所示。

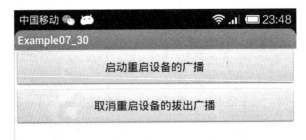

图 7.30 重启系统的广播

3. 实例程序讲解

在本实例中，首先显示两个按钮控件，当用户单击启动重启系统广播时开启广播监听，当用户单击取消重启系统广播时，取消对应的广播接收器。想要实现本实例效果，首先修改 res/layout/activity_main.xml 文件。代码省略。

然后修改 src/com.wyl.example/MainActivity.java 文件，代码如下：

```
01  //定义了本实例的主要Activity
02  public class MainActivity extends Activity {
03      //省略定义控件对象的代码
04
05      //声明RestartPhoneReceiver对象
06      private RestartPhoneReceiver restartPhoneReceiver;
07
08      @Override
09      public void onCreate(Bundle savedInstanceState) {
10          super.onCreate(savedInstanceState);
11          setContentView(R.layout.activity_main);
12          //给RestartPhoneReceiver对象赋值
13          restartPhoneReceiver=new RestartPhoneReceiver();
14          //得到布局中的所有对象
15          findView();
16          //设置对象的监听器
17          setListener();
18      }
19
20  //省略findview和setlistener方法定义
21
22  OnClickListener listener=new OnClickListener() {
23
24      @Override
25      public void onClick(View v) {
26          //TODO Auto-generated method stub
27          switch (v.getId()) {
28
29          //开启重启设备广播
30          case R.id.btn_startrestartphone:
31              //注册监听者
32              registerReceiver(restartPhoneReceiver,
33                  new IntentFilter(Intent.ACTION_REBOOT));
34              Toast.makeText(MainActivity.this,
35                  "设置重启设备广播启动已完成", Toast.LENGTH_SHORT).show();
36              break;
37
38          //取消监听者
39          case R.id.btn_cancelrestartphone:
40              if(restartPhoneReceiver!=null)
41              {
42                  //取消监听者
43                  unregisterReceiver(restartPhoneReceiver);
44                  Toast.makeText(MainActivity.this,
45                      "取消重启设备广播已完成", Toast.LENGTH_SHORT).
                        show();
46              }
47              break;
48          }
```

```
49      }
50    };
51 }
```

此文件是 Activity 的代码文件，在其中第 6 行定义了重启系统广播的广播接收器对象，在第 13 行进行了初始化，当用户单击启动广播按钮的时候调用第 32～33 行注册广播，当用户单击取消广播的时候，调用第 43 行取消广播接收器。

然后定义广播接收器类，新建 src/com.wyl.example/ RestartPhoneReceiver.java 文件，代码如下：

```
01 //定义广播接收器，接收重启完毕的广播
02 public class RestartPhoneReceiver extends BroadcastReceiver {
03
04    @Override
05    public void onReceive(Context arg0, Intent arg1) {
06        //Intent.ACTION_REBOOT;
07        //重启设备时的广播
08        if (arg1.getAction().equals(Intent.ACTION_REBOOT)) {
09            Toast.makeText(arg0,
10                "检测到重启设备广播", Toast.LENGTH_SHORT).show();
11        }
12    }
13 }
```

此文件为广播接收器类，用来接收重启系统的广播。

4．实例扩展

本实例主要应用在当系统重启前需要进行应用数据的保存功能。

范例 162 断开电源的广播

1．实例简介

在我们使用某些应用程序的时候，经常需要检测手机断开电源的广播。这样的功能，我们一般是通过广播接收器，接收手机断开电源的广播。本例子就带领大家使用广播接收器来接收手机断开电源的广播的实例。

2．运行效果

该实例运行效果如图 7.31 所示。

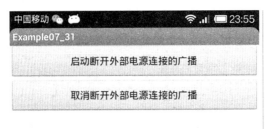

图 7.31 断开电源的广播

3. 实例程序讲解

在本实例中，首先显示两个按钮控件，当用户单击启动断开电源广播时开启广播监听，当用户单击取消断开电源广播时，取消对应的广播接收器。想要实现本实例效果，首先修改 res/layout/activity_main.xml 文件。代码省略。

然后修改 src/com.wyl.example/MainActivity.java 文件，代码如下：

```java
01 //定义了本实例的主要 Activity
02 public class MainActivity extends Activity {
03     //省略定义控件对象代码
04
05     //声明 BreakPowerReceiver 对象
06     private BreakPowerReceiver breakPowerReceiver;
07
08     @Override
09     public void onCreate(Bundle savedInstanceState) {
10         super.onCreate(savedInstanceState);
11         setContentView(R.layout.activity_main);
12         //给 breakPowerReceiver 对象赋值
13         breakPowerReceiver=new BreakPowerReceiver();
14         //得到布局中的所有对象
15         findView();
16         //设置对象的监听器
17         setListener();
18     }
19
20     //省略 findview 和 setlistener 方法定义
21
22     OnClickListener listener=new OnClickListener() {
23
24         @Override
25         public void onClick(View v) {
26             //TODO Auto-generated method stub
27             switch (v.getId()) {
28             //开启已断开外部电源连接广播
29             case R.id.btn_startbreakpower:
30                 //注册监听者
31                 registerReceiver(breakPowerReceiver,
32                     new IntentFilter(Intent.ACTION_POWER_DISCONNECTED));
33                 Toast.makeText(MainActivity.this,
34                     "设置已断开外部电源连接广播启动已完成", Toast.LENGTH_SHORT).
                    show();
35                 break;
36
37             //取消监听者
38             case R.id.btn_cancelbreakpower:
39                 if(breakPowerReceiver!=null)
40                 {
41                     //取消监听者
42                     unregisterReceiver(breakPowerReceiver);
43                     Toast.makeText(MainActivity.this,
44                         "取消已断开外部电源连接广播已完成", Toast.LENGTH_
                        SHORT).show();
45                 }
46                 break;
47             }
```

```
48        }
49    };
50  }
51
```

此文件是 Activity 的代码文件，在其中第 6 行定义了断开电源广播的广播接收器对象，在第 13 行进行了初始化，当用户单击启动广播按钮的时候，调用第 31～32 行注册广播，当用户单击取消广播的时候，调用第 42 行取消广播接收器。

然后定义广播接收器类，新建 src/com.wyl.example/ BreakPowerReceiver.java 文件，代码如下：

```
01  //定义广播接收器，接收断开外部电源的广播
02  public class BreakPowerReceiver extends BroadcastReceiver {
03
04      @Override
05      public void onReceive(Context arg0, Intent arg1) {
06          //Intent.ACTION_POWER_DISCONNECTED;
07          //已断开外部电源连接时发出的广播
08          if (arg1.getAction().equals(Intent.ACTION_POWER_DISCONNECTED)) {
09              Toast.makeText(arg0,
10                  "检测到已断开外部电源连接广播", Toast.LENGTH_SHORT).show();
11          }
12      }
13  }
```

此文件为广播接收器类，用来接收断开电源的广播。

4．实例扩展

本实例主要应用在当系统断开电源时提醒用户。

范例 163　墙纸改变的广播

1．实例简介

在我们使用某些应用程序的时候，经常需要检测手机中是否在进行墙纸改变的广播。这样的功能，我们一般是通过广播接收器，接收手机墙纸改变的广播。本例子就带领大家使用广播接收器来接收手机墙纸改变广播的实例。

2．运行效果

该实例运行效果如图 7.32 所示。

图 7.32　墙纸改变的广播

3. 实例程序讲解

在本实例中,首先显示两个按钮控件,当用户单击启动墙纸改变广播时开启广播监听,当用户单击取消墙纸改变广播时,取消对应的广播接收器。想要实现本实例效果,首先修改 res/layout/activity_main.xml 文件。代码省略。

然后修改 src/com.wyl.example/MainActivity.java 文件,代码如下:

```
01  //定义了本实例的主要Activity
02  public class MainActivity extends Activity {
03      //省略控件定义代码
04
05      //声明ChangeWallerReceiver对象
06      private ChangeWallerReceiver changeWallerReceiver;
07
08      @Override
09      public void onCreate(Bundle savedInstanceState) {
10          super.onCreate(savedInstanceState);
11          setContentView(R.layout.activity_main);
12          //给changeWallerReceiver对象赋值
13          changeWallerReceiver = new ChangeWallerReceiver();
14          //得到布局中的所有对象
15          findView();
16          //设置对象的监听器
17          setListener();
18      }
19
20      //省略findview和setlistener方法代码
21
22      OnClickListener listener = new OnClickListener() {
23          @Override
24          public void onClick(View v) {
25              //TODO Auto-generated method stub
26              switch (v.getId()) {
27              //开启更换壁纸广播
28              case R.id.btn_startchangewaller:
29                  //注册监听者
30                  registerReceiver(changeWallerReceiver, new IntentFilter(
31                      Intent.ACTION_WALLPAPER_CHANGED));
32                  Toast.makeText(MainActivity.this,"设置更换壁纸广播启动已完成",
33                      Toast.LENGTH_SHORT).show();
34                  break;
35              //取消监听者
36              case R.id.btn_cancelchangewaller:
37                  if (changeWallerReceiver != null) {
38                      //取消监听者
39                      unregisterReceiver(changeWallerReceiver);
40                      Toast.makeText(MainActivity.this,"取消更换壁纸广播已完成",
41                          Toast.LENGTH_SHORT).show();
42                  }
43                  break;
44              }
45          }
46      };
47  }
```

此文件是 Activity 的代码文件,在其中第 6 行定义了墙纸改变广播的广播接收器对象,在第 13 行进行了初始化,当用户单击启动广播按钮的时候,调用第 30～31 行注册广播,当用户单击取消广播的时候,调用第 39 行取消广播接收器。

然后定义广播接收器类,新建 src/com.wyl.example/ ChangeWallerReceiver.java 文件,代码如下:

```
01  //定义广播接收器,接收壁纸更换的广播
02  public class ChangeWallerReceiver extends BroadcastReceiver {
03
04  @SuppressWarnings("deprecation")
05  @Override
06  public void onReceive(Context arg0, Intent arg1) {
07      //Intent.ACTION_WALLPAPER_CHANGED;
08      //设备壁纸已改变时发出的广播
09      if (arg1.getAction().equals(Intent.ACTION_WALLPAPER_CHANGED)) {
10          Toast.makeText(arg0,
11              "检测到更换壁纸的广播", Toast.LENGTH_SHORT).show();
12      }
13  }
14  }
```

此文件为广播接收器类,用来接收墙纸改变的广播。

4.实例扩展

本实例主要应用在当应用和墙纸改变的事件有联系时。

范例 164 电话黑名单

1.实例简介

在我们使用某些应用程序的时候,经常需要检测手机中是否有电话呼入的事件。例如,当电话拨入时我可以根据设置拒接某些电话,这就是我们通常说的电话黑名单功能。这样的功能,我们一般是通过广播接收器,接收手机电话拨入的广播。本例子就带领大家使用广播接收器来接收手机电话拨入的广播的实例。

2.运行效果

该实例运行效果如图 7.33 所示。

图 7.33 电话黑名单

3. 实例程序讲解

在本实例中，首先显示一个输入框和两个按钮控件，当用户单击启动电话黑名单广播时，开启广播监听，当用户单击取消电话黑名单广播时，取消对应的广播接收器。想要实现本实例效果，首先修改 res/layout/activity_main.xml 文件。代码省略。

然后修改 src/com.wyl.example/MainActivity.java 文件，代码如下：

```java
01  //定义了本实例的主要 Activity
02  public class MainActivity extends Activity {
03      //省略控件定义代码
04      //定义电话状态改变的 ACTION
05      public final static String PHONE_STATE = TelephonyManager.ACTION_PHONE_STATE_CHANGED;
06  
07      //声明 HoldUpReceiver 对象
08      private HoldUpReceiver holdUpReceiver;
09  
10      @Override
11      public void onCreate(Bundle savedInstanceState) {
12          super.onCreate(savedInstanceState);
13          setContentView(R.layout.activity_main);
14          //得到布局中的所有对象
15          findView();
16          //设置对象的监听器
17          setListener();
18      }
19  
20      //省略 findview 和 setlistener 方法定义
21  
22      OnClickListener listener = new OnClickListener() {
23          @Override
24          public void onClick(View v) {
25              //TODO Auto-generated method stub
26              switch (v.getId()) {
27  
28              //开启拦截号码广播
29              case R.id.btn_startholdup:
30                  //给 HoldUpReceiver 对象赋值
31                  holdUpReceiver = new HoldUpReceiver(etHoldUpPhone.getText()
32                          .toString());
33                  //注册监听者，第一个参数是需要绑定的监听器，第二个是需要监听的广播
34                  registerReceiver(holdUpReceiver, new IntentFilter(PHONE_STATE));
35                  Toast.makeText(MainActivity.this, "设置拦截号码广播启动已完成",
36                          Toast.LENGTH_SHORT).show();
37                  break;
38              //取消监听者
39              case R.id.btn_cancelholdup:
40                  if (holdUpReceiver != null) {
41                      //取消监听者
42                      unregisterReceiver(holdUpReceiver);
43                      Toast.makeText(MainActivity.this, "取消拦截号码广播已完成",
44                              Toast.LENGTH_SHORT).show();
45                  }
46                  break;
```

```
47          }
48        }
49    };
50 }
```

此文件是 Activity 的代码文件,在其中第 8 行定义了电话黑名单广播的广播接收器对象,当用户单击启动广播按钮的时候,调用第 31～34 行注册广播,当用户单击取消广播的时候,调用第 42 行取消广播接收器。

然后定义广播接收器类,新建 src/com.wyl.example/ HoldUpReceiver.java 文件,代码如下:

```java
01 //定义电话广播接收器,接收电话广播
02 public class HoldUpReceiver extends BroadcastReceiver {
03   //声明的拦截号码
04   private String holdUpPhone;
05   //HoldUpReceiver 的构造函数
06   public HoldUpReceiver(String string) {
07       //TODO Auto-generated constructor stub
08       holdUpPhone = string;
09   }
10
11   @Override
12   public void onReceive(Context arg0, Intent arg1) {
13
14       //电话拨入时发出的广播
15       if (arg1.getAction().equals(MainActivity.PHONE_STATE)) {
16           Toast.makeText(arg0, "检测到来电的广播", Toast.LENGTH_SHORT).show();
17           //得到来电的号码
18           String phoneNumber = arg1
19               .getStringExtra(TelephonyManager.EXTRA_INCOMING_NUMBER);
20           //得到 TelephonyManager 的实例化
21           TelephonyManager telephony = (TelephonyManager) arg0
22               .getSystemService(Context.TELEPHONY_SERVICE);
23           //得到来电的状态
24           int state = telephony.getCallState();
25           //判断来电属于哪种状态
26           switch (state) {
27           //响铃
28           case TelephonyManager.CALL_STATE_RINGING:
29               //判断来电的号码和需要拦截的号码是不是一样
30               if (holdUpPhone.equals(phoneNumber)) {
31                   //来电的号码和需要拦截的号码一样,调用电话挂断函数
32                   holdupphone(telephony);
33               }
34               break;
35           //挂电话
36           case TelephonyManager.CALL_STATE_IDLE:
37               Toast.makeText(arg0,
38                   "idle: " + arg1.getStringExtra("incoming_number"),
39                   Toast.LENGTH_LONG).show();
40               break;
41           //无活动
42           case TelephonyManager.CALL_STATE_OFFHOOK:
43               Toast.makeText(arg0,
44                   "OffHook: " + arg1.getStringExtra("incoming_number"),
```

```
45                          Toast.LENGTH_LONG).show();
46              break;
47          }
48      }
49  }
50
51  public void holdupphone(TelephonyManager teleManager) {
52      Class c;
53      try {
54          //通过 forName 动态得到 teleManager 所属的类
55          c = Class.forName(teleManager.getClass().getName());
56          //得到类里面的方法
57          Method m = c.getDeclaredMethod("getITelephony");
58          //设置类中的方法可以得到
59          m.setAccessible(true);
60          //通过 invoke 反射实例化 ITelephony
61          ITelephony iTelephony = (ITelephony) m.invoke(teleManager);
62          //挂断电话
63          iTelephony.endCall();//结束通话
64      } catch (ClassNotFoundException e) {
65          //TODO Auto-generated catch block
66          e.printStackTrace();
67      } catch (SecurityException e) {
68          //TODO Auto-generated catch block
69          e.printStackTrace();
70      } catch (NoSuchMethodException e) {
71          //TODO Auto-generated catch block
72          e.printStackTrace();
73      } catch (IllegalArgumentException e) {
74          //TODO Auto-generated catch block
75          e.printStackTrace();
76      } catch (IllegalAccessException e) {
77          //TODO Auto-generated catch block
78          e.printStackTrace();
79      } catch (InvocationTargetException e) {
80          //TODO Auto-generated catch block
81          e.printStackTrace();
82      } catch (RemoteException e) {
83          //TODO Auto-generated catch block
84          e.printStackTrace();
85      }
86  }
87  }
```

此文件为广播接收器类,用来接收电话拨入的广播。在第 15 行进行广播状态判断,在第 18~22 行得到系统的电话管理服务,得到来电号码,根据电话服务的状态,判断当前广播为拨入、拨出或者挂断,如果为用户拨入则判断来电号码是否在黑名单中,如果在的话,调用 holduphone 方法挂断电话,这样就可以实现电话黑名单的功能。

4. 实例扩展

本实例在执行过程中需要代码挂断电话,这需要系统的 AIDL 支持,所以本实例在做的时候需要导入系统的 AIDL 文件才可以看到效果。

范例 165 短信接收的广播

1. 实例简介

在我们使用某些应用程序的时候，经常需要检测手机中是否有新的短信接收。例如，垃圾短信过滤器和短信自动回复器等。这样的功能，我们一般是通过广播接收器，接收手机短信接收的广播。本例子就带领大家使用广播接收器来接收手机短信的广播的实例。

2. 运行效果

该实例运行效果如图 7.34 所示。

图 7.34 短信接收的广播

3. 实例程序讲解

在本实例中，首先显示两个按钮控件，当用户单击启动手机短信广播时，开启广播监听，当用户单击取消手机短信广播时，取消对应的广播接收器。想要实现本实例效果，首先修改 res/layout/activity_main.xml 文件。代码省略。

然后修改 src/com.wyl.example/MainActivity.java 文件，代码如下：

```
01  //定义了本实例的主要 Activity
02  public class MainActivity extends Activity {
03      //省略控件定义代码
04
05      //声明 ReceiverMsgReceiver 对象
06      private ReceiverMsgReceiver receiverMsgReceiver;
07
08      @Override
09      public void onCreate(Bundle savedInstanceState) {
10          super.onCreate(savedInstanceState);
11          setContentView(R.layout.activity_main);
12          //给 ReceiverMsgReceiver 对象赋值
13          receiverMsgReceiver = new ReceiverMsgReceiver();
14          //得到布局中的所有对象
15          findView();
16          //设置对象的监听器
17          setListener();
```

```
18    }
19
20    //省略 findview 和 setlistener 方法定义
21
22    OnClickListener listener = new OnClickListener() {
23        @Override
24        public void onClick(View v) {
25            //TODO Auto-generated method stub
26            switch (v.getId()) {
27            //开启短信接收广播
28            case R.id.btn_startreceivemessage:
29                //注册监听者
30                registerReceiver(receiverMsgReceiver, new IntentFilter(
31                    "android.provider.Telephony.SMS_RECEIVED"));
32                Toast.makeText(MainActivity.this,
33                    "设置短信接收广播启动已完成",
34                    Toast.LENGTH_SHORT).show();
35                break;
36
37            //取消监听者
38            case R.id.btn_cancelreceivemessage:
39                if (receiverMsgReceiver != null) {
40                    //取消监听者
41                    unregisterReceiver(receiverMsgReceiver);
42                    Toast.makeText(MainActivity.this,
43                        "取消短信接收广播已完成",
44                        Toast.LENGTH_SHORT).show();
45                }
46                break;
47            }
48        }
49    };
50    }
51
```

此文件是 Activity 的代码文件，在其中第 6 行定义了手机短信广播的广播接收器对象，在第 13 行进行了初始化，当用户单击启动广播按钮的时候，调用第 30～31 行注册广播，当用户单击取消广播的时候，调用第 41 行取消广播接收器。

然后定义广播接收器类，新建 src/com.wyl.example/ ReceiverMsgReceiver.java 文件，代码如下：

```
01    //定义广播接收器，接收短信已送达的广播
02    public class ReceiverMsgReceiver extends BroadcastReceiver {
03
04    @Override
05    public void onReceive(Context arg0, Intent arg1) {
06        //android.provider.Telephony.SMS_RECEIVED
07        //短信到达时发出的广播
08        if (arg1.getAction().equals("android.provider.Telephony.SMS_RECEIVED")) {
09            Toast.makeText(arg0,
10                "检测到短信到达的广播", Toast.LENGTH_SHORT).show();
11        }
12    }
13    }
```

此文件为广播接收器类，用来接收手机短信的广播。

4．实例扩展

本实例主要应用在手机短信客户端应用中，其中包括接收短信、过滤短信和自动回复短信等。

范例 166　短信发送的广播

1．实例简介

在我们使用某些应用程序的时候，经常需要检测手机中短信是否发送成功。例如，发送成功提示等。这样的功能，我们一般是通过广播接收器，接收手机短信发送的广播。本例子就带领大家使用广播接收器来接收短信发送的广播的实例。

2．运行效果

该实例运行效果如图 7.35 所示。

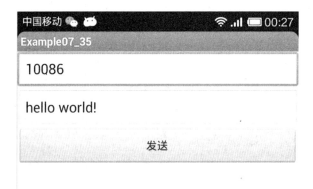

图 7.35　短信发送的广播

3．实例程序讲解

在本实例中，首先显示两个输入框，一个按钮控件，当用户单击短信发送按钮时，开启广播监听。想要实现本实例效果，首先修改 res/layout/activity_main.xml 文件。代码省略。然后修改 src/com.wyl.example/MainActivity.java 文件，代码如下：

```
01  /定义了本实例的主要 Activity
02  public class MainActivity extends Activity {
03  /* 自定义 ACTION 常数，作为广播的 Intent Filter 识别常数 */
04  public static String SMS_SEND_ACTIOIN = "SMS_SEND_ACTIOIN";
05  public static String SMS_DELIVERED_ACTION = "SMS_DELIVERED_ACTION";
06  /* 建立两个 mServiceReceiver 对象，作为类成员变量 */
07  private SendMsgReceiver sendReceiver, deliverReceiver;
08
09  //省略控件定义代码
10
11  @Override
12  public void onCreate(Bundle savedInstanceState) {
13      super.onCreate(savedInstanceState);
```

```java
14      setContentView(R.layout.activity_main);
15      //得到布局中的所有对象
16      findView();
17      //设置对象的监听器
18      setListener();
19  }
20
21  //省略 findview 和 setlistener 方法定义
22
23  OnClickListener listener = new OnClickListener() {
24      @Override
25      public void onClick(View v) {
26          //TODO Auto-generated method stub
27          switch (v.getId()) {
28          //开启短信接收广播
29          case R.id.btn_sendmessage:
30              /* 欲发送的电话号码 */
31              String strDestAddress = etSendPhone.getText().toString();
32              /* 欲发送的短信内容 */
33              String strMessage = etSendBody.getText().toString();
34              /* 建立 SmsManager 对象 */
35              SmsManager smsManager = SmsManager.getDefault();
36              try {
37                  /*建立自定义Action常数的Intent(给PendingIntent参数之用) */
38                  Intent itSend = new Intent(SMS_SEND_ACTIOIN);
39                  Intent itDeliver = new Intent(SMS_DELIVERED_ACTION);
40
41                  /* sentIntent 参数为传送后接收的广播信息 PendingIntent */
42                  PendingIntent mSendPI = PendingIntent.getBroadcast(
43                      getApplicationContext(), 0, itSend, 0);
44
45                  /* deliveryIntent 参数为送达后接收的广播信息 PendingIntent */
46                  PendingIntent mDeliverPI = PendingIntent.getBroadcast(
47                      getApplicationContext(), 0, itDeliver, 0);
48
49                  /* 发送 SMS 短信,注意倒数的两个 PendingIntent 参数 */
50                  smsManager.sendTextMessage(strDestAddress, null,
51                      strMessage, mSendPI, mDeliverPI);
52              } catch (Exception e) {
53                  e.printStackTrace();
54              }
55              break;
56          }
57      }
58  };
59  @Override
60  protected void onResume() {
61      //TODO Auto-generated method stub
62
63      /* 自定义 IntentFilter 为 SENT_SMS_ACTIOIN Receiver */
64      IntentFilter mFilter01;
65      mFilter01 = new IntentFilter(SMS_SEND_ACTIOIN);
66      sendReceiver = new SendMsgReceiver();
67      registerReceiver(sendReceiver, mFilter01);
68
69      /* 自定义 IntentFilter 为 DELIVERED_SMS_ACTION Receiver */
70      mFilter01 = new IntentFilter(SMS_DELIVERED_ACTION);
71      deliverReceiver = new SendMsgReceiver();
```

```
72          registerReceiver(deliverReceiver, mFilter01);
73          super.onResume();
74      }
75  }
```

此文件是 Activity 的代码文件,在其中第 7 行定义了短信发送广播的广播接收器对象,当用户单击发送按钮的时候,获取输入的电话号码和发送内容,然后调用第 35 行得到 SmsManager 短信管理器,再设置相应的参数后,调用第 50~51 行的 sendTextMessage 方法发送短信。此程序的第 64~72 行注册了两个广播接收器,用来接收短信发送成功与失败的广播接收器。

然后定义广播接收器类,新建 src/com.wyl.example/ SendMsgReceiver.java 文件,代码如下:

```
01  //定义广播接收器,接收短信发送的广播
02  public class SendMsgReceiver extends BroadcastReceiver {
03
04      private Context context;
05
06      @Override
07      public void onReceive(Context context, Intent intent) {
08          //TODO Auto-generated method stub
09          this.context = context;
10          if (intent.getAction().equals(MainActivity.SMS_SEND_ACTIOIN)) {
11              try {
12                  /* android.content.BroadcastReceiver.getResultCode()方法 */
13                  switch (getResultCode()) {
14                  case MainActivity.RESULT_OK:
15                      /* 发送短信成功 */
16                      mMakeTextToast("短信发送成功");
17                      break;
18                  case SmsManager.RESULT_ERROR_GENERIC_FAILURE:
19                      /* 发送短信失败 */
20                      mMakeTextToast("发送短信失败");
21                      break;
22                  }
23              } catch (Exception e) {
24                  e.getStackTrace();
25              }
26          } else if (intent.getAction().equals(MainActivity.SMS_DELIVERED_ACTION)) {
27              try {
28                  /* android.content.BroadcastReceiver.getResultCode()方法 */
29                  switch (getResultCode()) {
30                  case MainActivity.RESULT_OK:
31                      /* 短信 */
32                      mMakeTextToast("短信到达");
33                      break;
34                  case SmsManager.RESULT_ERROR_GENERIC_FAILURE:
35                      /* 短信未送达 */
36                      mMakeTextToast("短信未送达");
37
38                      break;
39                  }
40              } catch (Exception e) {
41                  e.getStackTrace();
42              }
```

```
43        }
44    }
45
46    private void mMakeTextToast(String string) {
47        //TODO Auto-generated method stub
48        Toast.makeText(context, string, Toast.LENGTH_SHORT).show();
49    }
50 }
```

此文件为广播接收器类,用来接收短信发送的广播。在第 10 行根据短信发送的状态显示对应的提示信息,其中常见的短信状态有短信发送成功、短信发送失败、短信已送达和短信未送达等状态,在本实例中仅仅做出了相应的提示信息。

4. 实例扩展

本实例主要针对短信发送状态的广播接收,根据短信发送的状态实例,做出相应动作。例如,短信发送失败的话定时再次发送;短信已送达的话,提示用户短信已送达等。

范例 167　电池电量低的广播

1. 实例简介

在我们使用某些应用程序的时候,经常需要检测手机中电池电量是否在正常范围内,当电池电量低的时候可以自动提醒,或者当电池电量低的时候把手机屏幕调暗等。这样的功能,我们一般是通过广播接收器,接收电池电量低的广播。本例子就带领大家使用广播接收器来接收电池电量低的广播的实例。

2. 运行效果

该实例运行效果如图 7.36 所示。

图 7.36　电池电量低的广播

3. 实例程序讲解

在本实例中,首先显示两个按钮控件,当用户单击启动电池电量低的广播时开启广播监听,当用户单击取消电池电量低的广播时,取消对应的广播接收器。想要实现本实例效果,首先修改 res/layout/activity_main.xml 文件。代码省略。

然后修改 src/com.wyl.example/MainActivity.java 文件,代码如下:

```
01  //定义了本实例的主要Activity
02  public class MainActivity extends Activity {
03  //省略控件对象的定义代码
04
05      //声明PowerLowReceiver对象
06      private PowerLowReceiver powerLowReceiver;
07
08      @Override
09      public void onCreate(Bundle savedInstanceState) {
10          super.onCreate(savedInstanceState);
11          setContentView(R.layout.activity_main);
12          //给PowerLowReceiver对象赋值
13          powerLowReceiver = new PowerLowReceiver();
14          //得到布局中的所有对象
15          findView();
16          //设置对象的监听器
17          setListener();
18
19      }
20
21  //省略findview和setlistener方法的定义
22
23  OnClickListener listener = new OnClickListener() {
24
25      @Override
26      public void onClick(View v) {
27          //TODO Auto-generated method stub
28          switch (v.getId()) {
29
30          //开启电量低的接收广播
31          case R.id.btn_startpowerlow:
32              //注册监听者
33              registerReceiver(powerLowReceiver, new IntentFilter(
34                      Intent.ACTION_BATTERY_LOW));
35              Toast.makeText(MainActivity.this,
36                      "设置电池电量低广播启动已完成",
37                      Toast.LENGTH_SHORT).show();
38              break;
39
40          //取消监听者
41          case R.id.btn_cancelpowerlow:
42              if (powerLowReceiver != null) {
43                  //取消监听者
44                  unregisterReceiver(powerLowReceiver);
45                  Toast.makeText(MainActivity.this, "取消电池电量低广播已完成",
46                          Toast.LENGTH_SHORT).show();
47              }
48              break;
49          }
50      }
51  };
52  }
```

此文件是Activity的代码文件，在其中第6行定义了电池电量低的广播接收器对象，在第13行进行了初始化，当用户单击启动广播按钮的时候，调用第33～34行注册广播，当用户单击取消广播的时候，调用第44行取消广播接收器。

然后定义广播接收器类，新建 src/com.wyl.example/ PowerLowReceiver.java 文件，代码如下：

```
01  //定义接收系统电量低的广播接收器
02  public class PowerLowReceiver extends BroadcastReceiver {
03
04      @Override
05      public void onReceive(Context arg0, Intent arg1) {
06          //Intent.ACTION_BATTERY_LOW;
07          //表示电池电量低
08          if (arg1.getAction().equals(Intent.ACTION_BATTERY_LOW)) {
09              Toast.makeText(arg0,
10                  "检测到电池电量低的广播", Toast.LENGTH_SHORT).show();
11          }
12      }
13  }
```

此文件为广播接收器类，用来接收电池电量低广播。

4．实例扩展

本实例主要应用在当系统电量低的时候进行提醒，或者进行省电的操作。

范例 168　音乐播放器

1．实例简介

在我们使用某些应用程序的时候，经常需要检测手机中是否在进行重启操作的广播。这样的功能，我们一般是通过广播接收器，接收手机重启的广播。本例子就带领大家使用广播接收器来接收手机重启的广播的实例。

2．运行效果

该实例运行效果如图 7.37 所示。

图 7.37　音乐播放器

3．实例程序讲解

在本实例中，首先显示一个进度条控件和两个按钮控件，当用户单击开始按钮时音乐开始播放，单击停止按钮时音乐停止。想要实现本实例效果，首先修改 res/layout/activity_main.xml 文件。代码省略。

然后修改 src/com.wyl.example/MainActivity.java 文件，代码如下：

```java
01 //定义了本实例的主要 Activity
02 public class MainActivity extends Activity {
03     //定义静态当前的播放进度变量
04     public static String ACTION_START_UPDATE_PROGRESS = "update_progress";
05     //定义播放进度的最大值的静态变量
06     public static String ACTION_START_TOTAL_PROGRESS = "total_progress";
07     //省略控件定义代码
08     //声明用于 service 的 intent 对象
09     private Intent serviceIntent;
10     //声明广播监听者 MusicBordcastReceiver 的对象
11     private MusicBordcastReceiver musicBordcastReceiver;
12
13     @Override
14     public void onCreate(Bundle savedInstanceState) {
15         super.onCreate(savedInstanceState);
16         setContentView(R.layout.activity_main);
17         //得到布局中的所有对象
18         findView();
19         //设置对象的监听器
20         setListener();
21         //启动 service
22         serviceIntent = new Intent(this, MusicService.class);
23         //实例化 musicBordcastReceiver 对象
24         musicBordcastReceiver = new MusicBordcastReceiver();
25
26     }
27
28     //省略 findview 和 setlistener 方法定义
29
30     OnClickListener listener = new OnClickListener() {
31
32         @Override
33         public void onClick(View v) {
34             //TODO Auto-generated method stub
35             switch (v.getId()) {
36
37             //开始播放音乐
38             case R.id.bt_start:
39                 //musicstate 来提示 service 进行什么操作，设置为 true，说明播放音乐
40                 serviceIntent.putExtra("musicstate", true);
41                 startService(serviceIntent);
42                 break;
43             //停止播放音乐
44             case R.id.btn_stop:
45                 //musicstate 来提示 service 进行什么操作，设置 false 说明暂停播放音乐
46                 serviceIntent.putExtra("musicstate", false);
47                 startService(serviceIntent);
48                 break;
49             }
50         }
51     };
52     //定义了广播接收器对象
53     private class MusicBordcastReceiver extends BroadcastReceiver {
```

```
54
55      @Override
56      public void onReceive(Context context, Intent intent) {
57          //TODO Auto-generated method stub
58          //判断接收的是哪个 action
59          if (intent.getAction().equals(ACTION_START_UPDATE_PROGRESS)) {
60              //设置音乐播放的当前的进度
61              sbMusicProgress.setProgress(intent.getIntExtra(
62                  "CurrentPosition", 0));
63          }
64          if (intent.getAction().equals(ACTION_START_TOTAL_PROGRESS)) {
65              //设置音乐播放的总进度,
66              sbMusicProgress.setMax(intent.getIntExtra("TotalPosition",
                    0));
67          }
68      }
69
70  }
71
72  @Override
73  protected void onResume() {
74      //TODO Auto-generated method stub
75      //判断 musicBordcastReceiver 是否为空,如果为空,则实例化
76      if (musicBordcastReceiver == null) {
77          musicBordcastReceiver = new MusicBordcastReceiver();
78      }
79      //初始化 intentFilter,并增加相应的 Action
80      IntentFilter intentFilter = new IntentFilter();
81      intentFilter.addAction(ACTION_START_UPDATE_PROGRESS);
82      intentFilter.addAction(ACTION_START_TOTAL_PROGRESS);
83      //注册广播
84      registerReceiver(musicBordcastReceiver, intentFilter);
85      super.onResume();
86  }
87
88  @Override
89  protected void onStop() {
90      //TODO Auto-generated method stub
91      //取消注册广播
92      unregisterReceiver(musicBordcastReceiver);
93      super.onStop();
94  }
95  }
96
```

此文件是 Activity 的代码文件,在其中第 6 行定义了音乐播放广播的广播接收器对象,在第 24 行进行了初始化,当用户单击启动播放按钮的时候,调用第 40~41 行启动服务播放音乐,单击停止按钮时,调用 46~47 行停止音乐播放服务。在音乐播放的过程中不断发送音乐播放的进度广播,这时候我们在第 53 行定义的音乐广播接收器进行此广播的接收,其中如果音乐在播放,则调用第 61~62 行更新音乐播放进度,如果音乐播放完毕,则调用第 66 行设置音乐播放进度为最大值。在第 73 行的 onresume 方法中注册广播接收者,在第 89 行 onStop 方法中取消广播接收者。

然后定义音乐播放服务类,新建 src/com.wyl.example/ MusicService.java 文件,代码如下:

```
01  //定义音乐播放服务
02  public class MusicService extends Service {
03
04      //声明音乐播放器对象
05      private MediaPlayer mPlayer;
06      //声明当前播放进度广播的 intent 对象
07      private Intent broadCastCurrentIntent;
08      //声明总进度广播的 inent 对象
09      private Intent broadCastTotalIntent;
10
11      @Override
12      public IBinder onBind(Intent arg0) {
13          //TODO Auto-generated method stub
14          return null;
15      }
16
17      //该服务不存在需要被创建时被调用,不管 startService()还是 bindService()都会启
    动时调用该方法
18      @Override
19      public void onCreate() {
20          Toast.makeText(this, "MusicSevice onCreate()", Toast.LENGTH_
            SHORT)
21                  .show();
22          //加载音乐文件
23          mPlayer = MediaPlayer.create(getApplicationContext(), R.raw.
            music);
24          //为 intent 对象赋值
25          broadCastCurrentIntent = new Intent(
26                  MainActivity.ACTION_START_UPDATE_PROGRESS);
27          broadCastTotalIntent = new Intent(
28                  MainActivity.ACTION_START_TOTAL_PROGRESS);
29          super.onCreate();
30      }
31
32      @Override
33      public int onStartCommand(final Intent intent, int flags, int startId) {
34          //TODO Auto-generated method stub
35          Toast.makeText(this, "MusicSevice onStart()", Toast.LENGTH_SHORT)
36                  .show();
37          //得到的 intent 值
38          if (intent.getBooleanExtra("musicstate", true)) {
39              //开始播放音乐
40              mPlayer.start();
41              //设置音乐总进度的 Extra 的值,并发送广播
42              broadCastTotalIntent.putExtra("TotalPosition",
43                      mPlayer.getDuration());
44              sendBroadcast(broadCastTotalIntent);
45              //声明并启动一个线程
46              new Thread(new Runnable() {
47
48                  @Override
49                  public void run() {
50                      //TODO Auto-generated method stub
51                      //当当前音乐播放进度小于总进度的时候循环
52                      while (mPlayer.getCurrentPosition() <= mPlayer
53                              .getDuration()) {
54                          //设置音乐当前进度的 Extra 的值,并发送广播
55                          broadCastCurrentIntent.putExtra("CurrentPosition",
56                                  mPlayer.getCurrentPosition());
```

```
57                         sendBroadcast(broadCastCurrentIntent);
58                         try {
59                             Thread.sleep(500);
60                         } catch (InterruptedException e) {
61                             //TODO Auto-generated catch block
62                             e.printStackTrace();
63                         }
64                     }
65
66                 }
67             }).start();
68         } else {
69             //暂停音乐播放
70             mPlayer.stop();
71         }
72         return super.onStartCommand(intent, flags, startId);
73     }
74 }
```

此文件为音乐播放服务类，用来在后台播放音乐。在其 onCreate 方法中，第 23 行初始化了 MediaPlayer 对象进行音乐播放。第 27 行初始化进度更新广播进行页面上的进度条更新，在启动命令的回调函数 onStartCommand 方法中进行服务状态的判断，如果用户发送播放音乐的状态，那么后台音乐播放，而其定时发送播放的进度广播，如果用户发送停止播放音乐的状态，调用 mPlayer 的 stop 方法停止播放音乐。

4．实例扩展

本实例主要结合了 BroadcastReceiver、Activity 和 Service 这三个组件之间的交互操作，在实际应用中基本都是采用这种方式来进行更新的。

7.3　小　　结

在本章节中主要介绍了 Android 中广播和服务。服务是在后台运行的程序，能够在用户看不到界面的情况下给用户提供一些功能；广播是系统通知应用程序某个事件发生的唯一方式，我们可以在应用程序中接收系统广播来做出对应的响应。本章的内容针对实际项目中的某些特殊功能提供支持，它可以使程序在看不到界面的情况下为用户提供服务，例如：后台音乐播放或垃圾短信的拦截等。下一章我们会讲述 Android 中的网络编程的例子。

第 8 章 Android 的网络编程

上一章了解了 Android 中广播和服务的使用方法。服务是 Android 系统中后台运行的程序，它没有界面，但是仍然可以为用户提供功能，如后台下载或后台音乐播放等。广播是 Android 系统中系统和应用程序交互的手段，当系统发生了某个操作的时候，系统就会发送相应的广播，对应的应用程序就会做出响应，如电话拨入的时候或短信发送的时候等。了解上一章的内容后，大家基本上就可以创建一个完整的单机 Android 应用了。但是对于我们现在常见的应用客户端来说，单机应用的市场还是有限的，一般的应用程序都是要与网络数据交互的。例如，QQ 网购，数据从服务器端获取，用户注册登录也要提交服务器审核等。这也就需要我们 Android 的客户端需要同网络的数据进行交互。那么本章就给大家介绍 Android 中的网络编程，通过不同的网络访问方式实现各种常见的功能。

Android 系统中的有关网络编程的内容，主要分为两部分：

第一部分是网络请求。这部分主要的功能就是把数据从网络上请求下来。

第二部分是根据请求下来的数据格式，解析出系统有用的数据资源进行使用。

本章主要通过各种实例来介绍 Android 中常用的网络访问方法。希望读者阅读完本章内容后，可以开发常见的 Android 网络应用，使自己的应用可以随时随地与网络进行交互。

8.1 网络请求

范例 169 在线天气查询

1. 实例简介

在使用一些应用的时候经常会需要请求网络的数据。因为对于我们手机来说，从手机的性能到软硬件都无法和电脑比拟，所以一般我们手机终端只作为展示的平台，而数据都存储在服务器上。例如，天气预报、在线读书和在线音乐等。对于这种情况，一般我们都是通过网络请求服务器端的数据，然后在客户端进行展示。我们通过本实例带领大家一起制作一个在线天气请求的应用。

2. 运行效果

该实例运行效果如图 8.1 所示。

图 8.1 在线天气查询

3. 实例程序讲解

上例中,用户在输入框中输入希望查询天气的地点,然后单击"查询"按钮,最后用户就可以看到服务器返回的天气信息了。想要实现如上效果,首先修改我们建立的工程下的 res/layout/activity_main.xml。代码省略。

然后修改 src/com.wyl.example/MainActivity.java 文件,代码如下:

```
01  //定义了本实例的主要 Activity
02  public class MainActivity extends Activity {
03      //省略定义控件对象的代码
04
05      //4.0 之后不允许在主线程中运行耗时的网络请求
06      private Handler mHandler = new Handler() {
07          //handler 接收到信息的回调函数
08          @Override
09          public void handleMessage(Message msg) {
10              //隐藏进度条,显示获取结果
11              mProgressBar.setVisibility(View.GONE);
12              mTvShow.setText(mReturnConnection);
13              super.handleMessage(msg);
14          }
15
16      };
17      private String mReturnConnection;
18
19      @Override
20      public void onCreate(Bundle savedInstanceState) {
21          super.onCreate(savedInstanceState);
            //设置当前页面的布局视图为 activity_main
22          setContentView(R.layout.activity_main);
            //设置标题文字
23          setTitle("POST 天气查询");
24
25          //得到布局中的控件
26          findView();
27          //绑定控件事件
28          setListener();
29      }
30
31  //省略 findview 方法的定义
32
33  private void setListener() {
34      //添加事件
35      mBtn.setOnClickListener(new OnClickListener() {
36
37          @Override
38          public void onClick(View arg0) {
39              //TODO Auto-generated method stub
40              mProgressBar.setVisibility(View.VISIBLE);
41              new Thread() {
42                  @Override
43                  public void run() {
44                      //发送 post 请求
45                      dopost(mEtInPut.getText().toString());
46                      mHandler.sendEmptyMessage(0);
47                      super.run();
```

```java
48              }
49          }.start();
50
51      }
52   });
53 }
54
55 private void dopost(String val) {
56     //http客户端
57     DefaultHttpClient client = new DefaultHttpClient();
58     //天气查询
59     HttpPost httpPost = new HttpPost(
60             "http://webservice.webxml.com.cn/WebServices/
              WeatherWS.asmx/getWeather");
61
62     //Post运作传送变量必须用NameValuePair[]数组储存
63     List<NameValuePair> params = new ArrayList<NameValuePair>();
64     params.add(new BasicNameValuePair("theCityCode", val));
65     params.add(new BasicNameValuePair("theUserID", ""));
66
67     try {
68         //编码格式
69         UrlEncodedFormEntity p_entity = new UrlEncodedForm
              Entity(params,
70                 "utf-8");
71         //将POST数据放入HTTP请求
72         httpPost.setEntity(p_entity);
73         //发出实际的HTTP POST请求
74         HttpResponse response = client.execute(httpPost);
75         //若状态码为200 ok
76         if (response.getStatusLine().getStatusCode() == 200) {
77             mReturnConnection = EntityUtils.toString
                  (response.getEntity());
78         } else {
79             mReturnConnection = "Error Response: "
80                     + response.getStatusLine().toString();
81         }
82     } catch (IllegalStateException e) {
83         e.printStackTrace();        //非法状态异常
84     } catch (IOException e) {
85         e.printStackTrace();        //输入输入异常
86     }
87 }
88 }
```

如上面中代码的第6行定义了handler对象，用来接收其他线程发送的消息，在第26行得到布局中的所有控件，在第28行设置所有的监听器，当用户单击提交请求按钮时，调用第45行的dopost方法，具体实现在第55～87行，首先定义HttpClient对象，然后定义HttpPost对象，传入请求地址，然后定义传入的参数列表并且设置参数，将参数传入httpPost对象，然后通过client的execute方法执行post请求，并且得到HttpResponse对象，如果返回的状态码为200代表请求成功，得到返回的结果，然后进行展示即可，否则提示请求错误。

4．实例扩展

此实例主要介绍了通过Post方式请求网络数据的方式，当然得到的数据没有进行处理，直接展示了一下，后面的实例中会介绍数据解析的方法。

范例 170　在线百度搜索

1．实例简介

在我们使用应用的过程中，经常需要得到网络数据。例如，查询某本图书的信息，需要提交图书编号，通过百度搜索关键字等。遇到这样的功能，我们一般是当用户进行某个操作的时候，例如，单击某个按钮的时候，发送本地 GET 请求到服务器端，然后服务器返回查询结果给客户端。本例子就带领大家来实现一个通过 GET 请求百度的搜索结果的实例。

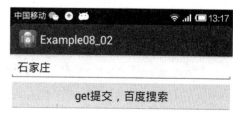

2．运行效果

该实例运行效果如图 8.2 所示。

图 8.2　百度搜索客户端

3．实例程序讲解

在本实例中，提供一个输入用户希望查找的关键字的输入框，然后用户单击按钮的时候，我们发送 GET 请求给百度，然后百度服务器将搜索结果返回给手机客户端进行展示。想要实现本实例效果，首先修改 res/layout/activity_main.xml 文件，定义如上布局。代码省略。

然后修改 src/com.wyl.example/MainActivity.java 文件，代码如下：

```
01  //定义了本实例的主要 Activity
02  public class MainActivity extends Activity {
03      //省略控件定义的代码
04
05      //4.0 之后不允许在主线程中运行耗时的网络请求
06      private Handler mHandler = new Handler() {
07
08          @Override
09          public void handleMessage(Message msg) {
10              //TODO Auto-generated method stub
11              mProgressBar.setVisibility(View.GONE);
12              mTvShow.setText(mReturnConnection);
13              super.handleMessage(msg);
14          }
15
16      };
17      private String mReturnConnection;
18
19      @Override
20      public void onCreate(Bundle savedInstanceState) {
21          super.onCreate(savedInstanceState);
22          //设置当前页面的布局视图为 activity_main
22          setContentView(R.layout.activity_main);
23          //得到布局中的控件
24          findView();
```

```java
25          //绑定控件事件
26          setListener();
27
28      }
29
30  //省略findview方法的定义
31
32      private void setListener() {
33          //添加事件
34          mBtn.setOnClickListener(new OnClickListener() {
35
36              @Override
37              public void onClick(View arg0) {
38                  //TODO Auto-generated method stub
39                  mProgressBar.setVisibility(View.VISIBLE);
40                  //开启线程进行 GET 请求
41                  new Thread() {
42
43                      @Override
44                      public void run() {
45                          //发送 GET 请求
46                          doGet();
                            //传递 handler 的消息
47                          mHandler.sendEmptyMessage(0);
48                          super.run();
49                      }
50
51                  }.start();
52
53              }
54          });
55      }
56
57      public void doGet() {
58          //定义接受的返回字节流对象
59          BufferedReader in = null;
60          String URL = "http://www.baidu.com/s?wd="
61                  + mEtInPut.getText().toString();
62
63          try {
64              //http 客户端
65              HttpClient client = new DefaultHttpClient();
66              //请求方式 get
67              HttpGet request = new HttpGet();
68              //设置请求地址
69              request.setURI(new URI(URL));
70              //定义相应对象
71              HttpResponse response = client.execute(request);
72              if (response.getStatusLine().getStatusCode() == 200) {
73                  //得到响应字符串
74                  mReturnConnection = EntityUtils.toString (response.getEntity());
75              } else {
76                  mReturnConnection = "Error Response: "
77                          + response.getStatusLine().toString();
78              }
79
80          } catch (Exception e) {
81              Log.e("wyl", e.toString());
```

```
82              } finally {
83                  //关闭字节流对象
84                  if (in != null) {
85                      try {
86                          in.close();                            //关闭流对象
87                      } catch (IOException ioe) {
88                          Log.e("wyl", ioe.toString());//输入输出异常
89                      }
90                  }
91              }
92          }
93      }
```

此文件是 Activity 的代码文件，在其中第 6～16 行定义了一个 Handler 对象，用来接收其他线程给主线程发送的 message 消息，当用户单击 GET 按钮的时候，执行第 41～51 行，用来启动一个线程，发送 GET 请求，然后发送 message 消息给 handler 对象。在第 57～92 行，首先定义了 httpclient 客户端，然后定义了 HttpGet 请求对象，最后设置 httpGget 的请求网址，通过客户端 client 的 execute 方法执行对应的 HttpGet 请求，得到返回结果为 response，通过 response 的状态得到是否请求成功，如果请求成功即可得到网络返回的数据。

4．实例扩展

在此实例中当用户单击 Button 的时候，就可以看到服务器返回的数据。本例中肯定是显示 html 的代码了，实际应用中要对返回的数据进行解析显示，后面的实例会介绍。上一个例子和本例子是最常见的网络请求方式 POST 和 GET，所以在 manifest 文件中一定要加入网络访问权限：

```
<uses-permission android:name="android.permission.INTERNET" />
```

范例 171　网络图片下载器

1．实例简介

在我们使用应用的过程中，经常会使用用户单击某按钮的时候下载网络的文件。例如，当用户看到某张图片的时候可以实现长按另存为效果，或者用户单击账单的时候，打印账单和账单保存等。遇到这样的功能，我们一般是当用户进行某个操作的时候，例如，单击某个按钮的时候，得到文件地址，然后通过连接此地址，得到文件流，然后进行保存。本例子就带领大家来实现一个通过网址下载网络图片的实例。

2．运行效果

该实例运行效果如图 8.3 所示。

3．实例程序讲解

在本实例中，提供一个用户希望下载网络图片的地址的输入框，然后用户单击下载按钮，开始后台下载文件，下载完毕

图 8.3　网络图片下载器

后保存在 sdcard 上,并且通过 Toast 提示用户下载完毕。想要实现本实例效果,首先修改 res/layout/activity_main.xml 文件,定义如上布局。代码省略。

然后修改 src/com.wyl.example/MainActivity.java 文件,代码如下:

```
001  //定义了本实例的主要 Activity
002  public class MainActivity extends Activity {
003
004      private Button mBtn;   //提交按钮
005      private EditText mEt;                          //下载网址输入框
006
007      //4.0 之后不允许在主线程中运行耗时的网络请求
008      private Handler mHandler = new Handler() {
009
010          @Override
011          public void handleMessage(Message msg) {
012              //根据 message 的状态做出不同提醒
013              if (msg.what == 0) {         //下载成功的状态提示
014                  Toast.makeText(MainActivity.this, "下载成功!",
015                          Toast.LENGTH_SHORT).show();
016              } else if (msg.what == 1) {//下载的文件已存在的状态提示
017                  Toast.makeText(MainActivity.this, "已有文件!",
018                          Toast.LENGTH_SHORT).show();
019              } else if (msg.what == -1) {     //下载失败的状态提示
020                  Toast.makeText(MainActivity.this, "下载失败!",
021                          Toast.LENGTH_SHORT).show();
022              }
023              super.handleMessage(msg);
024          }
025
026      };
027
028
029
030      @Override
031      public void onCreate(Bundle savedInstanceState) {
032          super.onCreate(savedInstanceState);
033          setContentView(R.layout.activity_main);
034          //得到布局中的控件
035          findView();
036          //绑定控件事件
037          setListener();
038
039      }
040
041      private void findView() {
042          //绑定控件
043          mBtn = (Button) findViewById(R.id.btn);
044          mEt = (EditText)findViewById(R.id.et);
045      }
046
047      private void setListener() {
048          //添加事件
049          mBtn.setOnClickListener(new OnClickListener() {
050
051              @Override
052              public void onClick(View arg0) {
053                  //定义新的线程去进行下载任务
```

```
054         new Thread() {
055             @Override
056             public void run() {
057                 //定义一个 Message 对象
058                 Message msg = new Message();
059                 //定义一个 HttpDownLoader 对象
060                 HttpDownloader httpDownLoader = new Http
                    Downloader();
061                 //把下载的结果返回给 message 对象
062                 msg.what = httpDownLoader
063                         .downfile(
064                                 mEt.getText().toString(),
065                                 "test/", "test1.jpg");
066                 //发送 message 消息给 handler
067                 mHandler.sendMessage(msg);
068                 super.run();
069             }
070         //开启此线程
071         }.start();
072     }
073 });
074 }
075
076 //定义文件下载类对象
077 public class HttpDownloader {
078     //定义 URL 对象
079     private URL url = null;
080     //定义文件操作对象
081     FileUtils fileUtils = new FileUtils();
082     //下载文件的方法
083     public int downfile(String urlStr,Stringpath,String fileName){
084         //判断是否本地已有此文件
085         if (fileUtils.isFileExist(path + fileName)) {
086             return 1;
087         } else {
088             //开始网络下载
089             try {
090                 InputStream input = null;
091                 input = getInputStream(urlStr);
092                 //写入 sdcard
093                 File resultFile = fileUtils.write2SDFrom
                    Input(path,
094                             fileName, input);
095                 //如果写入失败返回-1
096                 if (resultFile == null) {
097                     return -1;
098                 }
099             } catch (IOException e) {
100                 //TODO Auto-generated catch block
101                 e.printStackTrace();
102             }
103
104         }
105         return 0;
106     }
107
108     //由于得到一个 InputStream 对象是所有文件处理前必须的操作，所以将这个
        操作封装成了一个方法
109     public InputStream getInputStream(String urlStr) throws
```

```
           IOException {
110            //定义输入流对象
111            InputStream is = null;
112            try {
113                //得到根据url地址得到url对象
114                url = new URL(urlStr);
115                //URLConnection 连接网站
116                HttpURLConnection urlConn = (HttpURLConnection) url
117                        .openConnection();
118                //得到要下载文件的字节流对象
119                is = urlConn.getInputStream();
120
121            } catch (MalformedURLException e) {
122                //TODO Auto-generated catch block
123                e.printStackTrace();
124            }
125            return is;
126        }
127    }
128 }
```

此文件是 Activity 的代码文件,在其中第 8 行定义了一个 Handler 对象,用来接收其他线程发送的 message 消息,根据发送消息的 what 值判断是否成功,或失败。其中当用户单击下载按钮的时候执行第 54~71 行,首先定义一个 httpDownLoader 对象,然后传入下载地址,保存路径和图片名,就可以下载文件了,把返回状态通过 message 发送给主线程的 handler 进行处理。在第 77~127 行定义了 HttpDownloader 类,其中主要方法为第 83 行定义的 downfile 方法,第 109 行定义的 getInputStream 方法得到文件的字节流。

当然下载文件后用户希望保存到 SDCar 的目录下,这时候会涉及文件操作类,在我们的工程目录下新建 src/com.wyl.example/MainActivity.java 文件,代码如下:

```
01  //定义文件操作类
02  public class FileUtils {
03      //定义路径字符串
04      private String SDPATH;
05      //得到路径字符串
06      public String getSDPATH() {
07          return SDPATH;
08      }
09      //构造方法初始化路径字符串
10      public FileUtils() {
11          //得到当前外部存储设备的目录
12          ///SDCARD 路径
13          SDPATH = Environment.getExternalStorageDirectory() + "/";
14      }
15      /**
16       * 在SD卡上创建文件
17       *
18       * @throws IOException
19       */
20      public File creatSDFile(String fileName) throws IOException {
21          File file = new File(SDPATH + fileName);   //初始化一个文件对象
22          file.createNewFile();                      //创建这个文件
23          return file;                               //返回文件对象
24      }
25
```

```
26  /**
27   * 在SD卡上创建目录
28   *
29   * @param dirName
30   */
31  public File creatSDDir(String dirName) {
32      File dir = new File(SDPATH + dirName);      //创建一个目录对象
33      dir.mkdir();                                 //创建目录
34      return dir;                                  //返回目录对象
35  }
36
37  /**
38   * 判断SD卡上的文件夹是否存在
39   */
40  public boolean isFileExist(String fileName){
41      File file = new File(SDPATH + fileName);     //创建文件对象
42      return file.exists();                        //判断文件是否存在
43  }
44
45  /**
46   * 将一个InputStream里面的数据写入到SD卡中
47   */
48  public File write2SDFromInput(String path,String fileName,InputStream input){
49      File file = null;
50      OutputStream output = null;
51      try{
52          //在SD卡上创建目录
53          creatSDDir(path);                        //创建目录
54          file = creatSDFile(path + fileName);     //创建此目录下的文件
55          output = new FileOutputStream(file);     //创建此文件的流对象
56          byte buffer [] = new byte[4 * 1024];     //定义一个字节数组
57          while((input.read(buffer)) != -1){       //从流中读取字节
58              output.write(buffer);
59          }
60          output.flush();                          //清空输出流
61      }
62      catch(Exception e){
63          e.printStackTrace();
64      }
65      finally{
66          try{
67              output.close();
68          }
69          catch(Exception e){
70              e.printStackTrace();
71          }
72      }
73      return file;
74  }
75  }
```

此文件为文件管理类,用来管理文件操作的,在第 10～14 行定义了构造方法,在其中初始化了保存的根目录。在第 20～24 行定义了创建 sdcard 文件的方法,在第 30～35 行定义了创建 sdcard 目录的方法,在第 48～74 行定义了把文件写入 sdcard 的方法。这样关于文件的操作就可以在这个类中统一实现了。

4. 实例扩展

在此实例中我们需要用到网络访问、读取 SDCard 状态、创建 SDCard 文件和读写 SDCard 文件,所以我们需要在 manifest 文件中加入相应的权限,代码如下:

```xml
<!-- 授予访问互联网权限 -->
<uses-permission android:name="android.permission.INTERNET" />
<!-- 在SDCard中创建与删除文件权限 -->
<uses-permission
android:name="android.permission.MOUNT_UNMOUNT_FILESYSTEMS" />
<!-- 往SDCard写入数据权限 -->
<uses-permission
android:name="android.permission.WRITE_EXTERNAL_STORAGE" />
```

范例 172　文件上传

1. 实例简介

在我们使用应用的过程中,经常会使用文件上传的功能。例如,在一些社交客户端中,用户在客户端选择头像进行上传,或者用户在提交资料的时候需要上传电子文件等。遇到这样的功能,我们一般是当用户进行某个操作的时候,例如,单击某个按钮的时候,发送一个网络请求,进行文件的上传。本例子就带领大家来实现一个网络文件上传的实例。

2. 运行效果

该实例运行效果如图 8.4 所示。

图 8.4　网络文件上传

3. 实例程序讲解

在本实例中,提供两个 TextView 控件分别显示需要上传的文件目录和上传到的网络地址,然后用户单击上传文件按钮,就开始文件的上传,上传完毕后提醒用户已经上传完毕。想要实现本实例效果,首先修改 res/layout/activity_main.xml 文件,构造图 8.4 的控件布局。代码省略。

然后修改 src/com.wyl.example/MainActivity.java 文件,代码如下:

```
001  //定义了本实例的主要Activity
002  public class MainActivity extends Activity {
003      private String newName = "up.png";
004      private String uploadFile = "/sdcard/up.png";
005      private String actionUrl = "http://192.168.0.71:8086/HelloWord/myForm";
006      //省略控件定义代码
007
008      @Override
009      public void onCreate(Bundle savedInstanceState) {
010          super.onCreate(savedInstanceState);
011          setContentView(R.layout.activity_main);
```

```
012             //得到布局中的控件
013             findView();
014             //绑定控件事件
015             setListener();
016         }
017         //省略 findview 方法定义
018
019         private void setListener() {
020             /* 设置 mButton 的 onClick 事件处理 */
021             mButton.setOnClickListener(new View.OnClickListener() {
022                 public void onClick(View v) {
023                     //定义线程开始上传文件
024                     new Thread(){
025                         @Override
026                         public void run() {
027                             uploadFile();                    //上传文件
028                             super.run();
029                         }
030                     }.start();
031                 }
032             });
033         }
034
035         /* 上传文件至 Server 的方法 */
036         private void uploadFile() {
037             String end = "\r\n";
038             String twoHyphens = "--";
039             String boundary = "*****";
040             try {
041                 //定义 url 对象
042                 URL url = new URL(actionUrl);
043                 //建立 URLconnection 连接
044                 HttpURLConnection con =
045                     (HttpURLConnection) url.openConnection();
046                 /* 允许 Input、Output,不使用 Cache */
047                 con.setDoInput(true);
048                 con.setDoOutput(true);
049                 con.setUseCaches(false);
050                 /* 设置传送的 method=POST */
051                 con.setRequestMethod("POST");
052                 /* setRequestProperty */
053                 con.setRequestProperty("Connection", "Keep-Alive");
054                 con.setRequestProperty("Charset", "UTF-8");
055                 con.setRequestProperty("Content-Type",
056                     "multipart/form-data;boundary=" + boundary);
057                 /* 设置 DataOutputStream */
058                 DataOutputStream ds = new DataOutputStream
                    (con.getOutputStream());
059                 ds.writeBytes(twoHyphens + boundary + end);
060                 ds.writeBytes("Content-Disposition: form-data; "
061                     + "name=\"file1\";filename=\"" + newName + "\
                    "" + end);
062                 ds.writeBytes(end);
063                 /* 取得文件的 FileInputStream */
064                 FileInputStream fStream =
065                     new FileInputStream(uploadFile);
066                 /* 设置每次写入 1024bytes */
067                 int bufferSize = 1024;
```

```
068                byte[] buffer = new byte[bufferSize];
069                int length = -1;
070                /* 从文件读取数据至缓冲区 */
071                while ((length = fStream.read(buffer)) != -1) {
072                    /* 将资料写入 DataOutputStream 中 */
073                    ds.write(buffer, 0, length);
074                }
075                ds.writeBytes(end);
076                ds.writeBytes(twoHyphens + boundary + twoHyphens + end);
077                /* close streams */
078                fStream.close();
079                ds.flush();
080                /* 取得 Response 内容 */
081                InputStream is = con.getInputStream();
082                int ch;
083                StringBuffer b = new StringBuffer();
084                while ((ch = is.read()) != -1) {
085                    b.append((char) ch);
086                }
087                /* 将 Response 显示于 Dialog */
088                showDialog("上传成功" + b.toString().trim());
089                /* 关闭 DataOutputStream */
090                ds.close();
091            } catch (Exception e) {
092                showDialog("上传失败" + e);
093            }
094        }
095
096        /* 显示 Dialog 的 method */
097        private void showDialog(String mess) {
098            new AlertDialog.Builder(MainActivity.this).setTitle("Message")
099                    .setMessage(mess)
100                    .setNegativeButton("确定", new DialogInterface.OnClickListener() {
101                        public void onClick(DialogInterface dialog, int which) {
102                        }
103                    }).show();
104        }
105    }
```

此文件是 Activity 的代码文件,在其中第 3～4 行定义了一系列的字符串变量,用来表明上传文件和上传地址,当用户单击上传按钮的时候启动线程调用 uploadFile 方法。在第 36～93 行定义了 uploadFile 方法,在其中首先建立 URLConnection 对象,设置 input 和 output 参数,设置请求方式为 POST,然后通过 getOutputStream 方法得到数据流对象,在此数据流对象中写入我们需要上传的文件信息,通过我们传入的 inputstream 对象得到上传的结果。

4. 实例扩展

在此实例中当用户单击 Button 按钮的时候,进行文件上传,当然对于文件来说我这里是写死的文件路径,大家可以修改为通过选择目录中的某个文件然后进行上传,之前章节中写过选择系统文件的实例。

范例173 异步图片加载

1. 实例简介

在我们使用应用的过程中，经常会是看到用户首先打开了当前页面，但是页面中的某些网络图片还没有正常显示，然后用户在看页面其他内容的同时图片加载然后自动显示。例如，一些网上购物客户端，京东、淘宝和 QQ 网购等都采用的这些技术。遇到这样的功能，我们一般是在另一个线程中下载图片，然后下载完毕后显示，我们之前的例子都是这样的，这样有一个不好的地方就是需要开辟新的线程，然后通过 Handler+Message 的方式进行更新，还有一种方法可以实现异步的任务操作，就是异步任务。本例子就带领大家通过异步任务来实现异步图片的加载。

2. 运行效果

该实例运行效果如图 8.5 所示。

图 8.5 异步图片加载

3. 实例程序讲解

在本实例中，提供一个开始异步任务的按钮，还有一个显示异步任务加载结果的 ImageView 控件。想要实现本实例效果，首先修改 res/layout/activity_main.xml 文件，构造如图 8.5 所示的布局。代码省略。

然后修改 src/com.wyl.example/MainActivity.java 文件，代码如下：

```
01  //定义了本实例的主要 Activity
02  public class MainActivity extends Activity {
03      //网络图片地址
04      private static final String URL = "http://pica.nipic.com/
        2008-07-01/200871134114809_2.jpg";
05      private Button mBtnPicTask;
06      private ImageView mIvPic;
07
08      @Override
09      public void onCreate(Bundle savedInstanceState) {
10          super.onCreate(savedInstanceState);
11          setContentView(R.layout.activity_main);
12          //得到布局中的控件
13          findView();
```

```
14          //绑定控件事件
15          setListener();
16      }
17
18      private void findView() {
19          //绑定控件
20          mBtnPicTask = (Button) findViewById(R.id.btn_picTask);
21          mIvPic = (ImageView) findViewById(R.id.image);
22      }
23
24      private void setListener() {
25          //添加事件
26          mBtnPicTask.setOnClickListener(new OnClickListener() {
27
28              @Override
29              public void onClick(View arg0) {
30                  //定义异步任务,开启异步任务
31                  AsyncPicTask picTask = new AsyncPicTask();
32                  picTask.execute(mIvPic, URL);
33              }
34          });
35      }
36  }
```

此文件是 Activity 的代码文件,在第 13 行得到布局控件,在第 15 行设置监听器,当用户单击按钮的时候构造异步任务 AsyncPicTask,并且传入相应的 URL 和 ImageView 对象,然后执行此异步任务。

本实例中还需要定义一个异步任务类,新建 src/com.wyl.example/MainActivity.java 文件,代码如下:

```
01  //定义异步任务类
02  public class AsyncPicTask extends AsyncTask<Object, Integer, Bitmap> {
03      //定义 imageview 属性
04      private ImageView iv_pic;
05
06      //定义后台执行方法
07      @Override
08      protected Bitmap doInBackground(Object... params) {
09          //TODO Auto-generated method stub
10          Bitmap bitmap = null;
11          iv_pic = (ImageView) params[0];
12          try {
13              URL url = new URL((String) params[1]);
14              //调用 URL 对象 openConnection()方法来创建 URLConnection 对象
15              HttpURLConnection conn = (HttpURLConnection) url.
                    openConnection();
16              //设置该 URLConnection 的 doInput 请求头字段的值
17              conn.setDoInput(true);
18              //建立实际的连接
19              conn.connect();
20              //得到连接的字节流
21              InputStream inputStream = conn.getInputStream();
22              //解析图片
23              bitmap = BitmapFactory.decodeStream(inputStream);
24              //关闭字节流对象
25              inputStream.close();
26          } catch (MalformedURLException e) {
27              e.printStackTrace();
```

```
28          } catch (IOException e) {
29              e.printStackTrace();
30          }
31      return bitmap;
32  }
33
34  //执行获得图片数据后，更新UI:显示图片
35  @Override
36  protected void onPostExecute(Bitmap result) {
37      if (result != null) {
38          iv_pic.setImageBitmap(result);
39      }
40  }
41  }
```

此文件定义了一个异步任务，在其中 doInBackGround 是异步任务执行的主要方法，在其中定义了一个 URL 对象，然后建立了 URLConnection 对象，得到此连接的 InputStream 对象，然后通过 BimapFactory 的 decodeStream 方法得到 bitmap 对象。然后异步任务执行完毕，回调 onPostExecute 方法，在 ImageView 对象中显示请求下来的 Bitmap 对象。

4．实例扩展

在此实例中使用异步任务来加载图片，一般情况下异步任务比 Handler+Message 方法要轻松一些，不需要管理线程和处理 handler 操作，但是异步任务也有它的弊端，例如，只能在 view 中显示结果等。

范例 174　UDP 网络通信

1．实例简介

在我们使用应用的过程中，不但会用到 TCP 的传输协议，而且也可能会用到 UDP 的传输协议。例如，我在网络中进行在线授课则可能需要保证实时性。遇到这样的功能，我们一般是通过 UDP 来保证广播的效果的。本例子就带领大家来实现一个 DUP 的网络通信的实例。

2．运行效果

该实例运行效果如图 8.6 所示。

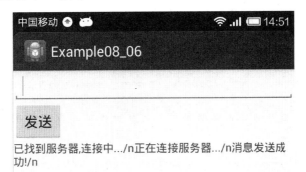

图 8.6　UDP 网络通信

3. 实例程序讲解

在本实例中，提供一个输入需要发送的文字输入框和一个发送按钮，然后当用户单击发送按钮时，广播发送用户输入的问题。想要实现本实例效果，首先修改 res/layout/activity_main.xml 文件，布局上述页面。代码省略。

然后修改 src/com.wyl.example/MainActivity.java 文件，代码如下：

```java
01  //定义了本实例的主要 Activity
02  public class MainActivity extends Activity {
03      //省略控件定义代码
04  
05      //定义 Handler 对象
06      private Handler mHandler = new Handler() {
07          @Override
08          //当有消息发送出来的时候就执行 Handler 的这个方法
09          public void handleMessage(Message msg) {
10              super.handleMessage(msg);
11              //处理 UI
12              mTvInfo.setText(mStrResult);
13          }
14      };
15  
16      @Override
17      public void onCreate(Bundle savedInstanceState) {
18          super.onCreate(savedInstanceState);
19          setContentView(R.layout.activity_main);
20          //省略得到控件对象的代码
21  
22          //开启服务器
23          ExecutorService exec=Executors.newCachedThreadPool();
24          //定义一个 UDPServer
25          UDPServer server=new UDPServer();
26          //执行此 Server
27          exec.execute(server);
28  
29          //发送消息
30          mBtnSend.setOnClickListener(new OnClickListener() {
31              @Override
32              public void onClick(View v) {
33                  new Thread(){
34                      public void run() {
35                          //TODO Auto-generated method stub
36                          //通过 UDPClient 发送 UDP 消息
37                          UDPClient client=new UDPClient(mEtMsg.getText().toString());
38                          mStrResult = client.send();
39                          //并且发送 handler 消息更新界面
40                          mHandler.sendEmptyMessage(0);
41                          super.run();
42                      }
43                  }.start();
44              }
45          });
46      }
47  }
```

此文件是 Activity 的代码文件，在其中第 6~14 行定义了一个 Handler 对象，接收用户发送的 message 消息，在第 23 行定义了一个服务器执行者，然后定义了一个 UDPServer 对象，然后执行此服务对象，开启 UDP 服务，用户单击按钮的时候启动一个线程，发送用户输入的文字，并且发送 Handler 消息。

在本实例中还定义了 UDPServer 类，新建 src/com.wyl.example/ UDPServer.java 文件，代码如下：

```java
01  //构建 Runnable 对象
02  public class UDPServer implements Runnable {
03  //端口号为 6000
04  private static final int PORT = 6000;
05  //定义缓存字节数组
06  private byte[] msg = new byte[1024];
07
08  private boolean life = true;
09
10  public UDPServer() {
11  }
12
13  public boolean isLife() {
14      return life;
15  }
16
17  public void setLife(boolean life) {
18      this.life = life;
19  }
20
21  @Override
22  public void run() {
23      //实例化 udp 广播类
24      DatagramSocket dSocket = null;
25      //要发送的信息
26      DatagramPacket dPacket = new DatagramPacket(msg, msg.length);
27      try {
28          //绑定端口
29          dSocket = new DatagramSocket(PORT);
30          while (life) {
31              try {
32                  //自我阻塞接受消息
33                  dSocket.receive(dPacket);
34                  Log.i("msg sever received", new String (dPacket.getData()));
35              } catch (IOException e) {
36                  e.printStackTrace();
37              }
38          }
39      } catch (SocketException e) {
40          e.printStackTrace();
41      }
42  }
43  }
```

本实例定义了一个 UDPServer 类，在此类中通过第 8 行定义 life 标记线程是否存活。第 22~42 行实现 run 方法，首先定义 Datagramsocket 类的对象，然后定义 DatagramPacket 的对象，然后绑定监听端口，当程序的 life 标记为真时，持续接收对应端口发来的信息。

在本实例中还定义了 UDPClient 类，新建 src/com.wyl.example/ UDPClient.java 文件，代码如下：

```java
01  //定义UDP客户端
02  public class UDPClient {
03      //定义UDP访问端口号
04      private static final int SERVER_PORT = 6000;
05      //UDP广播类
06      private DatagramSocket mDSocket = null;
07      //要发送的消息
08      private String mMsg;
09
10      /**
11       * @param msg
12       */
13      public UDPClient(String msg) {
14          super();
15          this.mMsg = msg;
16      }
17
18      /**
19       * 发送信息到服务器
20       */
21      public String send() {
22          //得到服务器的字符串构建对象
23          StringBuilder sb = new StringBuilder();
24          //得到网络地址对象
25          InetAddress local = null;
26          try {
27              //连接本机
28              local = InetAddress.getByName("localhost"); //本机测试
29              //设置状态对象
30              sb.append("已找到服务器,连接中...").append("/n");
31          } catch (UnknownHostException e) {
32              sb.append("未找到服务器.").append("/n");
33              e.printStackTrace();
34          }
35          try {
36              mDSocket = new DatagramSocket();//定义DatagramSocket对象
37              sb.append("正在连接服务器...").append("/n");
38          } catch (SocketException e) {
39              e.printStackTrace();
40              sb.append("服务器连接失败.").append("/n");
41          }
42          //判断消息
43          int msg_len = mMsg == null ? 0 : mMsg.length();
44          //向特定主机端口发送消息
45          DatagramPacket dPacket = new DatagramPacket(mMsg.getBytes(), msg_len,
46                  local, SERVER_PORT);
47          try {
48              mDSocket.send(dPacket);                         //发送数据包
49              sb.append("消息发送成功!").append("/n");
50          } catch (IOException e) {
51              e.printStackTrace();
52              sb.append("消息发送失败.").append("/n");
53          }
```

```
54        //关闭连接
55        mDSocket.close();
56        return sb.toString();
57    }
58 }
```

此类定义了 UDP 客户端的类。其中主要实现了第 21 行的 Send 方法，在此方法中首先定义 InetAddress 类的对象用来保存需要发送的地址，然后定义 DatagramSocket 对象，发送需要广播的内容。

4．实例扩展

在此实例中本例子既发送广播又接收广播，所以可能会有些不容易理解，在实际应用中也是这样的，你不但要学习如何发送 UDP 消息，还要学会如何接收 UDP 消息。

范例 175 在线音乐播放

1．实例简介

在我们使用应用的过程中，经常会使用在线音乐播放的效果。例如，千千静听和酷狗播放器，都提供在线音乐播放的功能。遇到这样的功能，我们一般是当用户进行某个操作的时候，例如：单击某个按钮的时候，开始加载网络的音频文件，一边加载一边播放。本例子就带领大家来实现一个在线音乐播放的实例。

2．运行效果

该实例运行效果如图 8.7 所示。

图 8.7 在线音乐播放

3．实例程序讲解

在本实例中，提供三个按钮，分别是播放网络音乐、暂停播放和停止播放，并且下面有一个播放的进度条，当用户单击对应按钮的时候实现对应的操作，例如：单击播放网络视频，则在线播放。想要实现本实例效果，首先修改 res/layout/activity_main.xml 文件，构造图 8.7 的布局效果，代码省略。

然后修改 src/com.wyl.example/MainActivity.java 文件，代码如下：

```java
01  //定义了本实例的主要Activity
02  public class MainActivity extends Activity {
03  //省略控件对象的定义代码
04
05  @Override
06  public void onCreate(Bundle savedInstanceState) {
07      super.onCreate(savedInstanceState);
08      setContentView(R.layout.activity_main);
09
10      //得到布局中的控件
11      findView();
12      //绑定控件事件
13      setListener();
14      mPlayer = new Player(mSkbProgress);        //定义一个player对象
15
16  }
17
18  //省略findview方法和setlistener方法的定义
19
20  class ClickEvent implements OnClickListener {
21
22      @Override
23      public void onClick(View arg0) {
24          if (arg0 == mBtnPause) {        //当用户单击暂停按钮时暂停播放
25              mPlayer.pause();
26          } else if (arg0==mBtnPlayUrl){//当单击播放按钮时播放指定音频文件
27              //在百度MP3里随便搜索到的,可以试试别的链接
28              String url = "http://m.zonse.net/music/data/upload
                  /12532622.mp3";
29              mPlayer.playUrl(url);
30          } else if (arg0 == mBtnStop) {
31              mPlayer.stop();
32          }
33      }
34  }
35
36  class SeekBarChangeEvent implements SeekBar.OnSeekBarChangeListener {
37      int progress;
38
39      @Override
40      public void onProgressChanged(SeekBar seekBar, int progress,
41              boolean fromUser) {
42          //设置播放进度为音乐播放的时间比例
43          this.progress = progress * mPlayer.mMediaPlayer.getDuration()
44                  / seekBar.getMax();
45      }
46
47      @Override
48      public void onStartTrackingTouch(SeekBar seekBar) {
49
50      }
51
52      @Override
53      public void onStopTrackingTouch(SeekBar seekBar) {
54          //seekTo()的参数是相对与音乐时间的数字,而不是与seekBar.getMax()相
                对的数字
55          mPlayer.mMediaPlayer.seekTo(progress);
```

```
56      }
57   }
58 }
```

此文件是 Activity 的代码文件,在其中第 11 行得到布局中的控件对象,第 13 行设置监听器,第 14 行初始化 Play 类的对象,当用户单击播放时调用第 28～29 行代码开始播放,同时停止时调用 mPlayer 的 stop 方法停止播放,用户单击暂停时调用 pause 方法暂停播放。在第 36～58 行定义了 SeekBar 改变的事件监听器,其中当进度改变时,通过播放的进度比例得到进度条的位置值。

本例还需要定义一个在线的音乐播放类,新建 src/com.wyl.example/ Player.java 文件,代码如下:

```
001 //定义在线播放音乐类
002 public class Player implements OnBufferingUpdateListener,
003         OnCompletionListener,
004         MediaPlayer.OnPreparedListener {
005
006     public MediaPlayer mMediaPlayer; //播放类
007     private SeekBar mSkbProgress; //进度条
008     //定时器对象
009     private Timer mTimer = new Timer();
010
011     //播放音乐的方法
012     public Player(SeekBar skbProgress) {
013         this.mSkbProgress = skbProgress;
014
015         try {
016             //定义 MediaPlayer 对象
017             mMediaPlayer = new MediaPlayer();
018             //设置 MediaPlayer 对象的参数
019             mMediaPlayer.setAudioStreamType(AudioManager.STREAM
                _MUSIC);
020             mMediaPlayer.setOnBufferingUpdateListener(this);
021             mMediaPlayer.setOnPreparedListener(this);
022         } catch (Exception e) {
023             Log.e("mediaPlayer", "error", e);
024         }
025         //设置执行的计划任务
026         mTimer.schedule(mTimerTask, 0, 1000);
027     }
028
029     //通过定时器和 Handler 来更新进度条
030     TimerTask mTimerTask = new TimerTask() {
031         @Override
032         public void run() {
033             if (mMediaPlayer == null)
034                 return;
035             if (mMediaPlayer.isPlaying()
036                     && mSkbProgress.isPressed() == false) {
037                 handleProgress.sendEmptyMessage(0);
038             }
039         }
040     };
041     //异步更新
042     Handler handleProgress = new Handler() {
043         public void handleMessage(Message msg) {
044
```

```java
            //当前播放的位置
            int position = mMediaPlayer.getCurrentPosition();
            //总体长度
            int duration = mMediaPlayer.getDuration();
            //当音频播放的长度大于0时，记录播放位置
            if (duration > 0) {
                //设置当前位置
                long pos = mSkbProgress.getMax() * (position / duration);
                mSkbProgress.setProgress((int) pos);
            }
        };
    };

    //播放
    public void play() {
        mMediaPlayer.start();
    }

    //在线播放音乐
    public void playUrl(String videoUrl) {
        try {
            //设备初始化
            mMediaPlayer.reset();
            //设置数据源
            mMediaPlayer.setDataSource(videoUrl);
            //prepare 之后自动播放
            mMediaPlayer.prepare();
            //mediaPlayer.start();
        } catch (IllegalArgumentException e) {
            //TODO Auto-generated catch block       //非法参数异常
            e.printStackTrace();
        } catch (IllegalStateException e) {
            //TODO Auto-generated catch block
            e.printStackTrace();                    //非法状态异常
        } catch (IOException e) {
            //TODO Auto-generated catch block
            e.printStackTrace();                    //输入输出异常
        }
    }

    //暂停
    public void pause() {
        mMediaPlayer.pause();
    }

    //停止
    public void stop() {
        if (mMediaPlayer != null) {
            mMediaPlayer.stop();            //播放停止
            mMediaPlayer.release();         //释放资源
            mMediaPlayer = null;
        }
    }

    @Override
    //通过 onPrepared 播放
    public void onPrepared(MediaPlayer arg0) {
```

```
102             arg0.start();
103             Log.e("mediaPlayer", "onPrepared");
104         }
105
106         @Override
107         public void onCompletion(MediaPlayer arg0) {
108             Log.e("mediaPlayer", "onCompletion");
109         }
110
111         //缓存更新的回调函数
112         @Override
113         public void onBufferingUpdate(MediaPlayer arg0, int buffering
            Progress) {
114
115             //设置缓存加载的长度
116             mSkbProgress.setSecondaryProgress(bufferingProgress);
117             //当前播放长度
118             int currentProgress = mSkbProgress.getMax()
119                     * mMediaPlayer.getCurrentPosition()
120                     / mMediaPlayer.getDuration();
121
122             Log.e(currentProgress + "% play", bufferingProgress + "%
            buffer");
123         }
124 }
```

此类主要功能是进行在线音乐的播放,其中主要实现的方法有 Player 构造方法,初始化 MediaPlayer 对象设置播放参数,设置定时任务 mTimer。TimerTask 类在第 30~40 行定义,主要功能是用来更新界面的进度条。在第 42~56 行定义了 Handler 对象,主要功能是时刻发送音乐播放的进度。在第 64~83 行定义了在线播放函数,设置 MediaPlayer 的预览效果。然后实现了 onBufferingUpdate 方法,用来缓冲加载的网络音频数据。这样一个在线播放软件就完成了。

4. 实例扩展

在此实例是在线音乐播放器的基本功能,当然一个完善的在线音乐播放器还有很多功能可以实现,例如:暂停时自动缓冲或者多点缓冲等。

范例 176 在线视频播放

1. 实例简介

在我们使用应用的过程中,经常会使用在线视频播放的效果。例如,优酷客户端和暴风影音客户端。遇到这样的功能,我们一般是当用户进行某个操作的时候,例如,单击某个按钮的时候,加载网络的视频,进行缓存并进行播放。本例子就带领大家来实现一个在线视频播放的实例。

2. 运行效果

该实例运行效果如图 8.8 所示。

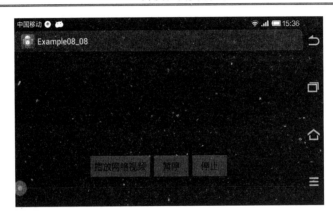

图 8.8 在线视频播放

3．实例程序讲解

在本实例中，提供播放网络视频按钮、暂停播放按钮和停止播放按钮。想要实现本实例效果，首先修改 res/layout/activity_main.xml 文件进行页面布局。代码省略。

然后修改 src/com.wyl.example/MainActivity.java 文件，代码如下：

```
01 //定义了本实例的主要Activity
02 public class MainActivity extends Activity {
03     //省略控件对象定义代码
04
05     private Player mPlayer;  //播放类
06
07     @Override
08     public void onCreate(Bundle savedInstanceState) {
09         super.onCreate(savedInstanceState);
10         setContentView(R.layout.activity_main);
11         //设置横向屏幕播放
12         setRequestedOrientation(ActivityInfo.SCREEN_ORIENTATION_
            LANDSCAPE);
13
14         //得到布局中的控件
15         findView();
16         //绑定控件事件
17         setListener();
18         mPlayer = new Player(mSurfaceView, mSkbProgress);
19
20     }
21     //省略findview和setlistener方法
22
23     class ClickEvent implements OnClickListener {
24
25         @Override
26         public void onClick(View arg0) {
27             //判断暂停播放视频
28             if (arg0 == mBtnPause) {
29                 mPlayer.pause();
30             }
31             //判断开始播放视频
32             else if (arg0 == mBtnPlayUrl) {
33                 String url = "网络视频地址";
```

```
34              mPlayer.playUrl(url);
35          }
36          //判断停止播放视频
37          else if (arg0 == btnStop) {
38              mPlayer.stop();
39          }
40      }
41  }
42  //定义 SeekBar 的改变监听器事件
43  class SeekBarChangeEvent implements SeekBar.OnSeekBarChangeListener {
44      int progress;
45
46      @Override
47      public void onProgressChanged(SeekBar seekBar, int progress,
48              boolean fromUser) {
49          //根据视频的播放进度更新 progressBar 的进度条
50          this.progress = progress * mPlayer.mMediaPlayer.getDuration()
51                  / seekBar.getMax();
52      }
53
54      @Override
55      public void onStartTrackingTouch(SeekBar seekBar) {
56
57      }
58
59      @Override
60      public void onStopTrackingTouch(SeekBar seekBar) {
61          //seekTo()的参数是相对与影片时间的数字,而不是与 seekBar.getMax()相
                对的数字
62          mPlayer.mMediaPlayer.seekTo(progress);
63      }
64  }
65  }
```

此文件是 Activity 的代码文件,在其中第 5 行定义了一个 Player 类。在第 12 行设置当前页面为横向显示,第 18 行初始 Player 对象,当用户单击播放网络视频的时候,调用 mPlayer 的 playUrl 方法播放在线视频,用户单击暂停按钮的时候,调用 mPlayer 的 pause 方法暂停播放,当用户单击停止播放的按钮时,调用 stop 方法停止播放。在第 43~64 行定义了 SeekBarChangeEvent 方法用来监听更新进度条的显示进度。

本例还需要定义一个在线的音乐播放类,新建 src/com.wyl.example/ Player.java 文件,代码如下:

```
001  //定义在线视频播放类
002  public class Player implements OnBufferingUpdateListener,
     OnCompletionListener,
003          MediaPlayer.OnPreparedListener, SurfaceHolder.Callback {
004      //定义播放视频的宽度
005      private int videoWidth;
006      //定义播放视频的高度
007      private int videoHeight;
008      public MediaPlayer mMediaPlayer;              //播放类
009      private SurfaceHolder mSurfaceHolder;
010      private SeekBar mSkbProgress;                 //进度条
011      private Timer mTimer = new Timer();
012
013      //构造方法,初始化播放的 Surfaceholder 类的对象
```

```
014    public Player(SurfaceView surfaceView, SeekBar skbProgress) {
015        this.mSkbProgress = skbProgress;
016        mSurfaceHolder = surfaceView.getHolder();
017        mSurfaceHolder.addCallback(this);
018        mSurfaceHolder.setType(SurfaceHolder.SURFACE_TYPE_PUSH_
           BUFFERS);
019        mTimer.schedule(mTimerTask, 0, 1000);
020    }
021
022    //通过定时器和Handler来更新进度条
023    TimerTask mTimerTask = new TimerTask() {
024        @Override
025        public void run() {
026            //如果没开始播放,就返回
027            if (mMediaPlayer == null)
028                return;
029            //如果开始播放就更新进度
030            if (mMediaPlayer.isPlaying() && mSkbProgress.isPressed()
               == false) {
031                handleProgress.sendEmptyMessage(0);
032            }
033        }
034    };
035
036    //根据消息更新进度条进度
037    Handler handleProgress = new Handler() {
038        public void handleMessage(Message msg) {
039            //当前播放的位置
040            int position = mMediaPlayer.getCurrentPosition();
041            //总体长度
042            int duration = mMediaPlayer.getDuration();
043
044            if (duration > 0) {
045                //设置当前位置
046                long pos = mSkbProgress.getMax() * position /
                   duration;
047                mSkbProgress.setProgress((int) pos);
048            }
049        };
050    };
051
052    //播放
053    public void play() {
054        mMediaPlayer.start();
055    }
056
057    //播放在线视频
058    public void playUrl(String videoUrl) {
059        try {
060            //设备初始化
061            mMediaPlayer.reset();
062            //设置数据源
063            mMediaPlayer.setDataSource(videoUrl);
064            //prepare之后自动播放
065            mMediaPlayer.prepare();
066            //mediaPlayer.start();
067        } catch (IllegalArgumentException e) {
068            //TODO Auto-generated catch block
```

```
069                e.printStackTrace();
070            } catch (IllegalStateException e) {
071                //TODO Auto-generated catch block
072                e.printStackTrace();
073            } catch (IOException e) {
074                //TODO Auto-generated catch block
075                e.printStackTrace();
076            }
077        }
078
079        //暂停
080        public void pause() {
081            mMediaPlayer.pause();
082        }
083
084        //停止
085        public void stop() {
086            if (mMediaPlayer != null) {
087                mMediaPlayer.stop();
088                mMediaPlayer.release();
089                mMediaPlayer = null;
090            }
091        }
092
093        @Override
094        public void surfaceChanged(SurfaceHolder arg0, int arg1, int arg2,
               int arg3) {
095            Log.e("mediaPlayer", "surface changed");
096        }
097
098        @Override
099        public void surfaceCreated(SurfaceHolder arg0) {
100            try {
101                //创建 surface 的回调方法
102                mMediaPlayer = new MediaPlayer();
103                mMediaPlayer.setDisplay(mSurfaceHolder);
104                mMediaPlayer.setAudioStreamType(AudioManager.STREAM_
                   MUSIC);
105                mMediaPlayer.setOnBufferingUpdateListener(this);
106                mMediaPlayer.setOnPreparedListener(this);
107            } catch (Exception e) {
108                Log.e("mediaPlayer", "error", e);
109            }
110        }
111
112        @Override
113        public void surfaceDestroyed(SurfaceHolder arg0) {
114            Log.e("mediaPlayer", "surface destroyed");
115        }
116
117        @Override
118        /**
119         * 通过 onPrepared 播放
120         */
121        public void onPrepared(MediaPlayer arg0) {
122            videoWidth = mMediaPlayer.getVideoWidth();
123            videoHeight = mMediaPlayer.getVideoHeight();
124
125            if (videoHeight != 0 && videoWidth != 0) {
```

```
126                arg0.start();
127            }
128            Log.e("mediaPlayer", "onPrepared");
129        }
130
131        @Override
132        public void onCompletion(MediaPlayer arg0) {
133            //TODO Auto-generated method stub
134
135        }
136
137        @Override
138        public void onBufferingUpdate(MediaPlayer arg0, int buffering
        Progress) {
139            //设置缓存加载的长度
140            mSkbProgress.setSecondaryProgress(bufferingProgress);
141            //当前播放长度
142            int currentProgress = mSkbProgress.getMax()
143                    * mMediaPlayer.getCurrentPosition()
144                    / mMediaPlayer.getDuration();
145            Log.e(currentProgress + "% play", bufferingProgress + "%
        buffer");
146        }
```

此类定义了在线视频的播放类，其实实现原理同上例的 Play 方法，在第 15~18 行初始化了播放视频的 SurfaceHolder 对象，同时开启定时任务更新进度条。第 22~34 行定义定时任务类。在 37~50 行定义了 Handler 类。第 58~77 行定义了 playUrl 方法，用来根据网络视频地址播放网络视频。第 99~110 行定义了 surfaceCreated 方法，用来创建 MediaPlayer 对象，并且设置播放参数。在第 121~129 行定义了 onPrepared 方法，用来播放 MediaPlayer 视频。

4．实例扩展

此实例是常见的视频播放软件的核心功能，当然还有很多可以完善的地方。例如，缓冲池的加入、预览播放的加入、用户快进和跳跃播放的功能加入等。如果想实现一个真正的视频播放器的话大家就去演技一下这些内容吧。

范例 177 应用程序在线更新

1．实例简介

在我们使用应用的过程中，经常会遇到应用程序升级的功能。例如，在一些软件市场的客户端中，都会有软件的更新功能，或者在一些成熟的客户端中都自带自动更新功能。遇到这样的功能，我们一般是当用户进行某个操作的时候，例如：单击某个按钮的时候，获取网络的应用程序的版本号与本地的版本号进行对比，如果需要升级则在线下载升级程序，下载完毕后自动安装。本例子就带领大家来实现一个在线更新应用程序的实例。

2．运行效果

该实例运行效果如图 8.9 所示。

第 8 章 Android 的网络编程

图 8.9 应用程序在线更新

3. 实例程序讲解

在本实例中，提供一个在线更新的按钮，当用户单击此按钮的时候程序检查网络版本开始更新。想要实现本实例效果，首先修改 res/layout/activity_main.xml 文件，定义如图 8.9 所示的布局控件。代码省略。

然后修改 src/com.wyl.example/MainActivity.java 文件，代码如下：

```java
01  //定义了本实例的主要 Activity
02  public class MainActivity extends Activity {
03
04      private Button mBtn; //请求更新的按钮
05
06      @Override
07      public void onCreate(Bundle savedInstanceState) {
08          super.onCreate(savedInstanceState);
09          setContentView(R.layout.activity_main);
10          //得到布局中的控件
11          findView();
12          //绑定控件事件
13          setListener();
14      }
15
16      private void findView() {
17          //绑定控件
18          mBtn = (Button) findViewById(R.id.btn);
19      }
20
21      private void setListener() {
22          //添加事件
23          mBtn.setOnClickListener(new OnClickListener() {
24
25              @Override
26              public void onClick(View arg0) {
27                  //显示更新对话框
28                  new UpdateManager(MainActivity.this).showDownload
                        Dialog();
```

```
29          }
30       });
31   }
```

此文件是 Activity 的代码文件，在其中第 11 行得到布局中的所有控件，第 13 行设置所有的监听器，当用户单击此按钮的时候，调用第 28 行定义一个 UpdateManager 类的对象并且显示下载对话框。

在本实例中，还需要新建一个更新类，新建 src/com.wyl.example/MainActivity.java 文件，代码如下：

```
001  //定义在线更新类
002  public class UpdateManager {
003      //定义三种更新状态
004      private static final int DOWN_NOSDCARD = 0;
005      private static final int DOWN_UPDATE = 1;
006      private static final int DOWN_OVER = 2;
007
008      private Context mContext;
009      //下载对话框
010      private Dialog downloadDialog;
011      //进度条
012      private ProgressBar mProgress;
013      //显示下载数值
014      private TextView mProgressText;
015      //进度值
016      private int progress;
017      //下载线程
018      private Thread downLoadThread;
019      //终止标记
020      private boolean interceptFlag;
021      //返回的安装包 url
022      private String apkUrl = "更新 APK 的地址";
023      //下载包保存路径
024      private String savePath = "";
025      //apk 保存完整路径
026      private String apkFilePath = "";
027      //临时下载文件路径
028      private String tmpFilePath = "";
029      //下载文件大小
030      private String apkFileSize;
031      //已下载文件大小
032      private String tmpFileSize;
033
034      public UpdateManager(Context context) {
035          this.mContext = context;
036      }
037
038      private Handler mHandler = new Handler() {
039          public void handleMessage(Message msg) {
040              switch (msg.what) {
041              case DOWN_UPDATE:
042                  //下载更新，更新进度条
043                  mProgress.setProgress(progress);
044                  mProgressText.setText(tmpFileSize + "/" + apkFileSize);
045                  break;
046              case DOWN_OVER:
```

```
047                    //下载完毕取消progressBar
048                    downloadDialog.dismiss();
049                    //安装apk
050                    installApk();
051                    break;
052                case DOWN_NOSDCARD:
053                    //检查SDcard是否装载,否则取消进度条
054                    downloadDialog.dismiss();
055                    Toast.makeText(mContext,
                           "无法下载安装文件,请检查SD卡是否挂载", 3000).show();
056                    break;
057                }
058            };
059        };
060
061    ...
062
063    private Runnable mdownApkRunnable = new Runnable() {
064        @Override
065        public void run() {
066            try {
067                String apkName = "wyl_" + ".apk";
068                String tmpApk = "wyl_" + ".tmp";
069                //判断是否挂载了SD卡
070                String storageState = Environment.getExternal
                   StorageState();
071                if(storageState.equals(Environment.MEDIA_MOUNTED)){
072                    savePath = Environment.getExternalStorage
                       Directory()
073                        .getAbsolutePath() + "/Update/";
074                    File file = new File(savePath);//得到文件对象
075                    if (!file.exists()) {   //判断文件是否存在
076                        file.mkdirs();                //创建文件目录
077                    }
078                    apkFilePath = savePath + apkName;//apk文件路径
079                    tmpFilePath = savePath + tmpApk;//临时文件路径
080                }
081
082                //没有挂载SD卡,无法下载文件
083                if (apkFilePath == null || apkFilePath == "") {
084                    mHandler.sendEmptyMessage(DOWN_NOSDCARD);
085                    return;
086                }
087
088                File ApkFile = new File(apkFilePath);
089
090                //是否已下载更新文件
091                if (ApkFile.exists()) {
092                    downloadDialog.dismiss();
093                    installApk();
094                    return;
095                }
096
097                //输出临时下载文件
098                File tmpFile = new File(tmpFilePath);
099                FileOutputStream fos = new FileOutputStream
                   (tmpFile);
100                //定义url对象
101                URL url = new URL(apkUrl);
```

```java
102            //定义url连接对象
103            HttpURLConnection conn = (HttpURLConnection) url
104                    .openConnection();
               conn.connect();                //建立连接
105            int length=conn.getContentLength();//得到连接内容的长度
106            InputStream is=conn.getInputStream();//得到连接的输入流
107
108            //显示文件大小格式：2个小数点显示
109            DecimalFormat df = new DecimalFormat("0.00");
110            //进度条下面显示的总文件大小
111            apkFileSize = df.format((float)length/1024/1024)+
                       "MB";
112            //定义下载量为0
113            int count = 0;
114            byte buf[] = new byte[1024];
115            ///循环下载文件
116            do {
117                int numread = is.read(buf);
118                count += numread;
119                //进度条下面显示的当前下载文件大小
120                tmpFileSize=df.format((float)count/1024/1024)+"MB";
121                //当前进度值
122                progress = (int) (((float) count / length) * 100);
123                //更新进度
124                mHandler.sendEmptyMessage(DOWN_UPDATE);
125                if (numread <= 0) {
126                    //下载完成 - 将临时下载文件转成APK文件
127                    if (tmpFile.renameTo(ApkFile)) {
128                        //通知安装
129                        mHandler.sendEmptyMessage(DOWN_OVER);
130                    }
131                    break;
132                }
133                fos.write(buf, 0, numread);
134            } while (!interceptFlag);       //单击取消就停止下载
135
136            fos.close();
137            is.close();
138        } catch (MalformedURLException e) {
139            e.printStackTrace();
140        } catch (IOException e) {
141            e.printStackTrace();
142        }
143
144    }
145 };
146
147 //下载apk
148 private void downloadApk() {
149     //开启线程下载apk
150     downLoadThread = new Thread(mdownApkRunnable);
151     downLoadThread.start();
152 }
153
154 //安装已下载的apk
155 private void installApk() {
156     //判断apk文件是否存在
157     File apkfile = new File(apkFilePath);
```

```
158              if (!apkfile.exists()) {
159                  return;
160              }
161              //如果存在，启动系统的 apk 安装流程
162              Intent i = new Intent(Intent.ACTION_VIEW);
163              i.setDataAndType(Uri.parse("file://" + apkfile.toString()),
164                      "application/vnd.android.package-archive");
165              mContext.startActivity(i);
166          }
167      }
```

此类定义了更新类的全部功能，主要代码在第 38～59 行定义了 Handler 对象，根据不同的 msg 信息设置不同的 downloadDialog 的文字。在第 63～154 行定义了一个 Runnable 对象，用来下载网络更新的 apk 文件，然后保存到 SDCard 下，然后在第 155～166 行定义了安装 apk 文件的函数，其中使用 intent 启动了系统的 apk 安装流程。

4．实例扩展

在此实例中主要的难点是版本的检查、apk 文件的下载和 apk 文件的安装。对于一些成熟的应用此功能是必备的，也希望大家在自己的程序中都加入在线更新的功能，方便自己新版本的发布和使用。

8.2 数据格式解析

范例 178 DOM 方式解析 XML

1．实例简介

我们在之前一部分主要讲解了如何从网络上获取数据，当然对于网络上获取的数据都是按照某种固定规则排列的字符串，我们统称为数据格式，常见的数据格式有 XML 和 JSON。拿到这些数据格式后我们要进行解析，得到我们想要的数据值。遇到这样的功能，我们一般是得到网络数据，使用固定方法解析。本例子就带领大家来实现一个使用 DOM 方法解析 XML 文件的实例。

2．运行效果

该实例运行效果如图 8.10 所示。

图 8.10 DOM 方式解析 XML

3. 实例程序讲解

在本实例中，提供一个显示解析结果的 TextView 控件和一个按钮控件，当用户单击此按钮时开始使用 DOM 方式解析 XML 文件。想要实现本实例效果，首先修改 res/layout/activity_main.xml 文件，布局如图 8.10 所示的效果。代码省略。

然后修改 src/com.wyl.example/MainActivity.java 文件，代码如下：

```
01  //定义了本实例的主要Activity
02  public class MainActivity extends Activity {
03
04      private TextView mTvShow;    //显示结果
05      private Button mBtnDom;  //Dom解析
06
07      @Override
08      protected void onCreate(Bundle savedInstanceState) {
09          super.onCreate(savedInstanceState);
10          setContentView(R.layout.activity_main);
11          //得到布局中的控件
12          findView();
13          //绑定控件事件
14          setListener();
15      }
16
17      private void setListener() {
18          //添加事件
19          mBtnDom.setOnClickListener(myListener);
20      }
21
22      private void findView() {
23          //绑定控件
24          mTvShow = (TextView) findViewById(R.id.textview);
25          mBtnDom = (Button) findViewById(R.id.btnDom);
26      }
27
28      private OnClickListener myListener = new OnClickListener() {
29
30          @Override
31          public void onClick(View v) {
32              //1、获取当前工程中的people.xml的文件流
33              InputStream inputStream = MainActivity.class.getClassLoader()
34                      .getResourceAsStream("people.xml");
35              //定义Person对象
36              List<Person> persons = null;
37              mTvShow.setText("");
38              switch (v.getId()) {
39              case R.id.btnDom:                       //DOM解析
40                  mTvShow.setText("DOM:");
41                  try {
42                      //使用DOMPersonService服务进行xml解析
43                      persons = DOMPersonService.readXml(inputStream);
44                  } catch (ParserConfigurationException e) {
45                      //TODO Auto-generated catch block
46                      e.printStackTrace();
47                  } catch (SAXException e) {
48                      //TODO Auto-generated catch block
49                      e.printStackTrace();
50                  } catch (IOException e) {
```

```
51                        //TODO Auto-generated catch block
52                        e.printStackTrace();
53                    }
54                    break;
55            default:
56                break;
57         }
58         //显示解析结果
59         for (Person person : persons) {
60             mTvShow.setText(mTvShow.getText().toString() + "\n" +
                   person.toString());
61         }
62     }
63 };
64 }
```

此文件是 Activity 的代码文件，在其中第 12 行得到页面的布局控件，第 14 行设置相应监听器，当用户单击解析按钮时，调用第 33～53 行的代码，首先得到系统的 xml 文件的内容，然后定义 persons 类表对象。然后第 43 行将需要解析的 XML 字符串传入 DOMPersonService 类中进行解析，得到结果。最后在 TextView 中显示解析结果。

本例还有一个 XML 的解析类，新建 src/com.wyl.example/ DOMPersonService.java 文件，代码如下：

```
01 //定义 dom 解析 xml 的类
02 public class DOMPersonService {
03   public static List<Person> readXml(InputStream inStream)
04         throws ParserConfigurationException, SAXException,IOException {
05     List<Person> persons = new ArrayList<Person>();
06     //实例化一个文档构建器工厂
07     DocumentBuilderFactory factory = DocumentBuilderFactory.newInstance();
08
09     //通过文档构建工厂构建文档构建器
10     DocumentBuilder builder = factory.newDocumentBuilder();
11
12     //通过文档构建器构建文档实例
13     Document document = builder.parse(inStream);
14
15     //得到文档的根节点
16     Element root = document.getDocumentElement();
17     //得到所有的 person 节点
18     NodeList nodes = root.getElementsByTagName("person");
19     //循环判断每个节点的内容
20     for (int i = 0; i < nodes.getLength(); i++) {
21         //得到节点的对象
22         Element personElement = (Element) nodes.item(i);
23         //定义 person 对象
24         Person person = new Person();
25         //得到 xml 的属性
26         person.setId(Integer.valueOf(personElement.getAttribute("id")));
27         //得到 xml 的子节点
28         NodeList childNodes = personElement.getChildNodes();
29         for (int j = 0; j < childNodes.getLength(); j++) {
30             //得到每个节点对象
31             Node childNode = (Node) childNodes.item(j);
32             //如果当前节点为内容节点，则取得相应的内容
```

```
33                if (childNode.getNodeType() == Node.ELEMENT_NODE) {
34                    Element childElement = (Element) childNode;
35                    if ("name".equals(childElement.getNodeName())) {
36                        person.setName(childElement.getFirstChild()
37                                .getNodeValue());
38                    } else if ("age".equals(childElement.getNodeName())) {
39                        person.setAge(new Short(childElement.
                            getFirstChild()
40                                .getNodeValue()));
41                    }
42                }
43            }
44            //添加到结果链表中
45            persons.add(person);
46        }
47        return persons;
48    }
49 }
```

此类定义了 DOM 类解析 XML 的方法，其中主要就是 readXML 方法，传入 InputStream 流，然后根据这个流构建 Document 文档，然后从文档的根目录依次得到每个节点的子节点，这样遍历每个节点的内容及属性。

4．实例扩展

DOM 方式解析 XML 文件的特点是一次性加载所有文件到内存中，然后根据 DOM 树的方式进行解析，优点是解析速度快，缺点是比较耗费内存，不能中途停止。

范例 179　SAX 方式解析 XML

1．实例简介

对于常见的 XML 格式解析我们还有另外一种方法-SAX。步骤同 DOM 解析步骤一样，都是拿到这些数据格式后我们要进行解析，得到我们想要的数据值。遇到这样的功能，我们一般是得到网络数据，使用固定方法解析。本例子就带领大家来实现一个使用 SAX 方法解析 XML 文件的实例。

2．运行效果

该实例运行效果如图 8.11 所示。

图 8.11　SAX 方式解析 XML

3. 实例程序讲解

在本实例中,提供一个显示解析结果的 TextView 控件和一个按钮控件。当用户单击此按钮时开始使用 SAX 方式解析 XML 文件。想要实现本实例效果,首先修改 res/layout/activity_main.xml 文件,布局如图 8.11 所示的效果。代码省略。

然后修改 src/com.wyl.example/MainActivity.java 文件,代码如下:

```
01  //定义了本实例的主要 Activity
02  public class MainActivity extends Activity {
03
04      private TextView mTvShow;              //显示结果
05      private Button mBtnSax;                //Sax 解析
06
07      @Override
08      protected void onCreate(Bundle savedInstanceState) {
09          super.onCreate(savedInstanceState);
10          setContentView(R.layout.activity_main);
11          //得到布局中的控件
12          findView();
13          //绑定控件事件
14          setListener();
15      }
16
17      private void setListener() {
18          //添加事件
19          mBtnSax.setOnClickListener(myListener);
20      }
21
22      private void findView() {
23          //绑定控件
24          mTvShow = (TextView) findViewById(R.id.textview);
25          mBtnSax = (Button) findViewById(R.id.btnSAX);
26      }
27
28      private OnClickListener myListener = new OnClickListener() {
29
30          @Override
31          public void onClick(View v) {
32              //1、获取当前工程中的 people.xml 的文件流
33              InputStream inputStream = MainActivity.class.getClassLoader()
34                      .getResourceAsStream("people.xml");
35              //定义 Person 对象
36              List<Person> persons = null;
37              mTvShow.setText("");
38              switch (v.getId()) {
39              case R.id.btnSAX:      //SAX 解析
40                  mTvShow.setText("SAX:");
41                  try {
42                      //使用 SAXPersonService 服务进行 xml 解析
43                      persons = SAXPersonService.readXml(inputStream);
44                  } catch (ParserConfigurationException e) {
45                      e.printStackTrace();
46                  } catch (SAXException e) {
47                      e.printStackTrace();
48                  } catch (IOException e) {
49                      e.printStackTrace();
```

```
50              }
51              break;
52          default:
53              break;
54          }
55          //显示解析结果
56          for (Person person : persons) {
57              mTvShow.setText(mTvShow.getText().toString() + "\n" +
                    person.toString());
58          }
59      }
60  };
61  }
```

此文件是 Activity 的代码文件,在其中第 12 行得到页面的布局控件,第 14 行设置相应监听器,当用户单击解析按钮时,调用第 33~59 行的代码,首先得到系统的 xml 文件的内容,然后定义 persons 类表对象,然后第 43 行将需要解析的 XML 字符串传入 SAXPersonService 类中进行解析,得到结果。最后在 TextView 中显示解析结果。

本例还有一个 XML 的解析类,新建 src/com.wyl.example/ SAXPersonService.java 文件,代码如下:

```
01  //使用SAX方式解析xml类
02  public class SAXPersonService {
03  public static List<Person> readXml(InputStream inStream)
04      throws ParserConfigurationException,
05      SAXException, IOException{
06      //得到SAX解析器的工厂对象
07      SAXParserFactory spf = SAXParserFactory.newInstance();
08      //让工厂对象去创建其对象
09      SAXParser saxParser = spf.newSAXParser();
10      //创建 DefaultHandler 对象
11      XMLContentHandler handler = new XMLContentHandler();
12      //使用parserr 的parser (inputstream in,DefaultHanderHandler
            handler);
13      saxParser.parse(inStream, handler);
14      //关闭字节流
15      inStream.close();
16      return handler.getPersons();
17  }
18  }
```

此类定义了 SAX 类解析 XML 的方法,其中主要就是 readXML 方法,传入 InputStream 流,然后根据固定的 Handler 来进行文档解析得到 persons 列表。

本例还有一个 XMLContentHandler 类。新建 src/com.wyl.example/ XMLContentHandler.java 文件,代码如下:

```
01  //定义SAX解析器的解析handle对象
02  public class XMLContentHandler extends DefaultHandler {
03
04  public static final String TAG = "XMLContentHandler";
05      //定义结果链表
06      private List<Person> persons;
07      private Person person;
08      //记录当前的节点名称
09      private String preTag;
```

```java
10
11   public List<Person> getPersons() {
12       return persons;
13   }
14
15   @Override
16   public void startDocument() throws SAXException {
17       //TODO Auto-generated method stub
18       super.startDocument();
19       //初始化结果链表
20       persons = new ArrayList<Person>();
21       Log.i(TAG, "开始解析...");
22   }
23
24   //uri 命名空间 localName:不带命名空间前缀的标签名 qName:
     //带命名空间前缀的标签名 attributes:属性集合
25   @Override
26   public void startElement(String uri, String localName, String qName,
27           Attributes attributes) throws SAXException {
28       //TODO Auto-generated method stub
29       super.startElement(uri, localName, qName, attributes);
30       //解析名称为 person 的节点
31       if ("person".equals(localName)) {
32           //初始化节点 person 对象
33           person = new Person();
34           //得到并设置 id 属性
35           person.setId(new Integer(attributes.getValue("id")));
36       }
37       //记录当前节点名称
38       preTag = localName;
39       Log.i(TAG, "解析元素: " + localName);
40   }
41
42   //ch[]内容 start 其实位置 length 长度
43   @Override
44   public void characters(char[] ch, int start, int length)
45           throws SAXException {
46       //TODO Auto-generated method stub
47       super.characters(ch, start, length);
48       if (person != null) {
49           //解析字段内容
50           String data = new String(ch, start, length);
51           //设置相应的属性
52           if ("name".equals(preTag)) {
53               person.setName(data);
54           } else if ("age".equals(preTag)) {
55               person.setAge(new Short(data));
56           }
57       }
58       Log.i(TAG, "解析的内容:" + new String(ch, start, length));
59   }
60
61   @Override
62   public void endElement(String uri, String localName, String qName)
63           throws SAXException {
64       //TODO Auto-generated method stub
65       super.endElement(uri, localName, qName);
66       //person 结束的节点
67       if ("person".equals(localName) && person != null) {
```

```
68          //添加 person 到结果 list 中
69          persons.add(person);
70          person = null;
71      }
72      preTag = null;
73      Log.i(TAG, localName + "解析完毕");
74   }
75
76   @Override
77   public void endDocument() throws SAXException {
78      //文档解析完毕
79      super.endDocument();
80      Log.i(TAG, "文档解析完毕");
81   }
82 }
```

此类主要定义了如何解析 XML，SAX 解析 XML 的方式是根据事件触发方式，第 16～22 行定义了文档开始解析的回调函数，第 26～40 行定义了开始一个节点的回调函数，第 44～59 行定义了取值的回调函数，第 62～74 行定义了节点结束的回调函数。第 77～81 行定义了文档解析结束的回调函数。

4．实例扩展

SAX 方式解析 XML 文件的特点是逐行进行加载，逐行解析，可以中途停止解析。

范例 180 PULL 方式解析 XML

1．实例简介

对于常见的 XML 格式解析，Android 提供了一种特殊的解析方法 PULL。思路和 SAX 解析步骤一样，都是拿到这些数据格式后进行解析，得到我们想要的数据值。遇到这样的功能，我们一般是得到网络数据，使用固定方法解析。本例子就带领大家来实现一个使用 PULL 方法解析 XML 文件的实例。

2．运行效果

该实例运行效果如图 8.12 所示。

图 8.12 PULL 方式解析 XML

3. 实例程序讲解

在本实例中，提供了一个显示解析结果的 TextView 控件和一个按钮控件。当用户单击此按钮时，开始使用 PULL 方式解析 XML 文件。想要实现本实例效果，首先修改 res/layout/activity_main.xml 文件，布局如图 8.12 所示的效果。代码省略。

然后修改 src/com.wyl.example/MainActivity.java 文件，代码如下：

```java
01 //定义了本实例的主要Activity
02 public class MainActivity extends Activity {
03
04     private TextView mTvShow;      //显示结果
05     private Button mBtnPull;       //Pull 解析
06
07     @Override
08     protected void onCreate(Bundle savedInstanceState) {
09         super.onCreate(savedInstanceState);
10         setContentView(R.layout.activity_main);
11         //得到布局中的控件
12         findView();
13         //绑定控件事件
14         setListener();
15     }
16
17     private void setListener() {
18         //添加事件
19         mBtnPull.setOnClickListener(myListener);
20     }
21
22     private void findView() {
23         //绑定控件
24         mTvShow = (TextView) findViewById(R.id.textview);
25         mBtnPull = (Button) findViewById(R.id.btnPull);
26     }
27
28     private OnClickListener myListener = new OnClickListener() {
29
30         @Override
31         public void onClick(View v) {
32             //1、获取当前工程中的 people.xml 的文件流
33             InputStream inputStream = MainActivity.class.getClassLoader()
34                     .getResourceAsStream("people.xml");
35             //定义 Person 对象
36             List<Person> persons = null;
37             mTvShow.setText("");                    //清空文本显示
38             switch (v.getId()) {                    //根据单击的按钮，进行 PULL 解析
39             case R.id.btnPull:  //PULL 解析
40                 //使用 PullPersonService 服务进行 xml 解析
41                 mTvShow.setText("PULL:");
42                 try { //通过 PULL 方式解析 xml 文件
43                     persons = PullPersonService.readXml(inputStream);
44                 } catch (XmlPullParserException e) {
45                     //TODO Auto-generated catch block
46                     e.printStackTrace();
47                 } catch (IOException e) {
48                     //TODO Auto-generated catch block
49                     e.printStackTrace();
```

```
50              }
51              break;
52          default:
53              break;
54          }
55          //显示解析结果
56          for (Person person : persons) {
57              mTvShow.setText(mTvShow.getText().toString() + "\n" +
                    person.toString());
58          }
59      }
60  };
61  }
```

此文件是 Activity 的代码文件,在其中第 12 行得到页面的布局控件,第 14 行设置相应监听器,当用户单击解析按钮时,调用第 33～50 行的代码,首先得到系统的 xml 文件的内容,然后定义 persons 类表对象,然后第 43 行将需要解析的 XML 字符串传入 PullPersonService 类中进行解析,得到结果。最后在 TextView 中显示解析结果。

本例还有一个 XML 的解析类,新建 src/com.wyl.example/ PullPersonService.java 文件,代码如下:

```
01  //Android 中 Pull 解析 xml 的类
02  public class PullPersonService {
03      public static List<Person> readXml(InputStream inStream)
04              throws XmlPullParserException, IOException {
05          //定义结果 List
06          List<Person> persons = null;
07          //创建 pull 解析类
08          XmlPullParser parser = Xml.newPullParser();
09          //设置数据读取格式
10          parser.setInput(inStream, "UTF-8");      //解析的数据格式
11          //得到每个节点的类型
12          int eventCode = parser.getEventType();
13          Person person = null;
14          while (eventCode != XmlPullParser.END_DOCUMENT) {
15              switch (eventCode) {
16              case XmlPullParser.START_DOCUMENT://文档开始
17                  //初始化结果链表
18                  persons = new ArrayList<Person>();
19                  break;
20              case XmlPullParser.START_TAG://开始元素
21                  //解析节点内容
22                  if("person".equals(parser.getName())){
23                      //初始化一个 person 对象
24                      person = new Person();
25                      //得到属性,设置 id
26                      person.setId(new Integer(parser.getAttributeValue(0)));
27                  }else if(null != person){
28                      //解析 person 的属性字段
29                      if("name".equals(parser.getName())){  //姓名
30                          person.setName(parser.nextText());
31                      }else if("age".equals(parser.getName())){  //年龄
32                          person.setAge(new Short(parser.nextText()));
33                      }
34                  }
```

```
35              break;
36          case XmlPullParser.END_TAG:
37              //文档结束
38              if("person".equals(parser.getName())&&person != null){
39                  persons.add(person);
40                  person = null;
41              }
42              break;
43          }
44          //移动下一个节点
45          eventCode = parser.next();
46      }
47      return persons;
48  }
49 }
```

此类定义了 Pull 类解析 XML 的方法，其中主要就是 readXML 方法，传入 InputStream 流，然后通过 XmlPullparser 得到解析对象，然后解析 inStream 流。Pull 的解析方式和 SAX 有些类似，也是基于事件的解析，当文档开始时会调用第 16 行的分支，当开始解析一个元素时，会调用第 20 行的分支，当节点结束时会调用第 36 行的分支，这样依次遍历 XML 文档中的所有节点取得想要的数据。

4. 实例扩展

PULL 方式解析 XML 文件的特点和 SAX 方式类似，而且 Android 提供的解析 xml 文档的方式也是官方推荐的解析方式。

范例 181 内置 JSON 解析

1. 实例简介

XML 是相对比较成熟的一种数据传输格式，但是其有一个缺点就是标签会成对出现，这无形中增大了数据传输的流量。近些年来网络上比较流行另一种传输格式——JSON，它的作用和 XML 是一样的，就是为了服务器和客户端进行数据传输。Android 中也自带了 JSON 的解析类。本例子就带领大家来实现一个使用 Android 内置类来解析 JSON 的实例效果。

2. 运行效果

该实例运行效果如图 8.13 所示。

图 8.13 内置 JSON 解析

3. 实例程序讲解

在本实例中,提供了五个按钮,分别通过代码来实现五种基本的 JSON 格式的字符串。想要实现本实例效果,首先修改 res/layout/activity_main.xml 文件,构造如图 8.13 效果的布局。代码省略。

然后修改 src/com.wyl.example/MainActivity.java 文件,代码如下:

```java
001 //定义了本实例的主要 Activity
002 public class MainActivity extends Activity {
003     //初始化需要解析的 JSON 字符串
004     private String json1 = "{\"url\":\"http://www.baidu.com\"}";
005     private String json2 = "{\"name\":\"android\",\"version\":\"2.3.1\"}";
006     private String jsonArray1 = "{\"namber\":[1,2,3]}";
007     private String jsonArray2 ="{\"number\":[[zhangsan,18],[lisi,20],[wangwu,23]]}";
008     private String jsonObj = "{\"mobile\":[{\"name\":\"android\"},{\"name\":\"iphone\"}]}";
009     //省略控件对象定义代码
010
011     @Override
012     protected void onCreate(Bundle savedInstanceState) {
013         super.onCreate(savedInstanceState);
014         setContentView(R.layout.activity_main);
015         //得到布局中的控件
016         findView();
017         //绑定控件事件
018         setListener();
019     }
020
021     //省略 findview 和 setlistener 方法的定义
022
023     private OnClickListener mylistener = new OnClickListener() {
024
025         @Override
026         public void onClick(View v) {
027             //TODO Auto-generated method stub
028             switch (v.getId()) {
029             case R.id.btn1: //解析一个属性的对象
030                 //定义 JSONObject 对象
031                 JSONObject demoJson;
032                 String url;
033                 try {
034                     //直接解析字符串得到 JSONObject 对象
035                     demoJson = new JSONObject(json1);
036                     //得到具体字段
037                     url = demoJson.getString("url");
038                     tv.setText("url:" + url);
039                 } catch (JSONException e) {
040                     //TODO Auto-generated catch block
041                     e.printStackTrace();
042                 }
043                 break;
044             case R.id.btn2: //解析多个属性的对象
045                 //定义 JSONObject 对象
046                 JSONObject demoJson2;
```

```
047         try {
048             //定义JSONObject对象
049             demoJson2 = new JSONObject(json2);
050             //得到相应的字段值
051             String name = demoJson2.getString("name");
052             String version = demoJson2.getString("version");
053             tv.setText("name:" + name + ",version" + version);
054
055         } catch (JSONException e) {
056             //TODO Auto-generated catch block
057             e.printStackTrace();
058         }
059         break;
060     case R.id.btn3: //解析一个普通数组
061         //定义JSONObject对象
062         JSONObject demoJson3;
063         try {
064             //定义解析json字符串得到JSONObject对象
065             demoJson3 = new JSONObject(jsonArray1);
066             //得到数组
067             JSONArray numberList = demoJson3.getJSONArray
                ("number");
068             tv.setText("");
069
070             for (int i = 0; i < numberList.length(); i++) {
071                 //因为数组中的数据类型是int 所以用getInt
072                 //其他getString ,getLong
073                 tv.setText(tv.getText() + "\n" + numberList.
                    getInt(i));
074             }
075
076         } catch (JSONException e) {
077             //TODO Auto-generated catch block
078             e.printStackTrace();
079         }
080
081         break;
082     case R.id.btn4://解析一个嵌套数组
083         //嵌套数组的遍历
084         JSONObject demoJson4;
085         try {
086             //解析json字符串
087             demoJson4 = new JSONObject(jsonArray2);
088             //得到嵌套数组
089             JSONArray numberList = demoJson4.getJSONArray
                ("number");
090             tv.setText("");
091             for (int i = 0; i < numberList.length(); i++) {
092                 //获取数组中的数据
093                 tv.setText(tv.getText() + "\n"
094                         + numberList.getJSONArray(i).
                            getString(0)
095                         + ";"
096                         + numberList.getJSONArray(i).
                            getInt(1));
097             }
098
099         } catch (JSONException e) {
100             //TODO Auto-generated catch block
```

```
101                         e.printStackTrace();
102                     }
103
104                     break;
105             case R.id.btn5:  //解析一个对象数组
106                 //定义JSONObject对象
107                 JSONObject demoJson5;
108                 try {
109                     //定义解析xml为JSONObject
110                     demoJson5 = new JSONObject(jsonObj);
111                     //得到数组对象
112                     JSONArray numberList = demoJson5.getJSONArray
                        ("mobile");
113                     tv.setText("");
114                     for (int i = 0; i < numberList.length(); i++) {
115                         tv.setText(tv.getText() + "\n"
116                                 + numberList.getJSONObject(i).
                                    getString("name"));
117                     }
118                 } catch (JSONException e) {
119                     //TODO Auto-generated catch block
120                     e.printStackTrace();
121                 }
122                 break;
123             default:
124                 break;
125             }
126         }
127     };
128 }
```

此文件是 Activity 的代码文件，在其中第 4～8 行定义了一系列需要解析的 JSON 字符串，在第 16 行得到了布局中的所有控件对象，在第 18 行设置了对应的监听器。然后主要的解析代码在第 23 行定义的监听器中，解析 JSON 对象的代码在第 31～42 行，首先通过 JSON 字符串得到 JSONObject 对象，然后再通过此对象的 getString 方法得到对应的字段值。解析多个 JSON 字段的方法同第一种方法类似，就是根据属性的名得到多个属性值即可。如果需要解析 JSON 数组，则使用第 62～79 行的代码，首先得到一个 JSONObject 对象，然后通过 getJSONArray 方法得到数组，然后遍历数组中的每个元素即可。同理，解析对象数组的代码在第 106～121 行，思路也是得到一个 JSONObject 对象，然后得到数组列表，遍历得到每个属性的值。

4．实例扩展

Android 内置的解析 JSON 的类库已经比较完善，可是对于对象的解析，需要根据每个字段的名称解析，然后放置在对应的对象属性中。

范例 182 Gson 解析 JSON

1．实例简介

由于 JSON 是现在比较流行的数据传输格式，所以 Google 开发了对应的解析库来解析 JSON，即 Gson。它的最大特点就是可以直接解析 JSON 对象，也可以把一个对象直接变成 JSON，这就简化了我们代码的复杂度。遇到这样的功能，我们一般是当用户进行某个操作的时候，例如，单击某个按钮的时候，使用 Gson 来解析 JSON。本例子就带领大家来实现一个使用 Gson 解析 JSON 的实例效果。

2. 运行效果

该实例运行效果如图 8.14 所示。

图 8.14 Gson 解析 JSON

3. 实例程序讲解

在本实例中，显示了一个 TextView 的结果显示框，然后定义了四个按钮，单击不同按钮的时候，在代码中进行不同格式的解析。想要实现本实例效果，首先修改 res/layout/activity_main.xml 文件，布局如图 8.14 所示。代码省略。

然后修改 src/com.wyl.example/MainActivity.java 文件，代码如下：

```
01  //定义了本实例的主要 Activity Gson
02  public class MainActivity extends Activity {
03      //省略控件对象的代码
04
05      //定义 Gson 的对象
06      private Gson gson = new Gson();
07
08      @Override
09      protected void onCreate(Bundle savedInstanceState) {
10          super.onCreate(savedInstanceState);
11          setContentView(R.layout.activity_main);
12
13          //得到布局中的控件
14          findView();
15          //绑定控件事件
16          setListener();
17      }
18
19      //省略 findview 和 setlistener 方法的定义
20
21      private OnClickListener mylistener = new OnClickListener() {
22
```

```
23      @Override
24      public void onClick(View v) {
25          //TODO Auto-generated method stub
26          switch (v.getId()) {
27          case R.id.btn1: //解析一个属性的对象
28              /**
29               * 将给定的 JSON 字符串转换成指定的类型对象
30               */
31              String json = "{\"name\":\"Wyl\",\"age\":30}";
32              Person person = gson.fromJson(json, Person.class);
33              tv.setText(person.toString());
34              break;
35          case R.id.btn2: //解析多个属性的对象
36              /**
37               * 将给定的目标对象转换成 JSON 格式的字符串
38               */
39              Person person1 = new Person("test", 20);
40
41              String json_Person = gson.toJson(person1);
42              tv.setText(json_Person);
43              break;
44          case R.id.btn3: //解析一个普通数组
45              /**
46               * 将给定的集合对象转换成 JSON 格式的字符串
47               */
48              ArrayList<Person> persons = new ArrayList<Person>();
49              Collections.addAll(persons, new Person("tom", 10),
50                  new Person(   "jon", 20));
51              String json_list = gson.toJson(persons);
52              tv.setText(json_list);
53              break;
54          case R.id.btn4://解析一个嵌套数组
55              /**
56               * 将给定的 JSON 格式字符串转换为带泛型的集合对象
57               */
58              String s = "[{\"name\":\"tom\",\"age\":10},
59                  {\"name\":\"jon\",\"age\":20}]";
                //定义解析的对象 list
60              List<Person> retList = gson.fromJson(s,
61                      new TypeToken<List<Person>>() {
62                      }.getType());
63              String t = "";
                //循环打印链表中的值
64              for (Person p : retList) {
65                  t += p.toString();
66              }
67              tv.setText(t);
68              break;
69          default:
70              break;
71          }
72      }
73  };
74  }
```

此文件是 Activity 的代码文件,在其中第 6 行定义了 Gson 的对象,在第 14 行得到布局中的所有控件。在第 16 行设置了所有对象的监听器对象,主要代码在第 21 行定义的监听器对象中，等需要把 JSON 字符串解析成一个对象的时候，调用第 31～33 行，首先根据需要解析的 JSON 字符串和解析的结果类，传入 gson 的 fromJson 方法中，结果就可以直接解析出对象了，解析的过程是根据 Person 类中的每个属性名来进行解析对应的值，如果用户希望直接把一个对象转换成 JSON 字符串，那就用第 41～42 行通过 gson 的 toJson 方法直接得到对象对应的 json 字符串了。同样道理，也可以直接把数组转换为 JSON 字符串。代码在第 48～52 行，也可以把 JSON 数组字符串转换为数组对象，代码在第 58～66 行。

4．实例扩展

Gson 为 Google 的开源库，所以在使用时候需要引入外部的 Jar 包，具体的实现方法及相关的 API 文档在 https://code.google.com/p/google-gson/。

8.3 小　　结

在本章节中主要介绍了 Android 中网络访问和数据解析这两部分内容。网络访问是为了把网络上的数据请求下来，数据解析是为了把请求下来的数据进行解析并进行处理和展示。本章的内容在实际项目开发中使用频率很高，它可以使用户随时随地与网络进行交互。例如，进行用户登录注册或进行微博分享等，这也就加大了我们应用的使用场景，增加用户体验。下一章我们会讲述在 Android 中常见的多媒体应用的开发。

第 9 章　Android 中的多媒体开发

上一章了解了 Android 中网络编程及数据解析的内容。有了上一章的内容讲解，大家基本上可以创建大多数的 Android 客户端应用了。其实对于我们一般的应用来说，之前的章节已经足够你制作一个近乎完美的应用了，但是基于手机应用平台的特点，Android 有一些与硬件相关的应用有时候也会被使用在应用中。例如，用户希望直接从摄像头得到程序头像，用户希望从手机的蓝牙中得到文件等。那么本章就给大家介绍 Android 中常见硬件相关的多媒体应用开发。

本章内容主要分为两部分：

第一部分是系统多媒体应用开发，这部分主要涉及系统的音频、视频和摄像头等系统硬件相关的应用开发。

第二部分是桌面插件开发，这部分主要是各种桌面插件的使用和开发。例如，桌面的联系人插件和桌面的时间插件等。

本章主要通过各种实例来介绍 Android 中常用的多媒体应用开发。希望读者阅读完本章内容后，可以在自己的应用程序中多加入一些硬件相关的小功能，增加自己应用的用户体验。

9.1　Android 中多媒体应用开发

范例 183　屏幕方向改变

1．实例简介

我们在使用一些应用的时候会进行屏幕方向的改变。例如，在一些视频播放器里面默认竖屏半屏播放，单击全屏播放按钮，屏幕方向旋转横屏播放。我们通过本实例带领大家一起来制作一个单击实现屏幕旋转的应用。

2．运行效果

该实例运行效果如图 9.1 所示。

图 9.1　控制屏幕方向改变

3. 实例程序讲解

在上例中提供了一个改变屏幕显示方向的按钮，当用户单击此按钮的同时屏幕旋转。想要实现如上效果，首先修改我们建立的工程下的 res/layout/activity_main.xml，添加图 9.1 所示的的控件。代码省略。

然后修改 src/com.wyl.example/MainActivity.java 文件，代码如下：

```
01  //定义了本实例的主要 Activity
02  public class MainActivity extends Activity {
03      //定义屏幕改变的按钮
04      private Button btnChangeScreen;
05
06      @Override
07      public void onCreate(Bundle savedInstanceState) {
08          super.onCreate(savedInstanceState);
09          setContentView(R.layout.activity_main);
10          //得到布局中的所有对象
11          findView();
12          //设置对象的监听器
13          setListener();
14      }
15      //省略 findview 和 setlistener 方法定义
16
17      OnClickListener listener = new OnClickListener() {
18
19          @Override
20          public void onClick(View v) {
21              //TODO Auto-generated method stub
22              switch (v.getId()) {
23
24              //屏幕变换的按钮
25              case R.id.btn_orientation:
26                  //无法进行画面的旋转
27                  if (MainActivity.this.getRequestedOrientation()
28                      == ActivityInfo.SCREEN_ORIENTATION_UNSPECIFIED) {
29                      Toast.makeText(MainActivity.this,
30                      "无法改变屏幕方向",
31                          Toast.LENGTH_SHORT).show();
32                  } else {
33                      //变为竖屏显示
34                      if (MainActivity.this.getRequestedOrientation()
35                          == ActivityInfo.SCREEN_ORIENTATION_
                            LANDSCAPE) {
36                          Toast.makeText(MainActivity.this,
37                          "已改变为竖屏方向",
38                              Toast.LENGTH_SHORT).show();
39                          setRequestedOrientation(
                            ActivityInfo.SCREEN_ORIENTATION_PORTRAIT);
40                          //变为横屏显示
41                      } else if (MainActivity.this.getRequested
                        Orientation()
42                          == ActivityInfo.SCREEN_ORIENTATION_PORTRAIT) {
43                          Toast.makeText(MainActivity.this, "已改变为横屏
                            方向",
44                              Toast.LENGTH_SHORT).show();
45                          setRequestedOrientation(
```

```
46                    }
47                }
48                break;
49            }
50        }
51    };
52 }
```

此文件为当前 Activity 的实现代码，在第 11 行得到了布局中的所有控件，在第 13 行设置了所有的监听器，在第 17 行定义了单击监听器，当用户单击按钮的时候，判断当前的屏幕方向，如果当前为横屏，就切换为竖屏，如果当前为竖屏，则切换为横屏。

4．实例扩展

由于此实例是通过按钮的单击来实现横屏和竖屏的切换的，当前在我们的手机中也可以通过手机的旋转来实现，这会用到我们后面的传感器相关的实例。

范例 184　调用系统相机拍照

1．实例简介

在我们使用应用的过程中，经常会启动系统照相应用的功能。例如，用户单击启动相机并且进行拍照显示。遇到这样的功能，我们一般是当用户进行某个操作的时候，例如，单击某个按钮的时候，发送 Intent 给系统的照相机应用进行拍照。本例子就带领大家来实现一个通过系统的照相机进行拍照的实例。

2．运行效果

该实例运行效果如图 9.2 所示。

图 9.2　调用系统相机拍照网页

3. 实例程序讲解

在本实例中，提供一个按钮，单击启动系统照相机进行拍照，然后在下面的 ImageView 中显示拍照结果。想要实现本实例效果，首先修改 res/layout/activity_main.xml 文件，布局成图 9.2 的样式。代码省略。

然后修改 src/com.wyl.example/MainActivity.java 文件，代码如下：

```
01  //定义了本实例的主要 Activity
02  public class MainActivity extends Activity {
03      //省略控件定义及 findview 和 setlistener 代码
04
05      @Override
06      public void onCreate(Bundle savedInstanceState) {
07          super.onCreate(savedInstanceState);
08          setContentView(R.layout.activity_main);
09          //得到布局中的所有对象
10          findView();
11          //设置对象的监听器
12          setListener();
13      }
14
15  OnClickListener listener = new OnClickListener() {
16
17      @Override
18      public void onClick(View v) {
19          //TODO Auto-generated method stub
20          switch (v.getId()) {
21          //拍照的按钮
22          case R.id.btn_takephoto:
23              //启动系统拍照
24              cameraMethod();
25              break;
26          }
27      }
28  };
29
30  //照相功能
31  private void cameraMethod() {
32      //实例化拍照的 Intent
33      Intent imageCaptureIntent =
34          new Intent(MediaStore.ACTION_IMAGE_CAPTURE);
35      //设置图片存放的路径
36      strImgPath = Environment.getExternalStorageDirectory().toString()
37              + "/CONSDCGMPIC/";//存放照片的文件夹
38      //给相片命名
39      String fileName = new SimpleDateFormat("yyyyMMddHHmmss")
40              .format(new Date()) + ".jpg";//照片命名
41      //检查存放的路径是否存在，如果不存在则创建目录
42      out = new File(strImgPath);
43      if (!out.exists()) {
44          out.mkdirs();
45      }
46      //在此目录下创建文件
47      out = new File(strImgPath, fileName);
48      //该照片的绝对路径
```

```
49              strImgPath = strImgPath + fileName;
50              //启动 ACITIVITY
51              startActivityForResult(imageCaptureIntent, RESULT_CAPTURE_
                IMAGE);
52      }
53
54      @Override
55      protected void onActivityResult(int requestCode, int resultCode, Intent
        data) {
56          super.onActivityResult(requestCode, resultCode, data);
57          switch (requestCode) {
58          case RESULT_CAPTURE_IMAGE://拍照
59              //如果返回为正确的结果
60              if (resultCode == RESULT_OK) {
61                  //intent.getExtras()得到intent所附带的额外数据,在这也就是
                        拍摄的图片
62                  Bundle extras = data.getExtras();
63                  //得到额外的数据的data字段,转化为bitmap类型
64                  Bitmap b = (Bitmap) extras.get("data");
65                  //实例化矩阵 Matrix
66                  Matrix matrix = new Matrix();
67                  //设置缩放
68                  matrix.postScale(5f, 4f);
69                  //创建bitmap对象,并设置bitmap的参数
70                  b = Bitmap.createBitmap(b, 0, 0, b.getWidth(), b.
                    getHeight(),
71                      matrix, true);
72                  //设置imageview的图片资源
73                  ivSurface.setImageBitmap(b);
74                  try {
75                      //把文件转化为outputstream
76                      FileOutputStream outStream = new FileOutputStream
                        (out);
77                      //把bitmap数据写入字符流中
78                      b.compress(CompressFormat.JPEG, 100, outStream);
79                      //关闭字符流
80                      outStream.close();
81                  } catch (Exception e) {
82                      e.printStackTrace();
83                  }
84              }
85          break;
86          }
87      }
88  }
```

此文件是 Activity 的代码文件,在其中第 10 行得到布局中的所有控件,第 13 行设置所有的监听器,当用户单击拍照按钮的时候,调用第 31~52 行的启动系统拍照按钮,在其中首先定义了一个 Intent 对象,然后在本地创建文件保存拍照的文件,再启动 Activity 进行拍照。在第 55~86 行,接收拍照后的返回结果,得到的 data 对象中包含了拍照的图片对象,将此对象写入本地文件,然后设置给 ImageView 进行显示。

4. 实例扩展

在此实例中主要是通过 Intent 启动系统的拍照界面,然后得到图片,当然系统的拍照界面还可以进行更多的设置,如闪光灯和曝光效果等。

范例 185 录音机

1. 实例简介

在 Android 的系统中一般都会有录音的功能。这样的功能，我们一般是当用户进行某个操作的时候，例如，单击某个按钮的时候，是调用系统的录音类来实现的。本例子就带领大家来实现一个系统的录音机的实例。

2. 运行效果

该实例运行效果如图 9.3 所示。

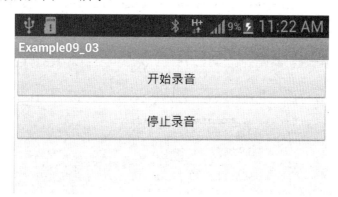

图 9.3 录音机

3. 实例程序讲解

在本实例中，提供了两个按钮，功能分别是开始录音和停止录音，在录音完毕后会在本地进行保存。想要实现本实例效果，首先修改 res/layout/activity_main.xml 文件，布局如图 9.3 所示。代码省略。

然后修改 src/com.wyl.example/MainActivity.java 文件，代码如下：

```
01  //定义了本实例的主要 Activity
02  public class MainActivity extends Activity {
03      //省略控件定义代码
04      //声明 MediaRecorder 对象
05      private MediaRecorder mr;
06  
07      @Override
08      public void onCreate(Bundle savedInstanceState) {
09          //省略
10      }
11  
12      private void setListener() {
13          //录音按钮单击事件
14          recordButton.setOnClickListener(new View.OnClickListener() {
15              @Override
16              public void onClick(View v) {
17                  //设置文件的路径
```

```java
18            File file = new File("/sdcard/"
19                    + "YY"
20                    + new DateFormat().format("yyyyMMdd_hhmmss",
21                        Calendar.getInstance(Locale.CHINA)) +
                        ".amr");
22
23            Toast.makeText(getApplicationContext(),
24                    "正在录音,录音文件在" + file.getAbsolutePath(),
25                    Toast.LENGTH_LONG).show();
26
27            //创建录音对象
28            mr = new MediaRecorder();
29            //从麦克风源进行录音
30            mr.setAudioSource(MediaRecorder.AudioSource.DEFAULT);
31            //设置输出格式
32            mr.setOutputFormat(MediaRecorder.OutputFormat.DEFAULT);
33            //设置编码格式
34            mr.setAudioEncoder(MediaRecorder.AudioEncoder.DEFAULT);
35            //设置输出文件
36            mr.setOutputFile(file.getAbsolutePath());
37
38            try {
39                //创建文件
40                file.createNewFile();
41                //准备录制
42                mr.prepare();
43            } catch (IllegalStateException e) {
44                e.printStackTrace();
45            } catch (IOException e) {
46                e.printStackTrace();
47            }
48
49            //开始录制
50            mr.start();
51            recordButton.setText("录音中……");
52        }
53    });
54
55    //停止按钮单击事件
56    stopButton.setOnClickListener(new View.OnClickListener() {
57        @Override
58        public void onClick(View v) {
59
60            if (mr != null) {
61                //停止录音
62                mr.stop();
63                //释放所占的资源
64                mr.release();
65                mr = null;
66                recordButton.setText("录音");
67                Toast.makeText(getApplicationContext(), "录音完毕",
68                        Toast.LENGTH_LONG).show();
69            }
70        }
71    });
72  }
73 }
```

此文件是 Activity 的代码文件,在其中第 5 行定义了一个 MediaRecorder 对象,当用户单击开始录音的时候调用第 18~51 行代码，首先定义一个 File 类的对象，然后设置 mr 的属性，包括输入输出格式等。然后调用 mr 的 start 方法开始录制，当录制完毕后单击停止录制按钮，调用第 60~69 行的代码停止录制并且释放录音器的资源。

4．实例扩展

在此实例中主要使用了系统的 MediaRecorder 类，在此类中进行录音，当然还可以进行一些其他的录音设置，如录音的频率和录音的输出格式等。

范例 186　录像机

1．实例简介

在 Android 系统中的照相机不但可以照相，而且可以进行视频的录制。这样的功能，我们一般是当用户进行某个操作的时候，例如，单击某个按钮的时候，开始进行了录制，然后再次单击停止录制。本例子就带领大家来实现一个录像机的实例。

2．运行效果

该实例运行效果如图 9.4 所示。

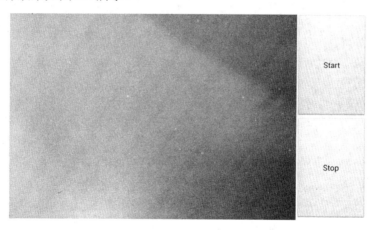

图 9.4　录像机

3．实例程序讲解

在本实例中，提供了两个按钮，用来开始录制视频和停止录制视频，还有一个 View 控件用来显示录制的效果。想要实现本实例效果，首先修改 res/layout/activity_main.xml 文件。代码省略。

然后修改 src/com.wyl.example/MainActivity.java 文件，核心代码如下：

```
001    //定义了本实例的主要 Activity
002    public class MainActivity extends Activity implements
       SurfaceHolder.Callback {
003        //布局控件代码省略
004        //录制视频的类
```

```
005     private MediaRecorder mediarecorder;
006     //显示视频的控件
007     private SurfaceView surfaceview;
008     //用来显示视频的一个接口
009     private SurfaceHolder surfaceHolder;
010
011     public void onCreate(Bundle savedInstanceState) {
012         super.onCreate(savedInstanceState);
013         //去掉标题栏
014         requestWindowFeature(Window.FEATURE_NO_TITLE);
015         //设置全屏
016         getWindow().setFlags(WindowManager.LayoutParams.FLAG_FULLSCREEN,
017                 WindowManager.LayoutParams.FLAG_FULLSCREEN);
018         //设置横屏显示
019         setRequestedOrientation(ActivityInfo.SCREEN_ORIENTATION_LANDSCAPE);
020         //选择支持半透明模式,在有 surfaceview 的 activity 中使用
021         getWindow().setFormat(PixelFormat.TRANSLUCENT);
022         setContentView(R.layout.activity_main);
023         //初始化控件
024         init();
025     }
026
027     //得到布局中的所有对象
028     private void init() {
029         surfaceview = (SurfaceView) this.findViewById(R.id.surfaceview);
030         //取得 holder
031         SurfaceHolder holder = surfaceview.getHolder();
032         //holder 加入回调接口
033         holder.addCallback(this);
034         //setType 必须设置
035         holder.setType(SurfaceHolder.SURFACE_TYPE_PUSH_BUFFERS);
036     }
037
038     class TestVideoListener implements OnClickListener {
039
040         @Override
041         public void onClick(View v) {
042             if (v == start) {
043                 //创建 mediarecorder 对象
044                 mediarecorder = new MediaRecorder();
045                 //设置录制视频源为 Camera(相机)
046                 mediarecorder.setVideoSource(MediaRecorder.VideoSource.CAMERA);
047                 //设置录制完成后视频的封装格式 THREE_GPP 为 3gp.MPEG_4 为 mp4
048                 mediarecorder
049                     .setOutputFormat(MediaRecorder.OutputFormat.THREE_GPP);
050                 //设置录制的视频编码 h263 h264
051                 mediarecorder.setVideoEncoder(MediaRecorder.VideoEncoder.H264);
052                 //设置视频录制的分辨率。必须放在设置编码和格式的后面,否则报错
053                 mediarecorder.setVideoSize(176, 144);
054                 //设置录制的视频帧率。必须放在设置编码和格式的后面,否则报错
055                 mediarecorder.setVideoFrameRate(20);
056                 //设置预览
```

```
057                 mediarecorder.setPreviewDisplay(surfaceHolder.
                        getSurface());
058                 //设置视频文件输出的路径
059                 mediarecorder.setOutputFile("/sdcard/wy1.3gp");
060                 try {
061                     //准备录制
062                     mediarecorder.prepare();
063                     //开始录制
064                     mediarecorder.start();
065                 } catch (IllegalStateException e) {
066                     //TODO Auto-generated catch block
067                     e.printStackTrace();
068                 } catch (IOException e) {
069                     //TODO Auto-generated catch block
070                     e.printStackTrace();
071                 }
072             }
073             if (v == stop) {
074                 if (mediarecorder != null) {
075                     //停止录制
076                     mediarecorder.stop();
077                     //释放资源
078                     mediarecorder.release();
079                     mediarecorder = null;
080                 }
081             }
082         }
083     }
084
085     @Override
086     public void surfaceChanged(SurfaceHolder holder, int format, int width,
087             int height) {
088         //这个holder为开始在oncreat里面取得的holder，将它赋给surfaceHolder
089         surfaceHolder = holder;
090     }
091
092     @Override
093     public void surfaceCreated(SurfaceHolder holder) {
094         //这个holder为开始在oncreat里面取得的holder，将它赋给surfaceHolder
095         surfaceHolder = holder;
096     }
097
098     @Override
099     public void surfaceDestroyed(SurfaceHolder holder) {
100         //surfaceDestroyed的时候同时对象设置为null
101         surfaceview = null;
102         surfaceHolder = null;
103         mediarecorder = null;
104     }
105 }
```

此文件是 Activity 的代码文件，在其中第 5 行定义了一个 MediaRecorder 对象，第 7 行定义了显示视频窗口的 View 对象，在 onCreate 方法中设置当前界面的显示样式。当用户单击视频录制按钮调用，第 44～72 行代码开始视频录制，首先设置视频源和设置视频的分辨率等信息，设置此处文件为 sdcard 上的固定文件，然后调用 prepare 方法开始录制，当用户单击停止的时候，调用 stop 方法停止录制，并且释放资源。

4．实例扩展

在此实例中仍然使用 MediaRecorder 类进行录制，这里需要注意一点就是在录制过程中要保证 SDCard 上有足够的可用空间存储录制的视频。

范例 187　手电筒应用

1．实例简介

在一些 Android 的系统中，会自带手电筒应用，其目的就是为了在晚上可以进行照亮。这样的功能，我们一般是当用户进行某个操作的时候，例如：单击某个按钮的时候，打开系统的闪光灯。本例子就带领大家来实现一个手电筒应用的实例。

2．运行效果

该实例运行效果如图 9.5 所示。

图 9.5　手电筒应用

3．实例程序讲解

在本实例中，提供一个开启闪光灯按钮。想要实现本实例效果，首先修改 res/layout/activity_main.xml 文件。代码省略。

然后修改 src/com.wyl.example/MainActivity.java 文件，代码如下：

```
01  //定义了本实例的主要 Activity
02  public class MainActivity extends Activity {
03      //声明 Camera 类
04      private Camera m_Camera;
05
06      @Override
07      public void onCreate(Bundle savedInstanceState) {
08          //代码省略
09      }
10
11      OnClickListener listener = new OnClickListener() {
12
```

```java
13      @Override
14      public void onClick(View v) {
15          //TODO Auto-generated method stub
16          switch (v.getId()) {
17          //打开/关闭手电筒
18          case R.id.btn_open:
19              if (isSupposeFlashLight()) {
20                  if (null == m_Camera) {
21                      //实例化camera
22                      m_Camera = Camera.open();
23                      //得到Camera.Parameters对象
24                      Camera.Parameters parameters = m_Camera.getPara
                        meters();
25                      //调用设置闪光灯
26                      parameters
27                          .setFlashMode(Camera.Parameters.FLASH_MODE_TORCH);
28                      //设置参数
29                      m_Camera.setParameters(parameters);
30                      //开启闪光灯
31                      m_Camera.startPreview();
32                      btnLight.setText("关闭手电筒");
33                      Toast.makeText(MainActivity.this, "手电筒已打开",
34                              Toast.LENGTH_SHORT).show();
35                  } else {
36                      //关闭闪光灯
37                      m_Camera.stopPreview();
38                      //释放所占的资源
39                      m_Camera.release();
40                      m_Camera = null;
41                      btnLight.setText("打开手电筒");
42                      Toast.makeText(MainActivity.this, "手电筒已关闭",
43                              Toast.LENGTH_SHORT).show();
44                  }
45              } else {
46                  Toast.makeText(MainActivity.this,"您的设备不支持闪光灯",
47                          Toast.LENGTH_SHORT).show();
48              }
49              break;
50
51          }
52      }
53  };
54
55  //判断是否支持摄像头设备
56  public Boolean isSupposeFlashLight() {
57      PackageManager pm = this.getPackageManager();
58      FeatureInfo[] features = pm.getSystemAvailableFeatures();
59      for (FeatureInfo f : features) {
60          if (PackageManager.FEATURE_CAMERA_FLASH.equals(f.name))
            //判断设备是否支持闪光灯
61          {
62              return true;
63          }
64      }
65      return false;
66  }
67  }
```

此文件是 Activity 的代码文件, 在其中第 4 行定义了 Camera 类的对象, 用来得到系统的摄像头。当用户单击打开手电筒时,调用第 20~31 行,首先判断系统是否支持摄像头功能,然后通过 Camera 的设置 FlashMode 的方法开启闪光灯,当用户单击按钮关闭闪光灯时,调用第 37~43 行代码,首先停止预览,然后清除占用的资源。第 56~66 行定义了 isSupposeFlashLight 方法,判断是否支持摄像头设备。

4. 实例扩展

在此实例中使用系统的 Camera 类,其中有闪光灯的模式,当然还有其他闪光模式,如间断频闪等。

范例 188 计时器

1. 实例简介

我们在日常生活中手机已经可以替代我们的一些常见的设备了。例如,音乐播放器、视频播放器和手电筒等,我们平时还常用一种设备叫做秒表,其最主要的功能就是计时。对于这样的功能,我们一般是当用户进行某个操作的时候,例如,单击某个按钮的时候,开始计时,再次单击按钮的时候,停止计时,得到一个时间差。本例子就带领大家来实现一个计时器的实例。

2. 运行效果

该实例运行效果如图 9.6 所示。

图 9.6 计时器

3. 实例程序讲解

在本实例中,定义了一个需要设置的计时时间输入框,还有一个计时时间显示框,还有两个按钮用来开始计时和重置计时器。想要实现本实例效果,首先修改

res/layout/activity_main.xml 文件。代码省略。

然后修改 src/com.wyl.example/MainActivity.java 文件，代码如下：

```java
01  //定义主要 Activity 的实现代码
02  public class MainActivity extends Activity {
03      //设置初始时间
04      private int startTime = 0;
05      //声明计时器 Chronomete 对象
06      private Chronometer chronometer;
07      //开始 boolOperate=true,停止 boolOperate=false
08      private Boolean boolOperate = true;
09
10      public void onCreate(Bundle savedInstanceState) {
11          //省略代码
12      }
13
14      //得到布局中的所有对象
15      private void setListener() {
16          //TODO Auto-generated method stub
17          btnRest.setOnClickListener(listener);
18          btnOperate.setOnClickListener(listener);
19          chronometer
20                  .setOnChronometerTickListener(new Chronometer.
                    OnChronometerTickListener() {
21                      @Override
22                      public void onChronometerTick(Chronometer chronometer){
23                          //如果开始计时到现在超过了 startime 秒
24                          if (SystemClock.elapsedRealtime()
25                                  - chronometer.getBase() > startTime *
                                    1000) {
26                              //计时器停止
27                              chronometer.stop();
28                              //给用户提示
29                              showDialog();
30                          }
31                      }
32                  });
33      }
34
35  OnClickListener listener = new OnClickListener() {
36
37      @Override
38      public void onClick(View v) {
39          //TODO Auto-generated method stub
40          switch (v.getId()) {
41              //开始按钮
42              case R.id.btntimeopp:
43                  //如果 boolOperate==true,说明计时开始
44                  if (boolOperate) {
45                      String ss = edtSetTime.getText().toString();
46                      //把到达时间转化为 Int 类型
47                      if (!(ss.equals("") && ss != null)) {
48                          startTime = Integer.parseInt(edtSetTime.getText()
49                                  .toString());
50                      }
```

```
51              //设置开始时间
52              chronometer.setBase(SystemClock.elapsedRealtime());
53              //开始计时
54              chronometer.start();
55              boolOperate = false;
56              //eidttext 不可点
57              edtSetTime.setClickable(false);
58              btnOperate.setText("点击停止");
59          }
60          //如果 boolOperate==false,说明计时结束
61          else {
62              //计时停止
63              chronometer.stop();
64              boolOperate = true;
65              //eidttex 可点
66              edtSetTime.setClickable(true);
67              btnOperate.setText("点击开始");
68          }
69
70          break;
71      //重置按钮
72      case R.id.btnReset:
73          chronometer.setBase(SystemClock.elapsedRealtime());
74          break;
75      default:
76          break;
77      }
78  }
79  };
80
81  protected void showDialog() {
82      AlertDialog.Builder builder = new AlertDialog.Builder(this);
83      builder.setIcon(R.drawable.ic_launcher);
84      builder.setTitle("警告").setMessage("时间到")
85          .setPositiveButton("确定", new DialogInterface.
             OnClickListener() {
86              @Override
87              public void onClick(DialogInterface dialog,int which) {
88              }
89          });
90
91      AlertDialog dialog = builder.create();
92      dialog.show();
93  }
94  }
```

此文件是 Activity 的代码文件,在其中第 6 行定义了 Chronometer 对象,其主要功能就是计时。在第 19~32 行定义了一个计时器的监听器,当时间还没有到达记录的时间时,定时更新时间显示控件。当用户单击开始计时按钮时,chronometer 设置基础时间,然后开始计时,当用户单击重置按钮时,计时器重新置零。

4. 实例扩展

在此实例中主要使用了一个 Android 中提供的 Chronometer 类,此类还有多种用法,如倒计时和秒表功能等。

范例189 语音识别功能

1. 实例简介

在现在及未来的应用中越来越多的用户操作手机的方式产生了,从最开始的基本按键操作到现在的屏幕操作,今后用户与手机的操作方式会越来越多,其中最主要的功能就是语音操作了。遇到这样的功能,我们一般是当用户进行某个操作的时候,例如,单击某个按钮的时候,用户说出自己的话,然后手机能够得到您想表达的含义。本例子就带领大家来实现一个语音识别的实例。

2. 运行效果

该实例运行效果如图9.7所示。

图9.7 语音识别功能

3. 实例程序讲解

在本实例中,提供了一个语音识别的按钮。想要实现本实例效果,首先修改 res/layout/activity_main.xml 文件。代码省略。

然后修改 src/com.wyl.example/MainActivity.java 文件,代码如下:

```
01  //定义了本实例的主要 Activity
02  public class MainActivity extends Activity {
03      //语音识别请求码
04      private static final int VOICE_RECOGNITION_REQUEST_CODE = 1234;
05
06      @Override
07      public void onCreate(Bundle savedInstanceState) {
08          //省略代码
09      }
```

```java
10
11  OnClickListener listener = new OnClickListener() {
12
13      @Override
14      public void onClick(View v) {
15          //TODO Auto-generated method stub
16          switch (v.getId()) {
17          //打开语音识别
18          case R.id.btn_open:
19              try {
20                  //通过 Intent 传递语音识别的模式,开启语音
21                  Intent intent = new Intent(
22                          RecognizerIntent.ACTION_RECOGNIZE_SPEECH);
23                  //语言模式和自由模式的语音识别
24                  intent.putExtra(RecognizerIntent.EXTRA_LANGUAGE_MODEL,
25                          RecognizerIntent.LANGUAGE_MODEL_FREE_FORM);
26                  //提示语音开始
27                  intent.putExtra(RecognizerIntent.EXTRA_PROMPT,
                            "开始语音");
28                  //开始语音识别
29                  startActivityForResult(intent,
30                          VOICE_RECOGNITION_REQUEST_CODE);
31              } catch (Exception e) {
32                  //TODO: handle exception
33                  e.printStackTrace();
34                  Toast.makeText(getApplicationContext(),
                        "找不到语音设备", 1)
35                          .show();
36              }
37              break;
38          }
39      }
40  };
41
42  @Override
43  protected void onActivityResult(int requestCode, int resultCode, Intent data) {
44      //TODO Auto-generated method stub
45      //回调获取从谷歌得到的数据
46      if (requestCode == VOICE_RECOGNITION_REQUEST_CODE
47              && resultCode == RESULT_OK) {
48          //取得语音的字符
49          ArrayList<String> results = data
50                  .getStringArrayListExtra(RecognizerIntent.EXTRA_RESULTS);
51          //定义已经识别的字符串
52          String resultString = "";
53          for (int i = 0; i < results.size(); i++) {
54              resultString += results.get(i);
55          }
56          Toast.makeText(this, resultString, 1).show();
57      }
58      super.onActivityResult(requestCode, resultCode, data);
59  }
60  }
```

此文件是 Activity 的代码文件,在其中当用户单击开始语音识别的按钮时,调用第 20~30 行代码,首先初始化语音识别的 Intent,设置语音模式,然后启动 Activity。当识别完成

后，系统回调 onActivityResult 方法，通过返回的 data 参数得到识别的结果，并且进行显示。

4．实例扩展

在此实例中使用了 Google 提供的语音识别功能，首先要保证你的手机支持此功能，而且此功能需要网络访问的权限，所以你的程序中需要添加访问网络的权限。

范例 190　语音转换文本

1．实例简介

在我们使用应用的过程中，我们不但希望能够通过语音控制手机，而且在某些环境下的时候，还希望能够通过语音直接转换为文字。例如，在演讲的时候，可以直接记录演讲笔记等。这样的功能，我们一般是当用户进行某个操作的时候，例如，单击某个按钮的时候，调用系统的语音识别功能来实现的。本例子就带领大家来实现一个语音转换文字的实例。

2．运行效果

该实例运行效果如图 9.8 所示。

图 9.8　语音转换文本

3．实例程序讲解

在本实例中，提供了一个开始转换的按钮。想要实现本实例效果，首先修改 res/layout/activity_main.xml 文件。代码省略。

然后修改 src/com.wyl.example/MainActivity.java 文件，代码如下：

```java
01  //定义了主要的activity
02  public class MainActivity extends Activity {
03
04  @Override
05  public void onCreate(Bundle savedInstanceState) {
06      //省略
07  }
08
09  //设置对象的监听器
10  private void setListener() {
11      //TODO Auto-generated method stub
12
13      btnSpeak.setOnClickListener(new View.OnClickListener() {
14          @Override
15          public void onClick(View v) {
16              //定义需要触发的意图
17              Intent intent = new Intent(
18                  RecognizerIntent.ACTION_RECOGNIZE_SPEECH);
19              //设置识别的语言
20              intent.putExtra(RecognizerIntent.EXTRA_LANGUAGE_MODEL,
                  "en-US");
21
22              try {
23                  //启动intent并返回结果
24                  startActivityForResult(intent, RESULT_SPEECH);
25                  txtText.setText("");
26              } catch (ActivityNotFoundException a) {
27                  Toast t = Toast.makeText(getApplicationContext(),
28                      "您的手机不支持Speech to Text功能",
29                      Toast.LENGTH_SHORT);
30                  t.show();
31              }
32          }
33      });
34
35  }
36
37  @Override
38  protected void onActivityResult(int requestCode, int resultCode, Intent data) {
39      super.onActivityResult(requestCode, resultCode, data);
40
41      switch (requestCode) {
42      case RESULT_SPEECH:
43          //如果返回的结果正确，并且data不为空，则从data里面提取出相应的文本，并显示出来
44          if (resultCode == RESULT_OK && null != data) {
45              //得到返回结果
46              ArrayList<String> text = data
47                  .getStringArrayListExtra(RecognizerIntent.EXTRA_
                  RESULTS);
48
49              txtText.setText(text.get(0));
50          }
51          break;
52      }
53  }
54  }
```

此文件是 Activity 的代码文件，当用户单击开始转换的时候调用第 16~31 行，首先初始化 Intent 对象，设置识别的语言类型，然后启动 Activity，当识别完毕后自动回调第 38~53 行的 onActivityResult 方法，其中的 data 参数包含了语音识别的结果，然后在 TextView 中进行展示即可。

4．实例扩展

语音识别功能在今后的应用中会越来越多，其中要求的识别率也越来越高，所以大家在使用这方面功能的时候要尽量提高识别的效率，这样才能提高程序的用户体验。

范例 191 TTS 文字朗读

1．实例简介

在我们使用应用的过程中，希望手机可以帮我们把需要看的内容朗读出来。例如，在一些听书软件和听报软件等。这样的功能，我们一般是当用户进行某个操作的时候，例如：单击某个按钮的时候，从系统中得到需要朗读的文字内容，然后调用 Google 提供的 TTS 服务。本例子就带领大家来实现一个 TTS 文字朗读的实例。

2．运行效果

该实例运行效果如图 9.9 所示。

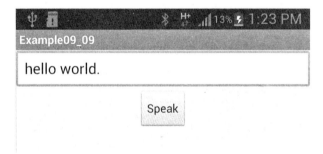

图 9.9 TTS 文字朗读

3．实例程序讲解

在本实例中，提供了一个输入框，在此输入框中输入你想让系统朗读的文字，然后提供一个朗读按钮。想要实现本实例效果，首先修改 res/layout/activity_main.xml 文件。代码省略。

然后修改 src/com.wyl.example/MainActivity.java 文件，代码如下：

```
001  //定义本例子主要的 Activity 代码
002  public class MainActivity extends Activity implements OnInitListener {
003      //省略控件定义代码
004      //定义文字转语音对象
005      private TextToSpeech mTts;
006  
```

```java
007    @Override
008    public void onCreate(Bundle savedInstanceState) {
009        super.onCreate(savedInstanceState);
010        setContentView(R.layout.activity_main);
011
012        //检查 TTS 数据是否已经安装并且可用
013        Intent checkIntent = new Intent();
014        checkIntent.setAction(TextToSpeech.Engine.ACTION_CHECK_TTS
           _DATA);
015        startActivityForResult(checkIntent, REQ_TTS_STATUS_CHECK);
016
017        findView();
018        setListener();
019    }
020    //省略 findview 实现代码
021
022    private void setListener() {
023        //设置单击按钮监听器
024        speakBtn.setOnClickListener(new OnClickListener() {
025
026            public void onClick(View v) {
027                //朗读输入框里的内容
028                mTts.speak(inputText.getText().toString(),
029                        TextToSpeech.QUEUE_ADD, null);
030            }
031        });
032    }
033
034    //实现 TTS 初始化接口
035    @Override
036    public void onInit(int status) {
037        //TODO Auto-generated method stub
038        //TTS Engine 初始化完成
039        if (status == TextToSpeech.SUCCESS) {
040            int result = mTts.setLanguage(Locale.US);
041            //设置发音语言
042            if (result == TextToSpeech.LANG_MISSING_DATA
043                    || result == TextToSpeech.LANG_NOT_SUPPORTED)
044            //判断语言是否可用
045            {
046                Log.v(TAG, "Language is not available");
047                speakBtn.setEnabled(false);
048            } else {
049                mTts.speak("This is an example of speech synthesis.",
050                        TextToSpeech.QUEUE_ADD, null);
051                speakBtn.setEnabled(true);
052            }
053        }
054    }
055
056    protected void onActivityResult(int requestCode, int resultCode,
       Intent data) {
057        if (requestCode == REQ_TTS_STATUS_CHECK) {
058            switch (resultCode) {
059            case TextToSpeech.Engine.CHECK_VOICE_DATA_PASS:
060                //这个返回结果表明 TTS Engine 可以用
061            {
062                mTts = new TextToSpeech(this, this);
063                Log.v(TAG, "TTS Engine is installed!");
064
```

```
065            }
066                break;
067            case TextToSpeech.Engine.CHECK_VOICE_DATA_BAD_DATA:
068                //需要的语音数据已损坏
069            case TextToSpeech.Engine.CHECK_VOICE_DATA_MISSING_DATA:
070                //缺少需要语言的语音数据
071            case TextToSpeech.Engine.CHECK_VOICE_DATA_MISSING_VOLUME:
072                //缺少需要语言的发音数据
073                {
074                    //这三种情况都表明数据有错,请重新下载安装需要的数据
075                    Log.v(TAG, "Need language stuff:" + resultCode);
076                    Intent dataIntent = new Intent();
077                    dataIntent
078                    .setAction(TextToSpeech.Engine.ACTION_INSTALL_TTS
                        _DATA);
079                    startActivity(dataIntent);
080
081                }
082                break;
083            case TextToSpeech.Engine.CHECK_VOICE_DATA_FAIL:
084                //检查失败
085            default:
086                Log.v(TAG, "Got a failure. TTS apparently not
                    available");
087                break;
088            }
089        } else {
090            //其他 Intent 返回的结果
091        }
092    }
093
094    @Override
095    protected void onPause() {
096        //TODO Auto-generated method stub
097        super.onPause();
098        //activity 暂停时也停止 TTS
099        if (mTts != null)
100        {
101            mTts.stop();
102        }
103    }
104
105    @Override
106    protected void onDestroy() {
107        //TODO Auto-generated method stub
108        //释放 TTS 的资源
109        super.onDestroy();
110        mTts.shutdown();
111    }
112 }
```

此文件是 Activity 的代码文件,在其中第 5 行定义了 TextToSpeech 对象用来进行语音识别,在第 13~15 行检查当前手机是否支持文字朗读功能,当用户单击朗读按钮的时候,执行第 28~29 行进行识别。识别的结果会在第 56 行的 onActivityResult 方法中进行返回。

4.实例扩展

在此实例中使用到了 Google 提供的 TTS 服务,所以想实现本实例一定要保证你的手

机中安装了 Google 服务才可以，当然如果大家的手机里安装了其他文字朗读的服务也可以执行。

范例 192　本地音频播放

1．实例简介

在我们使用应用的过程中，经常会使用播放本地音乐的效果。这样的功能，我们一般是当用户进行某个操作的时候，例如，单击某个按钮的时候，获取本地的音频资源，然后进行播放。本例子就带领大家来实现一个本地音乐播放器的实例。

2．运行效果

该实例运行效果如图 9.10 所示。

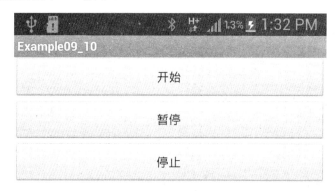

图 9.10　播放本地音乐

3．实例程序讲解

在本实例中，提供了三个按钮分别是开始播放、暂停播放和停止播放按钮。想要实现本实例效果，首先修改 res/layout/activity_main.xml 文件。代码省略。

然后修改 src/com.wyl.example/MainActivity.java 文件，代码如下：

```
01  //定义当前主 Activity 代码
02  public class MainActivity extends Activity {
03      //省略控件定义的代码，findview 和 setlistener 代码
04      //定义音乐播放对象
05      private MediaPlayer player = new MediaPlayer();
06  
07      @Override
08      public void onCreate(Bundle savedInstanceState) {
09          super.onCreate(savedInstanceState);
10          //设置当前 Activity 的布局文件
11          setContentView(R.layout.main);
12          try {
13              //加载本地的 mp3 文件
14              player.setDataSource("/sdcard/test.mp3");
```

```java
15          } catch (IllegalArgumentException e) {
16              //TODO Auto-generated catch block
17              e.printStackTrace();
18          } catch (IllegalStateException e) {
19              //TODO Auto-generated catch block
20              e.printStackTrace();
21          } catch (IOException e) {
22              //TODO Auto-generated catch block
23              e.printStackTrace();
24          }
25      //得到布局中的所有控件
26      findView();
27      //设置所有控件的监听器
28      setListener();
29  }
30
31  //自定义单击监听器对象
32  OnClickListener mylistener = new OnClickListener() {
33
34      @Override
35      public void onClick(View v) {
36          //TODO Auto-generated method stub
37          switch (v.getId()) {
38          case R.id.BtnStart:
39              try {
40                  //开始播放
41                  player.prepare();
42              } catch (IllegalStateException e) {
43                  //TODO Auto-generated catch block
44                  e.printStackTrace();
45              } catch (IOException e) {
46                  //TODO Auto-generated catch block
47                  e.printStackTrace();
48              }
49              player.start();
50              break;
51          case R.id.BtnPause:
52              //音乐暂停
53              player.pause();
54              break;
55          case R.id.BtnStop:
56              //重置音乐，重新播放
57              player.seekTo(0);
58              player.stop();
59              try {
60                  player.prepare();
61              } catch (IllegalStateException e) {
62                  e.printStackTrace();
63              } catch (IOException e) {
64                  e.printStackTrace();
65              }
66              break;
67          default:
68              break;
69          }
70      }
71  };
72  }
```

此文件是 Activity 的代码文件，在其中第 5 行定义了一个 MediaPlayer 对象，使用此对

象来进行音频播放。在第 12～24 行进行加载本地资源文件，当用户单击开始播放时，调用第 41 行准备播放，第 49 行开始音乐播放，当用户单击暂停播放时，调用第 53 行暂停播放。当用户单击重置时，调用 stop 方法让音乐从头播放。

4．实例扩展

在此实例中又使用了之前用过的 MediaPlayer 对象，此对象不但可以播放在线音乐，也可以播放本地音乐，前提是本地有对应的音乐文件。

范例 193 音效播放

1．实例简介

在我们使用应用的过程中，经常会使用音效的播放。例如，单击按钮的音效，游戏中人物的音效等。这样的功能，我们一般是当用户进行某个操作的时候，例如，单击某个按钮的时候，使用音效播放池来播放。本例子就带领大家来实现一个单击按钮播放音效效果。

2．运行效果

该实例运行效果如图 9.11 所示。

图 9.11 音效播放效果

3．实例程序讲解

在本实例中，提供了两个按钮，一个是播放音效，第二个按钮是同时在播放多个音效功能。想要实现本实例效果，首先修改 res/layout/activity_main.xml 文件，布局如上布局。代码省略。

然后修改 src/com.wyl.example/MainActivity.java 文件，代码如下：

```
01  //定义主要的Activity代码
02  public class MainActivity extends Activity
03      implements View.OnClickListener,
04          SoundPool.OnLoadCompleteListener {
05      static String TAG = "TestSoundPoolActivity";
06      //定义音乐池对象
07      private SoundPool sndPool;
08
```

```java
09  private static final int SOUND_LOAD_OK = 1;
10  private final Handler mHandler = new MyHandler();
11
12  /** Called when the activity is first created. */
13  @Override
14  public void onCreate(Bundle savedInstanceState) {
15      super.onCreate(savedInstanceState);
16      setContentView(R.layout.main);
17      findView();
18
19      //初始化音乐池对象
20      sndPool = new SoundPool(16, AudioManager.STREAM_MUSIC, 0);
21      sndPool.setOnLoadCompleteListener(this);
22  }
23  //省略findview方法定义代码
24
25  public void onDestroy() {
26      //释放音乐池播放对象
27      sndPool.release();
28      super.onDestroy();
29  }
30
31  private class MyHandler extends Handler {
32      public void handleMessage(Message msg) {
33          switch (msg.what) {
34              //发送播放短音乐的消息
35              case SOUND_LOAD_OK:
36                  //开始播放音频
37                  sndPool.play(msg.arg1, (float) 0.8, (float) 0.8, 16, 10,
38                          (float) 1.0);
39                  break;
40          }
41      }
42  }
43
44  public void onLoadComplete(SoundPool soundPool,int sampleId,int status){
45      //加载失败发送失败的消息
46      Message msg = mHandler.obtainMessage(SOUND_LOAD_OK);
47      msg.arg1 = sampleId;
48      mHandler.sendMessage(msg);
49  }
50
51  public void onClick(View v) {
52      int id = v.getId();
53      switch (id) {
54      case R.id.BtnPlay:
55          //单击播放音乐,如果音乐池不为空则进行加载播放
56          if (sndPool != null)
57              sndPool.load(this, R.raw.f, 1);
58          break;
59      case R.id.BtnMore:
60          //加载更多的音乐同时播放
61          sndPool.load(this, R.raw.f, 1);
62          Log.v(TAG, "more");
63          break;
64      }
65  }
66  }
```

此文件是Activity的代码文件,在其中第7行定义了一个音效播放对象用来播放音效,

在第 20～21 行初始化 sndPool 对象，并且设置对应监听器，当用户单击开始播放音效时，就在 sndPool 中设置 raw 目录下的 f 文件进行播放，这样几个音效就同时播放了。

4．实例扩展

在此实例中使用音效播放对象来播放音效，在 Android 播放音乐一般使用 MediaPlayer 对象，但是音效使用 SoundPool，因为一般 MediaPlayer 比较耗内存，而且只能同时播放一个文件，而音效可以同时播放多个声音，但是每个声音最好都是比较短小的。

范例 194 播放本地视频

1．实例简介

在我们使用应用的过程中，经常会使用播放本地视频的功能。例如，在一些视频播放客户端中、暴风影音和优酷客户端等。这样的功能，我们一般是当用户进行某个操作的时候，例如，单击某个按钮的时候，打开系统的视频文件进行播放。本例子就带领大家来实现一个本地视频播放器的实例效果。

2．运行效果

该实例运行效果如图 9.12 所示。

图 9.12 本地视频播放器

3．实例程序讲解

在本实例中，提供界面显示了一个 VideoView 控件，用来播放系统内置视频。想要实现本实例效果，首先修改 res/layout/activity_main.xml 文件。代码省略。

然后修改 src/com.wyl.example/MainActivity.java 文件，代码如下：

```
01  //定义主要的Activity代码
02  public class MainActivity extends Activity implements
```

```java
03         MediaPlayer.OnErrorListener,
04         MediaPlayer.OnCompletionListener {
05
06     public static final String TAG = "VideoPlayer";
07     //定义视频播放对象
08     private VideoView mVideoView;
09     //定义内部资源定位对象
10     private Uri mUri;
11     private int mPositionWhenPaused = -1;
12
13     //定义视频控制播放对象
14     private MediaController mMediaController;
15
16     @Override
17     protected void onCreate(Bundle savedInstanceState) {
18         //TODO Auto-generated method stub
19         super.onCreate(savedInstanceState);
20         //设置当前Activity的布局
21         setContentView(R.layout.activity_main);
22
23         //设置为横屏播放
24         this.setRequestedOrientation(ActivityInfo.SCREEN_ORIENTATION
            _LANDSCAPE);
25         mVideoView = (VideoView) findViewById(R.id.videoview);
26
27         //文件路径
28         mUri = Uri.parse(Environment.getExternalStorageDirectory()
29                 + "/video.3gp");
30
31         //创建音频控制对象
32         mMediaController = new MediaController(this);
33
34         //设置MediaController
35         mVideoView.setMediaController(mMediaController);
36     }
37
38     //监听MediaPlayer上报的错误信息
39     @Override
40     public boolean onError(MediaPlayer mp, int what, int extra) {
41         //TODO Auto-generated method stub
42         return false;
43     }
44
45     //Video播完的时候得到通知
46     @Override
47     public void onCompletion(MediaPlayer mp) {
48         this.finish();
49     }
50
51     //开始
52     public void onStart() {
53         //播放视频
54         mVideoView.setVideoURI(mUri);
55         mVideoView.start();
56         super.onStart();
57     }
```

```
58
59    //暂停
60    public void onPause() {
61        //暂停视频播放
62        mPositionWhenPaused = mVideoView.getCurrentPosition();
63        mVideoView.stopPlayback();
64        super.onPause();
65    }
66
67    public void onResume() {
68        //恢复视频播放
69        if (mPositionWhenPaused >= 0) {
70            mVideoView.seekTo(mPositionWhenPaused);
71            mPositionWhenPaused = -1;
72        }
73        super.onResume();
74    }
75 }
```

此文件是 Activity 的代码文件，在其中第 14 行定义了一个视频控制器。在第 28 行得到需要播放的视频路径，在第 32 行初始化 MediaController 对象，并且给 VideoView 设置了此控制对象，当视频播放出错时调用第 40 行的 onError 方法。当视频播放完毕时调用第 47 行的 onCompletion 方法，当视频开始播放时调用第 52 行的 onStart 方法。当视频暂停时调用第 60 行的 onPause 方法。当视频恢复播放时调用第 67 行定义的 onResume 方法。

4. 实例扩展

注意在此实例中主要使用 MediaController 控制视频播放，使用 VideoView 来播放视频。

范例 195 加速度传感器应用

1. 实例简介

在我们使用应用的过程中，有时候需要一些与传感器相关的应用。例如，在一些功能中需要知道手机在 X、Y 和 Z 方向的加速度。这样的功能，我们一般是需要得到传感器的服务类，然后通过监听器得到相应硬件的数据参数。本例子就带领大家来实现一个得到手机加速度的实例效果。

2. 运行效果

该实例运行效果如图 9.13 所示。

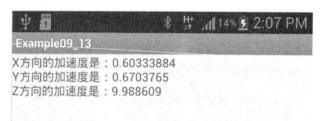

图 9.13 加速度传感器应用

3. 实例程序讲解

在本实例中，打开应用程序即可显示手机当前在 X、Y 和 Z 三个方向的加速度值。想要实现本实例效果，首先修改 res/layout/activity_main.xml 文件，加入 TextView 控件。代码省略。

然后修改 src/com.wyl.example/MainActivity.java 文件，代码如下：

```
01  //定义只要的Activity代码
02  public class MainActivity extends Activity {
03      //省略控件定义代码，及findview方法代码
04      //定义传感器管理对象
05      private SensorManager sm;
06
07      @Override
08      public void onCreate(Bundle savedInstanceState) {
09          super.onCreate(savedInstanceState);
10          //设置当前页面布局
11          setContentView(R.layout.activity_main);
12          //得到系统的传感器管理服务
13          sm = (SensorManager)getSystemService(Context.SENSOR_SERVICE);
14
15          findView();
16      }
17
18      @Override
19      protected void onResume() {
20          //TODO Auto-generated method stub
21          super.onResume();
22          //注册传感器监听器
23          sm.registerListener(sel,
24                  sm.getDefaultSensor(Sensor.TYPE_ACCELEROMETER),
25                  SensorManager.SENSOR_DELAY_UI);
26      }
27
28      @Override
29      protected void onStop() {
30          //取消注册传感器监听器
31          sm.unregisterListener(sel);
32          super.onStop();
33      }
34
35      //定义传感器监听器对象
36      SensorEventListener sel = new SensorEventListener() {
37
38          //当传感器的值发生改变时触发该方法
39          @Override
40          public void onSensorChanged(SensorEvent event) {
41              //TODO Auto-generated method stub
42              String str;
43              str = "X方向的加速度是："+event.values[0];
44              str += "\nY方向的加速度是："+event.values[1];
45              str += "\nZ方向的加速度是："+event.values[2];
46
47              Tv.setText(str);
48          }
49          //当传感器经度发生改变时触发该方法
```

```
50      @Override
51      public void onAccuracyChanged(Sensor sensor, int accuracy) {
52          //TODO Auto-generated method stub
53
54      }
55  };
56  }
```

此文件是 Activity 的代码文件,在其中第 5 行定义了 SensorManager 类的对象,通过它来得到系统的传感器服务,在第 13 行得到了系统的加速度传感器服务。在第 23~25 行设置了传感器数据变化的监听器,在第 36~55 行定义了一个 SensorEventListener 监听器,当传感器数据发生变化时,调用其中的 onSensorChanged 方法,在本实例中得到三个方向的加速度并且显示出来。

4. 实例扩展

注意在 Android 中提供了 11 种传感器,分别对应了一个固定参数及相应的功能,代码如下:

```
SENSOR_TYPE_ACCELEROMETER           1       //加速度
SENSOR_TYPE_MAGNETIC_FIELD          2       //磁力
SENSOR_TYPE_ORIENTATION             3       //方向
SENSOR_TYPE_GYROSCOPE               4       //陀螺仪
SENSOR_TYPE_LIGHT                   5       //光线感应
SENSOR_TYPE_PRESSURE                6       //压力
SENSOR_TYPE_TEMPERATURE             7       //温度
SENSOR_TYPE_PROXIMITY               8       //接近
SENSOR_TYPE_GRAVITY                 9       //重力
SENSOR_TYPE_LINEAR_ACCELERATION     10      //线性加速度
SENSOR_TYPE_ROTATION_VECTOR         11      //旋转矢量
```

这些都会在相应功能中有所体现。

范例 196 光强度查看器

1. 实例简介

在我们使用应用的过程中,经常会使用光强度检测的功能。例如,屏幕的自动亮度和闪光灯的自动开启等。遇到这样的功能,我们一般是通过得到系统的光传感器,然后得到光强度数据,最后进行相应的处理。本例子就带领大家来实现一个光强度查看器的实例效果。

2. 运行效果

该实例运行效果如图 9.14 所示。

图 9.14 光强度查看器

3. 实例程序讲解

在本实例中，打开应用，显示光强度。想要实现本实例效果，首先修改 res/layout/activity_main.xml 文件。代码省略。

然后修改 src/com.wyl.example/MainActivity.java 文件，代码如下：

```
01  //定义只要的Activity代码
02  public class MainActivity extends Activity {
03      //省略定义控件对象和findview代码
04
05      //定义传感器管理对象
06      private SensorManager sm;
07
08      @Override
09      public void onCreate(Bundle savedInstanceState) {
10          super.onCreate(savedInstanceState);
11          //设置当前页面布局
12          setContentView(R.layout.activity_main);
13          //得到系统的传感器管理服务
14          sm = (SensorManager)getSystemService(Context.SENSOR_SERVICE);
15
16          findView();
17      }
18
19      @Override
20      protected void onResume() {
21          //TODO Auto-generated method stub
22          super.onResume();
23          //注册传感器监听器
24          sm.registerListener(sel,
25                  sm.getDefaultSensor(Sensor.TYPE_LIGHT),
26                  SensorManager.SENSOR_DELAY_FASTEST);
27      }
28
29      @Override
30      protected void onStop() {
31          //取消注册传感器监听器
32          sm.unregisterListener(sel);
33          super.onStop();
34      }
35
36      //定义传感器监听器对象
37      SensorEventListener sel = new SensorEventListener() {
38
39          //当传感器的值发生改变时触发该方法
40          @Override
41          public void onSensorChanged(SensorEvent event) {
42              //TODO Auto-generated method stub
43              String str;
44              str = "当前的手机范围内的光强度为："+event.values[0];
45
46              Tv.setText(str);
47          }
48          //当传感器经度发生改变时触发该方法
49          @Override
50          public void onAccuracyChanged(Sensor sensor, int accuracy) {
51              //TODO Auto-generated method stub
```

```
52
53      }
54    };
55  }
```

此文件是 Activity 的代码文件，在其中第 6 行定义了 SensorManager 类的对象，通过它来得到系统的传感器服务，在第 14 行得到了系统的光传感器服务。在第 24～26 行设置了传感器数据变化的监听器，在第 37～54 行定义了一个 SensorEventListener 监听器，当传感器数据发生变化时调用其中的 onSensorChanged 方法，在本实例中得到光强度数据并且显示出来。

4．实例扩展

本实例一般不用在一些核心功能中，一般用在一些提高用户体验度的功能中。例如，自动屏幕亮度和自动开启闪光灯等功能中。

范例 197 微信摇一摇功能

1．实例简介

在我们使用应用的过程中，经常会使用摇一摇触发一个事件。例如，微信摇一摇找到好友，软件市场摇一摇找到软件等。这样的功能，我们一般是得到系统的传感器，然后判断是否进行了手机摇动。本例子就带领大家来实现一个微信摇一摇的实例效果。

2．运行效果

该实例运行效果如图 9.15 所示。

图 9.15 微信摇一摇功能

3．实例程序讲解

在本实例中，显示了 TextView 控件，当用户摇动手机的时候，提示用户摇到好友。想要实现本实例效果，首先修改 res/layout/activity_main.xml 文件。代码省略。

然后修改 src/com.wyl.example/MainActivity.java 文件，代码如下：

```
01  //定义主要的Activity代码
02  public class MainActivity extends Activity {
03    //省略控件对象定义代码和findview方法定义
04    //得到传感器的管理对象
05    private SensorManager sm;
```

```
06
07    @Override
08    public void onCreate(Bundle savedInstanceState) {
09        super.onCreate(savedInstanceState);
10        setContentView(R.layout.activity_main);
11        sm = (SensorManager) getSystemService(Context.SENSOR_SERVICE);
12
13        findView();
14    }
15
16    @Override
17    protected void onResume() {
18        //TODO Auto-generated method stub
19        super.onResume();
20
21        sm.registerListener(sel,
22                sm.getDefaultSensor(Sensor.TYPE_ACCELEROMETER),
23                SensorManager.SENSOR_DELAY_UI);
24    }
25
26    @Override
27    protected void onStop() {
28        //TODO Auto-generated method stub
29        sm.unregisterListener(sel);
30        super.onStop();
31    }
32
33    SensorEventListener sel = new SensorEventListener() {
34        //当传感器的值发生改变时触发该方法
35        @Override
36        public void onSensorChanged(SensorEvent event) {
37            //TODO Auto-generated method stub
38            int sensorType = event.sensor.getType();
39            float[] values = event.values;
40            if (sensorType == Sensor.TYPE_ACCELEROMETER) {
41                if (Math.abs(values[0]) > 14 || Math.abs(values[1]) > 14
42                        || Math.abs(values[2]) > 14) {
43                    Tv.setText("您摇到一位好友");
44                }
45            }
46        }
47        //当传感器经度发生改变时触发该方法
48        @Override
49        public void onAccuracyChanged(Sensor sensor, int accuracy) {
50            //TODO Auto-generated method stub
51
52        }
53    };
54 }
```

此文件是 Activity 的代码文件，在其中第 5 行定义了 SensorManager 类的对象，通过它来得到系统的传感器服务，在第 11 行得到了系统的加速度传感器服务。在第 21~23 行设置了传感器数据变化的监听器，在第 33~52 行定义了一个 SensorEventListener 监听器，当传感器数据发生变化时调用其中的 onSensorChanged 方法，在此方法中判断手机在三个方向上的加速度是否满足相应的值，然后显示用户得到好友。

4. 实例扩展

很多应用的功能可能都和传感器有关，合理使用传感器功能可以大大提高你的程序的可用性，吸引用户。

9.2 桌面插件开发

范例 198 切换壁纸插件

1. 实例简介

在我们使用应用的过程中，经常会希望在桌面上有快捷的程序应用。例如，快速切换背景的功能。这样的功能，我们一般是通过桌面插件的方式来实现的。本例子就带领大家来实现一个单击切换系统背景的桌面插件。

2. 运行效果

该实例运行效果如图 9.16 所示。

图 9.16 切换壁纸插件

3. 实例程序讲解

在本实例安装完成后不会有应用程序的提示，而是在手机的小工具中可以看到本插件，当单击本插件的时候会切换桌面背景。想要实现本实例效果，新建 src/com.wyl.example/AppWidet.java 文件，代码如下：

```
001  public class AppWidet extends AppWidgetProvider {
002      //定义intent的action
003      private static final String CLICK_NAME_ACTION = "com.wyl.
         action.widget.click";
004      //声明RemoteViews对象
005      private RemoteViews myRemoteViews;
006      //定义单击的次数
007      private static int clickTime;
008      //定义图片的资源
009      private int[] images = { R.raw.image_01, R.raw.image_02, R.raw.
         image_03,
010              R.raw.image_04, R.raw.image_05 };
011      //声明inputstream对象
```

```
012    private InputStream is;
013
014    @Override
015    public void onUpdate(Context context, AppWidgetManager appWidget
       Manager,
016            int[] appWidgetIds) {
017        //初始化信息
018        updateAppWidget(context, appWidgetManager);
019    }
020
021    public void updateAppWidget(Context context, AppWidgetManager
           appWidgeManger) {
022        myRemoteViews = new RemoteViews(context.getPackageName(),
           R.layout.main);
023        //创建单击意图对象
024        Intent intentClick = new Intent(CLICK_NAME_ACTION);
025        PendingIntent pendingIntent = PendingIntent.getBroadcast
           (context, 0,
026            intentClick, 0);
027        //绑定单击事件
028        myRemoteViews.setOnClickPendingIntent(R.id.btn, pending
           Intent);
029        ComponentName myComponentName = new ComponentName(context,
030            AppWidet.class);
031        //得到AppWidget管理器
032        AppWidgetManager myAppWidgetManager = AppWidgetManager
033            .getInstance(context);
034        //更新控件
035        myAppWidgetManager.updateAppWidget(myComponentName, myRemote
           Views);
036    }
037
038    @Override
039    public void onDeleted(Context context, int[] appWidgetIds) {
040        System.out.println("onDeleted");
041        super.onDeleted(context, appWidgetIds);
042    }
043
044    @Override
045    public void onEnabled(Context context) {
046        System.out.println("onEnabled");
047        super.onEnabled(context);
048    }
049
050    @Override
051    public void onDisabled(Context context) {
052        System.out.println("onDisabled");
053        super.onDisabled(context);
054    }
055
056    //接收到每个广播时都会被调用
057    @Override
058    public void onReceive(Context context, Intent intent) {
059        System.out.println("onReceive");
060        super.onReceive(context, intent);
061        //如果myRemoteViews==null,则实例化
062        if (myRemoteViews == null) {
063            myRemoteViews = new RemoteViews(context.
               getPackageName(),
064                R.layout.main);
```

```
065            }
066            //如果接收到意图的 action 是不是 CLICK_NAME_ACTION
067            if (intent.getAction().equals(CLICK_NAME_ACTION)) {
068
069                Log.e("CLICK_NAME_ACTION", clickTime + "");
070                //单击的次数小于资源的长度
071                if (clickTime < images.length) {
072                    //得到图片资源
073                    Resources resources = context.getResources();
074                    //可换成任意图片资源
075                    is = resources.openRawResource(images[clickTime]);
076                    try {
077                        /* 更换桌面背景 */
078                        context.setWallpaper(is);
079                        /* 用 Toast 来显示桌布已更换 */
080                        Toast.makeText(context, "桌面壁纸已切换", Toast.LENGTH_SHORT)
081                                .show();
082                    } catch (Exception e) {
083                        e.printStackTrace();
084                    }
085
086                    clickTime++;
087                    if (clickTime >= images.length ) {
088                        clickTime = 0;
089                    }
090
091                } else {
092                    Toast.makeText(context, "桌面壁纸已换完", Toast.LENGTH_SHORT).show();
093                }
094
095                //AppwidgetManager 实例，更新 appwidget
096                AppWidgetManager appWidgetManger = AppWidgetManager
097                        .getInstance(context);
098                int[] appIds = appWidgetManger.getAppWidgetIds(new ComponentName(
099                        context, AppWidet.class));
100                appWidgetManger.updateAppWidget(appIds, myRemoteViews);
101            }
102        }
103 }
```

此文件是本插件的代码文件，其第 12～36 行设置了当前插件的布局及监听器方式，得到了 AppWidget 管理器对象。当用户单击此桌面控件时发送内部广播，在第 58～102 行接收此广播，然后得到系统的资源图片，通过 context 的 setWallpaper 方法设置桌面布局。

4．实例扩展

本实例定义了桌面插件并且实现桌面壁纸的切换。

范例 199　倒计时插件

1．实例简介

在我们使用应用的过程中，经常会使用一些倒计时的插件。例如，放假倒计时和世界杯倒计时等。这样的功能，我们一般是通过桌面插件来实现的。本例子就带领大家来实现一个时间倒计时的桌面插件。

2．运行效果

该实例运行效果如图9.17所示。

图9.17　倒计时插件

3．实例程序讲解

在本实例安装完成后不会有应用程序的提示，而是在手机的小工具中可以看到本插件，当单击本插件的时候会显示倒计时效果。想要实现本实例效果，新建 src/com.wyl.example/AppWidet.java 文件，代码如下：

```
01  public class AppWidet extends AppWidgetProvider {
02    //定义intent的action
03    private static final String CLICK_NAME_ACTION = "com.wyl.action.widget.click";
04    //声明RemoteViews对象
05    private  RemoteViews myRemoteViews;
06
07      @Override
08      public void onUpdate(Context context, AppWidgetManager appWidgetManager, int[] appWidgetIds) {
09        //声明计时器对象
10        Timer timer=new Timer();
11        //每天定点执行某一任务
12        //第一个参数是需要触发的事件，第二个参数是触发的事件，第三个参数是每隔多长时间执行一次
13        timer.scheduleAtFixedRate(new MyTime(context, appWidgetManager), 1, 60000);
14        super.onUpdate(context, appWidgetManager, appWidgetIds);
15      }
16
17      @Override
18      public void onDeleted(Context context, int[] appWidgetIds) {
19          System.out.println("onDeleted");
20          super.onDeleted(context, appWidgetIds);
21      }
22
23      @Override
24      public void onEnabled(Context context) {
25          System.out.println("onEnabled");
26          super.onEnabled(context);
```

```java
27      }
28
29
30      @Override
31      public void onDisabled(Context context) {
32          System.out.println("onDisabled");
33          super.onDisabled(context);
34      }
35
36      @Override
37      public void onReceive(Context context, Intent intent) {
38          System.out.println("onReceive");
39          super.onReceive(context, intent);
40
41      }
42      //MyTime 的定义
43      private class MyTime extends TimerTask{
44       //RemoteViews 描述一个view,而这个view是在另外一个进程显示的
45       //声明 RemoteViews 对象
46          RemoteViews remoteViews;
47          //声明 AppWidget 管理器的对象
48          AppWidgetManager appWidgetManager;
49          //声明组件名字的对象
50          ComponentName thisWidget;
51
52          public MyTime(Context context,AppWidgetManager appWidget
        Manager){
53              this.appWidgetManager = appWidgetManager;
54              //实例化 remoteViews, 第一个参数是包的名字，第二个参数是XML 文件的名字
55              remoteViews = new RemoteViews(context.getPackageName(),
            R.layout.main);
56
57             //组件名称，第一个参数是包名,第二个是类名，要带上包名
58              thisWidget = new ComponentName(context,AppWidet.class);
59          }
60          //需要定期执行的方法
61          public void run() {
62           //得到当前的时间
63              Date date = new Date();
64              //设置到期时间
65              Calendar calendar = new GregorianCalendar(2013,07,01);
66              //转化成天数
67              long days = (((calendar.getTimeInMillis()-date.getTime())
                /1000))/86400;
68              remoteViews.setTextViewText(R.id.text, "距离期末考试还有" +
                days+"天");
69              appWidgetManager.updateAppWidget(thisWidget, remoteViews);
70          }
71      }
72  }
```

此文件是本插件的代码文件，其第 10～14 行设置了当前插件的布局及监听器方式，定义定时器对象间隔调用。在第 43～71 行定义了定时任务类，在此类中第 61～69 行定时得到当前时间，然后更新显示内容。

4. 实例扩展

本实例定义了桌面插件并且实现显示固定的倒计时功能。

范例 200　日期插件

1. 实例简介

在我们使用应用的过程中，经常希望在桌面上一眼就可以看到当前的日期。这样的功能，我们也可以在桌面插件中实现。本例子就带领大家来实现一个桌面的日期插件。

2. 运行效果

该实例运行效果如图 9.18 所示。

图 9.18　日期插件

3. 实例程序讲解

在本实例安装完成后不会有应用程序的提示，而是在手机的小工具中可以看到本插件，拖动到桌面显示当前日期。想要实现本实例效果，新建 src/com.wyl.example/AppWidet.java 文件，代码如下：

```
01  public class AppWidet extends AppWidgetProvider{
02  //定义12个月份
03      private String[] months={"一月","二月","三月",
04          "四月","五月","六月","七月",
05          "八月","九月","十月","十一月","十二月"};
06  //定义一周7天
07      private String[] days={"星期日","星期一","星期二",
08          "星期三","星期四","星期五","星期六"};
09      @Override
10      public void onUpdate(Context context, AppWidgetManager appWidget
    Manager,
11              int[] appWidgetIds) {
12          //TODO Auto-generated method stub
13          //声明 RemoteViews 对象
14          RemoteViews remoteViews=new RemoteViews(context.getPackageName(),
            R.layout.main);
15          //设置为当前的时间
16          Time time=new Time();
17          time.setToNow();
18          //得到当前的月份
```

```
19      String month=time.year+" "+months[time.month];
20
21      //设置textview的值
22      remoteViews.setTextViewText(R.id.txtDay, new Integer(time.
        monthDay).toString());
23      remoteViews.setTextViewText(R.id.txtMonth, month);
24      remoteViews.setTextViewText(R.id.txtWeekDay, days[time.week
        Day]);
25
26      //调用系统日历
27      ComponentName cn;
28          Intent i = new Intent();
29          cn = new ComponentName("com.android.calendar",
30            "com.android.calendar.LaunchActivity");
31          i.setComponent(cn);
32          //为PendingIntent设置intent的值
33      PendingIntent pendingIntent=PendingIntent.getActivity(context, 0,
        i, 0);
34
35          //当桌面控件单击时调用系统日历
36      remoteViews.setOnClickPendingIntent(R.id.layout, pendingIntent);
37      appWidgetManager.updateAppWidget(appWidgetIds, remoteViews);
38
39          super.onUpdate(context, appWidgetManager, appWidgetIds);
40  }
41
42  @Override
43  public void onReceive(Context context, Intent intent) {
44      //TODO Auto-generated method stub
45      super.onReceive(context, intent);
46  }
47  }
```

此文件是本插件的代码文件，其第 10～40 行定义了 onUpdate 方法，在此方法中得到了当前的日期，并且设置给桌面插件布局进行显示。

4．实例扩展

本实例定义了桌面插件并且实现日期显示效果。

范例 201　电池状态显示插件

1．实例简介

在我们使用应用的过程中，经常会希望在桌面上能看到电池的状态。本例子就带领大家来实现一个桌面的电池状态显示插件。

2．运行效果

该实例运行效果如图 9.19 所示。

第 9 章 Android 中的多媒体开发

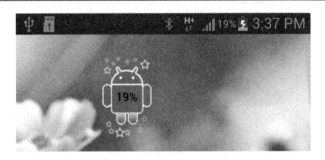

图 9.19　电池状态显示插件

3. 实例程序讲解

在本实例安装完成后不会有应用程序的提示，而是在手机的小工具中可以看到本插件，当单击本插件拖动桌面即可显示效果。想要实现本实例效果，新建 src/com.wyl.example/AppWidet.java 文件，代码如下：

```
01  //定义桌面widget 类
02  public class AppWidet extends AppWidgetProvider {
03  //监控电池电量
04  private static int currentBatteryLevel;
05  //存储电池状态
06  private static int currentBatteryStatus;
07
08  public void onUpdate(Context context, AppWidgetManager appWidget
    Manager,
09          int[] appWidgetIds) {
10      super.onUpdate(context, appWidgetManager, appWidgetIds);
11
12      //启动自动更新电池信息的service
13      context.startService(new Intent(context, updateService.class));
14
15      //为AppWidget 设置单击事件的响应，启动显示电池信息详情的activity
16      Intent startActivityIntent = new Intent(context,
17              BatteryInfoActivity.class);
18      PendingIntent Pintent = PendingIntent.getActivity(context, 0,
19              startActivityIntent, 0);
20      //设置远程显示view
21      RemoteViews views = new RemoteViews(context.getPackageName(),
22              R.layout.other_layout);
23      views.setOnClickPendingIntent(R.id.imageView, Pintent);
24      appWidgetManager.updateAppWidget(appWidgetIds, views);
25  }
26
27  /**自动更新电池信息的service,通过AlarmManager 实现定时不间断地发送电池信息 */
28  public static class updateService extends Service {
29      Bitmap bmp; //定义机器人图片
30
31      @Override
32      public IBinder onBind(Intent intent) {
33          //TODO Auto-generated method stub
34          return null;
35      }
36
37      /** 定义一个接收电池信息的broascastReceiver */
```

```java
38      private BroadcastReceiver batteryReceiver = new BroadcastReceiver() {
39          @Override
40          public void onReceive(Context context, Intent intent) {
41              //得到电量和电池状态
42              currentBatteryLevel = intent.getIntExtra("level", 0);
43              currentBatteryStatus = intent.getIntExtra("status", 0);
44          }
45      };
46
47      public void onStart(Intent intent, int startId) {
48          super.onStart(intent, startId);
49
50          //注册接收器
51          registerReceiver(batteryReceiver, new IntentFilter(
52                  Intent.ACTION_BATTERY_CHANGED));
53          //定义一个AppWidgetManager
54          AppWidgetManager manager = AppWidgetManager.getInstance(this);
55
56          //定义一个RemoteViews,实现对AppWidget界面控制
57          RemoteViews views = new RemoteViews(getPackageName(),
58                  R.layout.other_layout);
59          //定义当正在充电或充满电时,显示充电的图片
60          if (currentBatteryStatus == 2 || currentBatteryStatus == 5) {
61              //根据电量显示图片,代码省略
62          }
63          else //未在充电时,显示不在充电状态的系列图片
64          {
65              //根据电量显示图片,代码省略
66          }
67
68          //设置AppWidget上显示的图片和文字的内容
69          views.setImageViewBitmap(R.id.imageView, bmp);
70          views.setTextViewText(R.id.tv, currentBatteryLevel + "%");
71
72          ComponentName thisWidget = new ComponentName(this,
73                  AppWidet.class);
74
75          //使用AlarmManager实现每隔一秒发送一次更新提示信息,实现信息实时动态变化
76          long now = System.currentTimeMillis();
77          long pause = 1000;
78
79          Intent alarmIntent = new Intent();
80          alarmIntent = intent;
81
82          PendingIntent pendingIntent = PendingIntent.getService(this, 0,
83                  alarmIntent, 0);
84          //定时发送更新提示
85          AlarmManager alarm = (AlarmManager) getSystemService(Context.ALARM_SERVICE);
86          alarm.set(AlarmManager.RTC_WAKEUP, now + pause, pending Intent);
87
88          //更新AppWidget
89          manager.updateAppWidget(thisWidget, views);
90      }
91  }
92  }
```

此文件是本插件的代码文件，其第 8~25 行定义了 onUpdate 方法，得到插件的布局文件，设置监听器。在第 27~91 定义了一个服务器，在这个服务中定义了一个广播接收器用来接收电池电量的广播，并根据电池的状态设置桌面控件的显示图片。

4．实例扩展

本实例定义了实时显示电池电量的桌面插件，其实很多功能都可以在桌面控件中实现，只要你能想的到，就去做吧。

9.3 小　　结

在本章节中主要介绍了 Android 中常见的多媒体应用的开发实例，其中主要包含两部分，一部分是录音、照相和视频这部分的多媒体应用。另一部分是桌面插件的开发。本章的内容在实际项目开发中可以提高用户的体验性。例如，可以单击启动照相机获取照片，启动视频软件播放视频等。通过本章的讲解，Android 中常见的应用的开发方式已经全部介绍完了，相信大家都已经能够做出自己的应用了。